Keith J. Laidler

1989

Ottawa

ACS SYMPOSIUM SERIES 390

Electrochemistry, Past and Present

John T. Stock, EDITOR
University of Connecticut

Mary Virginia Orna, EDITOR
College of New Rochelle

Developed from a symposium sponsored
by the Division of the History of Chemistry
and the Division of Analytical Chemistry
of the American Chemical Society,
at the Third Chemical Congress of North America
(195th National Meeting of the American Chemical Society),
Toronto, Ontario, Canada,
June 5-11, 1988

American Chemical Society, Washington, DC 1989

Library of Congress Cataloging-in-Publication Data

Electrochemistry, past and present
John T. Stock, editor. Mary Virginia Orna, editor.

Developed from a symposium sponsored by the Division of the History of Chemistry and the Division of Analytical Chemistry at the 195th Meeting of the American Chemical Society, Toronto, Ontario, Canada, June 5–11, 1988.

p. cm.—(ACS Symposium Series, 0097–6156; 390).
Includes index.

ISBN 0–8412–1572–3
1. Electrochemistry—Congresses.
2. Electrochemistry–History—Congresses.

I. Stock, John T. (John Thomas), 1911– . II. Orna, Mary Virginia. III. American Chemical Society. Division of the History of Chemistry. IV. American Chemical Society. Division of Analytical Chemistry. V. Chemical Congress of North America (3rd: 1988: Toronto, Ont.) VI. American Chemical Society Meeting (195th: 1988: Toronto, Ont.). VII. Series

QD551.E492 1989
541.3'7—dc19

89–15
CIP

PRINTED IN THE UNITED STATES OF AMERICA

ACS Symposium Series

M. Joan Comstock, *Series Editor*

Foreword

The ACS SYMPOSIUM SERIES was founded in 1974 to provide a medium for publishing symposia quickly in book form. The format of the Series parallels that of the continuing ADVANCES IN CHEMISTRY SERIES except that, in order to save time, the papers are not typeset but are reproduced as they are submitted by the authors in camera-ready form. Papers are reviewed under the supervision of the Editors with the assistance of the Series Advisory Board and are selected to maintain the integrity of the symposia; however, verbatim reproductions of previously published papers are not accepted. Both reviews and reports of research are acceptable, because symposia may embrace both types of presentation.

Contents

Preface

NICHOLSON AND CARLISLE RECOGNIZED hydrogen and oxygen as products of the electrolysis of water in 1800. From this small beginning, the use of electrochemistry has grown tremendously, both in magnitude and in diversity. Electrochemistry affects everyone. Small stores carry arrays of that indispensable device, the battery. At one time, aluminum cost nearly as much as silver. Today fused-salt electrolysis produces the metal so cheaply that household items made from aluminum can be cheerfully discarded when damaged. Our health is less likely to suffer because of rapidly developing electrochemical approaches to the control of pollution.

Chemists have at their disposal many preparative and investigative techniques that are based on electrochemistry. The symposium on which this book is based presented not only the basic history of electrochemistry, but also its growth to a major branch of science and technology. *Electrochemistry, Past and Present* captures the major events and technologies of classical and fundamental electrochemistry, electrosynthesis, electroanalytical chemistry, industrial electrochemistry, electrode systems, and pH measurement. This volume contains an overview of the field, organized under the general headings of Foundations of Electrochemistry, Organic and Biochemical Electrochemistry, Electroanalytical Chemistry, and Industrial Electrochemistry.

Electrochemistry, Past and Present is a unique collection that will educate and engage the interested reader. An equivalent background understanding of electrochemistry could only otherwise be gleaned by many days of persistent searching through the primary literature.

JOHN T. STOCK
Department of Chemistry
University of Connecticut
Storrs, CT 06269–3060

MARY VIRGINIA ORNA
Department of Chemistry
College of New Rochelle
New Rochelle, NY 10801

August 23, 1988

Chapter 1

Electrochemistry in Retrospect

An Overview

John T. Stock

Department of Chemistry, University of Connecticut, Storrs, CT 06268

The foundations of electrochemistry are to be found in the late 18th-century investigations by Galvani and Volta. Galvani believed that electricity arose within the frog's leg that he was examining. He could not possibly have foreseen the great advances in bioelectrochemistry that have occurred during the present century. Volta, who saw the frog's leg as a mere indicator of electricity, produced the first source of direct current, a battery. Immediately, this led to the beginning of "physical" electrochemistry.

Electrochemical techniques and theories developed quite rapidly, especially in the hands of Ostwald and his associates. Towards the end of the 19th century, continued progress in these areas was accompanied by the development of applications, both in the laboratory and in industry. Progress in measurement, essential in most areas of physical and analytical chemistry, fostered the development of instrumentation. In turn, the availability of instruments fertilized the further development of electrochemistry.

When engaged in any field of science or technology, it is not a bad idea to pause occasionally and to look backwards a little. In the case of areas such as astronomy, surveying, weighing and other measurements, the view may extend to the horizon, i.e., to the earliest human records. However, not all history is "ancient". Sometimes development in a particular area is so fast that we have to confine ourselves to the past decade, or even less. Rapid advances have one supreme advantage from the historian's point of view; sometimes we can persuade those who have made some of the history to write about it!

The reading of a little history is an excellent backup to the usual literature searches that form part of any projected experimental investigation. An old idea or observation can sometimes trigger a new line of thought.

0097–6156/89/0390–0001$06.00/0

Electrochemistry appears in many guises that are often interconnected. An overall history of the subject would run to many volumes. Although by no means all-embracing, the present work is at least representative. Two major areas that receive no attention are corrosion and electroplating. The first of these causes billion-dollar losses, while the other is a billion-dollar industry. To treat even a few aspects of either would require a plethora of chapters.

Early Electrochemistry

Galvani's observations of the twitching of the detached leg of a frog are usually mentioned in texts that refer even briefly to the history of science. Galvani, an anatomist, believed that the twitching was due to electricity that arose in the muscle of the leg. The electric eel was a known producer of "animal electricity". Alessandro Volta (1745-1827), a physicist, concluded that the electricity arose from the junction of the copper hook that impaled the leg and the iron support upon which the hook hung. The twitching was seen as a mere indication of the externally-generated electricity. To augment the electrical effect, Volta built a "pile" consisting of alternate plates, such as of silver and zinc, separated by paper or cloth that was soaked in brine. Figure 1 shows the arrangement, described in 1800 (1). Volta may have felt that the wet separators merely acted as conductors.

With Volta's invention, the intensive study of areas such as electrochemistry that require a continuous source of electricity could begin. Galvani's ideas, temporarily put aside, contain more than an element of truth. Living systems are not only affected by electricity, but can also generate it. A common step in a medical checkup is the recording of an EKG. Bioelectrochemical studies may be said to take in all creatures (or parts of them!) great and small--even down to the bacterial level.

William Nicholson (1753-1815) read Volta's account before it appeared in print. Immediately, he and Anthony Carlisle (1768-1840) used a "pile" to electrolyze water and various solutions. They noted the liberation of hydrogen at one pole and of oxygen at the other (2). Systematic electrochemistry had begun with a process that would eventually grow to the kiloampere scale!

Similar experiments were carried out by others. In 1803, Jons Berzelius (1779-1848) and William Hisinger (1766-1852) observed that, in the electrolysis of salts, the solution becomes alkaline near the negative pole and acidic near the positive pole (3). This simple observation had at least two major consequences. It may have led Berzelius to develop his electrochemical theory of affinity. A practical consequence was the eventual development of the now massive chlor-alkali industry.

Figure 1. Volta's "pile". (Reproduced from Phil. Trans., 1800.)

Davy and Faraday. Humphry Davy (1778-1819) and Michael Faraday (1791-1867) were two of the scientists who brought great prestige to the Royal Institution. Davy made extensive electrolytic studies, sometimes using very large batteries. His aim was to isolate the components of alkali metal compounds. When all attempts using solutions failed, Davy turned to solid, slightly damp, potassium hydroxide. Mercury-like globules appeared, bursting into flame. Sodium hydroxide underwent a similar decomposition. Davy's 1807 account (4) marks the beginning of "fused salt" electrolysis, later to become important both in the laboratory and in industry.

The literature concerning Faraday is extensive. A most convincing account of his many investigations is contained in his own diary. The original volumes rest in the Royal Institution, where he lived and worked for nearly half a century. However, their contents are readily available (5). Professor Ronald King, who organized the Institution's Faraday Museum, has written an excellent short account of Faraday's life and activities (6).

It would be surprising to find a freshman chemistry text that fails to refer to Faraday's laws of electrolysis, or to terms such as "ion" and "electrode" that he introduced. The electrochemist remembers him every time the symbol "F" appears in a numerical expression. Although he did not commercialize his discoveries of the principles of the electric motor, the dynamo, and the transformer, these discoveries led to enormous changes in our daily lives. With the appearance of practical forms of these devices the era of "electricity for everyone" began. Batteries were not banished; today they are needed more than ever for standby supplies and for the vast amount of portable electrical equipment.

Electrolysis and Electrolytic Conductance. In 1805, Christian Grotthus (1785-1822) put forward a theory that attempted to explain why electrolysis products appear only at the electrodes (7). Although shown later to be open to grave objections, no doubt this theory caused others to consider the problem. On the experimental side, areas such as quantitative electrolytic conductance and the determination of transference numbers were beginning to open up.

In fact, Henry Cavendish (1731-1810) had examined the relative conductances of water and salt solutions as early as 1776 (8,9). Figure 2 is an attempt to illustrate his method. This involves the comparison of electric shocks when Leyden jars are discharged through the solutions and through the observer. Remarkably, Cavendish's results are not far from modern values.

After direct current became available, attempts were made to use it in electrolytic conductance measurements. In 1847, Eben Horsford (1818-1893) tried to allow for the effects of electrode polarization (10). He varied the length of the liquid path, then adjusted a calibrated series-connected resistor to restore the current to its original value. However, precise measurements of electrolytic conductance by d.c. methods had to await the perfection of the 4-electrode system by Gordon and his co-workers in the 1940's.

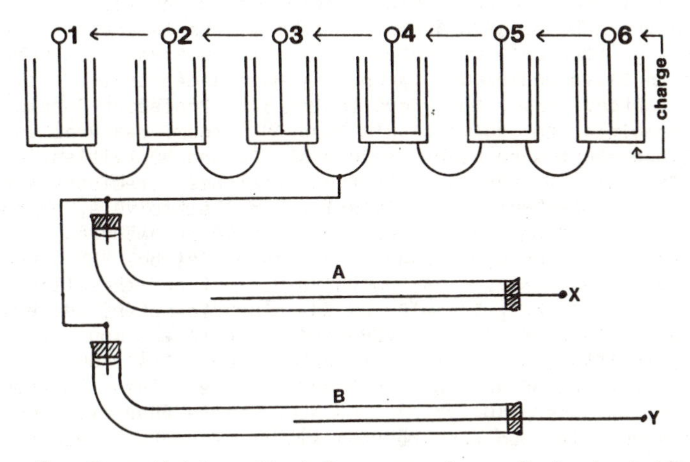

Figure 2. Cavendish's method for comparison of electrolytic conductances.
A,B, tubes containing solutions; X,Y, sliding wires to adjust length of liquid path; 1-6, Leyden jars carrying equal charges. The experiment involved grasping X with one hand and the knob of jar 1 with the other. Next, jar 2 was discharged while grasping Y, and so forth. (Reproduced from Ref. 11. Copyright 1984, American Chemical Society.)

The introduction of a.c. methodology by Friedrich Kohlrausch (1840-1910) in 1869 and his extensive work in the field of electrolytic conductance led to the high-precision measurements by Washburn and others around 1920 (11).

Even traces of dissolved electrolytes considerably increase the very small conductance of water. This effect is often used to continuously monitor the purity of distilled or demineralized water. The fundamental importance of conductance measurements became obvious when they were used by Svante Arrhenius (1859-1927) to develop his theory of electrolytic dissociation (12).

From experiments with dilute solutions of strong electrolytes, Kohlrausch developed a relationship that allowed the maximum (or "zero concentration") value of the conductance, Λ_0, to be estimated from measurements made at finite concentrations. He also showed that Λ_0 is the sum of the conductances of the anion and the cation. The additional information needed to find the separate ionic conductances from a knowledge of Λ_0 is the ratio of the mobilities of the two ionic species. The faster-moving ion will have the higher ionic conductance. In the period 1853 to 1859 Johann Hittorf (1824-1914) published the results of his experiments. These involved the analysis, after electrolysis, of the solutions around one or both of the electrodes. The "transference numbers" thus obtained are related to the required ratio.

It is interesting to note that the simple relationship used by Kohlrausch to obtain Λ_0 values has been subjected to considerable theoretical study and correction, notably by Peter Debye (1884-1966) and Erich Hückel (1896-1980) and by Lars Onsager (1903-1976).

Electrical Energy. The quantitative relationships between heat, electrical energy, and mechanical energy are cornerstones of any branch of electrical theory and practice. So basic were the investigations of these relationships carried out by James Joule (1818-1889) that his name is given to a unit (usual symbol, J).

Although Joule's major achievements were the accurate determinations of the mechanical and electrical equivalents of heat, his interests were wide. From electrolysis experiments he began to formulate theories of the heating power of an electric current. He developed various electromagnetic engines, noting that the force of the engine when driven by a given battery decreases as the speed increases. This is, of course, due to the progressive development of back-emf. He was forced to conclude that the "duty" of his engines, with zinc as "fuel" could not compete with that of the steam engine, using an equal weight of coal (13).

Ostwald, Father of Physical Chemistry

Nowadays, we would describe much of the early work in electrochemistry as physical chemistry, which was established as a significant discipline by Wilhelm Ostwald (1853-1932). This came about through his own studies, writings, founding of periodicals such as Zeitschrift für physikalische Chemie, and the work of others that he inspired (14).

In 1884, while Ostwald was Professor of Chemistry at Riga Polytechnicum, he received, and was impressed by, Arrhenius's dissertation on the conductance of electrolytes. Ostwald traveled to Sweden to meet Arrhenius; the two became lifelong friends and associates. The development of Arrhenius's theory and of Kohlrausch's conductance techniques allowed Ostwald to enunciate his "Dilution Law", relating the degree of ionization of a weak electrolyte to its concentration. He became interested in the emfs of cells, and hence in the problem of the potential of a single electrode system. He thought that it should be possible to employ a dropping mercury electrode as the basis of a system of absolute potentials.

In 1887 Ostwald became Professor of Physical Chemistry at Leipzig, remaining there until his retirement in 1906. He was fortunate in his choice of assistants; one, recommended by Arrhenius, was Walther Nernst (1864-1941). The work on the factors that govern the emf of cells, begun at Riga, pushed ahead. One result was Nernst's announcement in 1889 of the relationship that is the basis of electrolytic potentiometry. This relationship soon acquired the simple title "The Nernst Equation".

Ostwald rapidly built up a famous school of Physical Chemistry at Leipzig, often with a preponderance of English-speaking students. At that time, organic chemistry was the dominant area in Germany. Theodore Richards (1868-1928), the first American recipient of the Nobel Prize for Chemistry, studied with Ostwald in 1894. Richards, best known for his work on the determination of atomic weights, made major contributions to fields such as precise coulometry.

Two more of Ostwald's many associates may be mentioned. In 1891 Max Le Blanc (1865-1943) studied the decomposition voltages of solutions at platinum electrodes. Compounds as dissimilar as sulfuric acid and sodium hydroxide gave essentially the same results; the overall reaction in both cases is merely the decomposition of water. Hydrochloric acid and the other hydrogen halides gave lower values; here halogen can replace oxygen as an anodic product. By use of a third, or reference electrode, Le Blanc was able to separately determine the anodic and the cathodic potentials. A few years later, Wilhelm Böttger (1871-1949) used the hydrogen electrode to study the potentiometric titration curves of acids and bases (15).

Standardization of Potentials. The increasing interest in potentiometry stressed the need for standard values of electrode potentials. In 1890, Ostwald introduced the calomel electrode, calibrating this against a dropping mercury electrode. Nernst chose the normal hydrogen electrode, assigning to it a potential of zero. The upshot was a confrontation between the two scientists as the 19th century ended.

It was realized that chemical equilibria were governed not by stoichiometric concentrations, but by "active masses". The contributions to the theories made by various workers culminated in the concept of ionic activity by Gilbert Lewis (1875-1946) in 1907. Later on, Lewis used this concept to set up a listing of electrode potentials based on the standard hydrogen electrode. This was generally adopted, with one source of difficulty, that of sign. We had the so-called 'American' and 'European' sign conventions! In 1953, international agreement confirmed the hydrogen scale and settled the question of sign, specifying that half-cell reactions shall be written as reductions.

Although this agreement regularizes the values of the electrode potentials that we use, the problem of measuring or calculating the absolute values of these potentials, or of the activities of single ionic species still remains.

Preparative Electrochemistry

Electrochemistry was a precocious infant, enabling Humphry Davy to discover (i.e., to prepare) potassium, sodium, and other reactive metals in 1807. As already indicated, the now-massive chlor-alkali industry that began towards the end of the last century owes its origin to early observations during the electrolysis of salt solutions.

Although mineralogically abundant, aluminum was quite expensive when it had to be produced by purely chemical means. Nowadays electrolytically-produced aluminum allows the container for a "TV dinner" to be treated as a throwaway item. The almost simultaneous but independent reporting of the cryolite-alumina electrolytic process in 1886 by C.M. Hall in the U.S.A. and by P. Hérault in France is an amazing coincidence. An excellent account of the events that led Hall to his discovery is available (16).

Although aluminum production is the outstanding example of fused-salt electrochemistry, the laboratory-scale applications of this field are quite wide (17). A recent example is the electrolytic preparation of the spinel XV_2O_4 (X = Zn, Mg, or Cd) (18). A recent bibliography of the synthesis of inorganic compounds by electrochemical means contains more than 4000 references (19).

The development of any industrial chemical process inevitably involves chemical engineering, process control, and a close study of any means to improve overall efficiency. An illustration is the winning of bromine from brine that contains less than 0.2 per cent of bromine as bromide (20). When the process was started in 1889, bleaching powder was added to large batches of brine, then the liberated bromine was carried off in a current of air. The introduction of electrolytic oxidation allowed this batch process to be changed to a more economic continuous one. Over-oxidation, leading to contamination of the bromine with chlorine, was minimized by sampling, chemical analysis, and subsequent adjustment either of the brine flow rate or of the electrolysis current. During the next few years, sampling and analysis were supplanted by a continuously indicating galvanometer system. Eventually, the electrochemical indication was extended to provide automatic control of the current.

Organic Electrochemistry. Fichter's classic text names Faraday, Schoenbein, and Kolbe as the founders of organic electrochemistry (21). Although published nearly half a century ago, this text and its massive bibliography are valuable aids to the study both of reactions and of the history of electrochemistry.

The electroreduction of nitrobenzene received early attention, probably because of the need for dyestuffs precursors. A patent for making aniline by this procedure was granted over a century ago. Depending upon the conditions, nitrobenzene can yield a variety of reduction products. By his introduction of controlled-potential electrolysis near the end of last century, Fritz Haber (1868-1934) was able to elucidate the electrochemical and chemical steps that give rise to these products. The Luggin capillary, named for Haber's friend and colleague, dates from this period (22).

Interest in organic electrochemistry, never at a standstill has increased greatly during the past few decades. Apart from the carrying out of small-scale syntheses, much attention has been paid to the elucidation of reaction mechanisms leading, for example, towards an understanding of how various alkaloids are formed in living systems. Surprisingly, large-scale organic electrosynthesis was not commercialized until around 1965. The most obvious example is the manufacture of the nylon intermediate adiponitrile.

Some Other Aspects of Industrial Electrochemistry

Impurities in copper seriously decrease its value as a conductor of electricity. Accordingly, the purity of copper needed for cables became a matter of great concern as long-distance telegraphy developed. Electrorefining, first introduced over a century ago, provides a means for the production of high-purity copper.

Not much "off the shelf" equipment was available when early industrial processes were being developed. Nowadays the control engineer is called during the early stages of plant design. Before canning, tomatoes, peaches, and the like are "caustic peeled" in a continuous process that involves treatment with hot, quite concentrated NaOH solution, followed, of course, by a thorough water wash. A typical conveyer system is diagrammed in Figure 3. Temperature control, long practiced in industry, presents no special problems; the maintenance of the desired alkalinity is another matter (11). An effective approach makes use of the oldest electrochemical measurement, conductance. However, the modern electrodeless sensing system uses totally-enclosed coupled coils and is not fouled by suspended matter in the electrolyte.

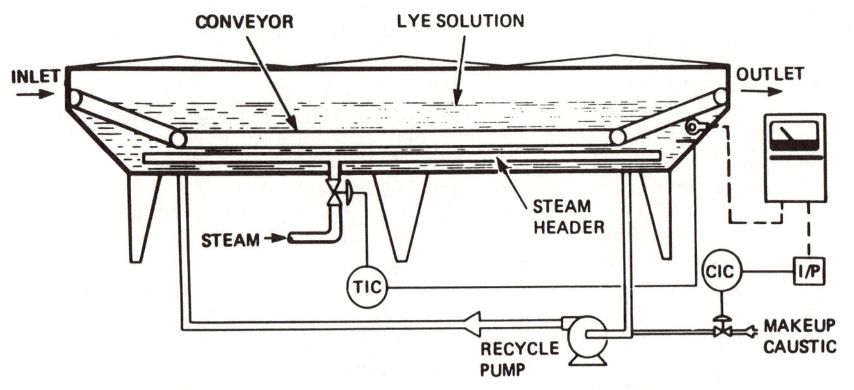

Figure 3. Control of temperature and of NaOH concentration in a fruit-peeling system.
TIC, temperature indicating controller; CIC, conductance-indicating controller; I/P, interface between conductance monitor and CIC. (Reproduced from Ref. 11. Copyright 1984, American Chemical Society.)

Electrochemical Machining. "Destructive electrolysis" conjures up grave visions of losses by corrosion. There is, however, at least one area in which such destruction is put to constructive, i.e., valuable, uses. This is the technique of electrochemical machining (23). The shaping of hard, brittle, or tough metals by conventional means has always been an engineering problem. Machining involves the controlled removal of metal; in many cases this can be done electrolytically. The flow of electrons ignores the hardness of a workpiece that can blunt the conventional cutting tool. All that is needed is susceptibility to anodic attack when flooded with a suitable electrolyte solution. Properly applied, electrochemical machining can produce objects of quite complex shapes and with a smooth finish. Quite small holes can be drilled by a modification of the technique.

Instrumentation

As amply demonstrated by the early workers, it is possible to carry out a preparation electrochemically with little other than a suitable electrolysis cell and a source of power such as a battery. The situation is quite different in fields such as analytical or physical chemistry; instrumentation becomes important, and may be a controlling factor (24). A basic requirement is the ability to measure electrical quantities such as potential, current, and resistance or conductance.

An early method for assessing the "strength" of a battery was to find the length of thin wire that the battery could cause to be heated to redness. Oersted's 1820 discovery of the deflection of a magnetic needle by an adjacent current-carrying wire provided a new approach to "strength" measurements. Within a year, Ampere had enunciated the basic principles of electromagnetism, while three independent workers showed that a multi-turn coil could greatly increase the sensitivity of Oersted's device (25). Here is the ancestry of the sensitive, and eventually accurate, ammeters and voltmeters of later times.

The laying of the Transatlantic cable in 1858 accentuated the need for sensitive and reliable instruments. A galvanometer designed by William Thomson (later Lord Kelvin) deflected when energized through 3700 miles of cable; the sole source of energy was a slip of zinc dipped in dilute acid contained in a silver thimble! In Thomson's telegraphic "siphon recorder", patented in 1867, pen-to-paper friction was eliminated by ink-jet printing (26). The same general idea is used in modern computer printers.

The null-point method introduced by Poggendorf in 1841 paved the way for precise potentiometry. Although the essentials of "Ohm's Law" were published in 1826, they received little attention until they were publicized in 1843 by Wheatstone, whose name is associated with the "bridge" method for comparing electrical resistances.

The Move Towards the Electronic Age. Progress during the past 100 years has been very rapid. Coulometry, the measurement of quantities of electricity, was initiated by Faraday. Eventually coulometry became so precise that, until 1948, the ampere was defined internationally in terms of the rate of deposition of silver under constant-current conditions.

Galvanometric pen recorders were commercialized in the late 1890's (26). The early 1800's ancestor of the "string galvanometer" gave rise to the electrocardiograph and the electromagnetic oscillograph, both making possible the observation or recording of rapidly-varying currents. From its birth in a home workshop, the modern type of X-Y recorder became a commercial item in 1951 (27). Photographic recording of current-voltage curves was a feature of the polarograph, invented in 1925.

Developments such as that of the vacuum triode, cathode ray tube and, later, the transistor and the solid-state integrated-circuit device, had an enormous influence on the design and construction of instruments. Electronic systems can respond with lightning speed to an almost-zero signal. The glass ion-sensitive electrode was discovered in the early 1900's, but its appearance in the pH meter had to wait for developments in electronics.

Automation, both in the laboratory (28) and in industry (29), became noticeable around 1900. Most, but not all of the early devices were entirely mechanical. Mention may be made of a water-softening system, patented in 1906, that claimed to electrophotometrically "analyze" the feed water and to automatically adjust the dosages of lime and soda (30). Nowadays computerization can control a single instrument, a fully-automated laboratory, or an entire factory.

Preservation of Historic Instruments. Equipment that has been superseded is put aside and then quite often scrapped; this is to be expected. However, if the item in question is the last of its kind, or is unique, perhaps a prototype, the scrapping becomes an act of extinction. Because of universal recognition and value as antiques, long-outdated microscopes, telescopes, timepieces and the like have quite good chances of survival. General laboratory equipment, recognizable by scientists but not by many others, has not fared so well. Historic industrial equipment seems to have a quite poor chance of survival. Factories have little space for discarded equipment which, in any case, lacks the "brass, mahogany, and glass" eye appeal of an old laboratory instrument. Only sheer luck saved the high-pressure flowmeter shown in Figure 4. This instrument once formed part of Britain's first synthetic ammonia plant.

Diligent and persistent searching for a particular historic instrument sometimes results in success. Unfortunately, failure is much more common; apparently the item in question no longer exists (31). The fate of instruments that will form part of the heritage of the history of science for those that are to follow us lies in our hands!

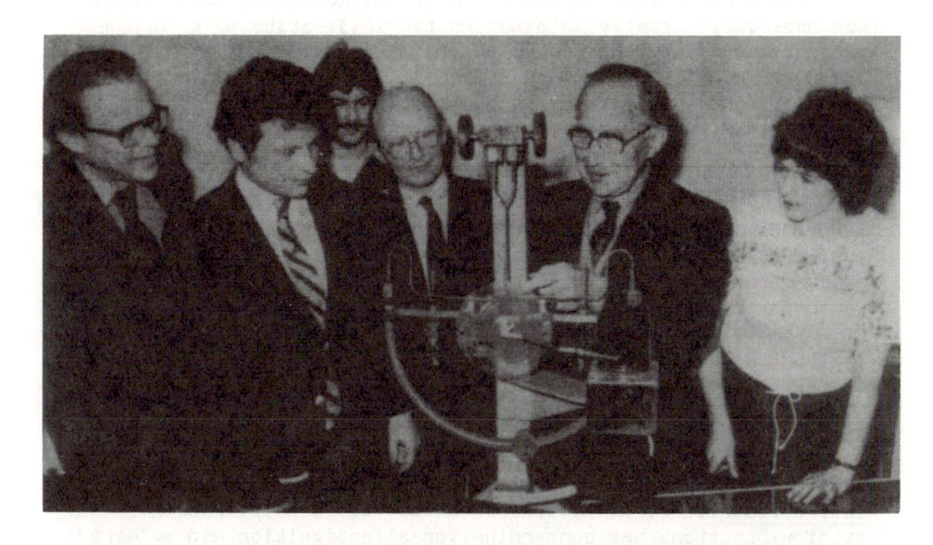

Figure 4. Jasper Clark (2nd from right) presents an historic flowmeter to the Science Museum, London. (Reproduced from Ref. 31. Copyright 1980, American Chemical Society.)

Electroanalytical Chemistry

For a concise overview of the influence of electrochemistry on the practice of analytical chemistry, the relevant section of the American Chemical Society monograph, A History of Analytical Chemistry, is hard to beat (32). The account deals both with techniques and with persons. One of the contributors is Emeritus Professor I. M. Kolthoff, very much a maker of history and also the scientific ancestor of generations of electroanalytical chemists. Through him we have, as a bonus, a direct linkage with Arrhenius, Haber, Nernst, and other scientific giants of the past.

The applications of electrochemistry to analytical chemistry are numerous. A table in a modern text (33) lists more than a dozen interfacial techniques; other fields such as conductometry are dealt with in separate chapters. Accordingly the present comments are limited to a few of the basic areas of electroanalytical chemistry.

Electrogravimetry. In his History of Analytical Chemistry, Szabadvary devotes an entire chapter to electrogravimetry (34). He points out that electrolytic deposition was recommended as a qualitative test for copper in 1800. The first electrogravimetric determinations were reported by Wolcott Gibbs in 1864. By a remarkable coincidence, the German industrial chemist C. Luckow reported similar investigations in 1865, indicating that he had practiced electrogravimetry since 1860. Szabadvary does not mention the major contributions of E. F. Smith in U.S.A., or of H. J. S. Sand in Britain. Once a major analytical technique, electrogravimetry has largely passed into history.

Potentiometry. As is obvious from the vast and continuing outflow of literature concerning ion-selective electrodes, direct potentiometry needs no emphasis. The development of the glass electrode, the pH concept, and pH measurement is a fascinating and important branch of this aspect of potentiometry. According to Szabadvary (35), the first potentiometric titrations were reported by R. Behrend in 1893. The technique became very popular, and has remained so; the first monograph, by E. Mueller, appeared in 1923. Automatic potentiometric titration, first performed in 1914, has an extensive literature (36).

Conductometry. Because conductometric titration curves are easily linearized, precise results can often be obtained at low titrand concentrations. According to I. M. Kolthoff, who made major contributions to this field, the first analytical determination by this method occurred in 1895 (37).

Because no liquid junctions are involved, conductometry can be useful in studies of nonaqueous systems, including solvents such as liquid ammonia or concentrated sulfuric acid. In fact, measurements can be made without metal-to-solution contact. The use of high-frequency a.c. permits the sensing system, sometimes merely a pair of metal rings, to be located outside of the solution container. The modern industrial electrodeless sensor is a completely-sealed immersible device.

Coulometry. Faraday's laws of electrolysis, enunciated in 1834, form the basis of coulometric techniques. By the beginning of the present century the silver coulometer had been shown to provide an accurate means for the measurement of quantities of electricity. An excellent survey of various chemical and other coulometers is available (38). The electronic digital coulometer, first described in 1962 (39), was a major practical advance.

Although the measurement of electricity through its chemical effects began quite early, the reverse of this process received little attention. The first major development occurred in 1938, when the Hungarian workers L. Szebelledy and Z. Somogyi described the titration of, for example, HCl with electrogenerated hydroxyl ion (40). The technique, coulometric titration, was rapidly extended, largely by American workers (41).

Considerable selectivity can often be obtained by keeping the working electrode at a suitable fixed potential. Until the invention of the potentiostat in 1942, the method required the constant attention of the operator. Sometimes a combination of two quite different disciplines can yield more information than is given by either one alone. A good example of this approach is the technique of spectroelectrochemistry.

Voltammetry. The discovery of polarography, the study of voltammetry at a dropping mercury electrode (DME) focused attention upon the analytical potentialities of voltammetric methods. Although excellent for cathodic processes, mercury is susceptible to anodic attack, especially in the presence of ions such as chloride. Platinum and other solid electrodes are useful as anodes, and can be used as cathodes if the potential is not too negative. A supreme advantage of the DME is the periodic renewal of the electrode surface; the surface condition of a solid electrode can greatly affect its performance. Attempts to electrochemically clean a platinum electrode led to the discovery of quantitative anodic stripping analysis during the then classified work on radionuclides. Later work, published in 1952, clearly demonstrated the value of stripping analysis as a means for trace determinations (42).

The sensitivity and versatility of polarography has been greatly extended by the introduction of pulse techniques. Although explored in the early days of polarography, the theory and practice of pulse polarography did not develop until the early 1950's. The first commercial electronic instrument appeared around 1958.

Cyclic voltammetry involves the scanning of the potential in one direction, followed by an immediate reverse scan to the starting potential. Often the cycle is repeated. The technique is now used extensively and is valuable for such purposes as the diagnosis of electrode reactions. A key paper of 1964 which essentially opened up the field refers to the limited earlier work (43).

Amperometric Titration. This analytical process is performed by observing changes in the cell current as the titration proceeds. Titration at one indicator electrode was first reported by Heyrovsky and Berezicky in 1929. A few years later, Kolthoff and his coworkers began their extensive studies of the method which, like conductometric titration, yields linear curves.

Biamperometric titration involves the application of a fixed, usually small, voltage across a pair of identical electrodes. First performed as early as 1897, the technique was rediscovered and applied in 1926. Although the titration curves are not linear, the end point is usually marked by a sharp change in the current. A text (44) and a succession of reviews (45) describe the history and subsequent development of these and related titration techniques.

Literature Cited

1. Volta, A. Phil. Trans. 1800, 90, 403.
2. Nicholson, W.; Carlisle, A. Nicholson's J. 1800, 4, 179.
3. Leicester, H. M. The Historical Background of Chemistry; Wiley: New York, 1956; p. 165.
4. Davy, H. Phil. Trans. 1808, 98, 1.
5. Martin, T., Ed.; Faraday's Diary; Bell: London, 1932.
6. King, R. Michael Faraday of the Royal Institution; London, 1973.
7. Grotthus, C. J. D. Ann. Chim. Phys. 1806, 58, 54.
8. Cavendish, H. Phil. Trans. 1776, 46, 196.
9. Maxwell, J. C.; Larmor, J., Eds.; The Scientific Papers of the Honourable Henry Cavendish, F.R.S.; Cambridge University Press: Cambridge, 1921; Vol. 1, pp. 23, 311.
10. Horsford, E. N. Ann. Physik. 1847, 70, 238.
11. Stock, J. T. Anal. Chem. 1984, 56, 561A.
12. Kauffman, G. B. J. Chem. Educ. 1988, 65, 437.
13. Crowther, J. G. British Scientists of the Nineteenth Century; Penguin: London, 1940; Vol. 1, p. 183.
14. Donnan, F. G. J. Chem. Soc. 1933, 316.
15. Böttger, W. Z. Phys. Chem. 1887, 24, 251.
16. Craig, N. C. J. Chem. Educ. 1986, 63, 557.
17. Wold, A.; Bellavance, D. In Preparative Methods in Solid State Chemistry; Hagenmuller, P., Ed.; Academic: New York, 1972; pp. 279-307.
18. Chamberland, B.; Wu, J. F. J. Electrochem. Soc. 1988, 135, 921.
19. Nagy, Z.; Electrochemical Synthesis of Inorganic Compounds; Plenum: New York, 1985.
20. Dow, H. H. Ind. Eng. Chem. 1930, 22, 113.
21. Fichter, F.; Organische Electrochemie; Steinkopff: Dresden, 1942.
22. Stock, J. T. J. Chem. Educ. 1988, 65, 337.

23. McGeough, J. A. Principles of Electrochemical Machining; Wiley: New York, 1974.
24. Stock, J. T.; Orna, M. V. The History and Preservation of Chemical Instrumentation; Reidel: Dordrecht, 1986.
25. Stock, J. T. J. Chem. Educ. 1976, 53, 29.
26. Stock, J. T.; Vaughan, D. The Development of Instruments to Measure Electric Current; Science Museum: London, 1983.
27. Moseley, F. L. Symposium on the History of Chemical Instrumentation; A.C.S. Annual Meeting, Washington, D.C., 1979, Paper No. 11.
28. Stock, J. T. Educ. Chem. 1983, 20, 7.
29. Stock, J. T. Am. Lab. (Fairfield, CT), 1984, 16 (6), 14.
30. Stock, J. T. Trends Anal. Chem. 1983, 2, 211.
31. Stock, J. T. Anal. Chem. 1980, 52, 1518A.
32. Laitinen, H. A.; Ewing, G. W., Eds. A History of Analytical Chemistry; American Chemical Society: Washington, D.C., 1977, Chapter 4.
33. Kissinger, P. T.; Heineman, W. R. Laboratory Techniques in Electroanalytical Chemistry; Dekker: New York, 1984, p. 6.
34. Szabadvary, F. History of Analytical Chemistry; Pergamon: New York, 1966, Chapter 10.
35. Ref. 34, Chapter 13.
36. Lingane, J. J. Electroanalytical Chemistry, 2nd Ed.; Interscience: New York, 1958, Chapter 8.
37. Kolthoff, I. M. Konduktometrische Titration; Steinkopff: Dresden, 1923.
38. Ref. 36, pp. 340, 452.
39. Bard, A. J.; Solon, E. Anal. Chem. 1962, 34, 1181.
40. Ref. 34, p. 316.
41. Ref. 32, p. 269.
42. Lord, S. S.; O'Neill, R. C.; Rogers, L. B. Anal. Chem. 1952, 24, 209.
43. Nicholson, R. S.; Shain, I. Anal. Chem. 1964, 36, 706.
44. Stock, J. T. Amperometric Titration; Interscience: New York, 1965.
45. Stock, J. T. Anal. Chem. 1984, 56, 1R, and earlier parts cited.

RECEIVED August 12, 1988

FOUNDATIONS OF ELECTROCHEMISTRY

Chapter 2

Wollaston's Microtechniques for the Electrolysis of Water and Electrochemical Incandescence

A Pioneer in Miniaturization

Melvyn C. Usselman

Department of Chemistry, University of Western Ontario, London, Ontario N6A 5B7, Canada

In 1801 W.H. Wollaston demonstrated that the electrolysis of water could readily be accomplished by a small current passing through a short distance of water between two exceedingly fine conducting wires. The similar chemical effects of frictional and voltaic electricity led him to conclude that they were identical, long before such a view became commonplace. Later, after perfecting the process of drawing extremely fine platinum wires, Wollaston constructed his "thimble battery", which was a very small galvanic cell capable of bringing a platinum filament to incandescence upon immersion in dilute vitriolic acid. Both demonstrations illustrate Wollaston's talent for exhibiting natural phenomena by means of inexpensive "shirt-pocket" devices.

The heuristic value of demonstration devices was much appreciated by our chemical forebears, and historians of science and technology prize those artifacts of scientific pedagogy which fame or circumstance has rescued from oblivion. The itinerant chemical lecturers of the 18th century depended for their livelihood on the appeal of their courses, and they supplemented their lectures whenever possible with demonstrations of novel chemical phenomena, often with the aid of appropriate apparatus. Instrument makers, in turn, were quick to meet the demand for popular instruments, many of which served both to demonstrate science and to aid in its acquisition. Devices produced in large numbers from durable materials are generally well known to us, but more obscure, or fragile, artifacts seldom survive to modern times. Often, only literature references remain, and the curious historian's knowledge of the devices themselves is dependent upon the detail of the original written record. Some natural philosophers were especially adept at demonstrating theoretical postulates by means of specially-designed devices. Such singular devices, although often well known at the time to a small circle of associates, rarely attract later historical scrutiny. The discovery and analysis of

0097–6156/89/0390–0020$06.00/0

the design and function of these rare devices often makes evident presuppositions of the inventors that are obscure or ambiguous in their theoretical writings. The electrochemical devices of William Hyde Wollaston are pertinent examples, for they were small contrivances designed to illustrate the principles of electrochemistry as simply and dramatically as possible. His devices had their effect magnified by comparison with the ever larger, more powerful galvanic devices that became so popular in early 19th century England.

The Birth of Electrochemistry

The seminal event in electrochemistry is Volta's letter of March 20, 1800 to Joseph Banks, president of the Royal Society. During the last decade of the 18th century Volta investigated the phenomena of animal electricity and galvanism and had developed two novel prototypes of an electric cell, one called an "artificial electric organ", and the other a "crown of cups" (1). Volta summarized his research in a two-part letter to Banks, and requested that the results be communicated to the Royal Society (2). After Banks received the second instalment of the paper, it was read to the Society on June 26, 1800 and published shortly thereafter in the Philosophical Transactions. Soon after Banks received the first part of the letter in April, he took the liberty of showing it to some of his colleagues in the Society. By the end of April, Sir Anthony Carlisle, chief surgeon in Westminster Hospital, had constructed a pile from 17 pairs of silver and zinc discs separated by pasteboard which had been soaked in salt water. His experiments with the pile quickly verified Volta's results (3). During the course of his experiments, Carlisle had sought to improve contact between the end plate of the pile and the connecting wire by placing a drop of water on the point of contact. He noted that gas was evolved from the water when the pile was in operation. His friend William Nicholson believed the gas was hydrogen and proposed experiments which soon proved this hypothesis correct. On May 2, Carlisle and Nicholson determined that when the pile was allowed to discharge through water hydrogen was evolved from the wire connected to the silver end and oxygen from the wire connected to the zinc end. The two gases were evolved in a ratio of nearly 2 : 1, "the proportions in bulk, of what are stated to be the component parts of water" (4). These experiments were, as Humphry Davy was later to claim, "the true origin of all that has been done in electrochemical science" (5).

Volta had focussed in his work on the source of the electricity in his devices, and had concluded that it emanated from the simple contact of two dissimilar metals; he took no note of the chemical changes that accompanied current flow. It was William Nicholson who first established the chemical phenomena associated with galvanism, and his work with Carlisle set off the explosion of interest in the chemical aspects of galvanism in the spring of 1800 (6). Although publication of Nicholson and Carlisle's results was withheld until after Volta's paper was read to the Royal Society, word of their discoveries spread rapidly and the assembly of voltaic piles became the rage of scientific inquiry. The natural tendency was to increase the power of the pile by increasing the size and number of the metallic pairs. Humphry Davy, still in Bristol in 1800, built a pile of 100 pairs of plates (7) and in 1803 William Pepys superintended the construction of a large galvanic apparatus containing 60 pairs of zinc and copper plates, each 6 inches square (8). This up-sizing culminated in John Children's behemoth of 21 sets of zinc and double copper plates, each of which measured 32 square feet (9). Clearly, for many investigators bigger was better, and the research advantage shifted to those with the resources to build larger, more powerful batteries. Among those who bucked the trend was Wollaston, who favoured miniature devices for his electrochemical work.

Wollaston and Electrochemistry

The year 1800 was the pivotal one in Wollaston's career. In that year, at the age of 34, he decided to turn his back on a promising medical career and to seek a new life as a chemical entrepreneur. By December, 1800, he had formed a financial partnership with Smithson Tennant for the production of chemical commodities, notably malleable platinum, and his scientific achievements over the next two decades would place him in the upper echelon of Europe's scientific elite. Wollaston's talent for doing science on a scale much smaller than normal was a trait which often impressed his contemporaries. William Brande, for example, wrote

> [Wollaston's] uncommon tact, neatness, and dexterity as an experimental chemist, will never be forgotten by those who had an opportunity of witnessing his performance of any analytical operation; he practised a peculiar method of microscopic research, in which he willingly instructed those who asked his information; and we owe to him numerous abbreviations of tedious processes, and a variety of improvements in the application of tests, which have gradually become public property, although he never could be induced to describe his manipulations in print, or to communicate to the world his happy and peculiar contrivances (10).

In a similar vein, John Paris commented

> Every process of [Wollaston's] was regulated with the most scrupulous regard to microscopic accuracy, and performed with the utmost neatness of detail..... By long discipline, Wollaston had acquired such power in commanding and fixing his attention upon minute objects, that he was able to recognise resemblances, and to distinguish differences, between precipitates produced by reagents, which were invisible to ordinary observers, and which enabled him to submit to analysis the minutest particle of matter with success... Wollaston appeared to take great delight in showing by what small means he could produce great results (11).

Wollaston regularly attended the meetings of the Royal Society and was fully acquainted with Nicholson and Carlisle's discoveries. In a letter dated June 6, 1800 to his close friend Henry Hasted, Wollaston reported

> I cannot write without a few words upon the most curious discovery (as it appears to me, unless we

> accept Cowpox) which has been made in our time....
> [After a description of Volta's pile, he continues]
> Nicholson and Carlisle a surgeon made the apparatus
> and in one week added some very important facts upon
> the decomposition of water by this apparatus.
>
> I too have been dabbling with so curious a
> subject and can tell you a few facts. 40 pieces are
> not necessary - 3 shillings are sufficient with
> similar pieces of zinc and pasteboard to decompose
> water, and if the pasteboard is wetted with salt-water
> even two pieces will do it. (12)
>
> ......
>
> [Galvanism] is a fresh hare just started; the Royal
> Society Hounds are now in full cry; which will run her
> down don't yet appear (13).

Wollaston was not present at the Royal Society the night Volta's paper was read, for on the previous day he had set off on a four week tour of the Lake District with Hasted. Before his departure he had reduced the scale of his electrolytic cell even further, for he took a small demonstration device with him. As Hasted was to recall

> He had a minute tube in his pocket which with a wire
> connecting thro' a few drops of muriatic acid the zinc
> and silver, shewed the whole principle, and first set
> Dr. Currie, whom he saw on his way back thro'
> Liverpool, if not Dr. Henry also, at Manchester, upon
> the right scent (14)

We cannot be sure what Hasted meant by "the whole principle" or "the right scent", but it is likely that Wollaston believed his device revealed the cause of electric current. One possibility, the so-called "contact theory" (favoured by Volta), built on the premise that electricity emanated from the mere contact of two dissimilar metals, with the liquid serving only as a conducting medium. This theory treated concurrent chemical changes only as ancilliary events. The alternative "chemical theory" which located the source of electrical current in the chemical changes themselves, was favoured by Nicholson and Carlisle and many of their English contemporaries (5,15). When Wollaston was to reveal the workings of his device in 1801, he strongly emphasized the chemical aspects.

> If a piece of zinc and a piece of silver have each one
> extremity immersed in the same vessel, containing
> sulphuric or muriatic acid diluted with a large
> quantity of water, the zinc is dissolved, and yields
> hydrogen gas, by decomposition of the water : the
> silver, not being acted upon, has no power of
> decomposing water; but, whenever the zinc and silver
> are made to touch, or any metallic communication is
> made between them, hydrogen gas is also formed at the
> surface of the silver (16).

Wollaston gives no details on the size or assembly of his small cell in the published paper, but a description is available from one who saw it. While visiting London in the summer of 1801, the Swiss natural philosopher Marc-Auguste Pictet witnessed Wollaston's demonstration of galvanism at the Royal Institution, and he described the device for the readers of the Bibliotheque Britannique.

> I also witnessed the [decomposition of water] effected by the following apparatus, which is of an admirable simplicity. It consists of a small glass tube, full of water slightly acidified by a common acid. In this tube is a silver wire, which passes through a cork from above, bends immediately, and extends upward along the outside of the tube until it bends again above the tube so that it can be made to touch, or not, the end of a zinc wire (itself protruding [from the aqueous solution]). As the two metals touch outside the tube, a stream of small ascending bubbles appears at once at the end of the silver wire; when they are separated, the chemical action in the liquid on the silver ends,and the stream ceases. (17)

This miniature cell, Figure 1, is undoubtedly the same as that referred to by Hasted. In it Wollaston has reduced the metallic components to two wires, and increased the current-carrying capacity of the water by slightly acidifying it. By reducing the galvanic apparatus to its simplest form, Wollaston wished to demonstrate that the flow of electricity between two different metals was dependent upon oxidation of the less noble metal, and the production of hydrogen gas on the surface of the more noble metal was a consequence of electrical flow. For Wollaston, galvanism was fundamentally a chemical process and for him "oxidation of the metal is the primary cause of the electric phaenomena observed" (18). (It is important to note here that Wollaston's cell does not fully decompose water into its elemental constituents, oxygen and hydrogen. The silver wire functions as the cathode for the reduction of hydrogen ion to molecular hydrogen. The added acid increase the hydrogen ion concentration to a level at which the evolution of hydrogen gas becomes visible. The zinc anode corrodes as metallic zinc is oxidized, and there is no molecular oxygen produced. Thus, the cell is not fully analogous to one which produces both hydrogen and oxygen. In order not to misconstrue the historical context, however, I refer to the process as Wollaston and his contemporaries did, as a decomposition of water).

Having established that the decomposition of water was an electrical phenomenon with his rudimentary galvanic apparatus, Wollaston then sought to achieve the same effects with common, or static, electricity. In order to counter arguments that the electrical effects of static electricity were due principally to its greater intensity, Wollaston wished to decompose water with a static charge of lowest possible intensity. In his words,

> It has been thought necessary to employ powerful machines, and large Leyden jars, for the decomposition of water; but, when I considered that the decomposition must depend on duly proportioning the strength of the charge of electricity to the quantity of water, and that the quantity exposed to its action at the surface of communication depends on the extent of that surface, I hoped that by reducing the surface of communication the decomposition of water might be effected by smaller machines, and with less powerful

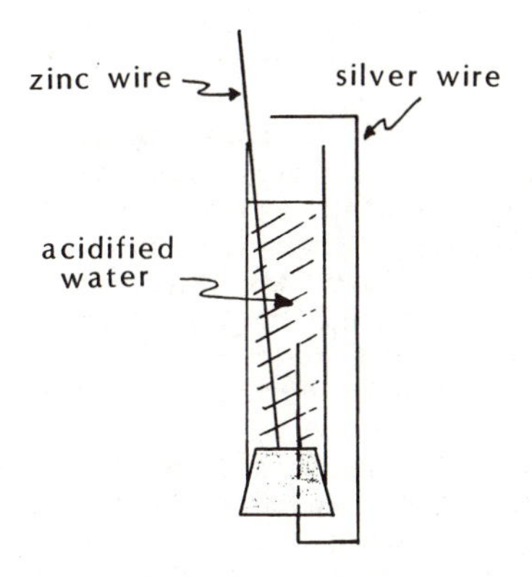

Figure 1. Wollaston's miniature galvanic cell (1800)

> excitation, than have hitherto been used for that purpose....(19).

The strength of the electrical charge generated by electrical machines was customarily measured as the maximum gap between two conductors over which a spark could be made to pass. By melting a glass capillary tube around a fine gold wire, and filing down the exposed end, Wollaston was able to fabricate wire conductors which had only a very small area of exposed metal. When these insulated conductors were placed in a small glass tube and connected to an electrical machine, Wollaston found that "a spark passing to the distance of 1/8 of an inch would decompose water, when the point exposed did not exceed 1/700 of an inch in diameter"(20). But even this was not enough, since a galvanic apparatus had the ability to decompose water even when it had insufficient power to produce any spark at all. Once again, Wollaston's aptitude for miniaturization reveals itself

> In order to try how far the strength of the electric spark might be reduced by proportional diminution of the extremity of the wire, I passed a solution of gold in aqua regia through a capillary tube, and, by heating the tube, expelled the acid. There remained a thin film of gold, lining the inner surface of the tube, which, by melting the tube, was converted into a very fine thread of gold, through the substance of the glass.
>
> When the extremity of this thread was made the medium of communication through water, I found that the mere current of electricity would occasion a stream of very small bubbles to rise from the extremity of the gold, although the wire, by which it communicated with the positive or negative conductor, was placed in absolute contact with them. Hence it appears, that decomposition of water may take place by common electricity, as well as by the electric pile, although no discernible sparks are produced.(21)

For Wollaston, then, the electrolysis of water by both galvanic and common electricity, under comparable conditions, was evidence for their identity. He showed in further experiments that the two classes of electricity produced analogous chemical effects, and therefore

> This similarity in the means by which both electricity and GALVANISM appear to be excited, in addition to the resemblance that has been traced between their effects, shews that they are both essentially the same, and confirms an opinion that has already been advanced by others, that all the differences discoverable in the effects of the latter, may be owing to its being less intense, but produced in much larger quantity (22).

Although Wollaston's published conclusions did little to promote a consensus on the topic, his live demonstrations were more effective. After witnessing the experiments at the Royal Institution, Pictet was convinced

> [The experiments of Wollaston] reveal more on Galvanism than has been discovered to date on the

> subject....[By them] is Galvanism fully reconciled with common electricity, and one might say, identified with it (23).

Twenty years later, after Arago had shown in 1820 that both common and galvanic electricities could induce a magnetic field in steel, the editor of the Annals de Chimie reprinted Wollaston's paper, with the comment that "it would be beneficial for investigators to study in detail the ingenious experiments by which Wollaston demonstrated the identical chemical effects of common and voltaic electricities"(24). It was, however, another decade before Faraday used Wollaston's results as a point of departure for his decisive demonstration of the equivalence of differently-generated electricities (25). (See also the paper in this volume by Frank James on Michael Faraday's First Law of Electrochemistry). The impact of Wollaston's electrochemical researches was lessened, no doubt, by the fact that he made no further contributions to the ensuing electrical debates. It may seem strange to us that Wollaston would fail to follow up on this early work, but we must remember that in the first decade of the 19th century, he subjugated his scientific curiosity to his entrepreneurial enterprises. He states his decision explicitly in letters to Hasted.

> I do not publish for the gratification of idle curiosity but to friends I make no mystery of my intentions. I am partial to Chemistry; I have here [in his new Buckingham Street house] room for a laboratory, and tho' many have spent fortunes in such amusements more have made fortunes by the same processes differently conducted (26)...
> [and] My business must be Chemistry. It is late to be beginning life entirely anew, and as efforts that I am not equal to may be requisite, I must not play with botany or think of Chemistry at present (27).

We also know that Wollaston had, in partnership with Smithson Tennant, invested heavily in the purchase of crude platina ore in late 1800, and from 1801 to 1805 he perfected the techniques of powder metallurgy that led to the production of malleable platinum in 1805 (28). Only when the processing of platinum became regularized about 1810 did Wollaston turn his attention again to electrochemistry, once more in connection with a small, demonstration device.

Wollaston's Thimble Battery

In his biography of Davy, Paris relates the following anecdote

> Shortly after he [Wollaston] had inspected the grand galvanic battery constructed by Mr. Children, and had witnessed some of those brilliant phenomena of combustion which its powers produced, he accidentally met a brother chemist in the street, and seizing his button, (his constant habit when speaking on any subject of interest,) he led him into a secluded corner; when taking from his waistcoat pocket a tailor's thimble, which contained a galvanic arrangement, and pouring into it the contents of a small phial, he instantly heated a platinum wire to a white heat(29).

This 'brother chemist' appears to have been Thomas Thomson, for it was he who persuaded Wollaston to publish a description of the device in Annals of Philosophy. In the published paper, Wollaston says that the small battery had been first constructed three years previously (in 1812) to demonstrate "the vast quantity of electricity evolved during the solution of metals"(30) by thermal luminescence of a platinum wire. At that time he had shown it to Berzelius, who was then visiting England, and Berzelius wrote of it to Berthollet

> Another interesting idea of Wollaston's is his galvanic Thimble. It is an ordinary tailor's thimble, made of copper.... This apparatus, which has a zinc surface of no more than one square inch, and only a single pair, produces an electric discharge between the two metals strong enough to bring to redness the small metal band that connects them. (31)

The device, Figure 2, consisted of a plate of zinc nearly one inch square mounted inside a topless, flattened copper thimble by sealing wax. Attached to the metals of the battery were two closely-spaced, parallel platinum wires, between which ran a very short length of extremely fine platinum wire about 1/3000 of an inch in diameter. When lowered into a dilute solution of sulphuric acid, the current generated is sufficient to heat the fine platinum wire to incandescence. Of the working of this remarkable little device, Wollaston only says

> though the ignition effected by this acid be not permanent, its duration for several seconds is sufficient for exhibiting the phenomenon, and for showing that it does not depend upon mere contact, by which only an instantaneous spark should be expected. (32)

Wollaston's invention of this device owed less to electrochemical principles than to his newly-discovered process for producing extremely fine platinum wires. These wires, as fine as 1/30000 of an inch in diameter, had been made for use as cross-hairs in the eye-pieces of astronomical instruments. To make them Wollaston hit upon the idea of encasing platinum in silver, drawing out the composite wire, and removing the silver coating with nitrous acid(33). Within weeks of first producing these fine platinum wires, Wollaston had constructed his 'thimble' battery, and was astonishing his friends with its operation. (Interestingly, Edison's first patents for an incandescent glow lamp in 1878 made use of a fine platinum wire encased in a partially evacuated glass globe).

The electrochemical devices discussed here are but two products of the fertile mind of Wollaston. He designed many other ingenious pieces of apparatus that combined pedagogical and practical attributes. He was especially fond of miniature 'shirt-pocket' devices and there was much to his character that presages a 20th century 'Mr. Wizard', one enthralled with science and whose pockets conceal the latest in scientific novelties. This was indeed just the way those who met him as children remembered him. As one of them was much later to recall

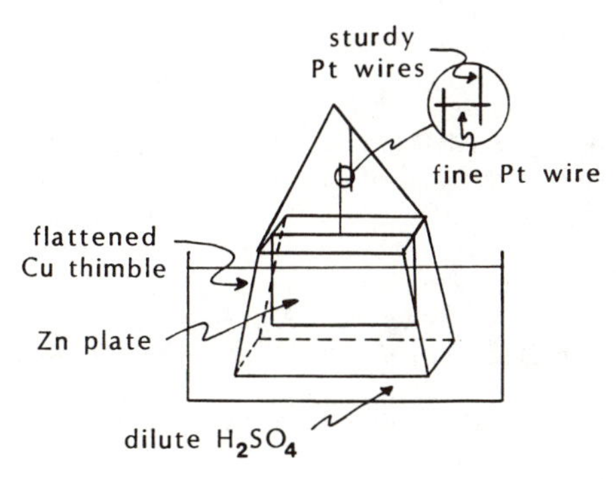

Figure 2. Wollaston's thimble battery (1812)

> You must have observed the great pleasure he [Wollaston] always derived from the company of children and young people, and the pains he took to amuse and at the same time to instruct them. Whilst he entertained them with little experiments, he appeared to be himself making experiments on the different degrees of intelligence they possessed and if they comprehended his explanations with tolerable quickness he seemed particularly pleased and interested in them..... I considered him as my playfellow and friend, he never came without bringing some puzzle or plaything (usually composed of cord and sealing wax) for the amusement of my brothers and myself and his arrival was always hailed as a joyful event (34).

Acknowledgements

This research was made possible by a grant from the Social Sciences and Humanities Research Council of Canada.

Literature Cited

1. Volta, A. Phil. Trans. 1800, 90, 403-431.
2. Partington, J.R. A History of Chemistry; Macmillan: London, 1972; Vol.4, p.19.
3. Nicholson, W. J. Nat. Phil. Chem. & Arts. 1800, 4, 179-187.
4. Ibid., p.182.
5. Davy, H. Phil. Trans. 1826, 116, p.383.
6. Sudduth, W.M. Ambix. 1980, 27, 26-35.
7. Davy, H. J. Nat. Phil. Chem. & Arts. 1801, 4, 275-281.
8. Anon. Phil. Mag. 1803, 15, 94-96.
9. Children, J.G. Phil. Trans. 1815, 105, 363-374.
10. Brande, W.T. A Manual of Chemistry; 1836, Parker: London, 1836, 4ed; p.103.
11. Paris, J.A. The Life of Sir Humphry Davy; Colburn and Bentley: London, 1831; Vol. 1, pp 146-147.
12. W.H. Wollaston to Henry Hasted, Friday 6 [June, 1800]. A copy of the Wollaston/Hasted correspondence is in the Gilbert papers, D.M.S. Watson library, University College, London.
13. W.H. Wollaston to Henry Hasted, Tuesday, 17 [June, 1800].
14. Hasted, H. Reminiscences of a Friend; privately printed, no date, p.7.
15. Russell, C.A. Ann. Sci., 1959, 15, 1-13.
16. Wollaston, W.H. Phil. Trans., 1801, 91, 427-434, p.427.
17. Pictet, M.A. Voyage de trois mois en Angleterre, en Ecosse et en Irlande pendant l'été de l'an IX, Geneva, an XI [1802]; pp.17-18.
18. Wollaston, op. cit. (ref.16), p.427.
19. Ibid., 430.
20. Ibid., 431.
21. Ibid., 431-432.
22. Ibid., 434.
23. Pictet, op. cit. (ref.17), p.17
24. Wollaston, W.H. Ann.Chim. 1821, 16, p.45.
25. Faraday, M. Phil. Trans., 1833, 123, 23-54.
26. W.H. Wollaston to Henry Hasted, Monday, November 16 [1801].
27. Ibid., Tuesday, November 24 [1801].

28. B.I. Kronberg; L.L. Coatsworth; M.C. Usselman. In Archaeological Chemistry - III; Lambert, J.B. Ed.; Advances in Chemistry Series No. 205; American Chemical Society: Washington, DC, 1984, 295-310.
29. Paris, op. cit. (ref.11), p.147.
30. Wollaston, W.H. Ann. Phil., 1815, 6, 209-211, p.209.
31. Berzelius to Berthollet, Oct., 1812, in Soderbaum, H.G. (ed), Jac. Berzelius Bref; Uppsala, 1912; Vol.1, p.42-43.
32. Wollaston, op. cit. (ref.30), p.211.
33. Coatsworth, L.L.; Kronberg, B.I.; Usselman, M.C. Hist. Tech., 1981, 6, 91-111.
34. Julia Hankey to Henry Warburton, June 30, 1829. Copy in Gilbert papers, University College, London.

RECEIVED August 12, 1988

Chapter 3

Michael Faraday's First Law of Electrochemistry

How Context Develops New Knowledge

Frank A. J. L. James

Royal Institution Centre for the History of Science and Technology, Royal Institution, 21 Albemarle Street, London W1X 4BS, United Kingdom

This paper analyses the path of experimental research that led Michael Faraday (1791-1867) to enunciate in 1834 the first law of electro-chemistry. The specific experiments that Faraday performed and his view of the results were influenced by a number of factors. This paper discusses these including his antipathy towards mathematics, his religious beliefs, his epistemology and the large number of other important experimental discoveries that he made between 1831 and 1834 which threatened to overwhelm him with empirical information which could not be quickly assimilated. This forms the context in which Faraday's new knowledge of electro-chemistry was developed.

If we think of some of Faraday's most important experimental discoveries: of electro-magnetic rotation, of electro-magnetic induction, of the magneto-optical effect and of diamagnetism, then his two laws of electro-chemistry stand out as his only statements of definitive quantitative relationships (1). It is then of great interest to understand why Faraday, in this case, chose to use quantitative methods in his experimental work.

Faraday's final version of the first law of electro-chemistry, enunciated in 1834, states: "The chemical power of a current of electricity is in direct proportion to the absolute quantity of electricity which passes" (2). I shall leave discussion of his work on the second law which states: "Electro-chemical equivalents coincide, and are the same, with ordinary chemical equivalents" (3) to another paper. Also, since most of the modern electro-chemical terminology was coined by Faraday during the experimental work which led to his enunciation of these laws, it would be inappropriate to apply these terms to period before he defined them (4).

Though these laws of electro-chemistry are quantitative statements, they do not quantify electricity - that is they do not in themselves provide an absolute measure of electricity. They do provide a quantitative relationship between different types of electricity using electro-chemical phenomena (5). David Gooding has

0097-6156/89/0390-0032$06.00/0

shown that Faraday's conceptual notions of force did not favour his developing an absolute measure of electricity (6) although he was aware of the possibility. In this paper I shall analyse the various factors (his antipathy towards mathematics, his religious views, his epistemology, the large number of discoveries he made between 1831 and 1834) that led him to the first law of electro-chemistry.

It is necessary to characterise briefly Faraday's complex attitude towards mathematics and quantification. He certainly did not believe in the superiority of the mathematical way. In a letter of 29 November 1831 Faraday wrote to Richard Phillips that "It is quite comfortable to me to find that experiment need not quail before mathematics, but is quite competent to rival it in discovery" (7). While this is an acknowledgement by Faraday of the value of mathematics in making scientific discoveries, it also states that experiment - in the fullest sense of the word - is equally capable of making scientific discoveries. To characterise Faraday as being simply anti-mathematical, or incapable, due to lack of education, of thinking mathematically would be inaccurate. James Clerk Maxwell acknowledged that Faraday had a well developed ability to think geometrically (8). This is strikingly demonstrated by the diagram (figure 1) that Faraday drew in his laboratory diary to illustrate his realisation that electricity, magnetism and motion were mutually orthogonal (9).

What Faraday did not do, or only very rarely, was to manipulate algebraically. Geoffrey Cantor has argued that Faraday's reluctance to perform such manipulation may well have stemmed from his Sandemanian beliefs (10). Sandemanians were a small sect of Christians who believed in interpreting the King James version of the bible as literally true - albeit they recognised that there was some metaphorical language in it. They believed that each individual in the sect should read the bible directly without the intervention of a priest - hence the sect had no priests, only elders of whom Faraday became one. Cantor has argued that since Sandemanians read the bible directly then Faraday believed that one should study nature directly without the mediation of theory and that to manipulate numbers derived from nature would be to destroy their natural meaning.

If Cantor is correct about Faraday's initial reluctance to work mathematically, especially algebraically, it was certainly reinforced be some of his life experiences as a scientist. To take one example: André-Marie Ampère's theory of electro-dynamics was extremely mathematical and yet failed completely to predict the electro-magnetic rotations which Faraday discovered experimentally in September 1821. This phenomenon he subsequently used to criticise severely Ampere's theories to such an extent that Pearce Williams has argued that Ampère refrained from ever again venturing physical hypotheses (11).

But, on the other hand, as Gooding has shown, Faraday was willing to accept the mathematical arguments of William Thomson about some aspects of diamagnetism (12). The point is that it is impossible to characterise fully Faraday's views on theory and on mathematics other than in very broad terms and even these are open to exceptions. There was also an element of pragmatism in his attitude to theory and to mathematics - perhaps more so in the former than the latter. What does need to be stressed is that for

Faraday experiment was the supreme method of investigation; results and theories arrived at by other methods had at some point to be submitted to experiment. Further, his attitude towards mathematics did not automatically preclude quantification or the use of measurement, but on the other hand it did not obviously encourage it. Let us then consider these aspects of Faraday's scientific character in his work on electro-chemistry and the way in which quantification came to play an ever more important role in his experiments.

On 29 August 1831 Faraday commenced experiments on an anchor ring that he had previously wound with two coils A and B (figure 2) (13). When he passed electricity from a voltaic pile into coil A, he found that he had induced, as a transient phenomenon, a current in coil B; this was indicated by the swing of a galvanometer needle. When he disconnected the pile from coil A, he found that the needle swung in the opposite direction. While this discovery of electro-magnetic induction and the invention of the first transformer has been rightly lauded, both at the time and since, the discovery did present Faraday with a number of very serious problems that required resolution. He had found that electricity could be produced from a source whence it was not previously known that it could be derived. Faraday, careful experimenter that he was, needed to be certain that the electricity he had produced by electro-magnetic induction was the same as that from other sources - common, voltaic, animal, thermal etc. Part of the reason for this need was probably due to his knowledge that some scientists were challenging the identity of electricities. In particular his colleague the Professor of Natural Philosophy at the Royal Institution, William Ritchie, and also the brother of the late Sir Humphry Davy, John Davy, were both arguing for differences in electricities (14). Faraday himself seems to have been convinced that he had produced electricity from magnetism (15). But as Gooding has shown, it was an essential part of Faraday's epistemology to demonstrate transparently his results to others (16). In this case Faraday had to show that electricity from magnetism was the same as other electricities and not some new force.

This point seems to have occurred to Faraday on the first day of his discovery of electro-magnetic induction for he tried to obtain electro-chemical decomposition of a solution of copper using induced electricity (17). By showing that electricity from whatever source produced the same effects, the identity of the causal electricities would be established. This had been the standard method during the first three decades of the nineteenth century and Faraday saw no need to alter it. He was quite elaborate in his arrangement to produce electro-chemical decomposition from induced electricity. He realised that when he broke the circuit the electric current flowed in the opposite direction and this would cancel out the electro-chemical effect of the original induced current. So before he broke the circuit he removed the pole from the solution of copper. He repeated this process many times, but found no electro-chemical decomposition.

Despite the fact that Faraday and other scientists had spent some considerable effort in the 1820s searching for electro-magnetic induction, he did not pursue his researches immediately, but went off to Hastings for a short period. When he returned to the

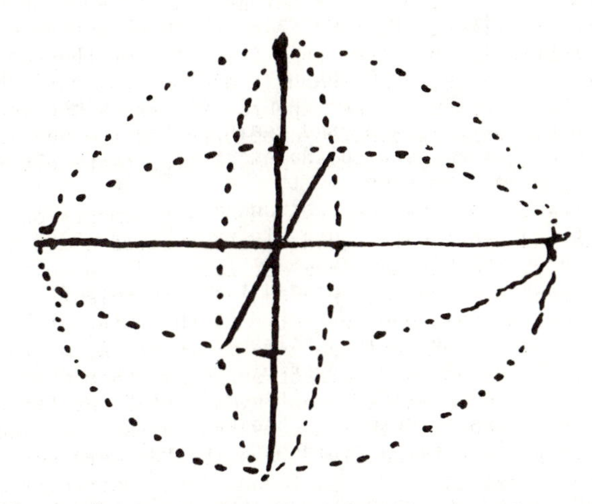

Figure 1. Faraday's diagram to show the mutual orthogonality of electricity, magnetism and motion. From Diary, 26 March 1832, 1, 403.

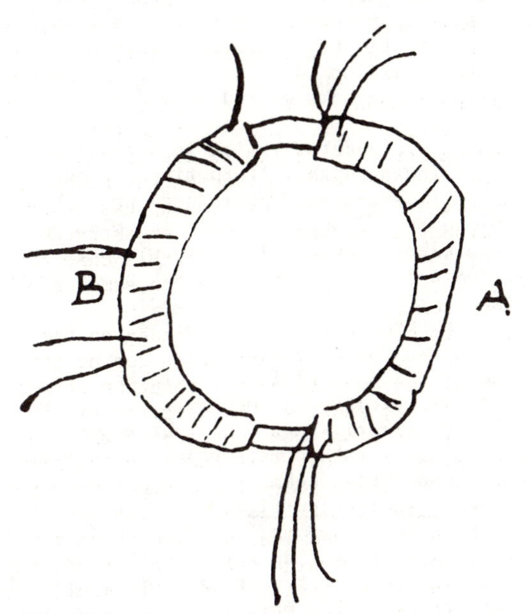

Figure 2. Faraday's drawing of his electro-magnetic induction ring. From Diary, 29 August 1831, 1, 1.

laboratory on 12 December, he set to work exploring the phenomenon, by discovering under what circumstances a current could be induced.

During this time Faraday worked with the hypothesis that a wave of electricity travelled across the ring, induced the current in the second coil and passed on to register in the galvanometer. On 1 October Faraday repeated, quite specifically, the experiment he had performed in 29 August, and found again that he could not produce electro-chemical decomposition. He ascribed this to the interference of the fluid conductors stopping the wave of electricity (18). He changed the ring to one which produced a more powerful effect on the galvanometer, but when he tried again to obtain electro-chemical decomposition he met with no success (19).

On 18 October Faraday concluded that the wave of electricity might be more powerful than indicated by the deflection of the galvanometer needle. He decided to see if he could "store" the wave in a solution of brine with copper plates in the circuit to retard the electric wave and so increase the deflection on the needle. He found that there was no change in the behaviour of the needle and also noted that electricity could pass through fluids but not decompose them (20).

As Ryan Tweney has pointed out, Faraday's rate of successful confirmation of experiments until 28 October was significantly lower than after that day (21). On that day he went to Samuel Christie's in Woolwich where he worked with the giant magnet of the Royal Society which had granted him permission to use it four days earlier. With this he was able to obtain much more powerful effects on the galvanometer. He again tried to obtain electro-chemical effects, apparently using the same method as before, but with no success (22).

On this day, 28 October, Faraday also invented the first direct current dynamo (he had invented the alternating current dynamo on 17 October). A week or so later Faraday again used the direct current dynamo - a copper plate turned between magnetic poles - as a source to attempt to produce electro-chemical decompositions from magnetically generated electricity. With this the most powerful arrangement he again failed to obtain any effect. Faraday then checked carefully to confirm that with voltaic electricity, decomposition did occur in the solution (23).

It is clear that as Faraday refined his methods of obtaining magnetically induced electricity, he sought to produce electro-chemical effects. It was clearly important to the prosecution of his work to obtain such efforts, but not essential as he spent the remainder of November and early December writing the first series of his Experimental Researches in Electricity which was read to the Royal Society on 24 November and 8 and 15 December 1831 (24). Here Faraday described his experiments and reported that he had been unable to obtain from induced electricity either electro-chemical effects from this electricity (25) or any of the other the standard effects of electricity - heating, production of the spark or physiological effects either with his tongue or with frogs legs.

Faraday then concentrated mainly on working with terrestrial electricity before returning to the problem of chemical decomposition using magneto-electricity. On 8 March 1832 Faraday used a direct current dynamo for five minutes to try to decompose copper sulphate but again with no success (26). However, for the first time, he was able to make a frog's leg contract (27). On 11 June he again tried to obtain electro-chemical effects using magneto-electricity but again with no success (28).

On 26 July Faraday returned to London to find a pseudonymous letter, signed PM, whom I have not been able to identify (29). This letter which Faraday promptly sent to the Philosophical Magazine, claimed, according to Faraday, to contain the first description of an experiment in which chemical decomposition was obtained by induced magneto-electricity. Faraday's reason for sending this letter to the Philosophical Magazine was that he had been involved in an unfortunate confusion of dates over the discovery of electro-magnetic induction with the Italian scientists Nobili and Antinori. His aim in forwarding this letter was to clarify dates. He did, however, find it difficult to decide whether PM had achieved what he claimed. There is no evidence that I have so far found which indicates that this letter had any effect on the course of Faraday's researches. He had certainly been searching for the phenomenon before-hand and would continue to do so.

By the time Faraday returned to the laboratory at the end of August 1832 he had quite clearly decided to establish as an experimental fact that electricity from all sources was the same phenomenon. His methodology was to show that the seven standard effects of electricity (physiological, magnetic deflection, the making of magnets, the production of the spark, the heating power of electricity, its chemical action, its attractive and repulsive characteristics and its discharge into hot air) were common to the five varieties of electricity (voltaic, common, magneto, thermo and animal).

Faraday concentrated initially on confirming the identity of voltaic and common electricity which had been challenged by Ritchie and by Davy. Faraday eventually repeated many of their experiments and those of others either to confirm that a particular effect was caused by a species of electricity, or to show that an earlier experimenter had been incorrect in stating that such an effect did not occur. In particular Faraday needed to confirm that magnetism was caused by common electricity (30). This he had failed to do in March, but in August he succeeded (31), by showing under what circumstances a galvanometer could be deflected by common electricity. However, in this work Faraday had to pass common electricity through water and he found that there was no decomposition. This contradicted received wisdom and in particular experiments which William Hyde Wollaston had described in 1801 (32). Here Wollaston had used electro-chemical effects to help establish that common electricity was the same as voltaic. These experiments produced similar, although as Wollaston pointed out, not identical electro-chemical effects, to those caused by voltaic electricity. These differences Wollaston ascribed to the higher tension of common electricity compared with voltaic. So although Faraday evidently thought that Wollaston had reached the correct conclusion, so far as the identity of electricity was concerned, his reasoning, combined with Faraday's own observations, led him to repeat Wollaston's work (33). In these experiments Faraday found that there was no need to have a direct physical contact of the pole with the substance undergoing decomposition, in this case turmeric paper dipped in sulphate of soda, for electro-chemical decomposition to occur. This led Faraday to consider the nature of the interactions between the matter that was decomposed and the electricity that decomposed the matter and to completely rethink electro-chemical action. On 6 September he wrote in his diary:

> Hence it would seem that it is not a mere repulsion of the alkali and attraction of the acid by the positive pole, etc. etc. etc., but that as the current of electricity passes, whether by metallic poles or not, the elementary particles arrange themselves and that the alkali goes as far as it can with the current in one direction and the acid in the other. The metallic poles used appear to be mere terminations of the decomposable substance
>
> The effects of decomposition would seem rather to depend upon a relief of the chemical affinity in one direction and an exaltation of it in the other, rather than to direct attractions and repulsions from the poles, etc. etc. (34)

It is typical of Faraday's work during this highly intensive period of 1832 and 1833 that one experimental investigation led directly to another as he discovered hitherto unknown phenomena, and inconsistencies and inadequacies in earlier work. In this case Faraday had also to raise the problem of what an electric current was. On 10 September Faraday defined what he meant by an electric current:

> By current I mean any thing progressive, whether it be a fluid of electricity or vibrations or generally progressive forces. By arrangement I understand a local arrangement of particles or fluids or forces, not progressive (35)

Faraday's work on the identity of common and voltaic electricity led to his examining the nature of electro-chemical decomposition and of his thinking about the nature of the transmission of electric force. Although Faraday published his work on the identity of electricity and on electro-chemistry in separate series of his *Experimental Researches in Electricity*, his examination of electro-chemical phenomena emerged within the general context of his work on the identity of electricities and not simply as a method of quantifying electricity.

Faraday in his experimental work had established that common and voltaic electricity were identical and that there was hence no need to invoke Wollaston's explanation of the higher tension of common electricity to account for their difference in electro-chemical decomposition. Nevertheless chemical effects were different for different quantities and intensities of electricity and this needed to be investigated. Thus it was necessary to discover the conditions under which decomposition could occur and to do this he needed to find the relationship between common and voltaic electricity was to be desired. Since Faraday had established their identity, it should therefore be possible to establish a quantifiable relationship between them.

It is worth pausing to discuss this decision of Faraday's to pursue a quantitative method. We must try and rid ourselves of the notion that Faraday had a theory that he wished to explicate and that this directed the course of his experiments. We must also rid ourselves of the notion that his methodological precepts were rigid and unbreakable. Faraday, as I indicated earlier, should be regarded as a theoretical pragmatist, willing to use experiments in any way that would further his active exploration of nature. In this case Faraday had by mid September 1832 realised that he must develop some quantitative method of comparing common and voltaic electricity. He did this not to provide further evidence of the identity of electricities, but because he had discovered something far more interesting than this. He had found that the theory of electro-chemical action which had been propounded for over thirty years was wholly inadequate to account for his experimental discoveries. Faraday well knew that theories could be a severe block to experimental discovery, as they could exclude phenomena which might exist. Here then was an inadequate theory which might be hiding much knowledge about electro-chemistry. If it required that Faraday take the quantitative road, then he was quite willing to follow it.

On 14 September Faraday commenced work on establishing a quantitative relationship between common and voltaic electricity (36). He charged eight jars of his great electric battery of 15 jars with thirty turns of an electric machine. He discharged them through a galvanometer which swung two and a half large divisions (of Faraday's making) to the left and then three to the right. He then connected the other seven jars, charged the whole battery with thirty turns from the machine and discharged into the galvanometer. It behaved as in the previous experiment. Faraday repeated this and concluded that:

> It would seem that if the same absolute quantity of electricity pass through the instrument, whatever may be its intensity, the deflecting force is the same (37)

He then charged the whole battery with sixty turns and although he observed that the galvanometer swing was much larger he could not tell the proportion but thought it was around double. Now while this is clearly an exercise in quantification and measurement, it does not, I think, go against Faraday's reluctance to manipulate numbers derived from natural sources. Both his measurements are separate - one number does not depend on the other - and both are direct readings from nature. However, Faraday did link them together as a proportion, albeit tentatively.

Faraday's next set of experiments was to obtain an equivalent for voltaic electricity (38). To this end he drew a piece each of platina and zinc wire 1/18th of an inch thick and mounted them on wood, 5/16ths of an inch apart, connected to a galvanometer. When he dipped the wires instantaneously into 4 ounces of water mixed with a drop of sulphuric acid to a depth of 5/8ths of an inch, he saw the galvanometer swung 3 of his divisions to the right. He concluded:

> that in that moment the two wires in such weak acid produced as much electricity as the battery contained after 30 turns of the machine (39).

When he left the wires in the acid solution, the needle swung up to the 6th division and settled at the third (40).

In the circuit he then placed a piece of paper coated with hydriodate of potassa which he had used previously in showing electro-chemical decomposition due to the passage of common electricity (41). When he passed electricity from his "standard voltaic arrangement" (42) through the hydriodate of potassa he observed that iodine went to one pole but that the deflection of the galvanometer needle was less, because he thought the paper was a bad conductor (43). However, when he used larger poles he observed that more electricity was indicated by the galvanometer and that much greater decomposition followed (44). This Faraday interpreted as being caused by the differences in intensities between the two sorts of electricity. Common electricity of high intensity passed easily through a small conductor into the hydriodate, while it required a large surface area to allow all the low tension voltaic electricity to pass.

The final set of experiments that Faraday performed that day involved moistening four filter papers with potassium hydriodate and then observing the effects of passing electricity through a platinum pole 1/12th of an inch in diameter (45). He observed that a different number of turns of the electric machine produced different intensities in the iodine produced due to decomposition. With electricity from his standard voltaic battery, the same amount of decomposition occurred but only after the passage of some time - "equal I think to that required to turn the machine" (46). He next included a "galvanometer in the circuit so as to give as it were a measure of the quantity of electricity passing" (47); he found that he had to use stronger acid in the battery and also had to keep the battery poles there for two or three seconds to produce the same decomposition as 30 turns of the machine. With that he finished experimentation for the day having found that time was needed for the passage of electricity to cause electro-chemical decomposition.

The following day, 15 September, he "resumed the experiments of yesterday on the comparative effects of the Voltaic pile and the Electrical machine" (48). In this he used the beats of his watch (150 beats per minute (49)) to measure the time it took for the standard voltaic cell to cause the galvanometer needle to swing the same distance as it did when the electricity from thirty turns of the electric machine was passed through it. Faraday was endeavouring to obtain a comparative measure of common and voltaic electricity. He concluded:

> Hence as an approximation, two wires, one of platina one of zinc, each 1/18 of an inch in diameter, when immersed each 5/8 of an inch into an acid consisting of 1 drop oil of vitriol and 4 oz. of water, being also 5/16 of an inch apart, give as much electricity in 8 beats of my watch as the Electrical battery charged with 30 turns of the large machine in excellent action (50)

For the remainder of the day he confirmed these results under different experimental conditions and concluded:

> Hence it would appear that both in magnetic deflection and in chemical effect the current of the standard voltaic battery for eight beats of the watch was equal to the electricity of 30 turns of the machine, and that therefore common and voltaic electricity are alike in all respects (51)

For most of the remainder of 1832 Faraday ceased laboratory work and concentrated on writing his paper on the identity of electricities - series three of Experimental Researches in Electricity (52). He was in danger of making discoveries at such a rate that one would overtake the other, and he really needed time to give order to his ideas and work by writing. Faraday had become distracted by his electro-chemical discoveries which clearly needed much more experimentation. In series three Faraday concentrated on writing up his experiments which he had done and established the identity of voltaic and common electricity. To establish the identity of the other forms of electricity Faraday referred to the work of other scientists which established that magneto-, thermo- and animal electricity were the same phenomena as voltaic and common electricity. For example in the case of magneto-electricity he noted that he had tried experimentally many times without success to establish that this caused electro-chemical decomposition (53). He then referred to the PM letter which, he said, did not really indicate that PM had produced the phenomenon experimentally. After references to other workers in the field whose results he also doubted, he then referred to a paper by J.N.P. Hachette who had "given decisive chemical results" (54) and to many other electrical workers who had established successfully that a particular electrical effect was caused by a species of electricity. After all this he wrote:

> The general conclusion which must, I think, be drawn from this collection of facts is, that electricity, whatever may be its source, is identical in its nature. (55)

and included in this Table I summarising his and others' experimental results. The blank spaces, particularly in thermo-electricity, Faraday occasionally returned to work on during the 1830s (56). But to his satisfaction he had established the identity of electricities whatever their source.

It may come as something of a surprise to find that Faraday was willing to rely on experiments that he had not tested himself, to make such a general statement. But if we are surprised then this says more about earlier biographical approaches to Faraday than it says about Faraday himself. Faraday, as a leading member of the scientific community, was not in ignorance of what others were doing, as his extensive collection of off-prints reveal. He knew whose experiments he could trust and whose he found dubious. While Faraday might have wished to test all these identities for himself, in his quantitative electro-chemical work he had now found something which was potentially far more interesting. Furthermore this was completely unsuspected and unknown to the scientific community, whereas the identity of electricities was a well established problem which could safely be left to others to work on.

In series three Faraday indicated that the effects of electro-chemical decomposition were dependent on the quantity of electricity rather than the tension (57). It was this that he set out to demonstrate in the second section of series three after he had made his general statement that electricity of all forms was identical. Faraday described his experiments on the measurement of voltaic and common electricity using the galvanometer and chemical measures (58). He stated as a general proposition his view that equal quantities of electricity produced equal deflections of the galvanometer and equal amounts of electro-chemical decomposition.

There Faraday left his quantitative work for a little. It must not be assumed that his work was theory driven or that throughout these months he was seeking a single goal. Each of his experimental discoveries led to new questions, unanticipated answers and theoretical interpretations which constantly had to be altered and modified. Thus for example when he returned to the laboratory in January 1833 he found, because his water had frozen in that bitterly cold winter, that ice would not conduct electricity (59). This led him to explore experimentally the effects of heating bodies as they transmitted electric currents and this formed the basis for his fourth series of Experimental Researches in Electricity (60).

In his fifth series he returned to studying electro-chemical decomposition (61). Here Faraday used common electricity of high tension to discover effects which could not be produced by voltaic electricity of low tension. This does not mean that Faraday thought there was a fundamental difference between the two electricities, only that no one had been able to develop a voltaic pile which produced high tension electricity. He confirmed that it was not necessary for the decomposing substance to be attached by a wire to the machine since air would also act as a pole and allow electricity to pass. He then reviewed the previous theories of electro-chemical decomposition, which mainly assigned an active role to the poles, to demonstrate their inadequacy for his experimental observations.

As a preamble Faraday criticised the two fluid theory of electricity. This he did so that instead electricity "may perhaps best be conceived of as an axis of power having contrary forces, exactly equal in amount, in contrary directions" (62). According to this view of electricity Faraday thought that electro-chemical decomposition was "produced by an internal corpuscular action" (63) which he went on to elucidate, analysing the passing of particles between the poles of the circuit. In this Faraday argued that the chemical affinity of the particles of the decomposing substance was disrupted by and changed the passage of electricity:

> The ordinary chemical affinity [he wrote] is relieved, weakened, or partly neutralised by the influence of the electric current in the one direction, parallel to the course of the latter [the electric current], and strengthened or added to in the opposite direction, that the combining particles have a tendency to pass in opposite courses (64) [figure 3]

Faraday spent the remainder of series five explaining and justifying this theory of electro-chemical decomposition. Although he had repeated his statement that the amount of electro-chemical decomposition was dependent on the quantity of electricity passing, and had even said that it was "probably true" for all cases (65), he did not expand on this here.

Faraday seems to have been reluctant to emphasise this quantitative relationship. For him this was not yet a law of nature, for laws were special, not to be lightly enunciated, and in the case of quantitative laws difficult to establish by experiment because there might always be exceptions. All his earlier discoveries - electro-magnetic rotations, induction and the dynamo etc - had been qualitative effects where it was possible to demonstrate these phenomena and thus make them accessible to an audience in such a way that they would accept the visible experimental reasoning. He had departed from this with the identity of electricities, as he had had to resort in many cases to using the experiments of others to establish this result. In his electro-chemical work Faraday had to make more departures from what had been his normal practice. He had proposed a quantitative relationship, a new theory of electro-chemical action and a theory of matter to support this - things which Faraday had never done before as a mature scientist. As I

Table I. Faraday's table showing the identity of electricities that had been confirmed experimentally. From Phil.Trans. 1833, 123, p.48.

	Physio-logical Effects.	Magnetic Deflection.	Magnets made.	Spark.	Heating Power.	True chemical Action.	Attraction and Re-pulsion.	Discharge by Hot Air.
1. Voltaic Electricity....	×	×	×	×	×	×	×	×
2. Common Electricity ..	×	×	×	×	×	×	×	×
3. Magneto-Electricity ..	×	×	×	×	×	×	×	
4. Thermo-Electricity ..	×	×	?		?			
5. Animal Electricity....	×	×	×	?		×		

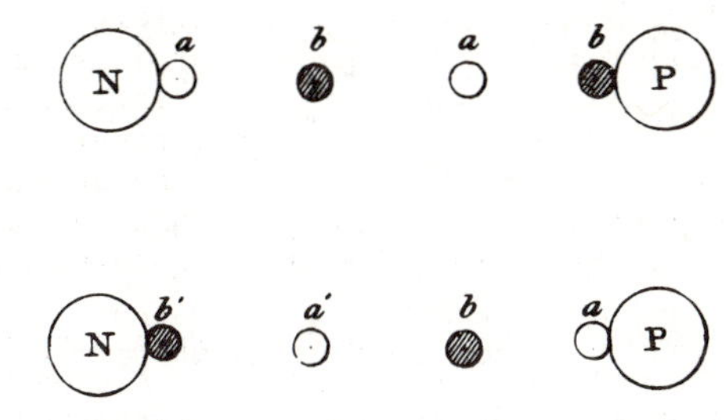

Figure 3. Faraday's illustration of his theory of electrochemical action. From Phil.Trans., 1833, 123, plate 19.

noted earlier Faraday's rate of discovery was such that he did not have sufficient time to digest his results before he made another discovery. He did not have time to cast his work in the experimental form which most suited his epistemological preferences. He needed some theoretical justification for the results of his experimental work which he could not justify in his normal manner. Hence in his fifth series Faraday proposed this theory of particle action ascribing to the particles the properties that were necessary for the explanation of the phenomena he had discovered.

Faraday, after developing this theory of electro-chemical action, turned in September 1833 to exploring the quantitative relationship he had found (66). He decided that this relationship might allow him to produce an instrument which would measure either relatively or absolutely, the quantity of electricity from a voltaic cell; in the end, although aware of the possibility (67), he did not pursue the absolute measure of electricity. Initially he continued to confirm his earlier observations that equal quantities of electricity produced equal amounts of electro-chemical decomposition. However, on 17 September he observed that the volume of gas in a tube that he had obtained by electrically decomposing water, had diminished (68). Judging by the manner in which Faraday immediately set to work to investigate this phenomenon, he must have felt that it potentially raised severe doubts about his quantitative relationship. Though he did not cease work on electro-chemistry, Faraday explored this phenomenon to ensure that it did not violate the quantitative electro-chemical relationship. The tube had a platinum pole; Faraday had observed what Berzelius was shortly to call the catalytic property of platinum which effected the combination of hydrogen and oxygen into water. At first Faraday thought that this might be connected with positive electricity as the combination seemed to be stronger there than at the negative pole (69). However, he found eventually that the strength of the combination was dependent on whether or not the platinum was clean or dirty. If clean then the combination proceeded well, if not then the contrary (70).

In his sixth series of Experimental Researches in Electricity Faraday described this work and reviewed the theories of platinum action on oxygen and hydrogen in a similar way to his earlier review of electro-chemical theories - that is he thought them inadequate to explain experimental phenomena (71). To explain how oxygen and hydrogen could combine under the influence of platinum Faraday had to take into account Dalton's work which showed that gases behaved as a vacuum towards each other (72). Faraday's problem was to explain how platinum overcame the resistance of oxygen and hydrogen in order to unite chemically. First he said that he admitted that he and probably others thought

> that the sphere of action of particles extends beyond those other particles with which they are immediately and evidently in union (73)

He argued that the platinum attracted the particles of hydrogen and oxygen and that this led to the diminution of elastic force on the side of the particles next to the platinum such that it would

> seem to result that the particles of hydrogen or any other gas or vapour which are next to the platina, &c., must be in such contact with it as if they were in the liquid state (74)

Hence it was possible for the gases to combine as they were now in their liquid form. This would, of course, be better with clean platinum, which Faraday had experimentally established, because its forces would have easier access to the gases. This brief description of Faraday's theory does not describe fully what was an extremely elaborate theory of action. However, by this time Faraday was beginning to be critical of this sort of theorising - one has to be especially careful with one's own theories to ensure that they do not become prejudices and thereby distort experimental knowledge. He wrote: "We have but very imperfect notions of the real and intimate conditions of the particles of a body existing in the solid, the liquid, and the gaseous state" (75). In this Faraday seems to have started to realise that he had gone too fast with interpreting his experimental work in terms of these unobservable entities. Despite this cautionary note Faraday interpreted this action in terms of particle properties, writing towards the end of the paper he commented

> The theory of action which I have given for the original phenomena appears to me quite sufficient to account for all the effects by reference to known properties (76)

What Faraday had established was that the recombination of oxygen and hydrogen was not an electrical phenomenon and hence did not affect his quantitative electro-chemical work to which he returned.

The initial experimental consequence of Faraday's discovery of the recombination of oxygen and hydrogen was to develop a new measuring device (which he later called a volta-electrometer) in which such combining action was not possible (77). With such instruments Faraday confirmed the quantitative relationship and established what became known as the second law of electro-chemistry.

He wrote up all these results in the seventh series of Experimental Researches in Electricity in late 1833 (78). He first noted that the theory of electro-chemical action he had advanced in the fifth series was greatly at variance with earlier theories and that his new theory of internal electro-chemical action within the decomposing body required a new nomenclature. He then proceeded to state the new electro-chemical terms with which we are now so familiar - electrode, electrolyte, cathode, anode etc. Faraday used these terms to emphasise that he was not taking any view of the nature of the electric current beyond what he had already stated in earlier papers. He described his volta-electrometers and described all the experiments by which he had tested all the possible instrumental differences which might have an effect on decomposition: the size of the electrodes, the intensity of electricity, variations in both, and the variation in the strength of the electrolytic solution. He showed that under all these different circumstances the same amount of water was decomposed when the same quantity of electricity was passed. He confidently stated:

> I consider the foregoing investigation as sufficient to prove the very extraordinary and important principle with respect to WATER, that when subjected to the influence of the electric current, a quantity of it is decomposed exactly proportionate to the quantity of electricity which has passed (79)

Which proposition he later graced with the title of law (80) and made general to other electrolytes.

At the end of this paper Faraday turned, once again, to the theory of the phenomena he had been investigating. This time he was much more circumspect and cautious than in his previous two series. When he discussed an atomic interpretation of his experimental results he wrote:

> I must confess I am jealous of the term atom; for although it is very easy to talk of atoms, it is very difficult to form a clear idea of their nature (81)

Presumably Faraday had in mind the uncritical ease with which he had introduced particle action in his previous two series. What thoughts led Faraday to this conclusion, which places series seven in very sharp contrast to five and six, is not clear and requires further research. When Faraday came to say what he would do next he emphasised that it was only by "a closer experimental investigation of the principles which govern the development and action of this subtile agent" (82) that further knowledge of electricity would be obtained. Perhaps Faraday realised that he had been theorising too much about matter and atoms and needed to experiment more. Perhaps he thought that the materialistic implications of the atomism he had been working with were undesirable. Certainly William Rowan Hamilton after meeting Faraday in mid 1834 wrote that he found him as anti-materialistic as himself (83). These are questions for future research.

What we can say now is that Faraday's electro-chemical work was intimately related to his electrical work and was not something separable from it. We can also say that the first law of electro-chemistry emerged naturally from his experiments and that he checked very carefully and very minutely that this quantitative relationship was in fact a real law of nature. Faraday had also to develop a theory of matter, at least initially, to cope with interpreting his work, with the wealth of discoveries he was making at this period. We see Faraday always willing to follow experiment, even if it led him towards quantification, and towards and away again from matter theory as his ideas developed and matured. Faraday repudiated the supremacy of mathematics in the exploration of nature, but he recognised that in the divinely-ordained economy of nature, constant proportions must apply (80). In his discovery of the first law of electro-chemistry, mathematics had a clear, but circumscribed, role.

Acknowledgments

I thank Dr David Gooding for discussions on Faraday generally and on this paper in particular. I also gratefully acknowledge the comments of three referees.

Literature Cited

1. However, this does not represent his only quantitative work. See M. Faraday, "Experimental Researches in Electricity - Eleventh Series. On Induction", Phil.Trans., 1838, 128: 1-40, section v for his work on specific inductive capacity.

2. M. Faraday, "Experimental Researches in Electricity - Seventh Series. On electro-chemical decomposition. On the absolute quantity of electricity associated with the particles or atoms of matter", Phil.Trans., 1834, 124: 77-122, paragraph 783
3. Ibid, paragraph 836
4. S. Ross, "Faraday consults the scholars: The origins of the terms of electrochemistry", Notes Rec.Roy.Soc.Lond., 1961, 16: 187-220
5. For the different interpretations of these laws that have been made see S.M. Guralnick, "The Contexts of Faraday's Electrochemical Laws", ISIS, 1979, 70: 59-75.
6. D. Gooding, "Metaphysics versus Measurement: The Conversion and Conservation of Force in Faraday's Physics", Ann.Sci. 1980, 37: 1-29, p.15-17.
7. Faraday to R.Phillips, 29 November 1831 in L.P. Williams, R. FitzGerald and O. Stallybrass (editors) The Selected Correspondence of Michael Faraday, 2 volumes, Cambridge, 1971, letter 122.
8. [J.C. Maxwell], "Scientific Worthies 1. Faraday", Nature, 1873, 8: 397-9.
9. Faraday's Diary. Being the various philosophical notes of experimental investigation made by Michael Faraday, DCL, FRS, during the years 1820-1862 and bequeathed by him to the Royal Institution of Great Britain. Now, by order of the Managers, printed and published for the first time, under the editorial supervision of Thomas Martin, 7 volumes and index, London, 1932-6. Hereafter this will be cited as "Faraday Diary", followed by volume and paragraph number. 26 March 1832, 1, 403.
10. G.N. Cantor, "Reading the Book of Nature: The Relation Between Faraday's Religion and his Science" in D. Gooding and F.A.J.L. James (editors), Faraday Rediscovered: Essays on the Life and Work of Michael Faraday, London, 1985, p.69-81.
11. L.P. Williams, "Faraday and Ampère: A Critical Dialogue" in Ibid, p.83-104
12. D. Gooding, "A Convergence of Opinion on the Divergence of Lines: Faraday and Thomson's Discussion of Diamagnetism", Notes. Rec.Roy.Soc.Lond., 1982, 36: 243-59.
13. Faraday Diary, 29 August 1831, 1, 1.
14. These they published in W. Ritchie, "Experiments in Voltaic Electricity and Electro-Magnetism", Phil.Trans., 1832, 122: 279-98. J. Davy, "An Account of Some Experiments and Observations on the Torpedo (Raia Torpedo, Linn)", Phil. Trans., 1832, 122: 259-78.
15. Faraday Diary, 29 August 1831, 1, 1.
16. D. Gooding, "'In Nature's School': Faraday as an Experimentalist" in Gooding and James op.cit. (10), p.103-35.
17. Faraday Diary, 29 August 1831, 1, 11.
18. Ibid, 1 October 1931, 1, 41.
19. Ibid, 45.
20. Ibid, 10 October 1831, 1, 65
21. R. Tweney, "Faraday's Discovery of Induction: A Cognitive Approach", in Gooding and James op.cit. (10), p.181-209.
22. Faraday Diary, 28 October 1831, 1, 92.
23. Ibid, 4 November 1831, 1, 158.

24. M. Faraday, "Experimental Researches in Electricity. On the induction of electric currents. On the evolution of electricity from magnetism. On Arago's magnetic phenomena", Phil.Trans., 1832, 122: 125-62. For the dates of reading see the Athenaeum, 26 November 1831, p.772, 10 December 1831, p.805, 17 December 1831, p.820.
25. Faraday op.cit. (24), paragraphs 22, 56, 133.
26. Faraday Diary, 8 March 1832, 1, 387.
27. Ibid, 1, 389
28. Ibid, 11 June 1832, 1, 429-41.
29. P.M., "An account of an Experiment in which Chemical Decomposition has been effected by the induced Magneto-electric current"; "Preceded by a letter from Michael Faraday", Phil.Mag. 1832, 1: 161-2.
30. L.P. Williams, Michael Faraday: A Biography, London, 1965, p.216.
31. Faraday Diary, 27 August 1832, 2, 16.
32. W.H. Wollaston, "Experiments on the Chemical Production and Agency of Electricity", Phil.Trans., 1801, 91: 427-34.
33. Faraday Diary, 31 August 1832, 2, 39 et seq.
34. Ibid, 6 September 1832, 2, 103 & 4.
35. Ibid, 10 September 1832, 2, 116
36. Ibid, 14 September 1832, 2, 122-7.
37. Ibid, 126.
38. Ibid, 129 et seq.
39. Ibid, 130.
40. Ibid, 131.
41. Ibid, 134.
42. Ibid, 129.
43. Ibid, 133.
44. Ibid, 134.
45. Ibid, 135.
46. Ibid, 136.
47. Ibid, 137.
48. Ibid, 15 September 1832, 2, 138.
49. Ibid, 147.
50. Ibid, 146.
51. Ibid, 159.
52. M. Faraday, "Experimental Researches in Electricity - Third Series. Identity of Electricities derived from different sources. Relation by measure of common and voltaic electricity", Phil.Trans., 1833, 123: 23-54.
53. Ibid, paragraph 346.
54. Ibid. J.N.P. Hachette, "De l'Action Chimique produite par l'Induction electrique; Decomposition de l'Eau", Ann.Chim., 1832, 51: 72-6.
55. Faraday, op.cit. (52), paragraph 360.
56. See "Henry's European Diary" in The papers of Joseph Henry, Washington, 1972- , 3, 315-20.
57. Faraday, op.cit. (52), paragraphs 320 & 9.
58. Ibid, paragraphs 366, 367, 377.
59. Faraday Diary, 23 January 1833, 2, 222 et seq.
60. M. Faraday, "Experimental Researches in Electricity - Fourth Series. On a new law of electric conduction. On conducting power generally", Phil.Trans., 1833, 123: 507-22.

61. M. Faraday, "Experimental Researches in Electricity - Fifth Series. On electro-chemical decomposition", Phil.Trans., 1833, 123: 675-710.
62. Ibid, paragraph 517.
63. Ibid, paragraph 518.
64. Ibid.
65. Ibid, paragraph 510.
66. Faraday Diary, 2 September 1833, 2, 628 et seq especially 638.
67. Faraday, op.cit. (2), paragraphs 736-9
68. Faraday Diary, 17 September 1833, 2, 714.
69. Ibid, 10 October 1833, 2, 882 et seq.
70. Ibid, 14 November 1833, 2, 1060 et seq.
71. M. Faraday, "Experimental Researches in Electricity - Sixth Series. On the power of metals and other solids to induce the combination of gaseous bodies", Phil.Trans., 1834, 124: 55-76.
72. Ibid, paragraph 626.
73. Ibid, paragraph 619.
74. Ibid, paragraph 627.
75. Ibid, paragraph 626.
76. Ibid, paragraph 656.
77. Faraday Diary, 21 September 1833, 2, 760.
78. Faraday op.cit. (2).
79. Ibid, paragraph 732.
80. Ibid, paragraphs 747 & 783.
81. Ibid, paragraph 869.
82. Ibid, paragraph 873.
83. W.R. Hamilton to S. Hamilton, 30 June 1834, in R.P. Graves, Life of Sir William Rowan Hamilton, 3 volumes, Dublin, 1882-9, 2, 95-6.
84. Gooding, op.cit. (6).

RECEIVED August 9, 1988

Chapter 4

The Universal Agent of Power

James Prescott Joule, Electricity, and the Equivalent of Heat

Stella V. F. Butler

Greater Manchester Museum of Science and Industry, Castlefield Manchester M3 4JP, United Kingdom

James Prescott Joule is best known for his determination of the mechanical equivalent of heat. Joule began his scientific career by investigating electrical phenomena. Joule was fascinated by the possibilities that electromagnets might become useful as sources of industrial power. He designed and built various electro-magnetic engines, trying to understand how to produce maximum efficiency. He began to link together electricity, heat and mechanical power by observing their transformations. His investigations culminated in his quantification of the dynamical equivalent of heat. Joule lived and worked in an industrial community; his concern about "power" was shared by many others who derived income from manufacturing. His success in developing a conceptual framework about energy owed much to his considerable experimental skills.

At the 1847 annual meeting of the British Association for the Advancement of Science (BAAS), in Oxford, James Prescott Joule, a young Salford-born brewer, presented a paper in which he demonstrated that heat must be regarded as an equivalent form of mechanical "force" or, what we could call, "energy". (1) Four years previously, at the Cork meeting of the BAAS, he had begun by showing that the work done to drive a magneto-electric engine could be converted by means of the electric current into the equivalent amount of heat. (2) Despite the implications of Joule's conclusions for the interpretation of nature and especially his assertion that heat was simply another form of "force" and not, as many still held, a material substance, his presentation in Oxford did not arouse much interest. However, William Thomson, who had been appointed Professor of Natural Philosophy at Glasgow University the year before, was impressed by Joule's experimental work and his ideas. (3) The two men discussed his paper and thus began one of the most important friendships in the history of science.

0097–6156/89/0390–0050$06.00/0

Joule's work provided important experimental proofs for Thomson's exposition of the laws of thermodynamics published between 1851-1855. (4) Indeed in the opening passage to his paper on the mechanism of electrolysis, Thomson maintained "certain principles discovered by Mr. Joule .... must ultimately become an important part of the foundation of the mechanical theory of chemistry". (5)

To a certain extent Joule's presentation to the BAAS in 1847 and his lecture at St. Ann's Church School the previous month represent the end point in his investigations concerning the relationships between various physical "forces" notably electricity, heat, and mechanical power. Joule was not, of course, alone in speculating about the nature of heat. As Thomas Kuhn has outlined, between 1842 and 1847 three other individuals, Mayer, Colding and Helmholtz also presented similar hypotheses. (6) Others also speculated on the concept of energy or "force" as a unifying heuristic tool for understanding natural phenomena. There were a number of reasons why these philosophers presented similar hypotheses at this period, including in many cases, a "Naturphilosophen" approach in which the organism is used as a fundamental metaphor for understanding the natural world. In this paper I focus specifically on the development of Joule's ideas by examining his experimental work. The sequence of investigations through which he formulated his theories underline the importance of his enthusiasm for electrical phenomena and, in particular, magneto-electricity. By examining both Joule's published work and his apparatus, part of which survives in the Joule Collection of the Greater Manchester Museum of Science and Industry, the considerable experimental skill involved in determining the equivalent values becomes clear. We are also able to chart the changing style of his scientific work, and can begin to relate his method of work to his developing theories about natural phenomena.

A Fascination for Magnets

James Prescott Joule was born in Salford in 1818. (7) His father owned a brewing business from which the family derived considerable wealth. James, a delicate child with a minor spinal deformity was educated at home with his elder brother Benjamin. Between 1833 and 1837, the two boys were taught maths and science by the venerable John Dalton.

The following year, 1838, Joule's first researches were published in the Annals of Electricity. He described different forms of electro-magnetic engines, his aim to design a more efficient and powerful machine. (8) Annals had been established as a scientific periodical in 1836 by William Sturgeon who moved to Manchester in 1838 to set up the Royal Victoria Gallery of Practical Science. (9) Joule's interest in electro-magnetism was stimulated by this self-taught man who, in 1825 had demonstrated the first soft-iron electromagnet at the Society of Arts in London. (10) The Royal Victoria Gallery was a speculative venture which did not enjoy long term success. Nevertheless, Sturgeon clearly drew around himself a group of enthusiasts for electro-magnets whose designs were exhibited at the Gallery. Joule also lectured there in February, 1841 "On a New Class of Magnetic Forces". (11)

Joule, like many others, envisaged electricity as a source of industrial power: in May 1839 he wrote "I can hardly doubt that electro-magnetism will ultimately be substituted for steam to propel machinery". (12) He therefore set about investigating the relationship between the current used to create the magnet and the strength of magnetic attraction. (13) He designed, made, and calibrated delicate galvanometers. He then measured the relationship between the strength of the magnet and the size of the current, establishing that the magnetic attraction is proportional to the square of the electric current and the square of the length of the wire. The similarities between the formula he deduced and the laws of gravitational attraction no doubt appealed to him. This work was typical of Joule's experimental style - he was always concerned to devise accurate and sensitive measuring instruments and always interested in the mathematical relationships between phenomena.

By 1841 however, Joule's optimism regarding the industrial potential of his electromagnets had diminished. In his lectures at the Victoria Gallery in February he described the simple rules he had found to hold true of his electro-magnetic engines - that the "duty" which, for Joule was a measure of the efficiency of the machine, is proportional to the intensities of the battery, and that what he called the "magneto-electric resistance" produced by the rotation of the bars, (the back emf) acts against the battery current which consequently reduces the magnetism of the bars. (14) At certain velocities therefore the force is so reduced that the bars are no longer accelerated. He concluded "I almost despair of the success of electro-magnetic attractions as an economical source of power" (15) - for, his electro-magnetic engines simply did not compare to the fuel efficiency of steam engines: without vastly more efficient batteries Joule could not see how to improve his machine.

Heat and Electricity

A few months before this lecture Joule had presented his first experiments on the relationship between heat and electricity. He simply placed coils of different kinds of wire in jars of water and measured the change in temperature. This work represents a shift away from his concern about improving and inventing gadgets and a growing interest in measurable phenomena, most notably electricity. Just as he had looked for simple mathematical relationships to understand magnetisation so he sought patterns in the quantities of heat and electricity which he was able to measure. In his first communication he maintained that "the calorific effects of equal quantities of transmitted electricity are proportional to the resistance opposed to its passage whatever may be the length, thickness, shape or kind of metal which closes the circuit". (16) Although, at this stage his work gave no opposition to the caloric theory which conceived of heat as a fluid, Joule in presenting ideas about magnetism, maintained that atoms were surrounded by atmospheres of electricity and magnetism; the skewing of these atmospheres explained electrical and magnetic phenomena. The vibration of the magnetic atmospheres "is called heat and will of course, increase in violence and extent with the increase of temperature of the bar". (17)

Although Joule used the term "caloric" value of heat he, like many of his contemporaries, never thought in earlier nineteenth century terms of "caloric" as a material substance. (18)

By August 1841, Joule could report on further investigations into heat and electricity. (19) This time he wrapped the conducting wire around a glass tube which was then placed in a jar of water. This apparatus has survived and is illustrated in Fig.1. Using sensitive galvanometers and thermometers he established that the heat evolved is proportional to the resistance of the conducting wire and the square of the "electric intensity" (current). Faraday had previously noted the heating effects of the electric current and had speculated on the relationship of the two forces suggesting that they are always "definite in amount". (20)

Joule went on in the same paper to look at the heat evolved in the cells of batteries. By the end of 1841 he was beginning to regard the electric current as, fundamentally, a force - he carried out experiments to demonstrate that the heat evolved in "ordinary chemical combination" is the "product of resistance to electric conduction". (21) He maintained that such "phenomena are easily understood if, with the great body of philosophers we keep in view the intimate relation which subsists between chemical affinity and the electric current". (22)

In 1832 Faraday had demonstrated the decomposition of water by magneto-electricity, and had gone on to suggest that equivalent weights of chemicals are associated with equal quantities of electricity. (23) For Faraday, as for Joule, electricity represented the combining force of atoms for one another.

Joule was elected to the Manchester Literary and Philosphical Society on 25th January, 1842. Later that year, he attended the BAAS meeting which Manchester hosted and presented a paper on his electrical experiments. The BAAS provided Joule with a national audience for his ideas and brought him into contact with others who shared his electrical interests. (24) These included Reverend William Scoresby, vicar of Bradford, who had carried out a number of experiments on electro-magnets. Joule subsequently worked with Scoresby, designing and building an electro-magnetic engine to test its "duty" or efficiency. (25)

At the BAAS meeting, Joule described his own combustion experiments before discussing his electrolysis results - he maintained that the heat of chemical reactions resulted from the electrical resistance between the atoms at the moment of their union. (26)

Following these experiments, and others on the heat evolved during the electrolysis of water, he began to formulate a theory of heat. He concluded that the heating power of the current produced per chemical equivalent of electrolyte in the battery is proportional to the electromotive force. (27) He had already established in his work on electro-magnetism that the mechanical power of a machine is also proportional to the intensities of the batteries or the electromotive force. He therefore suggested a constant mathematical relationship between the mechanical and heating powers of the current. Before testing this proposition, he maintained that "electricity may be regarded as the grand agent for carrying, arranging and converting chemical heat". (28) At this time he

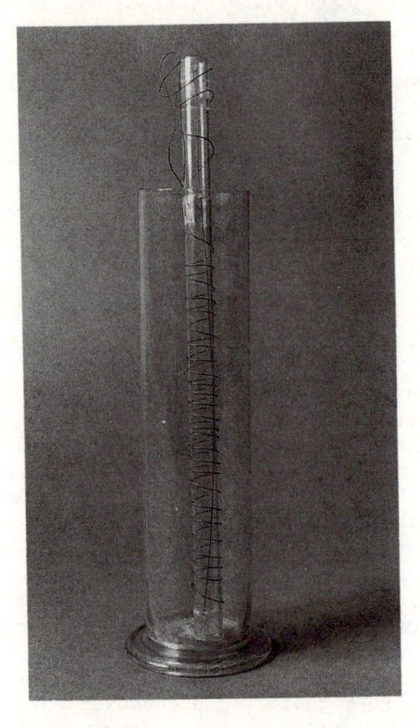

Figure 1. Apparatus used by Joule in his 1841 investigation concerning the relationship between heat and electricity. Courtesy of the Greater Manchester Museum of Science and Industry Trust.

still regarded heat as the "momentum of the atmospheres" of electricity.

He went on to explore the relationship between heat and mechanical "force" via the agency of electricity by immersing a small revolving electromagnet in water and measuring the changes in temperature for a given size of current induced in it by rotating it between the poles of another magnet. The little electromagnet was rotated by a system of falling weights. (29) He demonstrated that the heat evolved by the coil was proportional to the square of the current induced by the rotation of the electromagnet. All that had happened was that "force" had been expended and heat generated. He went on to demonstrate that the same effects are observed if the current through the electromagnet is produced by a battery instead of by electromagnetic induction. Having established the equivalence of voltaic and magneto electricity he went on to investigate what happens when you have both in a circuit. In one series the current induced in the electromagnet was used to oppose the current of the batteries, in another it was used to add to the effect of the battery. His results, he maintained were consistent with the "law" he had established earlier that the heat observed is proportional to the square of the current - i.e. there is no special effect observed because of the assistance or resistance which the magneto-electricity presents to the voltaic current.

For Joule this was further evidence for the fundamental nature of electricity as a force. Magneto-electricity could, he maintained, generate heat. (30)

It was a short step to measuring not only the temperature and the current observed in the electro-magnetic induction experiments but also the mechanical power necessary to turn the apparatus. He used his system of falling weights linked via pulleys and an axle to the magneto-electric machine.

He simply calculated the mechanical "force" expended measuring the distance fallen by his (known) weights. The falling weights drove the magneto-electric engine, the current from which generated heat. Having ascertained the force required to produce a particular temperature change, he equated it to the heat evolved, concluding that "the quantity of heat capable of increasing the temperature of a pound of water by one degree of Fahrenheit's scale is equal to and may be converted into a mechanical force capable of raising 838 lbs to the perpendicular height of one foot". (31) This was Joule's first full articulation of the dynamical theory of heat. In a post-script to his paper, presented at the Chemical Section of the British Association meeting in Cork, in August 1843, he elaborated upon his dynamical theory of heat by supporting Count Rumford's assertion that the heat observed when boring cannon results from friction not from a change in the "capacity" of the metal.

He went on to modify his views regarding the electrical nature of chemical heat. Previously he had believed that the heat evolved in chemical reactions was due to the electrical current involved in the chemical process. Instead he suggested that heat was the result of the "mechanical force expended by the atoms in falling towards one another." (32) He believed that his propositions would ultimately provide an explanatory framework for the whole of chemistry.

Further Work on the Mechanical Equivalent of Heat

These sets of experiments on the relationship between magneto-electricity and heat broadened Joule's view of the nature of heat and "force". Until this time, the summer of 1843, all his experiments had involved electrical phenomena. Although he did not abandon this interest, Joule's work began to focus on mechanical force and the possibility of demonstrating its direct equivalence with heat in the absence of the agency of electricity. He investigated the long-observed thermal effects produced when gases are compressed and allowed to expand. (33) He constructed two identical copper receivers into which he compressed air using a condensing pump he constructed for this purpose. From the observations of Erasmus Darwin, Dalton and Cullen before him, Joule realised that he could expect to observe an increase in temperature. He anticipated the effects to be slight and consequently commissioned from John Benjamin Dancer, perhaps the best known of Manchester's instrument makers, thermometers of "extreme sensibility and very great accuracy". (34) Dancer also made Joule a travelling microscope to enable the scale of each thermometer to be precisely etched. According to Joule these instruments represented a significant improvement in sensitivity and accuracy in thermometry in Britain. Previously Joule had bought thermometers from Fastré, the Parisian instrument maker. Joule showed that when compressed air from one cylinder was allowed to expand into an evacuated cylinder the net temperature change was zero. Under these circumstances no net external work is done so no heat is lost. He went on to calculate from the expansion of air a further set of values for the mechanical equivalent of heat which broadly agreed with the electromagnetic experiments. These experiments and, in particular, Joule's concern for very sensitive measuring instruments suggest that he was, to some extent at least, using experiments to prove his theory of the nature of heat. This experimental style contrasts with his earlier, more open-ended approach.

By 1847, Joule had further evidence for the generation of heat by mechanical work through the famous paddlewheel experiments. Using falling weights to turn a wheel in a bath of water, he equated the rise in temperature with the friction produced through the force of the falling weight. (35) Again he calculated a value for the mechanical equivalent of heat. Joule repeated these experiments with slight modifications in 1849 and 1878 establishing ever more accurate values for his equivalence. (36) His conclusions were received with sceptism by some. In 1849, the Royal Society refused to publish his assertion that friction was essentially the conversion of mechanical power into heat.

In 1847, Joule was able to outline in detail his views on the conservation of forces and their unity in his lecture at St. Ann's Church School. This was published subsequently in the *Manchester Courier*. (37) Joule maintained that "the most convincing proof of the conversion of heat into living force has been derived from my experiments with the electro-magnetic engine, a machine composed of magnets and bars of iron set in motion by an electrical battery". (38)

Joule continued to present his views both to his immediate peers within the Manchester scientific community and through both the Royal Society and the BAAS. As we have already noted, at the annual meeting of the BAAS in Oxford he met William Thomson. Although Thomson did not immediately accept Joule's propositions about heat they eventually became an important component in his synthesis of ideas concerning energy.

The Great Experimenter

Although by the 1850s he had completed his mechanical equivalent work, Joule continued to undertake a wide range of investigations. In 1861, at the urging of the telegraph engineers Sir Charles Bright and Mr. Latimer Clark, the British Association set up a committee to consider the need to establish a unit for electrical resistance and to decide the best form in which a standard could be kept. (39) Joule joined the committee the following year. Resistance was first measured in absolute units (cm and seconds) by James Clerk Maxwell, Balfour Stewart and Fleeming Jenkin. Joule's electrical heating law (current squared times resistance) coupled with his accurate determination of the mechanical equivalent of heat provided a vital check on the accuracy of this work. There was good, but not complete agreement. Joule therefore set out to determine the dynamical equivalent of heat. (40) The slight difference persisted. It was finally found by Lord Rayleigh and his group at Cambridge, that Maxwell, Balfour Stewart and Fleeming Jenkin were in error while H.A. Rowland, in the States, showed that Joule's final result (772.55 ft lbs per BThU) had to be increased slightly due to his reliance on glass thermometers. Joule demonstrated his genius for instrumentation during this investigation by developing a current balance - a horizontal flat coil was suspended by the current carrying wire between two fixed coils. The current could be determined by measuring the forces required to counterbalance the attractive or repulsive force experienced by the suspended coil. An instrument similar in design was used by Rayleigh in 1882 and also by the Laboratoire Central d'Eléctricité, Paris to determine the absolute value of current.

Although this current balance does not survive, the Joule Collection does contain a number of pieces of apparatus, some dating from the early part of Joule's career. His passion for electro-magnets is well demonstrated by the range of magnets. These include a pair made in 1839 to investigate whether any advantage could be gained by using a core made up of wires instead of a solid piece of iron (see Figure 2). The electromagnet constructed to give "great lifting-power" also survives. (41) From these relatively crude but, by Joule's accounts, effective pieces of apparatus we can also gain some insight into the process of his endeavour. For him, science clearly involved engineering skills. His "laboratory" was as much a workshop with furnace and lathe as an area for delicate experimentation.

In 1845 Joule could assure William Scoresby with whom he collaborated on a number of experiments on the "duty" or efficiency of electromagnets of the availability of steel wire, wrought iron tubing and copper wire which he sometimes ordered ready covered with

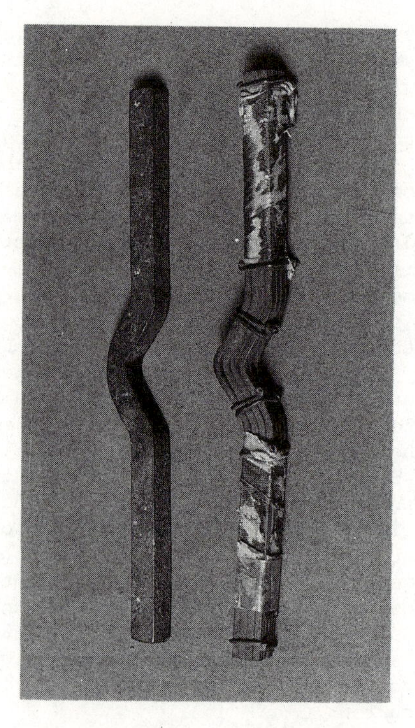

Figure 2. Electromagnets constructed by Joule to investigate the difference in effect between a solid core and a core made up of iron wires. Courtesy of the Greater Manchester Museum of Science and Industry Trust.

cotton. (42) Joule clearly suffered few financial constraints to his experimenting and could obtain all he needed from Manchester's engineering suppliers. He was very proud of his magnets - in 1851 he applied via the Royal Manchester Institution to exhibit an electro-magnet in the Great Exhibition. (43)

As well as the cruder pieces of apparatus, the collection also contains finely crafted instruments including a travelling microscope, specially constructed for Joule by John Benjamin Dancer. Dancer moved to Manchester from his home in Liverpool in 1841. (44) He had been apprenticed originally as an instrument maker to his father from whom he inherited the family business in 1835. Subsequently he went into partnership with A. Abraham, a very successful Liverpool instrument maker with the agreement that the young Dancer would move to Manchester to set up a branch of the firm. Dancer worked very closely with Joule. During 1844 Joule joined Dancer in his "workshop .... every morning for sometime until we completed the first accurate thermometers which were ever made in England". (45)

This close relationship between maker and customer was fostered in Manchester by a relatively open scientific community. Joule and Dancer undoubtedly met socially at the Lit and Phil as well as at lectures or conversaziones at the Royal Manchester Institution. Dancer also made several of the instruments Joule required for the long series of experiments on atomic volume he carried out in conjunction with Lyon Playfair. (46) Both Playfair and Dancer were close friends of John Mercer, (47) a manufacturing chemist from Oakenshaw, Lancashire, who was responsible for inventing the process of treating cotton with strong alkali to ensure a good uptake of dye, which continues to bear his name. This remarkable group of scientific talent shared a concern to investigate observable phenomena and demonstrated exceptional ability for inventing new gadgets and chemical processes.

However, the collection is not confined to objects originating from Manchester and Salford. Joule also purchased instruments from London - R. and G. Knight and Watkins and Hill. His papers demonstrate the breadth of his knowledge of scientific literature. He was clearly also well and critically informed about the availability of instruments.

Conclusions

The 1840 edition of the Encyclopaedia Britannica noted that "there is no branch of science more likely to reward the diligence of the young investigator than that which treats of the electric fluid in animal and vegetable life, its effects upon inorganic matter and its connection with the imponderable agents of light and heat". (48) The Voltaic pile had opened up a range of experimental possibilities after 1800; subsequently the work of Oersted and Faraday had indicated the possibilities of magneto-electricity. Faraday's work captured the imaginations of many enthusiastic investigators. Joule was, therefore, by no means alone in developing strong magnets and electro-magnetic machines. What set Joule apart from his fellow enthusiasts was his fascination with the connections between both magneto and voltaic electricity and the "imponderables" of heat and more especially mechanical force. Joule had grown up with, and drew his livelihood from industry. In Manchester, one of Britain's great manufacturing centres, sources of mechanical power were of tremendous importance. It is not surprising that Joule's interest in electricity and in particular in the design of electro-magnetic engines lay partly in their possibility as sources of industrial power.

Joule's main investigations on the relationship between electricity, heat and mechanical power were completed by 1847. His early descriptions of his electromagnets indicate the importance of William Sturgeon's influence and the lively group of inventors centred around Manchester's Victoria Gallery of Science. The electromagnets that survive emphasise these links. After 1842 his published work demonstrates his familiarity with the work of other investigators. His apparatus also indicates his knowledge of scientific instruments, both British and French, and his close relationship with John Benjamin Dancer. Joule probably directed the instrument maker with detailed specifications, while Dancer contributed his considerable technical expertise. Bennett has noted that this type of collaboration was relatively new in British science in this period. (49)

Close study of Joule's experimental procedures also indicate the meticulous skill which enabled him to obtain very accurate measurements. From his first experiments on the relationship between the strength of a magnet and the strength of the electric current, Joule was concerned to develop very sensitive measuring instruments. He devised galvanometers capable of detecting small differences in current and later thermometers which could measure temperature differences as small as 1/200°F. These concerns for accuracy, quantification using standard units and fine measurement, evident in his work from his first published papers, set Joule aside from his British peers and links him with the next generation of investigators, most notably his friend and colleague William Thomson who became principle advocate for modern thermodynamics based on Joule's ideas of the dynamical theory of heat.

Acknowledgments

I am very grateful to John Stock for his encouragement to prepare this paper.

Donald Cardwell, David Gooding and Neil Brown provided valuable and stimulating comments to earlier drafts. The British Council and the American Chemical Society provided financial support to make possible my involvement in the symposium on electrochemistry.

Literature Cited

1. Joule, J.P. Report of the Ann. Mtg. of the Brit, Assoc. Adv. of Sci. Chem. Sec., 1847, p 55.
2. Joule, J.P. "On the calorific effects of magneto-electricity and on the mechanical value of heat", in Joule, J.P. Collected Papers, Physical Society: London; 1884, vol, 1, pp 123-159.
3. Smith, C.W. "William Thomson and the Creation of Thermodynamics: 1840-1855", Archive Hist. Exact Sci., 1976, 16, 231-288.

4. Thomson, W. "On the dynamical theory of heat; with numerical results deduced from Mr. Joule's 'Equivalent of a thermal unit' and M. Regnault's 'Observations on steam'", Trans. Roy. Soc. Edin., 1853, 20, 261-288.
5. Thomson, W., Mathematical and Physical Papers, University Press: Cambridge, 1882-1911; vol. 1, pp 472-289.
6. Kuhn, T. "Energy conservation as an example of simultaneous discovery", in Critical Problems in the History of Science; Clagett, M.,Ed; Wisconsin Press: Wisconsin, 1959; Chapter 11.
7. Reynolds, O. "Memoir of James Prescott Joule", Proc. Manchester Lit. and Phil. Soc., 1892, 6.
8. Joule, J.P. "A Short Account of the Life and Writings of the Late Mr. William Sturgeon", Mem. Manchester Lit. and Phil. Soc, 1857, 14, 77-83.
10. Cardwell, D. "Science and Technology: The Work of James Prescott Joule", Technology and Culture, 1976, 17, 674-687.
11. Joule, J.P. Collected Papers, Physical Society: London, 1884; vol.1, pp 46-53.
12. Ibid., p 14.
13. Ibid., pp 10-14.
14. Ibid., pp 47.
15. Ibid., p 48.
16. Ibid., pp 59-60.
17. Ibid., p 53.
18. S. Brush, The kind of motion we call Heat, New York, 1976; pp 31-32.
19. Joule, J.P. Collected Papers, Physical Society: London 1884; vol. 1, pp 60-81.
20. Faraday, M. Experimental Researches in Electricity (1855), reprinted in Great Books of the Western World: Chicago, 1952; para. 1625.
21. Joule, J.P. Collected Papers, Physical Society: London, 1884; vol. 1, p 82.
22. Ibid., p 90.
23. Report of the Ann. Mtg. of the Brit. Assoc. Adv. Sci., 1832.
24. Morrell, J.; Thackray, A. Gentlemen of Science. Early Years of the British Association for the Advancement of Science; Clarendon Press: Oxford, 1981; p 411.
25. Stamp, T.; Stamp, C. William Scoresby, Arctic Scientist; Caedman of Whitby Press: Whitby, 1976; pp 175-185.
26. Joule, J.P. Collected Papers; Physical Society: London, 1884; vol. 1, pp 102-107.
27. Ibid., p 120.
28. Loc. Cit.
29. Ibid., pp 123-159.
30. Ibid., p 146.
31. Ibid., p 156.
32. Ibid., p 158.
33. Ibid., pp 172-189.
34. Ibid., p 174.
35. Ibid., p 202.
36. Ibid., pp 298-328, 632-657.
37. Ibid., pp 265-276.
38. Ibid., p 279.

39. Reports of Committee on Electric Standards appointed by the British Association for the Advancement of Science; Cambridge University Press: Cambridge, 1913; p xvii.
40. Ibid., p xxi.
41. Joule, J.P. Collected Papers; Physical Society: London, 1884; vol. 1, p 40.
42. Stamp, T.; Stamp, C. William Scoresby, Arctic Scientist; Caedmon of Whitby Press: Whitby, 1976; p 180.
43. Royal Manchester Institution Archives, Letter from J.P. Joule to Manchester Great Exhibition Committee, 30th April, 1850. Manchester Central Reference Library, M6/3/11/9.
44. "John Benjamin Dancer, FRAS 1812-1887, An Autobiographical Sketch", reprinted in Proc. Manchester Lit. and Phil. Soc., 1964-5, 107, 1-27.
45. Ashworth, J.R. "A list of apparatus now in Manchester which belonged to J.P. Joule, FRS with remarks on his Mss, letters and autobiography", Mem. Manchester Lit. and Phil. Soc., 1930-1, 75, 105-117, p 112.
46. Joule, J.P.; Playfair, L., in Joule's Collected Papers; Physical Society: London, 1884; vol. 2, pp 11-215.
47. Parnell, E.A. The Life and Labours of John Mercer, FRS, FCS; Longmans: London, 1886.
48. "Electricity" in Encyclopaedia Britannica; A. and C. Black: Edinburgh, 1840; 7th edition, vol. 8.
49. Bennett, J. "Instrument makers and the decline of science in England". In Nineteenth Century Scientific Instruments and their Makers; P.R. de Clercq, Ed.; Rodopi: Amsterdam, 1985; pp 13-27.

RECEIVED August 3, 1988

Chapter 5

The Contribution of Electrochemistry to the Development of Chemical Kinetics

Keith J. Laidler

Department of Chemistry, University of Ottawa, Ottawa, Ontario K1N 9B4, Canada

Arrhenius's theory of electrolytic dissociation, proposed in 1887, gave the first great impetus to chemical kinetics, and led to an understanding of catalysis by acids and bases. Later the Debye-Hückel theory of strong electrolytes and G.N. Lewis's definition of ionic strength led to a satisfactory interpretation of the kinetic effects of added foreign salts. These effects showed that, when transition-state theory is applied, the rate of a second-order reaction between A and B must involve a kinetic activity factor $y_A y_B / y_{\neq}$, where y_A and y_B are the activity coefficients of the reactants and $y_{\neq}$ is the activity coefficient of the activated complex. The detection of general acid-base catalysis was consistent with the Brønsted-Lowry extended definition of general acids and bases.

Electrochemistry and kinetics were born within a few years of each other, and grew up in close association. Unlike their sibling thermodynamics, which had a sickly and confused childhood, theirs was a healthy and vigorous one, largely as a result of the friendly cooperation between them. Most of those who made substantial contributions to either electrochemistry or kinetics made significant contributions to the other; names that come at once to mind are Faraday, Arrhenius, van't Hoff, and Ostwald.

The year 1887 is conveniently regarded as the year of birth of physical chemistry, by which is meant that the subject was then first recognized as a separate branch of chemistry. It was in that year that the first journal of physical chemistry, the *Zeitschrift für physikalische Chemie*, was founded, and that Wilhelm Ostwald (1853-1932) published the second and final volume of his *Lehrbuch der allgemeinen Chemie*, the first textbook of physical chemistry. In that same year Ostwald was appointed professor of physical chemistry at Leipzig, and Svante Arrhenius (1859-1927) published his famous theory of electrolytic dissociation.

Work in physical chemistry had, of course, been done before 1887, but it was somewhat spasmodic. Faraday's masterly work on

0097-6156/89/0390-0063$06.00/0

electrolysis (1), carried out in the 1830s, may perhaps be regarded as the first work in electrochemistry. Wilhelmy's work on the rate of inversion of sucrose (2), published in 1850, is perhaps the first work in kinetics although some interesting rate measurements had been made much earlier (3, 4). However, the work that gave the greatest impetus to both electrochemistry and kinetics was done in the early 1880s. It was then that Arrhenius was making the conductance measurements that led to his theory of electrolytic dissociation, and that J.H. van't Hoff (1852-1911) was working at the University of Amsterdam on osmotic pressure and was carrying out investigations in chemical kinetics that led in 1884 to his famous book *Etudes de dynamique chimique*, the first textbook on kinetics. Also at that time, at the Riga Polytechnic Institute, Ostwald was investigating the behaviour of acids and was studying the acid-catalyzed hydrolysis of various esters. By a happy chance these three men, dubbed by their contemporaries 'Die Ioner' - 'the Ionists' - came together in their endeavours; they corresponded and met over a period of years, and there is no doubt that the friedly relationship that existed between them had a strong influence on the development of both electrochemistry and kinetics. (For more detailed accounts see refs. 5 and 6).

Electrolytic Dissociation

Ironically, the happy chance was that Arrhenius's dissertation at the University of Uppsala, submitted in 1884, was not well received by the examiners, who gave it such a poor rating that his prospects for an academic career were seriously in jeopardy. Fortunately, instead of accepting defeat and retiring into oblivion, Arrhenius sent copies of his dissertation to several prominent scientists, including Ostwald and van't Hoff. Both of them, while not at once appreciating the significance of what Arrhenius had done, realized that it was probably of great importance and gave him much encouragement. Ostwald, in fact, took the unusual step of travelling to Stockholm to meet Arrhenius and discuss his work, a surprising compliment for a young but already distinguished professor to pay to a younger man who had nearly failed to obtain his degree. Ostwald also offered Arrhenius an appointment at the Riga Polytechnic Institute, but for personal reasons Arrhenius was unable to accept it.

Arrhenius had not had much help in carrying out his research, and in writing his dissertation, from the professors at the University of Uppsala. His dissertation was not clearly written, and his interpretations of his results were far from satisfactory; there was no suggestion in it that there can be dissociation into ions. The low rating the dissertation received was not altogether surprising. The discussions with Ostwald were valuable in leading Arrhenius to the idea of dissociation, although Ostwald himself, ever cautious, did not accept the idea until a little later. Before receiving a copy of Arrhenius's dissertation in June, 1984, Ostwald had himself made measurements of the electrical conductances of solutions of a number of acids, and had made kinetic measurements of their catalytic activities. He had observed a marked parallelism between the conductances, measured at a fixed concentration of acid, and the rate constants, and had also noted that the nature of the anion had little effect on the rates. On receiving Arrhenius's dissertation Ostwald at once prepared a short paper (7), submitted in July, 1884, in which he described his own kinetic and conductance results and discussed them on the basis of Arrhenius's work, to which he made full acknowledgement. At the time he was not, of course, able to explain the rate-conductance correlation in a very satisfactory way. It is easy today to understand this correlation. Hydrogen ions have much higher mobilities than the anions of the acids, so that the conductance of a solution of an acid provides an approximate measure of the hydrogen ion concentration in the solution; Ostwald was thus really observing a correlation between the rates of the acid-catalyzed reactions and the concentrations of hydrogen ions.

The support that van't Hoff provided to the idea of electrolytic dissociation came not from his kinetic work but from his osmotic pressure investigations. The German botanist Wilhelm Pfeffer (1845-1920) had used semipermeable membranes to make numerous osmotic pressure measurements, and van't Hoff noted that for a number of solutions the osmotic pressure Π is related to the concentration by an equation of the same form as that for an ideal gas:

$$\Pi V = nRT \tag{1}$$

Here R is the gas constant and T the absolute temperature; V is the volume and n the molar amount of solute present. Since n/V is the concentration c this equation can be written as

$$\Pi = cRT \tag{2}$$

However, van't Hoff noticed that in the case of certain solutes the osmotic pressure was greater than given by this formula, and for them he introduced a factor *i* that came to be called the 'van't Hoff factor':

$$\Pi = icRT \tag{3}$$

After reading Arrhenius's dissertation and discussing the matter with him van't Hoff realized that electrolytic dissociation could account for the result, since dissociation leads to an increase in the number of solute species present in the solution. It was in 1885-8 that van't Hoff published his classical papers (8-10) on osmotic pressure.

Arrhenius was awarded a travelling fellowship by the University of Uppsala, the authorities having realized belatedly, as a result in particular of the interest shown by Ostwald, that the dissertation could not have been so bad after all. Arrhenius made a very profitable use of his fellowship. He spent some months in 1886 with Ostwald in Riga; later that year and early in 1887 he worked in Würzburg with the German physicist Friedrich Kohlrausch (1840-1910). Later in 1887 he worked with the Austrian physicist Ludwig Boltzmann (1844-1906) at the University of Graz. Early in 1888 he visited van't Hoff in Amsterdam and then went to work again with Ostwald, who by that time had moved to the University of Leipzig. By 1887 Arrhenius had been able, as a result of some of this collaboration, to formulate in an explicit form his famous theory of electrolytic dissociation (11).

Shortly afterwards Ostwald published a paper (12) in which, by applying equilibrium theory to the dissociation process, he obtained the relationship between conductivity and concentration that we now refer to as the 'Ostwald dilution law'.

From then on progress in both electrochemistry and kinetics proceeded more smoothly and rapidly. The theory of electrolytic dissociation did not, however, at once gain universal assent, and even as late as the 1930s there were a few who did not accept it. Particularly vociferous opponents of the idea were the organic chemist Henry E. Armstrong (1848-1937) and the electrochemist and fruit farmer S.F.U. Pickering (1858-1920), both of them eccentric and controversial individuals. In the 1930s Louis Kahlenberg was giving, at the University of Wisconsin, a course on electrochemistry which took no account of the existence of ions in solution. It is of interest also to note that although Ostwald had played such an important role in the development of the theory of electrolytic dissociation, he believed for most of his career that atoms, molecules and ions were no more than a convenient fiction; he was not convinced of their real existence until 1909, after J.J. Thomson and J. Perrin had carried out their crucial experiments.

Strong Electrolytes

Arrhenius himself was recalcitrant on one point; he throughout insisted that his interpretation of conductivity in terms of a dissociation equilibrium applies to strong electrolytes as well as to weak ones; for an account of this controversy see ref. 13. Suggestions that strong electrolytes are completely dissociated and that ionic interactions must be invoked to explain the conductivities of their solutions were made by G.N. Lewis (14-16), Niels Bjerrum (17, 18), W. Sutherland (14) and S.R. Milner (20-24). Bjerrum provided spectroscopic evidence for this point of view, and Sutherland and Milner did important work on the theory of ionic interactions. Finally, in 1923, Debye and Hückel (25, 26) developed their comprehensive treatment of strong electrolytes. Arrhenius, however, rejected these ideas and for the most part refused to discuss them. At a meeting of the Faraday Society held in January, 1919, Arrhenius commented that Bjerrum's idea "seems not to agree very well with experiment" (27), and he maintained this position until his death in 1927.

Acid-Base Catalysis

It was work on catalysis by acids and bases, carried out from about 1890 until well into the present century, that particularly brought electrochemistry and kinetics together and led to great advances in both fields. The interpretation of the kinetics of these catalyzed reactions presented some difficulty. In 1889 Arrhenius suggested (28) that in acid catalysis a free hydrogen ion in solution adds on to a molecule of substrate, the intermediate so formed then undergoing further reaction. This theory, known as the 'hydrion theory', is along the right lines, but it did not lead at once to a quantitative interpretation of the kinetic behaviour. Arrhenius realized that certain 'activity' effects had to be taken into account but was not clear as to how this should be done. Later he showed (29) that the kinetics of acid-catalysed reactions could be better understood if one arbitrarily used osmotic pressures instead of concentrations, although the reason for this was not clear. Today we can understand it in terms of activity effects in kinetics.

Salt Effects. The solution to the problem came largely from investigations of salt effects in kinetics. The addition of 'foreign salts', which do not have an ion in common with any ion directly involved in the reaction, is often found to have an important kinetic effect. At first the explanations were given in terms of positive or negative catalysis, but this idea proved to be unhelpful. It is best instead to regard salt effects as environmental effects and to explain them in terms of the activity coefficients of the reacting species.

As understanding of the thermodynamics of solutions developed, it became clear that equilibrium constants must be expressed as ratios of activities rather than of concentrations. Thus for a general reaction

$$aA + bB + \ldots \rightleftharpoons \ldots yY + zZ$$

the equilibrium constant is

$$K = \frac{\ldots a_Y^y a_Z^z}{a_A^a a_B^b \ldots} = \frac{\ldots [Y]^y [Z]^z}{[A]^a [B]^b \ldots} \quad \frac{\ldots y_Y^y y_Z^z}{y_A^a y_B^b \ldots} \tag{4}$$

where the a's are the thermodynamic activities and the y's the activity coefficients, which multiply concentrations to give activities. The treatment of equilibrium constants in this way was straightfoward, but with rate equations it was not at first clear how to proceed.

According to the first proposal made, which came to be called the 'activity-rate theory', the rate of a second-order reaction between A and B should be expressed as

$$v = k_o \; [A][B] \; y_A y_B \tag{5}$$

where k_o is the rate constant. According to this theory the rate constant simply multiplies the product of the activities of the reactants rather than the product of their concentrations. This idea seems to have originated (30) with the Scottish organic chemist Arthur Lapworth (1872-1941), and it gained support (31-35) from the American electrochemist H.S. Harned (1885-1969) and the British kineticist W.C. McC. Lewis (1885-1956).

It soon became apparent, however, that this theory was not consistent with the experimental results. At low concentrations foreign ions always diminish the activity coefficients of ions, a result that is explained by the Debye-Hückel theory (25, 26) in terms of the effect of the ionic atmosphere. The activity-rate theory therefore always predicts a decrease in rate when a foreign salt is added at low concentrations. However, for some reactions a marked increase in rate is observed.

A completely satisfactory solution to the problem did not come until after transition-state theory (36-38) had been formulated in 1935. In terms of that theory, the rate should be expressed not as eq. (5), but as

$$v = k_o[A][B]\frac{y_A y_B}{y_{\neq}} \tag{6}$$

where $y_{\neq}$ is the activity coefficient of the activated complexes (transition states) formed when A and B come together. However, even before the transition-state theory had been formulated, important steps in the right direction had been taken, mainly through the efforts of three Danish physical chemists, J.N. Brønsted (1870-1947), N.J. Bjerrum (1879-1958) and J.A. Christiansen (1888-1969). Again there was a valuable contribution of electrochemistry leading to an important fundamental relationship in chemical kinetics.

The work of the Danish chemists was based on Lewis and Randall's introduction of the concept of the ionic strength and on the Debye-Hückel theory. Two years before Debye and Hückel had formulated their theory Lewis and Randall (39) had introduced the idea of the ionic strength of a solution, which they defined as

$$I \equiv \tfrac{1}{2}\Sigma z_j^2 c_j \tag{7}$$

where c_j is the concentration of an ion j and z_j is its charge number (positive for positive ions, negative for negative ions). The summation in eq. (7) is taken over all of the ions present in the solution. From various lines of evidence Lewis had realized that the effect of dissolved salts on certain properties, such as solubility, is mainly determined by the ionic strength of the solution. This conclusion was later supported when the Debye-Hückel theory was developed.

Shortly after that theory was published Brønsted and the American physical chemist V.K. La Mer (1895-1966) applied it to obtain a relationship (40) between the activity coefficient y_i of an ion and the ionic strength I. The relationship that they obtained was

$$\log_{10} y_i = -z_i^2 B\sqrt{I} \tag{8}$$

where z_i is the charge number of the ion i, and the positive quantity B can be expressed in terms of certain properties of the solution such as the dielectic constant. For water at 25°C the constant B has the value of 0.51 $mol^{-\frac{1}{2}}\ dm^{3/2}$. Equation (8) is obeyed satisfactorily in the limit of very low concentrations and is referred to as the 'Debye-Hückel limiting law'. Considerable deviations occur at higher concentrations, and modifications to the equation lead to better agreement.

Two years before this equation was obtained, Brønsted had made the suggestion (41) that in considering a reaction between two species A and B one should focus attention on a critical complex X that is formed during the course of the reaction:

$$A + B \longrightarrow X \longrightarrow \text{products}$$

This complex X is the most unstable of the various collision complexes formed and corresponds to what later became called an activated complex, or transition state. On the basis of arguments the validity of which is considered later, Brønsted arrived at the equation

$$v = k_o[A][B]\,\frac{y_A y_B}{y_X} \qquad (9)$$

This differs from the activity-rate equation (5) in having the activity coefficient y_X, relating to the intermediate X, in the denominator. Brønsted referred to the ratio $y_A y_B/y_X$ as the 'kinetic activity factor'.

Later, when equation (8) had been obtained, Brønsted introduced it into eq. (9), as follows (42). From eq. (9)

$$\log_{10}(v/[A][B]) = \log_{10} k_o + \log_{10} y_A + \log_{10} y_B - \log_{10} y_X \qquad (10)$$

and the use of eq. (8) for y_A, y_B and y_X leads to

$$\log_{10} k = \log_{10} k_o - B(z_A^2 + z_B^2 - z_X^2)\sqrt{I} \qquad (11)$$

where $k \equiv v/[A][B]$. Since $z_X = z_A + z_B$ this reduces to

$$\log_{10} k = \log_{10} k_o + 2Bz_A z_B\sqrt{I} \qquad (12)$$

Brønsted and others investigated the effect of ionic strength on the rates of reactions of various ionic types, and found that at sufficiently low ionic strengths plots of $\log_{10} k$ are quite linear, the plots having slopes that are consistent with eq. (12). For reactions between ions of the same sign ($z_A z_B$ is positive), eq. (12) predicts that increasing the ionic strength increases the rate, and this is found to be true experimentally. If $z_A z_B$ is negative, however, the rate decreases if the ionic strength is increased, in agreement with eq. (12). The activity-rate equation (5), however, fails for reactions between ions of the same sign, since it leads not to eq. (11) but to an equation with the term z_X^2 missing:

$$\log_{10} k = \log_{10} k_o - B\,(z_A^2 + z_B^2)\sqrt{I} \qquad (13)$$

Since $-B\,(z_A^2 + z_B^2)$ is bound to be negative, a plot of $\log_{10} k$ against $\sqrt{I}$ would always be negative, in disagreement with the results for ions of the same sign.

The experimental evidence thus favours Brønsted's equations (9) and (12) rather than the activity-rate equation (5). The justification for equations (9) and (12), however, remained a difficulty for a number of years. Brønsted's original derivation (41) was by no means explicit or convincing. He suggested that the rate should depend in some way on the difference between the chemical potential of X and that of the reactants A and B, and this difference is related to the ratio $y_A y_B/y_X$; however, it is by no means clear that this ratio should simply multiply the rate expression as in eq. (9). A more satisfactory derivation of eq. (12), although a somewhat complicated one, was suggested by Christiansen (43). His derivation dealt with the frequency of collisions between ions A and B and took account of the electrostatic interactions. By making use of the Debye-Hückel expression for the electric potential as a function of the distance from an ion and making some approximations, Christiansen was able to arrive at eq. (12). He did not obtain eq. (9), since no specific intermediate complex X was postulated in his derivation.

Bjerrum's derivation of eq. (9) was based on the idea of an intermediate S and seemed simple and straightforward at first, but on second thoughts it appeared to have a fatal flaw. Bjerrum postulated (44) that an equilibrium is first established between the reactants A and B and a purely physical collision complex S (Stosskomplex):

$$A + B \underset{k_{-1}}{\overset{k_1}{\rightleftharpoons}} S$$

Bjerrum's complex S is to be distinguished from Brønsted's critical complex X, which is the most unstable of the various complexes found during the course of a reaction and is what we now call an activated complex or transition state. Bjerrum expressed the equilibrium constant for the formation of the complex S as

$$K = \frac{[S]}{[A][B]} \cdot \frac{y_S}{y_A y_B} \qquad (14)$$

The concentration of S is thus

$$[S] = K\,[A][B]\,\frac{y_A y_B}{y_S} \qquad (15)$$

He then postulated that the rate is proportional to the <u>concentration</u> of S, so that

$$v = k'[S] = k'\,K[A][B]\,\frac{y_A y_B}{y_S} \qquad (16)$$

which is equivalent to Brønsted's equation (9).

There is, however, a difficulty with this derivation. The fact that the equilibrium equation (4) involves activity coefficients requires that the rate constants k_1 and k_{-1} for reaction in forward and reverse directions also involve activity coefficients, since $K = k_1/k_{-1}$. But if the rate constant k_{-1} for the reverse reaction involves activity coefficients, how can one say that $v = k'[S]$? It is unreasonable to suppose that the intermediate S follows a different type of kinetic equation according to whether it is reverting to reactants or forming products. Brønsted (45) raised this objection to the Bjerrum formulation in 1925, and in an extensive review of salt effects (46) published in 1928 he made only a passing reference to it. The difficulty was further discussed and elaborated by La Mer (47) in a review which appeared in 1932.

The answer to the difficulty with Bjerrum's derivation came with the advent of transition-state theory and was first pointed out by R.P. Bell (48). Transition-state theory (36-38) lays emphasis on the activated complexes, which are species that correspond to the col or saddle-point in a potential-energy surface for a reaction. Unlike an ordinary reaction intermediate, such as the Stosskomplex S postulated by Bjerrum, an activated complex has reached a point of no return; it is bound to pass into products and cannot revert to reactants. The process is thus represented as

$$A + B \longrightarrow X^{\neq} \longrightarrow \text{products}$$

with a single and not double arrow leading to $X^{\neq}$. The species $X^{\neq}$ is not in the ordinary sense in equilibrium with A and B; instead it is formed in a state of equilibrium, and is said to be in quasi-equilibrium. The concentration of $X^{\neq}$ is therefore correctly given by an equation analogous to eq. (15):

$$X^{\neq} = K^{\neq}[A][B]\,\frac{y_A y_B}{y_{\neq}} \qquad (17)$$

Because $X^{\neq}$ is bound to form products, the rate of their formation is given by transition-state theory as the concentration of $X^{\neq}$ multiplied by a frequency.

Thus, if Bjerrum's derivation is modified by replacing the intermediate S by the activated complex $X^{\neq}$, the logical dilemma disappears. Equation (6) is the correct transition-state formula to use, and the ionic-strength effects have been important in leading to this fundamental equation.

General Acid-Base Catalysis

Another important contribution of electrochemistry to kinetics has been in connection with what is referred to as general acid-base catalysis. It was originally supposed that catalysis by acids and bases is catalysis by hydrogen and hydroxide ions, but evidence was later obtained for catalysis by other acidic species such as undissociated acids and by basic species such as anions. According to the ideas of Brønsted (49) and Lowry (50), an acid is any species that can donate a proton and a base is one that can accept one. Thus CH_3COOH and NH_4^+ are acids, while CH_3COO^- and NH_3 are bases.

Conclusive evidence for catalysis by species other than hydrogen or hydroxide ions was first obtained by J.W. Dawson (51-52) of the University of Leeds. His most important work was on the conversion of the keto form of acetone into its enol form, a process that can be studied by measuring the rate of iodination of acetone, the iodine adding rapidly to the enol form. Dawson first showed that undissociated acids bring about acid catalysis and spoke of the 'dual theory' of acid catalysis, by which he meant acid catalysis by both hydrogen ions and by undissociated acids. Evidence for general base catalysis was later obtained by Brønsted and Pedersen (53) for the decomposition of nitramide. Somewhat later Brønsted and Guggenheim (54), simultaneously with Lowry and Smith (55), showed that the mutarotation of glucose was subject to general catalysis both by acids and by bases; in other words, there was catalysis by hydrogen and hydroxide ions, by undissociated acids, by the anions of acids, and by species such as NH_4^+. Both Brønsted and Lowry realized the significance of these facts when they suggested their extended definition of acids and bases. Dawson's term 'dual theory' of catalysis had to be changed to 'multiple theory'.

Concluding Remarks

At the outset it was said that both kinetics and electrochemistry, unlike thermodynamics, had a healthy and vigorous childhood. At the same time it should be recognized that at first investigators in physical chemistry tended to be treated as pariahs by the majority of chemists. Van't Hoff suffered a period of unemployment at the beginning of his career, and Ostwald at first had to work as a schoolteacher with no facilities for research. Van't Hoff's theory of the tetrahedral carbon atom was derided by the organic chemist Hermann Kolbe (1818-1884). Ostwald's important early work in physical chemistry was not at all appreciated by many chemists, who complained that Ostwald was not a true chemist since he had discovered no new compounds; to this Ostwald cheerfully retorted that the number of compounds that he had discovered was minus one, since he had shown by physical methods that an alleged new compound was identical with a substance that was already well known.

Such hostility to physical chemistry carried well into the present century. In the 1920s W.H. Perkin, Jr., the professor of organic chemistry at Oxford, enjoyed commenting that physical chemistry was all very well, but of course it does not apply to organic compounds, which constitute the vast majority of all chemical compounds! This attitude of Perkin created some discomfort for N.V. Sidgwick, a member of his department.

Fortunately these attitudes survive no longer. Physical chemistry is regarded as respectable, and even sometimes useful, by the vast majority of chemists.

Literature Cited

1. Faraday, M. Phil. Trans. Roy. Soc. 1834, 124, 77-122.
2. Wilhemy, L. Pogg. Ann. 1850, 81, 413-433,499-526.
3. Wenzel, C.F. Lehre von der Verwandschaft der Körper; Dresden, 1777.
4. Thenard, J. Ann. Chim. Phys. 1818, 9, 314-317.
5. Root-Bernstein, R.S. "The Ionists: Founding Physical Chemistry, 1872-1890", Ph.D. Thesis, Princeton University, 1980.
6. Laidler, K.J. "Chemical Kinetics and the Origins of Physical Chemistry", Arch. Hist. Exact Sci. 1985, 32, 43-75.
7. Ostwald, W. J. Prakt. Chem. 1884, [2], 30, 93-95.
8. van't Hoff, J.H. Arch. Neerlandaises 1885, 20, 239-302.
9. van't Hoff, J.H. Z. Physick. Chem. 1887, 1, 481-508. A

translation of the latter paper, by J. Walker, appears in The Foundations of the Theory of Dilute Solutions; Alembic Club, London, 1929; no. 19, pp. 5-42.
10. van't Hoff, J.H. Phil. Mag. 1888, 26, 81-105.
11. Arrhenius, S. Z. Physik. Chem. 1887, 1, 631-648.
12. Ostwald, W. Z. Physik. Chem. 1888, 2, 36-37.
13. Wolfenden, J.H. Ambix, 1972, 19, 175-196.
14. Lewis, G.N. Z. Physik. Chem., 1908, 61, 129-165.
15. Lewis, G.N. Z. Physik. Chem., 1910, 70, 212-219.
16. Lewis, G.N. J. Am. Chem. Soc., 1912, 34, 1631-1644.
17. Bjerrum, N. Proc. Seventh Int. Congress on Applied Chemistry, 1909, 55, (1909); a translation of this paper is included in Niels Bjerrum, Selected Papers; Copenhagen, 1949, p. 56.
18. Bjerrum, N. Seventh Int. Congr. Appl. Chem., 1909, 9, 58-60.
19. Sutherland; Phil. Mag., 1902, 3, 161-177; 1906, 12, 1-20; 1907, 14, 1-35.
20. Milner, S.R. Phil. Mag., 1912, 23, 551-578.
21. Milner, S.R. Phil. Mag., 1913, 25, 742-751.
22. Milner, S.R. Phil. Mag., 1918, 35, 214-220.
23. Milner, S.R. Phil. Mag., 1918, 35, 352-364.
24. Milner, S.R. Trans. Faraday Soc., 1919, 15, 148-151.
25. Debye, P.J.W.; Hückel, E. Physikal. Z., 1923, 24, 185-206.
26. Debye, P.J.W.; Hückel, E. Physikal. Z., 1923, 24, 305-325.
27. Arrhenius, S.A. Trans. Faraday Soc., 15, 10-17.
28. Arrhenius, S.A. Z. Physik. Chem., 1889, 4, 226-248.
29. Arrhenius, S.A. Z. Physik. Chem., 1899, 28, 319-335.
30. Lapworth, A. J. Chem. Soc., 1908, 93, 2187-2203.
31. Jones, W.J.; Lapworth, A.; Lingford, H.M. J. Chem. Soc., 1913, 103, 252-263.
32. Harned, H.S. J. Am. Chem. Soc., 1918, 40, 1461-1481.
33. Harned, H.S.; Pfenstiel, R. J. Am. Chem. Soc. 1922, 44, 2193-2205.
34. Jones, C.M.; Lewis, W.C.McC. J. Chem. Soc. 1920, 117, 1120-1133.
35. Moran, T.; Lewis, W.C.McC. J. Chem. Soc. 1922, 121, 1613-1624.
36. Eyring, H. J. Chem. Phys. 1935, 3, 107-115.
37. Evans, M.G.; Polanyi, M. Trans. Faraday Soc., 1935, 31, 875-895.
38. Evans, M.G.; Polanyi, M. Trans. Faraday Soc., 1937, 33, 448-452.
39. Lewis, G.N.; Randall, M. J. Am. Chem. Soc., 1921, 43, 1112-1154.
40. Brønsted, J.N.; La Mer, V.K. J. Am. Chem. Soc. 1924, 46, 555-573.
41. Brønsted, J.N. Z. Physik. Chem. 1922, 102, 169-207.
42. Brønsted, J.N. Z. Physik. Chem. 1925, 115, 337-364.
43. Christiansen, J.A. Z. Physik. Chem. 1924, 113, 35-52.
44. Bjerrum, N. Z. Physik. Chem. 1924, 108, 82-100.
45. Brønsted, J.N. Z. Physik. Chem. 1925, 115, 337-364.
46. Brønsted, J.N. Chem. Revs., 1928, 10, 179-212.
47. La Mer, V.K. Chem. Revs., 1932, 10, 179-212.
48. Bell, R.P. Acid-Base Catalysis; Clarendon Press, Oxford, 1941; pp. 28-31.
49. Brønsted, J.N. Rec. trav. chim., 1923, 42, 718-728.
50. Lowry, T.M. J. Soc. Chem. Ind., 1923, 42, 43-47.
51. Dawson, J.W.; Powis, F. J. Chem. Soc. 1913, 103, 2135-2146.
52. Dawson, J.W.; Carter, J.S. J. Chem. Soc., 1926, 2282-2296.
53. Brønsted, J.N.; Pederson, K.J. Z. Physik. Chem., 1924, 108, 185-235.

54. Brønsted, J.N.; Guggenheim, E.A. J. Am. Chem. Soc. 1927, 49, 2554-2584.
55. Lowry, T.M.; Smith, G.F. J. Chem. Soc. 1927, 130, 2539-2554.

RECEIVED August 9, 1988

Chapter 6

A Reappraisal of Arrhenius' Theory of Partial Dissociation of Electrolytes

R. Heyrovska

Na Stahlavce 6, 160 00 Praha 6, Czechoslovakia

The actual 'ionic concentrations' and hydration numbers of over fifty univalent and multivalent strong electrolytes have been presented (for the first time). The degrees of dissociation and hydration numbers calculated from vapor pressures correlate quantitatively with the properties of dilute as well as concentrated solutions of strong electrolytes. Simple mathematical relations have been provided for the concentration dependences of vapor pressure, e.m.f. of concentration cells, solution density, equivalent conductivity and diffusion coefficient. Non-ideality has thus been shown to be mainly due to solvation and incomplete dissociation. The activity coefficient corrections are, therefore, no longer necessary in physico-chemical thermodynamics and analytical chemistry.

A century ago, van't Hoff's (1) pioneering work on the gas-solution analogy was followed by Arrhenius' (2) theory of partial dissociation of electrolytes in solutions. Later, electrolytes came to be classified as weak or strong with the supposition that the former are partially dissociated whereas the latter are completely dissociated in the given solvent (3,4). However, with the advance of experimental and theoretical knowledge, it has become increasingly evident that many multivalent and even some univalent strong electrolytes (especially those with bulky anions or cations) are incompletely dissociated not only in solvents of low dielectric constant but also in water. On the other hand, association of ions of simple strong electrolytes like NaCl in water is considered negligible (3-7). This report, which summarizes several years' research work (8,9) shows that all electrolytes including alkali halides and halogen acids are in association/dissociation equilibrium with their ions in aqueous solutions, thereby confirming Arrhenius' views. A short survey of the development of the ideas in the literature about the state of dissociation of electrolytes in solutions is as follows:

0097-6156/89/0390-0075$06.00/0

The basic analogy of the gas and solution ideal laws, established by van't Hoff (1), was found to be valid only for very low gas pressures P and osmotic pressures π. Thus, at temperature T and molar volume V,

$$PV = RT \tag{1}$$
$$\pi V = RT \tag{2}$$

At higher P and π, the ratios (PV/RT) and (πV/RT) called, respectively, the compression factor (10-12), z, and the van't Hoff factor (1,13), i, deviated from unity.

Arrhenius (2) interpreted i as the total number of moles of solute actually present in the solution due to association/dissociation of one mole of solute (B) dissolved in the given solvent (A) at the given concentration. Thus, for one mole of an electrolyte B_{+-} dissociating into ν_+ cations of charge z_+ and ν_- anions of charge z_- according to

$$\begin{array}{llll} B_{+-} & \rightleftharpoons & \nu_+ B_+^{z_+} + & \nu_- B_-^{z_-} \\ 1-\alpha & & \nu_+ \alpha & \nu_- \alpha \end{array} \tag{3}$$

where α is the degree of dissociation, $\nu = \nu_+ + \nu_-$ and $\nu_+ z_+ = \nu_- z_-$, i is given by

$$i = 1 + (\nu - 1)\alpha \tag{4}$$

Thereby, an excess number, $(\nu-1)\alpha$, of moles are created by the dissociation of B_{+-}. At infinite dilution, an electrolyte is completely dissociated and therefore, $\alpha = 1$ and $i = \nu$. For a non-associating/dissociating solute, like sucrose in aqueous solution, $\alpha = 0$.

The above idea gained a wide support for nearly three decades, but was eventually given up mainly because there was no exact way of determining α from experiments like electrical conductivity. The use of the Arrhenius conductivity ratio (Λ/Λ_o), or other modifications, for the calculation of α, approximately satisfied the law of mass action or Ostwald's dilution law (2,14)

$$K = \alpha^{\nu} c/(1-\alpha) = (\Lambda/\Lambda_o)^2 c/ [1-(\Lambda/\Lambda_o)] \tag{5}$$

for weak acids. On the other hand, it gave neither the correct values of i nor the dissociation constant, K_c, independent of concentration for electrolytes like simple alkali halides. Strong electrolytes, by virtue of their 'anomalous' behavior (15), were assumed to be completely dissociated in water. This led to the introduction of the formal notations and conventions of 'activity and activity coefficients for a unified representation of non-ideality (15). From then on, the deviations from ideality came to be expressed commonly by the molal osmotic coefficient, ϕ, and the mean molal ionic activity coefficient, $\gamma_\pm$ (3,15,16).

ϕ is evaluated (3) usually from the vapor pressure measurements from the relation,

$$(p_A / p_A^\circ) = a_A = \exp(-\nu m \phi / RT) \tag{6}$$

where p_A^o is the vapor pressure of water at temperature T, p_A is that over the solution of molality m and a_A is defined (3,15) as the activity of the solvent. $\phi = 1$ corresponds to complete dissociation at infinite dilution. ϕ can also be obtained from measurements of osmotic pressure, changes in freezing and boiling points, etc.(3).

$\gamma_\pm$ is obtained directly from the e.m.f. measurements (3,14-16), say, of concentration cells without transference, from the deficit free energy, ΔG(non-id), attributed to non-ideality,

$$\begin{aligned} nF(E-E^o) + \nu RT\ln m &= -\nu RT\ln\gamma_\pm \\ \Delta G(\text{expt}) - \Delta G(\text{id}) &= \Delta G(\text{non-id}) \end{aligned} \quad (7)$$

where ΔG(id) is the supposed free energy due to complete dissociation at any molality m. $\gamma_\pm$ is also evaluated from ϕ through the Gibbs-Duhem relation by integration:

$$\phi = (1/m)\int_0^m m\,d\ln m \quad (8)$$

where $(m\gamma_\pm)^\nu = a_B$ is the activity of the solute B (3,15).

Subsequent theories of non-ideality have been mainly concerned with explaining the concentration and temperature dependences of γ and ϕ (3,16). For a comparison with various other theories for the non-ideal part of free energy of solutions, see (14). The interionic attraction theory (3,5,16-18) formulated on the assumption of complete dissociation of strong electrolytes, predicted the $\ln\gamma$ vs $\sqrt{m}$ linear dependence and explained the $\sqrt{c}$ dependence of Λ found empirically by Kohlrausch (3,14) for dilute solutions. Since the square-root laws were found to hold for dilute solutions of many electrolytes in different solvents, the interionic attraction theory gained a wide acceptance. However, as the square root laws were found to be unsatisfactory for concentrations higher than about 0.01m, the equations were extended or modified by the successive additions of more terms, parameters and theories to fit the data for higher concentrations. See e.g., (3,16) for more details.

At the same time, theories of ionic association were worked out by Bjerrum and others (3,5,14,16,19). According to these, free ions of opposite charge getting closer than a certain critical distance form separate associated entities. Thereby, the total number of moles of solute in the solution becomes lower than that expected on the basis of complete dissociation. These theories show that ion pairs can be formed, although to a small extent, even in aqueous 1:1 electrolytes where the critical distance is 3.57Å at 25 °C (3); for higher valent ions in solvents of lower dielectric constant, associated ions are more likely since the predicted critical distance is larger. In fact, the literature (3,5,16,20-32) provides growing evidence for ion association not only in aqueous and non-aqueous solutions of multivalent electrolytes and 1:1 strong electrolytes composed of bulky ions, but also in HCl and NaOH (29). Thus, in cases where incomplete dissociation and formation of associated ions were evident, α was incorporated (3,5,16,30-32) into the equations for ϕ and γ while retaining the Debye-Huckel-Onsager terms for the free ions

Solvation of the dissolved solutes as one of the important causes of non-ideality has long been recognized (14,33). Accordingly, there has been an increasing awareness of its influence on the properties of solutions (3,5,14,16,19,34,35). However, there is no concordance (19,36) in the reported values of the solvation numbers obtained by different methods, mostly due to the use of unsatisfactory theories of non-ideality.

On the whole, one finds that the existing interpretations of the non-ideal behavior of solutions are fairly complicated and that there is no simple, meaningful and unified explanation of the properties of dilute and concentrated solutions. Therefore, the present author decided to interpret directly, without presupposed models, the actual experimental data as such rather than their deviations from 'ideality' (or complete dissociation) represented by formal coefficients like ϕ and γ. Attention is paid here mainly to aqueous solutions of strong electrolytes, since these are considered anomalous (15). Extensive work on univalent and multivalent electrolytes has shown (8,9a-i) that when allowance is made for the solvation of solutes, Arrhenius' theory of partial dissociation of electrolytes explains the properties of dilute as well as concentrated solutions. This finding is in conformity with the increasing evidence for ion association of recent years mentioned above.

The method of determination of the actual degree of dissociation, α, and the hydration number, n_h, from the existing data on vapor pressures is outlined below. This is then followed by the quantitative correlation of α and n_h with various properties of electrolyte solutions.

α and n_h From Vapor Pressure

The method of obtaining α and n_h is briefly thus: The vapor pressures of solutions were found (8,9g) to obey Raoult's (37) law, on correcting it for hydration (34,38) and incomplete dissociation (2,34) of the electrolyte. Thus, the vapor pressure ratio (data stored in the form of a_A and ϕ in (3)) gives the mole fraction N_A of 'free' water,

$$(p_A/p_A^\circ) = N_A = n_A/(n_A + n_B) \qquad (9)$$

where $n_A = (55.51 - mn_h)$ is the number of moles of 'free' water, 55.51 is the number of moles of water in one kg and $n_B = im$ is the total number of moles of solute. In Figure 1 are shown two examples of the general dependence of $mp_A/(p_A^\circ - p_A)$ on m found for over fifty electrolytes compared with that for sucrose for which $\alpha = 0$. It can be seen that the graphs are linear over a considerable range of concentrations, e.g., 0 - 2m sucrose, 1.8 - 4m NaCl and 1.8 - 4.5m KBr. This implies that i and n_h are constant in these range of concentrations. From the slopes ($=n_h/i_m$) and the intercepts ($=55.51/i_m$), the values of the constants, i_m and n_h were obtained. For sucrose, i_m and n_h were found to be 0.999 and 5.01 (lit: (14,38), n_h=5) respectively. The values of n_h and α_m $[=(i_m-1)/(\nu-1)]$ obtained for the electrolytes are given in Table I and Figures 2 and 3.

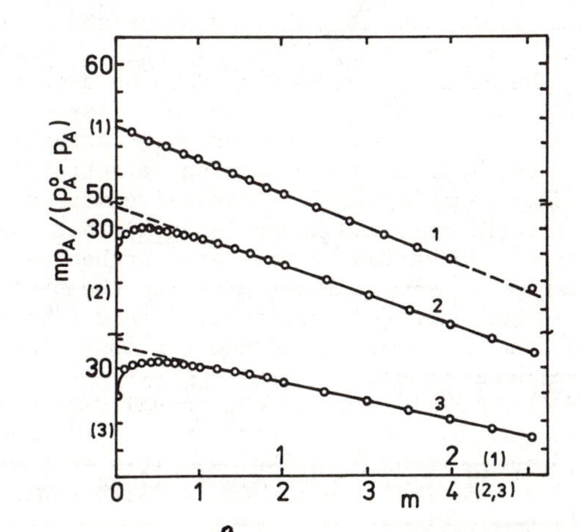

Figure 1. Plots of $mp_A/(p_A^o - p_A)$ vs. m for aqueous solutions at 25 °C, 1) sucrose, 2) NaCl and 3) KBr (Reproduced with permission from Ref. 8.

Table I. Degrees of Dissociation (α) of Some Multivalent Electrolytes in Aqueous Solutions at 25°C Hydration Numbers (n_h) are Given in the Last Row

m	$MgCl_2$	$CaCl_2$	$SrCl_2$	$BaCl_2$	$AlCl_3$	$ScCl_3$	YCl_3	$CrCl_3$
0.1	0.759	0.753	0.750	0.747	0.705	0.682	0.675	0.698
0.2	0.749	0.736	0.731	0.720	0.678*	0.671*	0.657*	0.675*
0.3	0.741*	0.728	0.720	0.711	0.677	0.670	0.655	0.672
0.4	0.739	0.724*	0.717*	0.707	0.681	0.672	0.658	0.673
0.5	0.741	0.725	0.717	0.704*	0.680	0.671	0.657	0.674
0.6	0.742	0.725	0.715	0.704	0.676*	0.670*	0.655*	0.673*
0.7	0.739	0.723	0.713	0.702	-	-	-	-
0.8	0.739	0.722	0.713	0.702	-	-	-	-
0.9	0.740	0.725	0.715	0.701	-	-	-	-
1.0	0.741*	0.725	0.717	0.702	-	-	-	-
1.2	-	0.724*	0.717	0.703	-	-	-	-
1.4	-	-	0.715*	0.704*	-	-	-	-
α_m ±0.002	0.740	0.724	0.715	0.703	0.678	0.671	0.656	0.673
n_h ±0.02	15.30	13.38	12.08	9.08	28.82	26.20	24.65	27.17

* the upper and lower limits of m between which $\alpha = \alpha_m$

Table I (continued)

m	$LaCl_3$	$CeCl_3$	$PrCl_3$	$NdCl_3$	$SmCl_3$	$EuCl_3$	$Th(NO_3)_4$
0.1	0.679	0.670	0.674	0.671	0.678	0.684	0.562
0.2	0.655	0.660	0.656	0.654	0.663	0.664	0.542
0.3	0.655*	0.655*	0.651*	0.651*	0.659*	0.657*	0.532*
0.4	0.657	0.655	0.652	0.653	0.659	0.658	0.530
0.5	0.659	0.657	0.652	0.655	0.659	0.659	0.532
0.6	0.659	0.654	0.649	0.651	0.656	0.657	0.532
0.7	0.654	0.658	0.653	0.654	0.660	0.660	0.530*
0.8	0.656*	0.654*	0.651*	0.650*	0.659*	0.657*	-
α_m	0.657	0.656	0.651	0.652	0.659	0.658	0.531
n_h	21.92	22.24	21.87	22.38	22.79	23.25	22.37

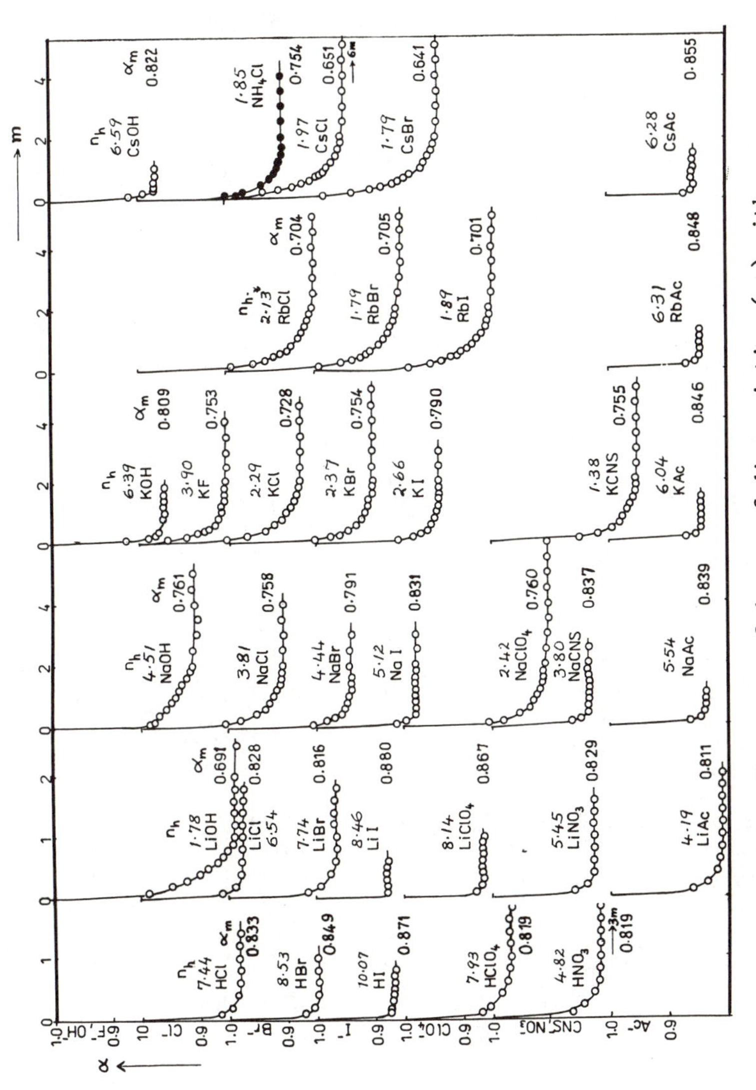

Figure 2. Variation of degree of disssociation (α) with concentration (m) for 1:1 strong electrolytes in aqueous solutions at 25°C.

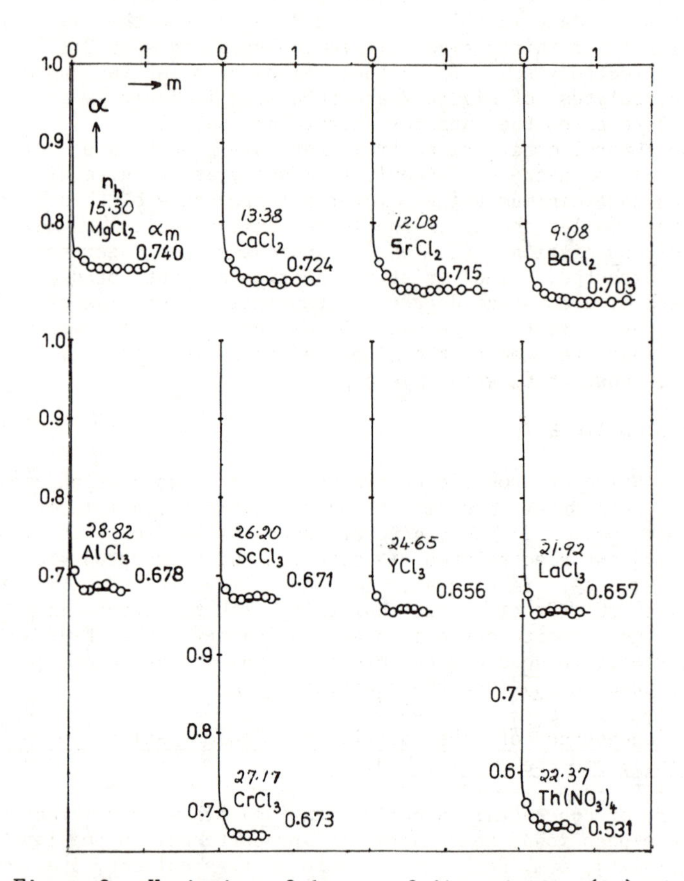

Figure 3. Variation of degree of dissociation (α) with concentration (m) for multivalent electrolytes in aqueous solutions at 25°C.

Assuming n_h to be the constant (maximum) hydration number in the concentration range from m = 0 to m corresponding to the end of linearity in Figure 1, as in (3), the values of i and hence of α [=(i-1)/(ν-1)] were then calculated from Equation 9 using the vapor pressure data in (3). Figures 2 and 3 show the variation of α with m for thirty five 1:1 electrolytes and some 2:1, 3:1, and 4:1 electrolytes. The actual values of α at various m for the electrolytes of Figure 2 are tabulated in Table I of (8). Table I here gives the data for multivalent salts.

The general observations from the above results are: 1) As m increases, α decreases steeply from unity at infinite dilution to a constsnt minimum value α_m over a large range of concentrations (the smaller the n_h, the larger this range). The maximum degree of association, $(1 - \alpha_m)$, increases as n_h decreases, as can be expected. The existing theories do not predict the attainment of a constant degree of association, or dissociation, in the given solvent. The equilibrium represented by Equation 3 can be characterized by the dissociation ratio (8), K, which becomes a constant K_m when $\alpha = \alpha_m$.

$$K = \alpha^{\nu} / (1-\alpha) = K_c / c \qquad (10)$$

For a detailed treatment of the electrolytic association/dissociation equilibrium in terms of Lange's inner potentials of the ions, see (9e). 2) The degree of association, $1-\alpha$, is higher for multivalent electrolytes, in agreement with accepted views. 3) n_h decreases, in general, for the halides in the order, $H^+ > ... > Cs^+$, $Cl^- < Br^- < I^-$, in conformity with the trend in (3), whereas the opposite cationic order is found for the hydroxides and acetates. In general, the hydration numbers are close to the values fitted in (3) into the equation for γ.

<u>Linear Dependence of the e.m.f. of Concentration Cells on $\log(\alpha m/n_A)$; and the Significance of pH:</u>

The e.m.f. of concentration cells with (9e) and without transport (8,9e) has been found to be directly proportional to the log of ionic molality, αm, reckoned per mole of free solvent, i.e., the mole ratio, $(\alpha m/n_A)$. Thus, for a cell without transport,

$$\Delta E = E - E^{o} = -\delta_A (\nu RT/\nu_{\pm} z_{\pm} F) \ln[(\alpha m/n_A)/(\alpha m/n_A)^{o}] \qquad (11)$$

where $\nu_+ z_+ = \nu_- z_- = \nu_{\pm} z_{\pm}$, $(\alpha m/n_A)^{o}$ pertains to the solution in the reference cell and δ_A is a constant (interperted tentatively as the factor by which the e.m.f. is affected by the polarization effect of the solvent). Figure 4 shows the linear dependence of ΔE (back-calculated from the data in (3) using Equation 7) on $\log(\alpha m/n_A)$ for dilute as well as concentrated solutions, i.e., for the whole concentration range for which α has been tabulated in (8) and plotted as a function of m in Figure 2. From the slopes, δ_A was found to be 0.970, 0.953, and 0.915 for HCl, NaCl, and KBr respectively. Similar graphs were obtained for the other electrolytes mentioned in this paper. For very dilute solutions, ΔE becomes proportional to $\log(\alpha c)$ as proposed by Nernst (39).

As Equation 11 now replaces the existing Equation 7 that defines the activity coefficient, the present correlation of e.m.f. with α and n_h is the most decisive proof in favor of solvation and incomplete dissociation as the main causes of non-ideality.

Significance of pH. An important consequence of Equation 11 concerns the definition and magnitude of pH. The conventional definition of pH is (3),

$$pH - (pH)^{o} = -\log a_{H^+} = -\log m\gamma_{\pm} = (F/2.303RT)\Delta E \tag{12}$$

but now from Equation 11, one has (8,9f,g),

$$pH - (pH)^{o} = (F/2.303RT)\Delta E = -\delta_A \log[(\alpha m/n_A)/(\alpha m/n_A)^{o}] \tag{13}$$

In Figure 4, line 3 for HCl, the ordinate is proportional to pH. If on the other hand, one follows the original definition of pH by Sorensen,

$$p[H^+] = -\log[H^+] = -\log \alpha c \tag{14}$$

where αc is the actual molar concentration of hydrogen ions. Since α can be obtained from the vapor pressure data, $p[H^+]$ for the strong acids can be calculated using Equation 14; see Table I of (8) for the α data.

Dependence of Solution Density on α

The conventional interpretation of the density d of strong electrolyte solutions is through parametrical equations like (16),

$$d = (1000 + mM_B)/V = d_A + A'c - B'c^{3/2} \tag{15}$$

where M_B is the molar mass of the electrolyte, V is the molal volume of the solution of molality m, d_A is the density of the solvent and A' and B' are constants.

From the above equation, it can be seen that an explanation of d amounts to that of the measured molal volume V. Since the solution contains ions, ion-pairs and solvent, it was recently demonstrated (8,9f,g) that,

$$V = V_A + mV_B = V_A + m(1-\alpha)V_B^{o} + \alpha(V_+ + V_- \pm \delta V) \tag{16}$$

where V_A is the volume of 1 kg of water in the solution, V_B is the volume of one mole of electrolyte B in the solution, V_B^o is the volume per mole of the electrolyte before it was dissolved in the solvent, V_+ and V_- are the volumes per mole of the cation and the anion and $\pm\delta V$ is the sum of the volume changes caused by the dissociation and interactions of the ions in the given medium (40-2).

Figure 5 shows the linear dependence of $(V-mV_B^o)$ on the ionic molality, αm, for NaCl (0-1.4m) and for KBr (0-0.8m); V was obtained from the tables for the (c/m) ratio in (16) and the data in (40) were used for V_B^o. From the value V_A of the intercept, it

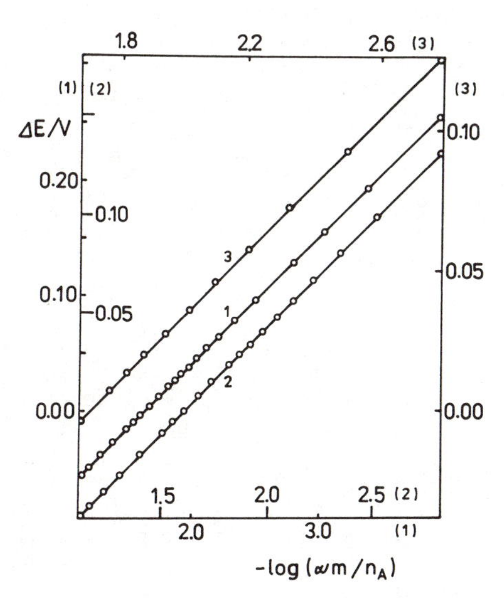

Figure 4. Linear dependence of ΔE on $-\log(\alpha m/n_A)$ for aqueous solutions at 25°C, 1) NaCl, 2) KBr and 3) HCl; (full line: least square fit). (Reproduced with permission from Ref. 8.

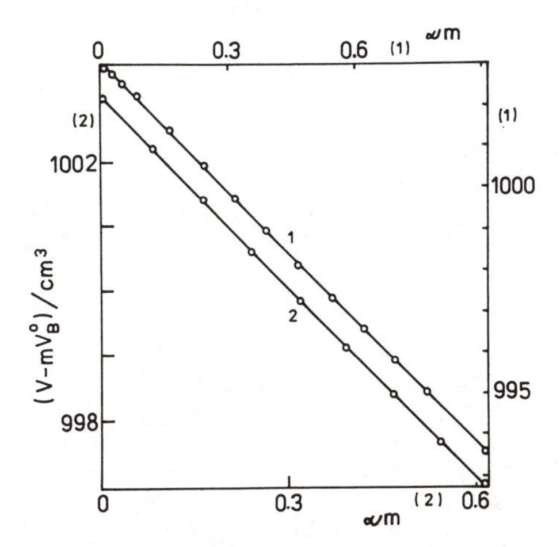

Figure 5. Linear dependence of $(V-mV_B^o)$ on ionic molality, αm, for aqueous solutions at 25 C, 1) NaCl and 2) KBr; (full line: least square fit). (Reproduced with permission from Ref. 8.

is found that the volume of water in the solution is the same as that of pure water, i.e., 1003.01(±0.04) cm^3/kg at 25 °C. This implies that, in the above concentration range, the solvent does not suffer any volume change. Hence, hydration is probably similar to adsorption (9d,43). From the slopes, the values of $(V_+ + V_- \pm \delta V)$ were found to be 16.35 and 32.94 cm^3/mol for NaCl and KBr respectively. These values are nearly the same as the apparent molal volumes at infinite dilution reported in (42): 16.3 and 33.6 cm^3/mol for NaCl and KBr respectively. The linearity also shows that the ion-pair, or the undissociated electrolyte, has the same molar volume in the solution as in the solid state before dissolution. Similar results were found for KCl and KI. At higher concentrations, the graphs are not linear, but that will not be discussed here. The above demonstration that V depends on α confirms Heydweiller's (44) results for many electrolytes, although he used, erroneously, the Arrhenius conductivity ratio for α. The validity of Equation 16 has been tested (8,9f) for solutions as dilute as 0.0064m NaCl and 0.00025m KBr.

Thus, the quantitative correlation of the molal volumes (and hence densities) with α is in itself a simple and sufficient proof for the correctness of the idea of partial dissociation of strong electrolytes in water.

Correlation of α and n_h with Equivalent Conductivity

An example of the existing equation (3) for the concentration dependence of equivalent conductivity of strong electrolytes, based on the idea of complete dissociation,

$$\Lambda = (\eta/\eta_A)\left[\Lambda_o - \frac{B_2\sqrt{c}}{1+B\dot{a}\sqrt{c}}\right]\left(1 - \frac{B_1\sqrt{c}\,F}{1+B\dot{a}\sqrt{c}}\right) \quad (17)$$

consists of collections of parameters, B, B_1, B_2 and F, see (3), and the coefficients of viscosity, η and η_A of the solution and solvent respectively.

In contrast, preliminary investigations in support of partial dissociation have given a simple relation,

$$\Lambda_o - \Lambda = \lambda_{+-}(1-\alpha) = K_{\Lambda,\pi}\,\pi \quad (18)$$

which holds for dilute as well as concentrated solutions (9c-e). In this equation, π calculated as $\nu m \phi RT d_{Af} = (-RT/V_{Af}) \ln N_A$ was used, where V_{Af} and d_{Af} are the molar volume and density of 'free' water. The first term in Equation 18 for $\Lambda_o - \Lambda$ gives the reduction in equivalent conductivity due to the formation of ion-pairs The second term shows the effect of osmotic pressure (crowding effect). Figure 6 shows the linear relation between $(\Lambda_o - \Lambda)/\pi$ and $(1-\alpha)/\pi$ for KBr from 0 to 3.75M; data are taken from (45). The constants, λ_{+-}, charecteristic of the ion-pair, and $K_{\Lambda,\pi}$ are obtained from the intercept and slope, respectively. Similar results were obtained for NaCl (9e). The correlation seems good despite the use for V_{Af} and d_{Af} the values for pure water.

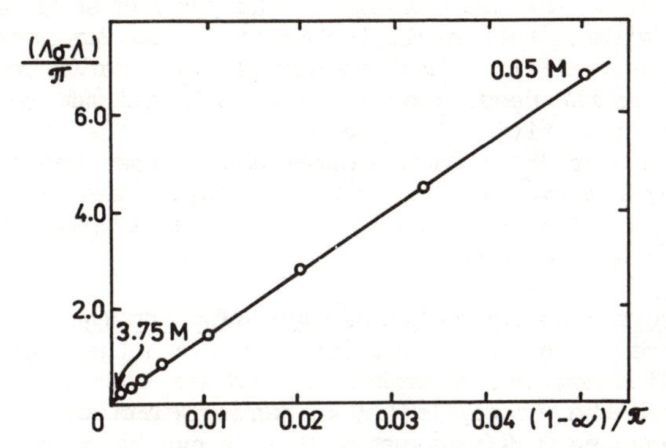

Figure 6. Linear dependence of $(\Lambda_0-\Lambda)/\pi$ on $(1-\alpha)/\pi$ for aqueous solutions of KBr at 25°C; π in atm; slope= $\lambda_{\pm}$=132, intcpt= $K_{\Lambda,\pi}$ =0.096; (full line: least square fit).

Limiting Law for Dilute Solutions. In this case, since the second term on the right becomes negligible, Equation 18 approximates to (9e),

$$\Lambda_o - \Lambda \simeq (1-\alpha)\lambda_{+-} \qquad (19)$$

which shows that the reduction in Λ from Λ_o is primarily due to ion association (as a result of interionic attraction in the given dielectric medium).

According to Arrhenius, $(\Lambda/\Lambda_o) = \alpha$, and therefore,

$$\Lambda_o - \Lambda = (1-\alpha)\Lambda_o \qquad (20)$$

The Λ, $\sqrt{c}$ empirical relation found by Kohlrausch can be explained on the basis of the analogy of the gas and solution properties. According to the simple kinetic theory of gases, the root-mean-square velocity $\sqrt{\bar{v}^2}$ is related (46) to $\sqrt{P}$ by $\sqrt{\bar{v}^2} = \sqrt{3P/\rho}$, where ρ is the density of the gas. In dilute solutions, therefore, the conductivity (or the mobility) of the ions is proportional to $\sqrt{\pi}$ or $\sqrt{c}$. On the other hand, the Debye-Huckel-Onsager (D-H-O) limiting law,

$$\Lambda_o - \Lambda \simeq (B_1\Lambda_o + B_2)\sqrt{c} \qquad (21)$$

which interprets Kohlrausch's observation, attributes the conductivity decrease to interionic forces between the ions, assuming complete dissociation.

In Figure 7 the Λ vs $(1-\alpha)$ dependence is compared with the Λ vs $\sqrt{c}$ plot and the D-H-O Equation 21. It can be seen that Equation 19 gives the best fit, whereas the D-H-O relation is in fact a tangential law for complete dissociation at infinite dilution.

Thus, Figures 6 and 7 give further support for the idea of partial dissociation of strong electrolytes.

Interpretation of Diffusion Coefficient in Terms of π

Based on the assumption of complete dissociation, the concentration dependence of diffusion coefficient D is described e.g., by (3)

$$D = (D^o + \Delta_1 + \Delta_2)\left(1 + m\frac{d\ln\gamma}{dm}\right)\left[1 + 0.036m\left(\frac{D^*_{H_2O}}{D^o} - n\right)\right](\eta_A/\eta) \qquad (22)$$

However, preliminary research (9c,d) has given a much simpler interpretation of the concentration dependence of D for dilute as well as concentrated solutions, in terms of the osmotic pressure π (see p. 78 and note that N_A, defined by Equation 9, depends on i and n_h),

$$\pi = D\eta c N_{Av}\Delta \qquad (23)$$

where N_{Av} is the Avogadro constant (N_{Av} is used here to distinguish it from the mole fraction of 'free' solvent, N_A) and $\eta\Delta$ (=$6\pi r\eta$, π = 3.1416) is the Stokes factor. Figure 8 shows the linear dependence of $D\eta c$ on π for the whole concentration range of

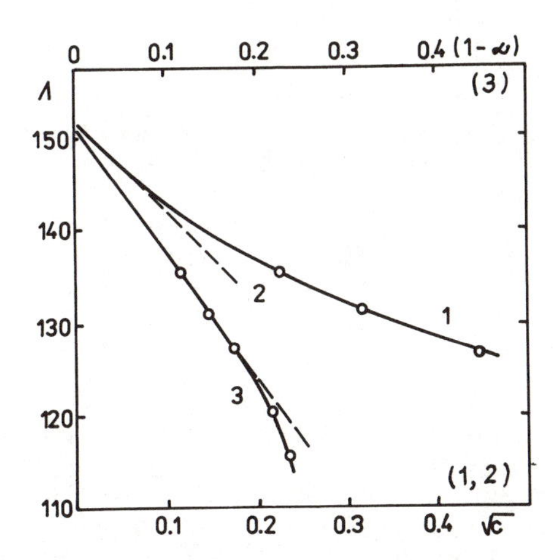

Figure 7. Equivalent conductivities of dilute solutions of KBr at 25°C; 1) Λ vs $\sqrt{c}$, $\Lambda_0 = 151.5$; 2) D-H-O, $\sqrt{c}$ law (Equation 21), $\Lambda_0 = 151.7$ and 3) Λ vs. $(1-\alpha)$, $\Lambda_0 = 151.15$ and slope = 134.7 $\simeq \lambda_{+-}$ (Equation 19), note: slope is not Λ_0 as required by Equation 20.

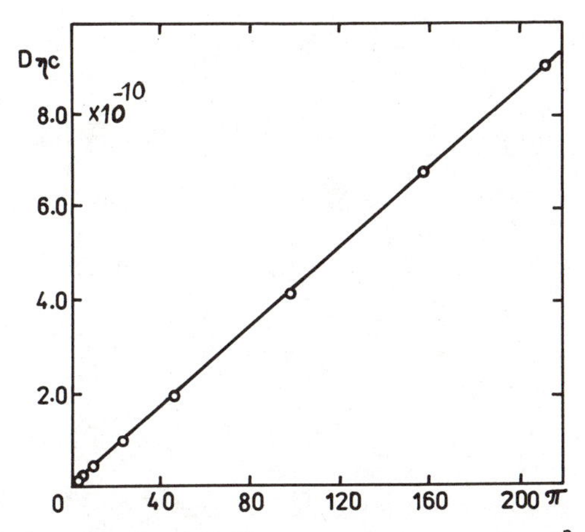

Figure 8. Linear dependence of Dηc (dyne.mol.cm^{-3}) on Π (atm) for aqueous solutions of KBr at 25°C; (full line: least square fit).

0 to 3.75M KBr; data for D and η were taken from (3) and (45,47) respectively. From the slope, r was found to be 2.40Å, which compares well with the value of 2.38Å (sum of the ionic radii) from conductivity (14), Similar results were obtained for HCl (0-1.5M), NaCl (0-3M) and KCl (0-4M). Equation 23 supports the original idea due to Nernst (48) that $D = \Pi/c\bar{K}N_{Av}$; and shows that $\bar{K}=\eta\Delta$ is not a constant since it depends on η. Also, since Π is proportional to $RT\ln N_A$, Equation 23 conforms with the view that chemical potential is the driving force for diffusion (3). In the Stokes-Einstein relation for Brownian motion, (14), $D\eta\Delta = kT$, Einstein replaced the (Π/cN_{Av}) of Nernst's equation with kT. After correcting for non-ideality this resulted in Equation 22.

At infinite dilution, Equation 23 reduces to (14),

$$D^o \eta_A = \nu kT/\Delta \qquad (24)$$

where D^o is the value at infinite dilution. On using the value of Δ from the slope of Figure 8, Equation 24 gives $D^o = 2.017\times10^{-5}$ $cm^2 s^{-1}$ (lit:(3,16),2.016 and 2.018).

This is the final evidence provided here in support of Arrhenius' idea of partial dissociation of electrolytes in solutions.

Literature Cited

1. van't Hoff, J. H. Z. physik. Chem. 1887, 1, 481; Phil. Mag. 1888, 26, 81.
2. Arrhenius, S. Z. physik. Chem. 1887, 1, 631.
3. Robinson, R. A.; Stokes, R. H. Electrolyte Solutions; Butterworths: London, 1955.
4. Guggenheim, E. A.; Stokes, R. H. Equilibrium Properties of Aqueous Solutions of Single Strong Electrolytes; in The International Encyclopedia of Physical Chemistry and Chemical Physics; Robinson, R. A., Ed.; Pergamon Press: Glasgow, 1969; Vol. 1, Topic 15.
5. Erdey-Gruz, T. Transport Phenomena in Aqueous Solutions; Akademiai Kiado': Budapest, 1974.
6. Conway, B. E. In Comprehensive Treatise of Electrochemistry; Vol. 5: Thermodynamic and Transport Properties of Aqueous and Molten Electrolytes; Conway, B. E.; Bockris; J. O'M; Yeager, E., Eds; Plenum Press: New York, 1983; Ch. 2.
7. Horvath, A. L. Handbook of Aqueous Electrolyte Solutions; Ellis Horwood: Chichester, 1985.
8. Heyrovska, R. Coll. Czech. Chem. Commun. 1988, 53, 686.
9. Heyrovska, R. Extd. Abstr. No. Mtgs. of the Electrochem. Soc., U.S.A.; a) 1981, 81-1, 487; b) 1982, 82-1, 649; c) 1984, 84-1, 425, 426; d) 1984, 84-2, 652, 653; e) 1985, 85-2, 442-4; f) 1987, 87-1, 463, 472; g) 1987, 87-2, 1454; h) Abs. 8th Int. Symp. SSSI Regensburg 1987, 49, 130; i) Abs. Int. Symp. MADATES Tokyo 1988, p. 44.
10. Eyring, H.; Henderson, D.; Jost, W. Physical Chemistry, An Advanced Treatise; Academic Press: New York, 1967; Vol. II.
11. Hala, E. Uvod do Chemicke Termodynamiky; Akademia: Praha, 1975.
12. Atkins, P.W. Physical Chemistry: Freeman: San Francisco, 1982.

13. Glasstone, S. Textbook of Physical Chemistry; MacMillan: London, 1951.
14. Moelwyn-Hughes, E. A. Physical Chemistry; Pergamon Press: London, 1957.
15. Lewis, G. N.; Randall, M. J. Am. Chem. Soc. 1921, 43, 1112 and the literature cited therein.
16. Harned, H. S.; Owen, B. B. The Physical Chemistry of Electrolyte Solutions; Rheinhold: New York, 1958.
17. Debye, P.; Huckel, E. Phys. Z. 1923, 24, 185.
18. Onsager, L. Phys. Z. 1927, 28, 277.
19. Kortum, G. Lehrbuch der Elektrochemie; Verlag Chemie: Weinheim, 1966.
20. Gray, J. L.; Mariel, G. E. J. Phys. Chem. 1983, 87, 5290.
21. Stellinger, F. H.; White, R. J. J. Chem. Phys. 1971, 54, 3395, 3405.
22. Gill, J. B. Pure & Appl. Chem. 1987, 59, 1127.
23. Pitzer, K. S.; Mayoraga, G. J. Solution Chem. 1974, 3, 539 and the literature cited therein.
24. Chang, T. G.; Irish, D. E. J. Solution Chem. 1974, 3, 175.
25. James, D.; Frost, R. L. Faraday Disc. Chem. Soc. 1977, 64, 48.
26. Davies, C. W. and others. Disc. Faraday Soc. 1957, 24, 83--108.
27. Hindman, J. C. J. Chem. Phys. 1962, 36, 1000.
28. Hinton, J. F.; Amis, E. S. Chem. Rev. 1967, 67, 367.
29. Chau-Chyun, C. AIChE Journal 1986, 32, 444.
30. Davies, W. G.; Otter, R. J.; Prue, J. E. Disc. Faraday Soc. 1957, 24, 103.
31. Justice, J. C.; Justice, M. C. Faraday Disc. Chem. Soc. 1977, 64, 263.
32. Salomon, M. J. Electrochem. Soc. 1982, 129, 1165.
33. Trans. Faraday Soc. 1907, 3, the discussion section.
34. Bousfield, W. R. Trans. Faraday Soc. 1917, 13, 141.
35. Faraday Disc. Chem. Soc. 1977, 64, "Ion-ion and Ion-solvent Interactions".
36. Bockris, J. O'M.; Conway, B. E. Modern Aspects of Electrochemistry, No. 5; Plenum Press: New York, 1969; chapter 1.
37. Raoult, F. M. Z. physik. Chem. 1888, 2, 353.
38. Porter, A. W. Trans. Faraday Soc. 1917, 13, 123.
39. Nernst, W. Z. physik. Chem. 1889, 4, 155.
40. Gurney, R. W. Ionic Processes in Solution; Dover: New York, 1953.
41. Millero, F. J. Chem. Rev. 1971, 71, 147.
42. Conway, B. E. Ionic Hydration in Chemistry and Biophysics; Elsevier: Amsterdam, 1981.
43. Stokes, R. H.; Robinson, R. A. J. Solution Chem. 1973, 2, 173.
44. Heydweiller, A. Z. physik. Chem. 1910, 70, 128.
45. Jones, G.; Bickford, C. F. J. Am. Chem. Soc. 1934, 56, 602.
46. Partington, J. R. Advanced Treatise on Physical Chemistry; Longmans: 1949; Vol. I.
47. Jones, G.; Talley, S. K. J. Am. Chem. Soc. 1933, 55, 4124.
48. Nernst, W. Z. physik. Chem. 1888, 2, 613.

RECEIVED August 9, 1988

Chapter 7

Historical Highlights in Transference Number Research

Michael Spiro

Department of Chemistry, Imperial College of Science and Technology, London SW7 2AY, United Kingdom

Transference numbers had a controversial birth 135 years ago. Their determination and interpretation have intrigued solution electrochemists ever since. Several ingenious methods of measuring transference numbers will be described because even today one cannot buy off-the-shelf transference kits suitable for research. Some of these methods have been developed and adapted to make them suitable for determinations under extreme conditions of concentration, temperature and pressure while others have remained historical curiosities. The absolute values of transference numbers and their variations with concentration have provided essential insight into the structure of ionic solutions. The triad of conductance, transference number and diffusion coefficient now furnishes a valuable basis for understanding the flow properties of electrolytes.

The story begins in 1853, the year in which Johann Wilhelm Hittorf, recently appointed professor at the University of Münster, published his seminal paper in Poggendorf's Annalen entitled "Ueber die Wanderung der Ionen während der Elektrolyse" (1). We must remember that Hittorf did not mean a free ion in the sense in which it is used today but the moving part of an electrolyte. The mechanism of electrolysis accepted in his time was that proposed by Theodor Grotthuss in 1805. It is illustrated in Figure 1 where each molecule of electrolyte (called atom in Hittorf's paper) is composed of a cation (coloured white) and an anion (coloured black). The first effect of the electric field was taken to be the orientation of the molecules into positions in which the cations all faced the cathode, the anions the anode (row 1 in Figure 1). The two ions in each molecule then separated and moved in opposite directions until they met and combined with the moving ions from the neighbouring molecules (row 2 in Figure 1).

0097–6156/89/0390–0092$06.75/0

Now, however, each cation faced the anode and each anion the cathode. Every molecule therefore had to rotate into the positions shown in row 3 of Figure 1 to allow the electrolysis process to continue.

Hittorf preferred a representation used by Berzelius in which the two ions in a molecule were arranged underneath each other, a representation which did not require the periodic rotation of the molecules (Figure 2). The ions were again pictured as moving horizontally in the electric field. Hittorf now pointed out a tacit but quite unnecessary assumption in the model, namely that the wandering ions met halfway between the molecules from which they originated. This implied that cations and anions moved with equal velocities. Figure 3, adapted from Hittorf's paper, illustrates a different situation, one in which the cations (empty circles) move twice as fast as the anions (filled circles). Here each new electrolyte molecule would form at a position two-thirds of the way from the original right-hand molecule (compare rows 1 and 2 in Figure 3). The experimental consequences can be seen by drawing a vertical line through the solution in each diagram. In Figure 2, per faraday of electricity passed, the anode compartment contains 1/2 equivalent more of anions and 1/2 equivalent less of cations; the reverse applies to the cathode compartment. In the anode compartment of Figure 3, on the other hand, each faraday of electricity produces a gain of 1/3 equivalent of anions and a loss of 2/3 equivalent of cations. The reverse again holds for the cathode section. Hittorf thus concluded that the true situation could be discovered by analysing the solutions around the electrodes after electrolysis.

The apparatus which Hittorf employed consisted essentially of a glass cylinder with a cathode near the top and an anode near the bottom. The anode was chosen to be of the same metal as in the salt used (e.g., copper in work with copper sulphate solutions) in order to maintain the chemical nature of the electrolyte. The cathode was of platinum, gold or silver on which the metal ions plated out. The upper catholyte solution therefore became more dilute during electrolysis and the lower anolyte more concentrated, so ensuring gravitational stability. After the passage of a known quantity of electricity (measured with a silver coulometer) the upper half of the cell was slid sideways by means of a glass plate and the catholyte solution was analysed. In the next paper (2) Hittorf analysed the anolyte solution also and introduced middle sections but unfortunately separated the compartments by means of intestinal membranes. He did make it plain, however, that the results should be calculated with respect to the mass of water in the final solution.

At first Hittorf expressed the amount of cation or anion transferred as a percentage; later (2) this was changed to a fraction and only in the last paper of the series (3) did he employ the term Ueberführungszahl (transference number). However, he realised from the beginning that this quantity was a measure of the relative velocities of the ions so that, in modern symbols,

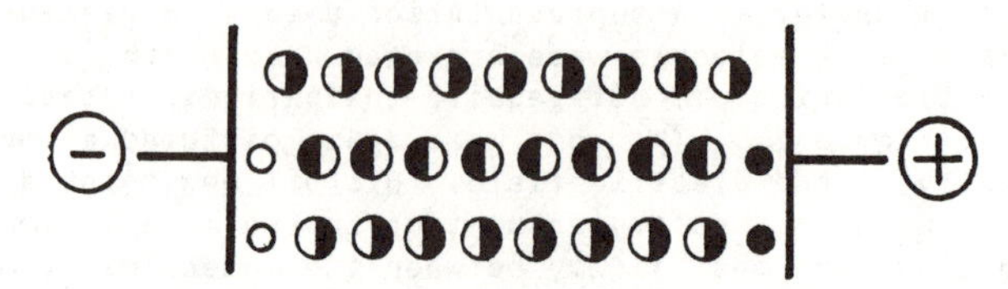

Figure 1. The Grotthuss model of electrolysis. White circles represent cations and black circles, anions.

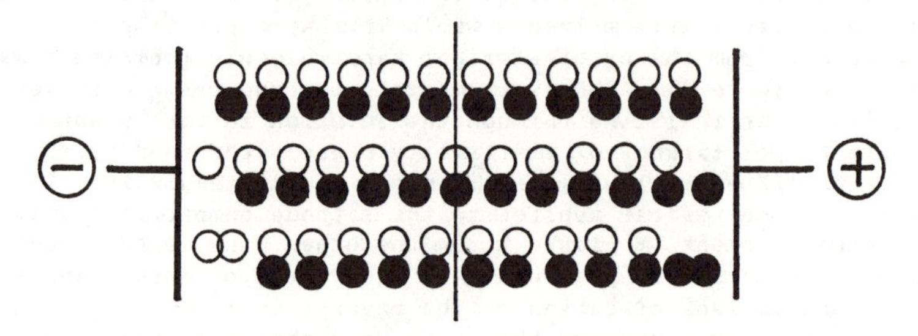

Figure 2. The Berzelius variant assuming equal ionic velocities.

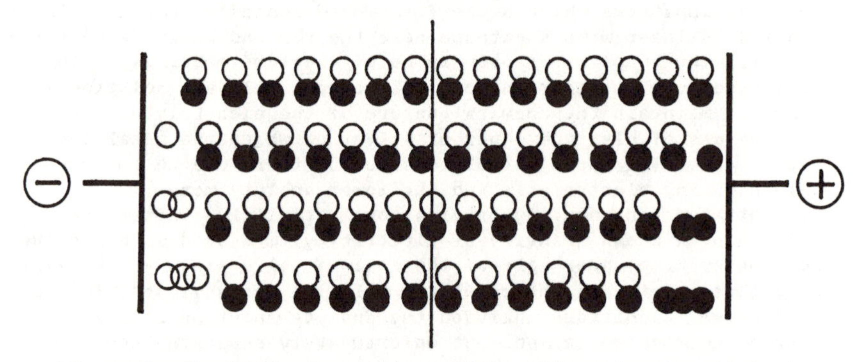

Figure 3. The Hittorf model assuming cations (empty circles) move twice as fast as anions (filled circles).

$$T_{\pm} = \frac{u_{\pm}}{u_{+} + u_{-}} = \frac{\lambda_{\pm}}{\lambda_{+} + \lambda_{-}} \quad (1)$$

where u is mobility and λ the equivalent conductance. The results he obtained did indeed prove that transference numbers were not 0.5 as had been assumed previously but varied from electrolyte to electrolyte. His figures for dilute solutions agree quite well with modern values at comparable concentrations as Table I shows.

Table I. Comparison of Hittorf's Cation Transference Numbers with Present Accepted Values

Electrolyte	$CuSO_4$	$AgNO_3$	KCl
Hittorf T_+ value	0.356 (4°C)	0.474 (19°C)	0.485 (7°C)
Modern T_+ value	0.365 (25°C)	0.468 (25°C)	0.492 (15°C)
Reference	4,5	6	7

The $CuSO_4$ transference numbers vary little with temperature (1,4). The agreement with recent values remains within 0.01 even up to $CuSO_4$ concentrations approaching 1 M, a fact which prompted Pikal and Miller to describe Hittorf's work as a remarkable achievement (4). Hittorf studied a number of different silver salts (nitrate, sulphate and acetate) and he noted that the amount of silver ions transported per faraday varied from salt to salt. He also discovered, to his evident astonishment, that the cation transference number for silver nitrate dissolved in water was not the same as for silver nitrate dissolved in ethanol (0.427 at 4°C). This unexpected result, he wrote, warns us to take great care in the interpretation of the data.

Hittorf was not actually the first to have carried out this type of experiment. Daniell and Miller (8), 8 years earlier, had electrolysed and analysed several electrolyte solutions but their results were vitiated by large experimental uncertainties (2). Hittorf himself took considerable care in his research. On reading his papers one is struck by the precautions he took to avoid disturbing influences; for instance, he checked that his results were independent of the current used. His theoretical analysis, too, is very clear and his calculations are described in great detail. Nevertheless, his papers aroused strong opposition from several quarters. Wiedemann could not understand how an electric current (sic) could produce a different effect on the cation and anion of a given electrolyte (9) or how transference numbers could differ in water and in ethanol (10). Another critic, Magnus, was more partial to Daniell and Miller's results and implied that Hittorf's work had been influenced by diffusion (11). Hittorf was forced to devote large parts of two subsequent long articles (12,13) to rebutting these charges. The attacks on his work must have been painful for a man of his reserved nature. In fact, Hittorf never married and he lived with his younger unmarried sister who kept house for him (14).

In the final paper of this series (3), courageously subtitled "Schluss" (End), Hittorf investigated so-called double salts like potassium ferri- and ferro-cyanides and potassium silver cyanide. He discovered that in solutions of these salts the heavy metals formed part of the anion and were transported to the anode. The formulae he used to describe these complex anions may appear strange to us now, e.g. potassium silver cyanide was written (CyAg + Cy) K. The transference number of (CyAg + Cy) in this solution was found to be 0.404. In a similar way he recognized the existence of a stable chloroplatinum anion, writing the formula of the sodium salt as ($PtCl_2$ + Cl) Na on the basis of the equivalent weight of Pt (98.7). In the modern molar formulation the anion clearly becomes $PtCl_6^{2-}$. Other complex halides were found to be less stable. In their concentrated solutions, Hittorf wrote, the double salt is not decomposed but on addition of water it is more and more decomposed into the two simple salts. Thus only for concentrated solutions was he able to represent potassium cadmium iodide as (JCd + J) K (in equivalent weight symbols with iodine given the symbol J; nowadays it would be written $2K^+[CdI_4^{2-}]$). Further experiments with cadmium and zinc halides yielded positive transference numbers for the metals in dilute solutions and negative ones in concentrated solutions. With aqueous CdI_2 (written JCd), for example, the cadmium transference number decreased from 0.387 to -0.258 as the concentration increased. Hittorf therefore represented the salt in concentrated solution as (JCd + J) + Cd, or Cd^{2+} + $[CdI_4^{2-}]$ in the modern formulation. Concentrated ethanolic solutions of CdI_2 gave an even more negative cadmium transference number of -1.102, suggesting to Hittorf that the solute should be written (3JCd + J) + Cd. The most negative cadmium transference number was obtained for CdI_2 in amyl alcohol, -1.3, making the iodine transference number +2.3. The interpretation of these remarkable results is still under discussion today, as will be mentioned at the end of this article.

Modern Developments

Rather surprisingly, another century was to pass before a general transference formula was published (15,16) that applied to simple and complex electrolytes alike. It depended upon the recognition that transference experiments measured the net transport, not of ions, but of ion-constituents. Only in solutions of completely dissociated electrolytes are the two terms synonymous. The ion-constituent concept was assiduously fostered in Duncan MacInnes' classic textbook "The Principles of Electrochemistry" (17) and elaborated by Harry Svensson (18) but its consequences had not been fully realised. Thus for a solution containing ions 1, 2,

...i, ..., the transference number of the ion-constituent R given by (15)

$$T_R = \frac{\sum_i (z_R/z_i) N_{R/i} u_i c_i}{\sum_i u_i c_i} = \frac{\sum_i (z_R/z_i) N_{R/i} \lambda_i c_i}{\sum_i \lambda_i c_i} \qquad (2)$$

where z is the algebraic charge number of the subscripted species, c_i the equivalent concentration of ion i, u_i its mobility and λ_i its equivalent conductance, and $N_{R/i}$ is the number of moles of ion-constituent R in each mole of ion i. In a concentrated solution of aqueous CdI_2, for example,

$$T_{Cd} = \frac{\lambda_{Cd^{2+}} c_{Cd^{2+}} + 2\lambda_{CdI^+} c_{CdI^+} - 2\lambda_{CdI_3^-} c_{CdI_3^-} - \lambda_{CdI_4^{2-}} c_{CdI_4^{2-}}}{\lambda_{Cd^{2+}} c_{Cd^{2+}} + \lambda_{CdI^+} c_{CdI^+} + \lambda_{CdI_3^-} c_{CdI_3^-} + \lambda_{CdI_4^{2-}} c_{CdI_4^{2-}} + \lambda_{I^-} c_{I^-}}$$

$$= 1 - T_I \qquad (3)$$

Exactly the same equations apply if c_i and λ_i represent the molar concentration and ionic conductance of i, respectively, since $(\lambda_i c_i)_{molar} = (\lambda_i c_i)_{equiv}$. For a simple electrolyte like KCl, Equation 2 turns into Equation 1.

Interest in transference numbers increased greatly around the turn of the 19th century. McBain (19) listed more than 100 experimenters up to the end of 1905 in his valuable critical review of the subject. The review revealed disagreements between the results of different workers which often amounted to several units in the second decimal place. One reason for this was the frequent failure to recognize that concentration changes had to be evaluated with reference to the mass of solvent, a point which Hittorf had clearly stated from the beginning. Moreover, most of the early forms of Hittorf apparatus possessed objectionable features and it is unfortunate that diagrams of some early cells were for years faithfully reproduced in physical chemistry textbooks. Hittorf himself (20) realised in time that membranes introduced to separate compartments frequently gave rise to selective transmission of ions. Not until the very careful researches of Washburn in 1909 (21) and subsequently of MacInnes and Dole in 1931 (22) were Hittorf cells and working practices developed which consistently produced transference numbers accurate to ± 0.0002. These scientists paid particular attention to the use of uniform tubing containing right-angled bends to reduce the effects of local heating and stirring, to the provision of more than one middle compartment to check on the absence of intermixing, to the incorporation of specially designed stopcocks and to the design and preparation of large non-gassing electrodes. The only significant refinement in the following 50 years was

Steel and Stokes' (23) idea of internal conductometric analysis. With these modifications Hittorf's original method reached its apogee.

The Moving Boundary Method

In 1886, just over a century ago, Professor (later Sir) Oliver Joseph Lodge of Liverpool University read a seminal paper to the British Association for the Advancement of Science at its meeting in Birmingham (24). Lodge was not known for his interest in electrolytes. His scientific reputation was made mainly as a pioneer of wireless telegraphy and he also developed an active interest in psychical research. Nevertheless, his paper entitled "On the Migration of Ions and an Experimental Determination of Absolute Ionic Velocity" laid the foundations for a brand new transference method even though it became clear later that Lodge had not properly understood the significance of his own experiments.

Figure 4 illustrates the arrangement of Lodge's first series of experiments. Two electrode vessels containing respectively strong Na_2SO_4 and $BaCl_2$ solutions of equal density were joined by a long tube of dilute HCl. After current had been passed for some time a precipitate of $BaSO_4$ appeared in the tube and increased in thickness. Lodge deduced from the position of the precipitate that the barium ions travelled three times as fast as the sulphate ions. In the second series of experiments, schematically drawn in Figure 5, he filled the central tube with jelly containing acetic acid and silver sulphate. A precipitate of $BaSO_4$ formed at its right-hand end and gradually migrated towards the cathode while a precipitate of AgCl appeared at the left-hand end and moved to the anode. In the last set of runs with acid solutions he incorporated some phenolphthalein into the jelly and observed the movement of colour change.

W.C.D. Whetham, the next worker to take up the method (25), criticized Lodge for employing jellies instead of solutions and for using precipitates since these changed the electrolyte concentrations. Later, however, he reluctantly returned to these very techniques (26) In his first paper Whetham introduced the use of a vertical tube in which a lighter solution was placed on top of a heavier solution. One of these solutions was coloured so that the boundary movement could be easily followed. He noticed that "... when travelling in one direction the boundary got vague and uncertain, and ... in the other (i.e, with current reversed) hard and sharp ", and he recorded the mean velocity.

The full explanation of this phenomenon was furnished 6 years later in 1899 by Professor (later Sir) David Orme Masson of the University of Melbourne (27). He employed a horizontal tube containing a KCl jelly with a yellow K_2CrO_4 solution at the cathode end and a blue $CuSO_4$ solution at the anode end (Figure 6). After current had passed for some time, the tube contained "...blue ($CuCl_2$) jelly at one end, colourless (KCl) jelly in the middle, and yellow (K_2CrO_4) jelly at the other end.... The blue and yellow boundaries remain quite clear cut ..." Masson found

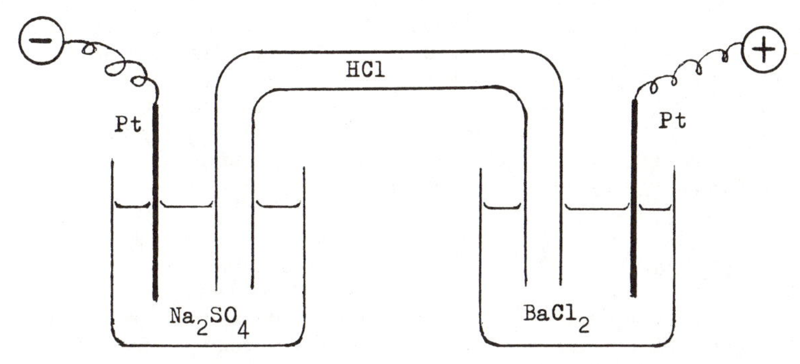

Figure 4. The apparatus used by Lodge in his first series of experiments.

cathode	NaBr solution	HOAc, Ag_2SO_4 jelly	$BaCl_2$ solution	anode

Figure 5. Schematic illustration of Lodge's second series of experiments.

cathode	K_2CrO_4 solution	KCl gelatine jelly	$CuSO_4$ solution	anode

Figure 6. Schematic illustration of Masson's moving boundary experiments.

that the ratio velocity of blue boundary/velocity of yellow boundary was constant throughout an experiment even though the current slowly decreased, and was independent of the dimensions of the tube. Moreover, the ratio was constant for the same salt jelly when different coloured indicators were used but changed when the jelly contained a different salt even when the coloured indicators were the same. The velocities of the boundaries were therefore determined by the properties of the <u>leading</u> solution only. Lodge had thus misinterpreted his own experiments: in Figure 4, for example, he had actually been measuring the relative velocities of the H^+ and Cl^- ions and not those of the following barium and sulphate ions. By the same reasoning, Whetham had been wrong to reverse the current flow and take the mean velocity.

Masson's conclusion agreed with that obtained from theoretical reasoning by Friedrich Kohlrausch two years earlier (28). Kohlrausch had derived a "beharrliche Function" (persistence or regulating function)

$$F = \sum_i (c_i/u_i) \tag{4}$$

which, at a given point in a tube, maintained a constant value independent of any concentration changes produced by ionic migration. The properties of the leading solution therefore controlled the concentration of the following solution that replaced it. In the KCl⟵ $CuCl_2$ system ,for instance, the $CuCl_2$ concentration behind the boundary would automatically adjust itself to the value

$$c_{CuCl_2}/c_{KCl} = T_{Cu}^{CuCl_2} / T_K^{KCl} \tag{5}$$

where c is the equivalent concentration of the named salt. This so-called Kohlrausch equation was derived independently by Masson.

In his experiments Masson showed that the boundaries were sharp only if the leading ion was the faster one. When the following ion was faster, the boundary slowly faded. This explained Whetham's observations on reversing the current. Both workers understood the reason for this phenomenon. Taking the KCl⟵ $CuCl_2$ system again as an illustration, at the boundary the velocities of the two non-common ions must be equal so that

$$v_{mb} = v_{K^+} = u_{K^+} X_{KCl} = v_{Cu^{2+}} = u_{Cu^{2+}} X_{CuCl_2} \tag{6}$$

where X is the potential gradient in the subscripted solution.

Since $u_{K^+} > u_{Cu^{2+}}$, $X_{KCl} < X_{CuCl_2}$. A copper cation straying ahead into the KCl solution would then find itself in a region of lower potential gradient and be eventually overtaken by the boundary whereas a K^+ ion diffusing into the copper chloride solution would experience a greater driving force and shoot back into the KCl solution. It is this electrical restoring effect that keeps the boundary in existence.

It was not appreciated for some time that all these equations should be formulated in terms of ion-constituents and not of ions. Thus in aqueous solutions where $u_{H^+} >> u_{Na^+}$ one expected to find a sharp HCl⟵NaCl boundary but not a sharp NaOAc⟵HOAc boundary. Yet Lash Miller in Toronto (29) predicted the existence of the latter boundary and then observed it. In the acetic acid solution only a small fraction, α, of the hydrogen ions is dissociated and the mobility of the hydrogen ion-constituent is therefore given by

$$\bar{u}_{H^+} = \alpha u_{H^+} \qquad (7)$$

Hence $\bar{u}_{H^+} < \bar{u}_{Na^+}$ so rendering the acetate system stable. Over half a century later a more detailed analysis by the present author of the stability of boundaries in weak electrolyte systems showed that the acetate boundary would be stable as long as the sodium acetate concentration was greater than 0.0005 M (30). The possibility of creating stable boundary systems even with faster following ions provided the following electrolyte is sufficiently weak has considerably enlarged the scope of the moving boundary method, especially in solvents of low permittivity.

Around the turn of the century discrepancies were noted between transference numbers obtained by the Hittorf method and those obtained by the moving boundary (m.b.) method. Explanations proposed in terms of complex ions were clearly unsatisfactory. Lash Miller (29) then pointed out that m.b. results needed a correction for the expansion and contraction caused by electrolysis and a year later, in 1910, G.N. Lewis (31) published a short definitive paper that clarified the problem. He explained that each m.b. transference number had been determined from the velocity of the boundary with respect to the glass tube whereas it should have been determined with respect to the solvent as in Hittorf experiments. A volume correction was therefore required, its calculation being greatly facilitated if one of the m.b. electrodes was kept closed. Both Hittorf and m.b. experiments then yielded the same quantity. Nevertheless, since the volume correction increases with increasing electrolyte concentration, uncertainties in its evaluation can limit the precision of m.b. results at higher concentrations. Several suggestions have recently been made for minimising this difficulty (32). The volume correction also affects the Kohlrausch equation. Hartley (33) showed (and Smits and Duyvis (34) later independently rediscovered) that this can be allowed for very simply by using molal instead of molar concentrations in the equation. A solvent correction is also required for all m.b. work (35) as well as for Hittorf experiments.

Moving Boundary Techniques in the Twentieth Century

Two papers significant in the experimental development of the m.b. method appeared right at the beginning of the century. In 1901 Bertram Steele (36), one of Masson's students, confirmed earlier observations that boundaries between colourless solutions could be seen by placing a light behind them (he used a small gas jet). Coloured indicator ions were therefore no longer necessary. Although Steele formed his initial boundaries by use of a layer of gelatine containing the following indicator electrolyte, the actual boundary movement took place in aqueous solution. He employed vertical tubes and studied both rising and falling boundaries. Optical observation was also used by Franklin and Cady in 1904 (37) in a courageous m.b. study with liquid ammonia solutions at $-33^{o}C$. The problem of initially forming the boundary was cleverly overcome by letting the anodic electrode reaction produce the following solution. With a KNO_3 solution over a mercury anode, for instance, a boundary automatically formed between the potassium and mercury nitrates upon passage of current. This so-called autogenic method has been widely employed since, especially under conditions such as high pressure or low or high temperatures where manipulation of the cell from the outside is difficult.

Normally the initial junction between the leading and following solutions has to be formed by some mechanical means (16). Among the devices employed during the last 65 years have been a plunger (38), sliding glass plates (39), displacement of an intervening air bubble (33), hollow (7) or PTFE (40) stopcocks and, most recently, the use of a flowing junction (41). New and often ingenious methods of observing the boundary have also been put forward. Chief among these are electrical methods. Many workers (33, 41-45) sealed platinum microprobes into the m.b. tube and followed changes in resistance or potential as the boundary advanced. Unfortunately these probes suffered from structural weaknesses and from electrical polarization problems. They are being gradually superseded by a technique invented in Robert Kay's group whereby sets of metallic ring probes are attached to the outside of the m.b. tube, the boundary passage being detected by radio frequency signals (46,47). Electrical methods of sensing the boundary movement are well suited to the remote control operations necessary in transference determinations at high pressures (43,48,49) and at high (50) and low (45) temperatures. Two quite different methods, which monitor the movement of specific constituents, are also now available. In the radioactive tracer method developed in the 1960's in Günter Marx's group (51), the leading or the following ion-constituent is radioactively labelled. Various types of detector are employed according to whether the tracer is an α, β, or γ emitter. Solutions as dilute as 10^{-4} M can be studied in this way as can mixed or hydrolysed systems (52). The latest idea, suggested and tested by Manfred Holz and his coworkers (53,54), has been to follow boundaries by NMR signals from a suitable nucleus in one of the solutions.

The m.b. method is generally recognized to be the most precise tool for determining transference numbers. It is for this reason that so much thought and effort has been devoted to improving and extending the technique. The measurements cannot be carried out with any commercially available "black box" : the appropriate cell design, the method of boundary detection, and even the constant current regulator must fit the problem in hand and be built "in house". Not surprisingly, then, m.b. research has been largely restricted to a few groups which have acquired the necessary expertise. Pre-eminent among these was the group led by Duncan MacInnes and Lewis Longsworth at the Rockefeller Institute for Medical Research in New York City. Their work from the late 1920s to the 1940s was instrumental in placing the m.b. method on a sound footing and their 1932 review (55) is required reading even today for anyone venturing into this field. They were also the first to publish accurate transference numbers for several strong aqueous electrolytes over a concentration range and so were able to test some important predictions of the Debye-Hückel-Onsager theory (56,57). The scene then shifted to the University of Toronto where Andrew Gordon and his group started to measure transference numbers over a range of temperatures (7) and went on to determine the first accurate transference numbers in alcoholic solvents (58). Robert Robinson and Robin Stokes, in their famous 1955 textbook on electrolyte solutions (59), aptly described Gordon's results in pure methanol as "the sole oasis of exact knowledge in a desert of ignorance". Gordon's laboratory, moreover, was a fertile training ground where both Robert Kay and I learnt the rudiments of the craft and eventually set up our own m.b. groups. Lately the centre of gravity has shifted to western Europe where Günter Marx in Berlin, Perie and Perie in Paris, Miguel Esteso in La Laguna and especially Josef Barthel in Regensburg have made and are making prominent contributions to the m.b. literature.

All the groups mentioned have measured transference numbers in non-aqueous or mixed media. Each solvent, after all, is a new world; each one offers the experimenter the opportunity of creating boundary systems never before seen by man. I still remember vividly the boundaries my student David Sidebottom produced in 100% sulphuric acid, boundaries of systems such as $NaHSO_4 \leftarrow Ba(HSO_4)_2$ (40). Sharp and black they certainly were as they crawled up the tube at the infuriatingly slow rate of only millimeters a day. The resulting transference numbers were all less than 0.01 because in this viscous medium the sodium ions moved so much more slowly than the proton-jumping hydrogensulphate ions. But the prize for the most fascinating system must surely go to Dye, Sankuer and Smith (60). Like Franklin and Cady more than half a century before, they chose liquid ammonia as their solvent but this time used metallic sodium as their leading electrolyte. The sharp boundaries $Na^+e^- \leftarrow Na^+Br^-$ were rendered easily visible through the high absorption of light by the leading solution. The transference numbers, combined with literature conductances, showed that the mobility of the e^- ion- constituent passed through a sharp minimum at 0.04 M while its limiting mobility was more than 6 times greater than that of the sodium ion.

This section would not be complete without mention of the complementary indirect m.b. method. Here the concentration of the adjusted Kohlrausch solution behind the boundary is measured: the transference number of the following ion-constituent can then be calculated from the Kohlrausch equation and a knowledge of the transference number of the leading ion-constituent. G.S. Hartley and his co-coworkers (33) were the first to develop this approach which has proved valuable for determining transference numbers of very slow ion-constituents such as cetyltrimethylammonium (61) and other surfactant species and, quite recently, of cadmium in dilute aqueous $CdCl_2$ solutions (62).

During the present century m.b. theory has been further generalised by the inclusion of acid-base reactions at boundaries (63), and this now forms the basis of isotachophoresis (64).

The Analytical Boundary Method

In 1948 A.P. Brady (65,66) invented a new technique with features akin to both the Hittorf and m.b. methods. In this analytical boundary method the solution under investigation was separated from an indicator solution by a porous glass frit: the amount of relevant ion-constituent migrating through the frit upon passage of current was then analysed. Results in satisfactory agreement with literature values were obtained by Spiro and Parton (67) who improved the experimental technique. In essence the method worked well because glass forms a fairly non-selective membrane material and net electroosmosis can be prevented by closing one side of the cell. Brady also suggested an ingenious variant in which the compositions of the two solutions were identical except that the ion-constituent migrating through the frit was tagged by a radioactive tracer. These tagged systems were enthusiastically adopted in the 1950s and 1960s by workers studying surfactant solutions (although some made the mistake of using two gassing electrodes). It was soon realised that the rate of exchange between free tracer ions like ^{22}Na and those associated with the micelles was finite and theory was developed to allow the rate of exchange to be calculated from a planned series of analytical boundary experiments (68,69). An alternative method of tagging, suggested by Hoyer, Mysels, and Stigter (70), was to add some water-insoluble dye solubilized by the micelles in the solution. These workers also carefully analysed the factors important to good cell design. Tagging allowed the use of horizontal cells since the densities of the two solutions barely differed; later variations included cells with two frits enclosing a central tagged solution which was magnetically stirred (71) and the replacement of frits by suitable stopcocks or capillaries (72). The analytical boundary method has been found most useful in its tagged form, especially for determining transference numbers in surfactant solutions and in mixed electrolytes like seawater (73).

The Significance of Transference Numbers

Transference numbers have formed one of the cornerstones in our understanding of electrolyte solutions. Hittorf's discovery in 1853 that transference numbers depended on the ion, the co-ion, and the solvent proved that each ion in a given solvent possesses its own individual mobility. Even today ionic mobilities must be determined by a combination of transference and conductance experiments for we still cannot predict their values accurately from first principles! The importance of ionic mobilities can hardly be overemphasized since they are the only properties of individual ions that can be unambiguously measured (either directly or via trace diffusion coefficients). They therefore provide unique insight into ion-solvent interactions. Hittorf's later transference experiments also revealed the existence and composition of a variety of complex ions in solution. His approach has been followed in more recent structural investigations, for instance in studying the complex ions present in aluminium plating solutions (74).

The next major transference contribution came in the 1930s when Longsworth and MacInnes accurately determined the transference numbers of several strong 1:1 electrolytes over a range of low concentrations. As Figure 7 shows, they found that transference numbers greater than 0.5 increased with increasing concentration, those smaller than 0.5 decreased, and those close to 0.5 hardly changed at all. These results nicely confirmed the predictions of the limiting Debye-Hückel-Onsager theory according to which

$$T_+ = T_+^o + (T_+^o - 0.5)\ \beta \sqrt{c}/\Lambda^o \qquad (8)$$

where superscript ^{o} denotes limiting values at zero ionic strength, Λ is the equivalent conductance and β the D-H-O electrophoretic parameter. The experimental transference numbers also approached the theoretical limiting slopes (shown as straight lines) as the concentration decreased. Only $AgNO_3$ deviated from the general pattern; so too does $AgClO_4$ (75). Much effort has been devoted in recent years to extending the theory to higher concentrations, particularly by the French school of Justice, Perie and Perie in Paris (76). More will be said about concentration dependence in the last section of this article. Transference numbers also play a role in theories of the concentration variation of certain other properties such as the permittivity of electrolyte solutions (77), to take a recent example.

It is sometimes forgotten that during the first half of this century the interpretation of transference data was affected by an unfortunate controversy. It began innocently enough in 1900 when Walther Nernst (78) pointed out that the solvent is not completely stationary for some of it solvates and travels with the migrating ions. This led later workers to the concept of "true" or absolute

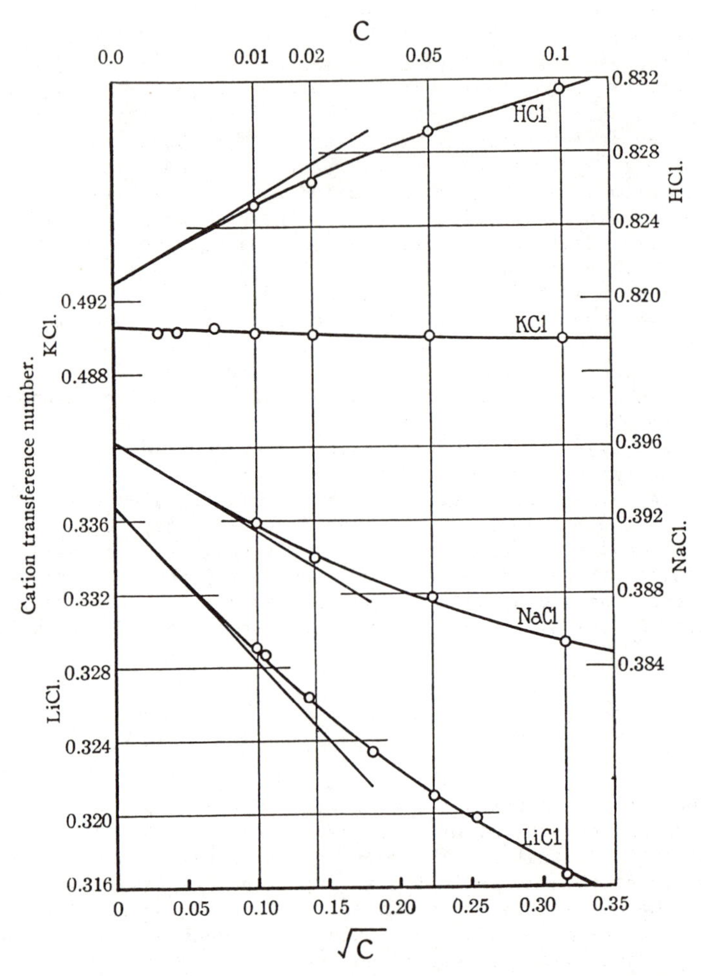

Figure 7. The variation with concentration of the transference numbers of several 1:1 electrolytes in water at 25°C (Reproduced from Ref. 56. Copyright 1932 American Chemical Society.)

transference numbers determined with respect to the "free" part of the solvent only, as compared with normal Hittorf transference numbers measured w.r.t. the solvent as a whole. Nernst suggested that solvation could be measured by adding a small amount of a non-electrolyte to act as an inert reference substance assumed to remain stationary in the electric field. The movement of solvent and of the various ion-constituents could then be determined relative to the inert non-electrolyte. Buchböck (79) and other workers subsequently used a variety of non-electrolytes such as sucrose and urea to determine "true" transference numbers (17). Around 1910 the idea gained ground that the m.b. method gave "true" transference numbers directly, an idea firmly put to rest by G.N. Lewis (31). Not until the late 1940s did incontrovertible evidence appear from a m.b. study by Longsworth (80) and Hittorf experiments by Hale and De Vries (81) that the net movement of solvent per faraday depended on the non-electrolyte used. The reason was simple - the polar reference substances partially solvated the ions and so were no more inert than the solvent itself. The solvation numbers and "true" transference numbers calculated from these experiments did not therefore have the significance which their names imply (82). Nevertheless, the concept itself remains attractive. Carman, for example, has suggested (83) that the Onsager-Fuoss equation applies to the true rather than to the Hittorf transference number although the numerical consequences of this hypothesis do not look favourable (84). No reference substance is needed for information on relative ion solvation in Hittorf experiments with mixed solvents. Here the composition of the solvent itself changes in the electrode compartments during electrolysis. Strehlow and his coworkers (85) have carried out a number of studies along these lines while Feakins' group has performed analogous experiments using emf cells with transference (86,87).

Cells with Transference

The Hittorf and m.b. methods both involve the passage of current and observation of the resulting changes in the system. No current flow takes place in the third major method of determining transference numbers, the measurement of the emf of concentration cells with transference such as

$$Zn|ZnSO_4(c_1)|ZnSO_4(c_2)|Zn.$$

Its theory was given by Hermann von Helmholtz, the great physiologist turned physicist, in an 1878 paper entitled "Ueber galvanische Ströme, verursacht durch Concentrationsunterschiede; Folgerungen aus der mechanischen Wärmetheorie" (88). His derivation, based on vapour pressures and transference numbers, led to equations which look strange to modern eyes but are in fact quite correct. They possessed three important properties. First, the emf was shown to be proportional to the transference number of the co-ion of the ion to which the electrodes were reversible. Second, the magnitude of the emf depended only on the concentrations (activities) of the solutions around the two electrodes, not on the concentration distribution in the middle of the liquid. This particular point was stressed by Helmholtz. And third, his equations did not depend upon solution ideality in contrast to those derived 11 years later by Walther Nernst (89). Nernst used osmotic pressure arguments to obtain equations that closely resemble those found in current textbooks. It is interesting that Nernst omitted to cite Helmholtz's paper although he several times referred to his work. Nernst admitted that Helmholtz's equations were more general than his own and showed that the two sets of equations were equivalent for dilute solutions. Both workers found satisfactory agreement between the predictions of their formulas and experimental data.

Nernst's equations were soon adopted by other workers although they often multiplied the ratio of concentrations by the ratio of molar conductances to allow for incomplete dissociation (even for strong electrolytes). Only in 1920 did MacInnes and Beattie (90) replace concentrations by activities and use the emf equation in its proper differential form. A more general equation in terms of ion-constituent transference numbers and applicable also to electrodes reversible to a complex ion was later derived by the present author (91). In 1935 Brown and MacInnes (92) initiated the converse procedure of calculating activity coefficients from the accurate m.b. transference numbers then available and the emfs of cells with transference, an approach that required only one type of reversible electrode.

The method of determining transference numbers from the emfs of cells with transport is particularly useful for obtaining data over a wide temperature and concentrations range. It has therefore gained increasing popularity in recent years when the emphasis has swung away from highly dilute solutions to more concentrated ones. Research workers have also become more adventurous in their choice of electrodes: the use of chlorine (93) and hydrogen (94) gas electrodes and especially the availability of new ion-selective electrodes have greatly extended the scope of the method. Hermann Weingärtner and his group in Karlsruhe have been very active in this latter area, a recent example being their determination of the transference numbers and activity coefficients of concentrated aqueous KF solutions using fluoride and potassium ion-selective electrodes (95).

Gravitational and Centrifugal Cells

In 1876, just prior to the advent of cells with transference, R. Colley in Moscow published two studies of a different type of cell (96,97). It consisted of a long vertical tube 1 m or more high filled with a uniform electrolyte solution and with identical electrodes at the ends. By calculating the work needed to move both the electrode metal and the ions through the gravitational field, Colley predicted that such vertical cells as

$$\mathrm{Ag}|\mathrm{AgNO_3}|\mathrm{Ag} \qquad (\mathrm{I})$$

$$\mathrm{Cd(Hg)}|\mathrm{CdI_2}(\text{concentrated})|\mathrm{Cd(Hg)} \qquad (\mathrm{II})$$

would exhibit tiny potentials of the order of microvolts. He succeeded in measuring these provided he got rid of bias potentials by also taking readings with the cell inverted. The results were reported as mm of galvanometer deflection. Since Hittorf had found negative cadmium transference numbers in concentrated cadmium iodide solutions, Colley forecast and then verified that the emf of cell II would be several times larger than that of cell I, and of opposite sign. In subsequent work carried out in Berlin (98) to test a prediction of Maxwell's, Colley dropped a vertical CdI_2 cell through 1.72 m from ceiling to floor into a deep layer of sand and recorded the small momentary galvanometer deflections at the beginning and end of the experiment. Both deflections were proportional to the emf of the cell and confirmed the effect of the inertia of the ions.

The study of gravity cells was taken up again 20 years later by Des Coudres (99). His theoretical treatment included a buoyancy correction, the necessity for which had been pointed out to him by Nernst in a letter. In fact, the need for it had already been recognised by Colley (97). Des Coudres measured the potentials of cell II and also of cells of type III

$$Hg|Hg_2Cl_2|MCl|Hg_2Cl_2|Hg \qquad (III)$$

whose emfs were a factor of ten smaller. In order to remove bias potentials and to increase the difference in height between the two electrodes he placed them in separate vessels joined by a rubber tube, full of cell solution, going over a pulley. The effective height difference - and hence the emf - could be doubled by reversing the position of the electrodes. The transference numbers calculated from Des Coudres' measurements agree reasonably well with 20th century values (100). Only one other group has ever tackled gravity cells. In 1942 Grinnell and Koenig (101) selected the system

$$Pt|MI + I_2|Pt \qquad (IV)$$

whose electrodes behaved well. They made various experimental improvements including better temperature control and the resulting transference numbers of KI are consistent with other literature data.

The effect of a force field can be greatly increased by placing the cell into a centrifuge. Des Coudres in 1893 (102) was the first person to realise this : his cell II rotating at 5.8 Hz developed 77.5 μV. More powerful centrifugal forces were employed by Tolman in 1911 (103) whose steam turbine was able to spin the rotator up to 80 Hz and produce emfs of several mV. With cell IV he determined the transference numbers of various iodides in 1 M solution. Cell IV was also the only one used by MacInnes and his coworkers who from 1949 to 1958 (104-107) made a determined effort to overcome the many practical difficulties of the centrifugal method. The effect of heat generated during the acceleration period was analysed in detail and procedures devised to compensate for it. Irreproducibility in the emf caused by the centrifuging of particles of dust on to the outer electrode was one of the major problems and was cured by changing the shape and the position of the platinum electrodes. In its modernised form the centrifugal emf method has become a feasible transference tool that should be applicable to a wider range of electrolyte/electrode systems. The theory for these cases has already been developed by Miller (108) and Schönert (109). Perhaps a future transference white knight will rouse the technique from its present state of dormancy.

The Current Scene

The Debye-Hückel-Onsager theory of the late 1920s, eagerly grasped by electrolyte workers, gradually became an addictive drug. For several decades afterwards researchers vied with one another in carrying out measurements at ever lower concentrations where the theory held better. The study of concentrated ionic solutions was simply ignored although it had been quite respectable in the late 19th and early 20th centuries. More recently, however, there has occurred a marked renewal of interest in concentrated electrolytes where a different type of theoretical framework was required. The most fruitful approach has been to combine all transport properties of a given system - conductances, transference numbers, diffusion coefficients - and to calculate from them certain other parameters which reflect the various ion-ion and ion-solvent interactions. Two major sets of representation exist. The first is provided by the thermodynamics of irreversible processes in which the flows of the ion-constituents are related to electrochemical potential gradients by linear "Onsager" phenomenological or transport coefficients l_{ij} (110). For a binary electrolyte there are four such coefficients (l_{++}, l_{+-}, l_{-+}, l_{--}) at any given molarity and four independent experimental properties are thus needed to evaluate them. These four coefficients can be reduced to three if the Onsager reciprocal relations (ORR) hold

$$l_{+-} = l_{-+} \tag{9}$$

If Equation 9 is true then the transference number measured by the Hittorf or the m.b. method should be identical to that obtained from the emfs of cells with transport (111). Moreover, the sensitivity of this test increases with increasing electrolyte concentration. During the 1970s three transference studies with concentrated solutions were undertaken to test the equation: Miller and Pikal's with $AgNO_3$ (6,112), McQuillan's with $CdCl_2$ (113) and Agnew and Paterson's with $ZnCl_2$ (114). Their results proved conclusively that the ORR were valid within the experimental uncertainties of 1-2%. The vindication of this important postulate is another highlight in the long history of transference research.

The parameter l_{++} reflects interactions of the cation-constituent with the solvent and with itself, l_{--} likewise for the anion-constituent, while l_{+-} is a measure of the degree of coupling between the motions of cation- and anion-constituent. The l_{ij} approach has been most actively fostered by Don Miller (110, 111, 115). An alternative method of representation, introduced by Gerhard Hertz (116,117), is based on velocity correlation coefficients. There are six of these for a binary electrolyte with self-diffusion coefficients being required to evaluate them. Sets of vcc values have recently been published for concentrated solutions of electrolytes which exhibit strong complexation like CdI_2 (118). The main problem in calculating these turned out to be a deficiency in suitable transference number information!

The new fundamental interest in concentrated and in mixed electrolyte solutions has coincided with the need to understand better the complex mixtures found in physiological and geological fluids. The measurements of transference numbers in seawater by Poisson et al.(73) and in wet timber (Simons, P.J; Spiro, M., to be published), have been attempts to shed light on transport in natural systems. Increasingly, too, demands for information on electrolytes are being made by the chemical and electricity-generating industries and by those working to improve batteries and fuel cells (119). In these cases the desire to help is strongly reinforced at the present time by the growing need of university scientists to seek external research funding. The study of transference numbers, for so long an area of pure research, is now inexorably moving out into the marketplace.

Literature Cited

1. Hittorf, W. Ann. Physik u. Chemie 1853, 89, 177.
2. Hittorf, W. Ann. Physik u. Chemie 1856, 98, 1.
3. Hittorf, W. Ann. Physik u. Chemie 1859, 106, 513.
4. Pikal, M.J.; Miller, D.G. J. Chem. Eng. Data 1971, 16, 226.
5. Miller, D.G.; Rard, J.A.; Eppstein, L.B.; Robinson, R.A. J. Solution Chem. 1980, 9, 467.
6. Miller, D.G; Pikal, M.J. J. Solution Chem. 1972, 1, 111.
7. Allgood, R.W.; Le Roy, D.J.; Gordon, A.R. J. Chem. Phys. 1940, 8, 418.
8. Daniell, J.F.; Miller, W.A. Ann. Physik u. Chemie 1845, 64, 18.
9. Wiedemann, G. Ann. Physik u. Chemie 1856, 99, 177.
10. Wiedemann, G. Ann. Physik u. Chemie 1858, 104, 162.
11. Magnus, G. Ann. Physik u. Chemie 1857, 102, 1.
12. Hittorf, W. Ann. Physik u. Chemie 1858, 103, 1.
13. Hittorf, W. Ann. Physik u. Chemie 1859, 106, 337.
14. Drennan, O.J. In Dictionary of Scientific Biography; Gillispie, C.C.; Ed,; Charles Scribner's Sons : New York, 1972; Vol. VI, pp. 438-440.
15. Spiro, M. J. Chem. Educ. 1956, 33, 464.
16. Spiro, M. In Physical Methods of Chemistry, 2nd edition, Vol. II Electrochemical Methods; Rossiter, B.W.; Hamilton, J.F., Eds.; Wiley: New York, 1986; Chapter 8.

17. MacInnes, D.A. The Principles of Electrochemistry Reinhold: New York, 1939; Chapter 4.
18. H. Svensson, Sci. Tools 1956, 3, 30.
19. McBain, J.W. Proc. Wash. Acad. Sci. 1907, 9, 1.
20. Hittorf, W. Z. Physik. Chem. 1902, 39, 613.
21. Washburn, E.W. J. Am. Chem. Soc. 1909, 31, 322.
22. MacInnes, D.A.; Dole, M. J. Am. Chem. Soc. 1931, 53, 1357.
23. Steel, B.J.; Stokes, R.H. J. Phys. Chem. 1958, 62, 450.
24. Lodge, O. Report of the British Association for the Advancement of Science, Birmingham, 1886, 389.
25. Whetham, W.C.D. Phil. Trans. Roy. Soc. (London) 1893, A184, 337.
26. Whetham, W.C.D. Phil. Trans. Roy. Soc. (London) 1895, A186, 507.
27. Masson, O. Phil. Trans. Roy. Soc. (London) 1899, A192, 331.
28. Kohlrausch, F. Ann. Physik u. Chemie 1897, 62, 209.
29. Miller, W.L. Z. Physik. Chem. 1909, 69, 436.
30. Spiro, M. Trans. Faraday Soc. 1965, 61, 350.
31. Lewis, G.N. J. Am. Chem. Soc. 1910, 32, 862.
32. King, F.; Spiro, M. J. Solution Chem. 1981, 10, 881.
33. Hartley, G.S.; Drew, E.; Collie, B. Trans. Faraday Soc. 1934, 30, 648.
34. Smits, L.J.M.; Duyvis, E.M. J. Phys. Chem. 1966, 70, 2747; 1967, 71, 1168.
35. Gwyther,J.R.; Spiro, M.; Kay, R.L.; Marx, G. J. Chem. Soc. Faraday Trans. 1 1976, 72, 1419.
36. Steele, B.D. J. Chem. Soc. 1901, 79, 414.
37. Franklin, E.C.; Cady, H.P. J. Am. Chem. Soc. 1904, 26, 499.
38. MacInnes, D.A.; Smith, E.R. J. Am. Chem. Soc. 1923, 45, 2246.
39. MacInnes, D.A.; Brighton, T.B. J. Am. Chem. Soc. 1925, 47, 994.
40. Sidebottom, D.P.; Spiro, M. J. Phys. Chem. 1975, 79, 943.
41. Kay, R.L.; Vidulich, G.A.; Fratiello, A. Chem. Instrum. 1969, 1, 361.
42. Cady, H.P.; Longsworth, L.G. J. Am. Chem. Soc. 1929, 51, 1656.
43. Wall, F.T.; Gill, S.J. J. Phys. Chem. 1955, 59, 278.
44. Lorimer, J.W.; Graham, J.R.; Gordon, A.R. J. Am. Chem. Soc. 1957, 79, 2347.
45. Barthel, J.; Ströder, U.; Iberl, L., Hammer, H. Ber. Bunsenges. Phys. Chem. 1982, 86, 636.
46. Pribadi, K.S. J. Solution Chem. 1972, 1, 455.
47. Lee, K.; Kay, R.L. Aust. J. Chem. 1980, 33, 1895.
48. Kay, R.L.; Pribadi, K.S.; Watson, B. J. Phys. Chem. 1970, 74, 2724.
49. Ueno, M.; Shimizu, S.I.; Shimizu, K. Bull. Chem. Soc. Japan 1983, 56, 846.
50. Smith, J.E. Jr.; Dismukes, E.B. J. Phys. Chem. 1963, 67, 1160.
51. Marx, G.; Fischer, L.; Schulze, W. Radiochim. Acta 1963, 2, 9.
52. Marx, G.; Bischoff, H. J. Radioanal. Chem. 1976, 30, 567.
53. Holz, M.; Müller, C. J. Magn. Reson. 1980, 40, 595.
54. Holz, M.; Radwan, J. Z. Physik. Chem. N.F. 1981, 125, 49.
55. MacInnes, D.A.; Longsworth, L.G. Chem. Rev. 1932, 11, 171.

56. Longsworth, L.G. J. Am. Chem. Soc. 1932, 54, 2741.
57. Longsworth, L.G. J. Am. Chem. Soc. 1935, 57, 1185.
58. Davies, J.A.; Kay, R.L.; Gordon, A.R. J. Chem. Phys. 1951, 19, 749.
59. Robinson, R.A.; Stokes, R.H. Electrolyte Solutions Butterworth: London, 1955; p.125.
60. Dye, J.L.; Sankuer, R.F.; Smith, G.E. J. Am. Chem. Soc. 1960, 82, 4797.
61. Hartley, G.S.; Collie, B.; Samis, C.S. Trans. Faraday Soc. 1936, 32, 795.
62. Indaratna, K.; McQuillan, A.J.; Matheson, R.A. J. Chem. Soc. Faraday Trans. 1, 1986, 82, 2763.
63. Alberty, R.A. J. Am. Chem. Soc. 1950, 72, 2361.
64. Bocek, P.; Deml, M.; Gebauer, P.; Dolnik, V. Analytical Isotachophoresis Verlag Chemie: Weinheim, 1988.
65. Brady, A.P. J. Am. Chem. Soc. 1948, 70, 911.
66. Brady, A.P.; Salley, D.J. J. Am. Chem. Soc. 1948, 70, 914.
67. Spiro, M.; Parton, H.N. Trans. Faraday Soc. 1952, 48, 263.
68. Wall, F.T.; Grieger, P.F. J. Chem. Phys. 1952, 20, 1200.
69. Gill, S.J.; Ferry, G.V. J. Phys. Chem. 1962, 66, 995.
70. Hoyer, H.W.; Mysels, K.J.; Stigter, D. J. Phys. Chem. 1954, 58, 385.
71. Mysels, K.J.; Hoyer, H.W. J. Phys. Chem. 1955, 59, 1119.
72. Perie, M.; Perie, J.; Chemla, M. Electrochim. Acta 1974, 19, 753.
73. Poisson, A.; Perie, M.; Perie J.; Chemla, M. J. Solution Chem. 1979, 8, 377.
74. Reger, A.; Peled, E.; Gileadi, E. J. Phys. Chem. 1979, 83, 869.
75. Kumarasinghe, S.; Spiro, M. J. Solution Chem. 1978, 7, 219.
76. Justice, J-C.; Perie, J.; Perie, M. J. Solution Chem. 1980, 9, 583.
77. Hubbard, J.B.; Colonomos, P.; Wolynes, P.G. J. Chem. Phys. 1979, 71, 2652.
78. Nernst, W. Nachr. Kgl. Ges. Wiss. Göttingen 1900, 68.
79. Buchböck, G. Z. Physik. Chem. 1906, 55, 563.
80. Longsworth, L.G. J. Am. Chem. Soc. 1947, 69, 1288.
81. Hale, C.H.; De Vries, T. J. Am. Chem. Soc. 1948, 70, 2473.
82. Spiro, M. J. Inorg. Nucl. Chem. 1963, 27, 902.
83. Carman, P.C. J. Phys. Chem. 1969, 73, 1095.
84. Spiro, M. In Physical Chemistry of Organic Solvent Systems; Covington, A.K.; Dickinson, T.; Eds.; Plenum: London, 1973; Chapter 5.
85. Strehlow, H.; Koepp, H.-M. Z. Elektrochem. 1958, 62, 372.
86. Feakins, D.; Lorimer, J.P. J. Chem. Soc. Faraday Trans. 1 1974, 70, 1888.
87. Feakins, D.; O'Neill, R.; Waghorne, E. Pure Appl. Chem. 1982, 54, 2317.
88. Helmholtz, H. Ann. Physik u. Chemie 1878, 3, 201.
89. Nernst, W. Z. Physik. Chem. 1889, 4, 129.
90. MacInnes, D.A.; Beattie, J.A. J. Am. Chem. Soc. 1920, 42, 1117.
91. Spiro, M. Trans. Faraday Soc. 1959, 55, 1207.

92. Brown, A.S.; MacInnes, D.A. J. Am. Chem. Soc. 1935, 57, 1356.
93. King, F.; Spiro, M. J. Solution Chem. 1983, 12, 65.
94. Davies, A.; Steel, B. J. Solution Chem. 1984, 13, 349.
95. Weingärtner, H.; Brown, B.M.; Schmoll, J.M. J. Solution Chem. 1987, 16, 419.
96. Colley, R. Ann. Physik u. Chemie 1876, 157, 370.
97. Colley, R. Ann. Physik u. Chemie 1876, 157, 624.
98. Colley, R. Ann. Physik u. Chemie 1882, 17, 55.
99. Des Coudres, Th. Ann. Physik u. Chemie 1896, 57, 232.
100. MacInnes, D.A. The Principles of Electrochemistry Reinhold: New York, 1939; Chapter 9.
101. Grinnell, S.W.; Koenig, F.O. J. Am. Chem. Soc. 1942, 64, 682.
102. Des Coudres, Th. Ann. Physik u. Chemie 1893, 49, 284.
103. Tolman, R.C. J. Am. Chem. Soc. 1911, 33, 121.
104. MacInnes, D.A.; Ray, B.R. J. Am. Chem. Soc. 1949, 71, 2987.
105. MacInnes, D.A.; Dayhoff, M.O. J. Chem. Phys. 1952, 20, 1034.
106. Kay, R.L.; MacInnes, D.A. J. Phys. Chem. 1957, 61, 657.
107. Ray, B.R.; Beeson D.M.; Crandall, H.F. J. Am. Chem. Soc. 1958, 80, 1029.
108. Miller, D.G. Am. J. Phys. 1956, 24, 595.
109. Schönert, H. Z. Physik. Chem. N.F. 1961, 30, 52.
110. Miller, D.G. J. Phys. Chem. 1966, 70, 2639.
111. Miller, D.G. Chem. Rev. 1960, 60, 15.
112. Pikal, M.J.; Miller, D.G. J. Phys. Chem. 1970, 74, 1337.
113. McQuillan, A.J. J. Chem. Soc. Faraday Trans. 1 1974, 70, 1558.
114. Agnew, A.; Paterson, R. J. Chem. Soc. Faraday Trans. 1 1978, 74, 2896.
115. Miller, D.G. Faraday Discuss. Chem. Soc. 1977, 64, 295.
116. Hertz, H.G. Ber. Bunsenges. Phys. Chem. 1977, 81, 656.
117. Hertz, H.G.; Harris, K.R.; Mills, R.; Woolf, L.A. Ber. Bunsenges. Phys. Chem. 1977, 81, 664.
118. Hertz, H.G.; Edge, A.V.J.; Mills, R. J. Chem. Soc. Faraday Trans.1 1983, 79, 1317.
119. Barthel, J.; Gores, H-J. Pure Appl. Chem. 1985, 57, 1071.

RECEIVED August 9, 1988

Chapter 8

Genesis of the Nernst Equation

Mary D. Archer

Newnham College, University of Cambridge, Cambridge CB3 9DF, United Kingdom

When Nernst started work alongside Arrhenius in Ostwald's Leipzig laboratory in 1887, there was uncertainty as to the location and quantification of the electromotive force (emf) of galvanic cells. Nernst's three seminal contributions of 1888 and 1889 (refs. 5-7) provided for the first time clear, atomistic explanations of, and quantitative expressions for, potential differences across:
i) liquid junctions between two solutions of different concentrations of the same strong electrolyte;
ii) electrode/electrolyte interfaces. Combined, these gave exact expressions for the emf of concentration cells and galvanic cells without liquid junction, in terms of the osmotic pressure p of each component in solution (proportional to concentration for very dilute solutions) and the 'dissolution pressure' P characteristic of each ion in a solid component. Nernst's early view that the separate terms of his whole-cell equations gave the absolute potential difference across each individual interface was later modified by the recognition that the condition of zero interfacial charge is not necessarily the condition of zero potential difference. His eponymous equations for whole cells in any case relate to relative not absolute differences, and remain the crucial cornerstone of equilibrium electrochemistry.

Hermann Walther Nernst, the son of a provincial judge, was born on 25th. June 1864 in the West Prussian town of Briesen (now Wabrzezno in Poland). He graduated *primus omnium* from the gymnasium at Graudenz, a fortified frontier town on the left bank of the Vistula. In the manner typical of German students of that time, he pursued his undergraduate studies at several universities. He spent his first (1883) and third semesters at Zürich, where Heinrich Weber had some years before discovered the strangely low heat capacities of certain elements at low temperatures. For his second semester, Nernst went

0097-6156/89/0390-0115$06.00/0

to Berlin to hear Helmholtz on thermodynamics, and for his fourth to Graz to hear Boltzmann, by whose atomistic ideas he was much impressed.

Nernst's first publications, which came out in 1886-7 (1-4), derive from work undertaken in Graz in collaboration with Ettingshausen, one of Boltzmann's former pupils, on the combined effect of temperature differences and magnetic fields in electromagnetic induction. (The Nernst-Ettingshausen effect now forms part of the phenomenological foundation for the free electron theory of metals.) After a year's work, this new effect had provided Nernst with enough material for his graduation thesis, adjudged *summa cum laude* and published in condensed form as reference (3).

In the autumn of 1886, Nernst moved to Würzburg to work on electrolyte conductivity with Friedrich Kohlrausch who, that same autumn, was visited by Svante Arrhenius. Arrhenius, five years older than Nernst, had been working with Ostwald, his early and important Ionist champion, at the Polytechnic Institute in Riga. After initial scepticism, Nernst was persuaded by Arrhenius's evidence for the complete dissociation of strong electrolytes and, excited by the range of data on ionic solutions that might thereby be explained, decided to go to Riga to work with Ostwald once he had obtained his Würzburg doctorate. However, before he could do so, Ostwald accepted in September 1887 the chair of Physical Chemistry in Leipzig; he offered Nernst the position of chief assistant in his laboratory, which Nernst gladly accepted. Arrhenius too moved to Leipzig, where Ostwald's group of *Ioner* also included van't Hoff. Work in the group centered around thermodynamics (Ostwald's 'energetics'), the colligative properties of gases and liquids and experimental evidence for Arrhenius's new hypothesis.

The two papers by Nernst containing the first formulations of the equations which rapidly acquired and retained the generic title of *Nernst equation* were published, after a precursor paper on liquid junction potentials (5), from Leipzig in 1889 (6,7). The Nernst equation expresses the equivalence of the electrical work $n\mathcal{F}E$ obtainable by the reversible (*i.e.* infinitely slow) discharge of a galvanic cell of cell emf E with the free energy change $-\Delta G$ of the cell reaction, and further quantifies E in terms of the activities of the cell components, equivalent to concentrations (c) in very dilute solutions.

In modern notation, the Nernst equation for the general cell reaction $\nu_R R \rightarrow \nu_P P$, in which n moles of electrons are transferred, is written

$$E = -\frac{\Delta G}{n\mathcal{F}} = E^{\ominus} - \frac{RT}{n\mathcal{F}} \ln \frac{c_P^{\nu_P}}{c_R^{\nu_R}} \qquad (1a)$$

where $E^{\ominus}$ and E are the standard and actual cell emf respectively, $\mathcal{F}$ is Faraday's constant, R the gas constant, T the temperature and ν_R and ν_P the number of reacting P(roduct) and R(eagent) molecules respectively. For example, for the Daniell cell

$$Zn \mid ZnSO_4\ (c_1) \ \vdots\ CuSO_4\ (c_2) \mid Cu$$

the Nernst equation, ignoring the small liquid junction potential and the difference between concentration and activity, is

$$E = E^{\ominus}_{2} + \frac{RT}{2\mathcal{F}} \ln c_2 - \{ E^{\ominus}_{1} + \frac{RT}{2\mathcal{F}} \ln c_1 \} \qquad (1b)$$

where $E^{\ominus}_{2}$ and $E^{\ominus}_{1}$ are the standard electrode potentials of the $Cu^{2+}|Cu$ couple and the $Zn^{2+}|Zn$ couple respectively.

A Nernst equation may be written for every galvanic cell at equilibrium. Therefore the equation is the basis of all thermodynamic applications of potentiometry (*i.e.* measurement of open-circuit cell potentials by means of a potentiometer or other zero-current device). Cells may be constructed and appropriate Nernst equations written to find, for example, the dissociation constant of water, and many electrolyte activity coefficients and stability and solubility constants. Potentiometric titration curves are also interpreted by means of the appropriate Nernst equation.

When current is drawn from a galvanic cell, or passed through an electrolytic cell, the electrodes generally depart from their equilibrium potentials and manifest *charge-transfer overpotential* due to a significant activation barrier to the faradaic process. However, if the kinetics of charge transfer at the electrode/electrolyte interface are so rapid that the electrochemical reactants and products stay in equilibrium at the electrode surface even though a current passes, the Nernst equation still applies to the surface concentrations. Such a process is said to be electrochemically *reversible* or *Nernstian* - sometimes written with a lower case n, a mark of distinction also accorded to the adjectives coulombic, ohmic and faradaic.

So powerful is the Nernst equation in respect of equilibrium electrochemistry that even now those whose electrochemistry has not progressed to the study of electrode kinetics sometimes misapply it to current-carrying systems which are not at equilibrium. Bockris has referred to the 1950's as the Nernstian Hiatus (8) because of the reluctance of electrochemists of the time to grapple with the branch of kinetics he has termed *electrodics*. In fact, although the Nernst equation is still to be found misapplied to irreversible systems, neglect of kinetic aspects of electrochemistry is not forgivable even on historical grounds, for the non-equilibrium concept of overpotential (*Uberspannung)* was introduced by Caspari in 1899 (9) and was quickly investigated by others, including Nernst (10).

At the time Nernst started his Leipzig work, there was still uncertainty as to the location or locations of the emf of a galvanic cell. The old controversy between the contact theory of Volta and the chemical theory of Davy had subsided with the realization that chemical changes necessarily accompanied the passage of current through the cell but not the mere manifestation of an open-circuit potential. Several reliable galvanic cells, including the Plante and Leclanche cells, existed. Faraday's Laws had established that the passage of current through an electrolyte produced stoichiometric chemical changes at the electrodes of an electrolytic cell, and it was recognized that the drawing of current from a galvanic cell produced entirely equivalent chemical changes in the cell which caused it gradually to discharge. Moreover, the electrochemical version of the Berthelot hypothesis, due to Thomsen, namely that the electrical work obtainable from a galvanic cell should be equal to the heat of the cell reaction, was recognised as incorrect and was obsolete. Helmholtz's work had established that the maximum electrical work obtainable was equal to the chemical free energy change of the cell reaction, *i.e.* that $\Delta G = - n\mathcal{F}E$.

However, there remained the question of whether the emf of a galvanic cell arose at the two electrode/electrolyte interfaces, at the contact between the two dissimilar metal electrodes or, in a cell with a liquid junction, at the boundary between the two electrolytes. It was to this problem, and to the quantification of the emf in terms of the solution composition, that Nernst turned his attention.

His three early Leipzig papers (5-7) represent a synthesis of concepts that he was well qualified to make. Working in Ostwald's laboratory, he must have absorbed some of the mass of electrochemical information which appeared a few years later in Ostwald's two-volume work on the history and theory of electrochemistry (11). He was thoroughly familiar with the second-law thermodynamics of Thomson and Clausius, and with the more recent pronouncements of van't Hoff and Helmholtz. Nernst was also imbued with the atomism of Dalton and Boltzmann, in which respect he differed from Ostwald and Helmholtz, and he had accepted Arrhenius's recently published (12,13) hypothesis of the complete dissociation of strong electrolytes in solution. However, his conductance work in Kohlrausch's laboratory had given him a lively appreciation of the effects of incomplete ionization of weak electrolytes.

In July 1888, Nernst published in *Zeitschrift für physikalische Chemie*, which had been started by van't Hoff and Arrhenius the previous year, an important paper entitled *Zur Kinetik der in Lösung befindlichen Körper* (Pt.I; Pt.II is not relevant in this context) on the diffusion theory of substances dissolved in solution (5). Having dealt quickly in this paper with the relatively simple case of diffusion of non-electrolytes using van't Hoff's recently published analogy (14) between gas pressure and osmotic pressure, Nernst moves on to electrolyte diffusion, a problem already considered by Wiedemann (15), Long (16) and Lenz (17). Drawing on the work of Hittorf and Kohlrausch on ionic transport numbers, mobilities and conduction, and that of Clausius, Ostwald and Arrhenius on the constitution of electrolytes, Nernst gives a clear atomistic description of how a liquid junction potential arises at the boundary between two solutions of different concentrations of the same strong electrolyte. He considers the case of a boundary formed between a dilute and a more concentrated solution of HCl. Since the ions in solution have an individual and separate existence, they must also have individual rates of diffusion. Initially, H^+ diffuses faster from the concentrated to the dilute solution than does the less mobile Cl^- but this produces an electric double layer (*Doppelschicht*) in which the field retards the cations and speeds the anions so that in the steady state both move by a combination of diffusion and migration at the same speed. Thus Nernst obtains his equation (5)

$$u\left(\frac{1}{c}\frac{dp}{dx} - \frac{dP}{dx}\right) = v\left(\frac{1}{c}\frac{dp}{dx} + \frac{dP}{dx}\right) \qquad (2)$$

where u = cation mobility, v = anion mobility, p = osmotic pressure and P = electric potential. By equating the quantities of cations and anions moving across the boundary in the yz-plane, Nernst is then able to evaluate Fick's previously empirical diffusion coefficient k in terms of the ion mobilities (measurable by means of Kohlrausch's law of the independent migration of ions), yielding his equation (10)

$$k = 9.689\ p_0 \frac{2uv}{u + v} \tag{3a}$$

where k is the diffusion coefficient in $cm^2\ day^{-1}$ and p_0 is the Boyle factor, *i.e.* the pressure of a perfect gas containing one mole in 1 cm^3 (p_0 = RT). Eq. 3a is the original form of the Nernst-Einstein equation for the diffusion coefficient $D_{\pm}$ of a 1:1 electrolyte, written in modern notation as

$$D_{\pm} = \frac{2\Lambda_+ \Lambda_-}{\Lambda_+ + \Lambda_-} \frac{RT}{\mathfrak{F}^2} = \frac{2u_+ u_-}{u_+ + u_-} \frac{RT}{\mathfrak{F}} \tag{3b}$$

where Λ_+ and Λ_- are the equivalent conductances of the cation and anion respectively, and u_+ and u_- their mobilities ($u_i = \Lambda_i/\mathfrak{F}$).

By integrating eq. (2) between the limits of the bulk concentrations of the two solutions, Nernst obtains for the liquid junction potential E

$$E = 0.0235 \frac{u - v}{u + v} \log \frac{p_1}{p_2}\ \text{Volt} \tag{4a}$$

which in modern terminology (essentially adopted by Nernst by the early 1890's (18)) is

$$E = \frac{RT}{\mathfrak{F}} (t_+ - t_-) \ln \frac{c_1}{c_2} \tag{4b}$$

where t_+ and t_- are the transport numbers of the cation and anion respectively ($t_+ = u_+/(u_+ + u_-)$, $t_- = u_-/(u_+ + u_-)$). Equations (4) are independent of the detailed nature of the boundary. Liquid junction potentials for two solutions containing $\geq$3 ions depend on the concentration profiles in the boundary region and awaited later treatment by Planck (19) and Henderson (20,21).

Nernst points (5) out that, according to eq. (4a), liquid junction potentials depend only on the ratio of the two osmotic pressures p_1 and p_2 and not on their absolute values. So if one makes a cell such that in one all the solutions are n times more concentrated than in the other, both must give the same liquid junction potential. Nernst called this the *superposition principle* and prefaced many of his discussions with it; its validity (only approximate, in that osmotic coefficients are ignored) is powerful evidence for the complete dissociation of strong electrolytes.

No means is available for measuring liquid junction potentials without introduction of additional interfaces through electrodes or other probes, so for experimental verification of his formula Nernst was led to construct cells containing liquid junctions. In paper (6), entitled *Zur Theorie umkehrbarer galvanischer Elemente*, published some six months after paper (5), Nernst considered concentration cells with transport, *i.e.* cells such as

$$Ag \mid AgNO_3\ (c_1) \,\vdots\, AgNO_3\ (c_2) \mid Ag \qquad \text{(Cell 2)}$$

in which there is no salt bridge and the two half-cells are identical save for the concentrations of the two solutions; the liquid junction in them is of the type considered by Nernst in paper (5) so the potential difference across it could be calculated. But to find the net emf of such a cell, the electrode processes and the potential difference at each electrode/electrolyte interface had also to be considered.

It was known that charge was carried across the electrode/electrolyte interface in such a cell solely by the metal ions. The *difference* in the work done in transporting a given quantity of ions into $AgNO_3$ solutions of osmotic pressure p' and p'' could be expressed on Second Law grounds as $K \ln p'/p''$, where K is a factor dependent on the units of measurement. But Nernst could not calculate the *absolute* work involved in transporting ions from electrolyte to electrode because there was no measure of the state of the ions in the electrode. Using vapor pressures and van't Hoff's theory of osmotic pressure as analogies, he therefore introduced, effectively as a constant of integration, the quantity P, the dissolution pressure (*Lösungstension*) characteristic of the metal. Although P was neither known nor calculable from other properties of the metal, this enabled Nernst to write the absolute work as $K \ln P/p$.

To find the potential difference across the electrode/electrolyte interface, Nernst then adapted his previous expression for a liquid junction potential by noting that all the electricity is carried across the interface by the moving metal ion so that u effectively becomes 1 and v zero. Thus Nernst obtains his equation (5) for the potential E_m of the metal with respect to the potential E_e of the electrolyte:

$$E = E_m - E_e = p_o \ln \frac{P}{p} \qquad (5)$$

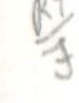

which by 1896, had evolved into the rather more familiar format $\mathcal{E} = RT \ln C/c$ (22). As evidence for eq. (5), Nernst adduced some unpublished data of Ostwald's on the $Zn|Zn^{2+}$ half-cell, taking into account the valence of zinc. Ostwald later commented (ref. (11), p 1136) that, considering the accumulation of experimental error, the numerical agreement could be regarded as satisfactory although it left something to be desired.

It follows from eq. (5) that, if it so happens that $P = p$ at the moment of first contact of electrode and electrolyte, the two are in equilibrium with no potential difference between them. However, $P > p$, metal ions will dissolve even at open circuit, taking with them a certain quantity of positive charge, leaving on the electrode the equivalent negative charge and creating an electric double layer at the interface. Conversely, if $P < p$, ions deposit on the electrode, causing it to become positively charged and leaving the equivalent negative charge at the solution boundary in the form of excess anions. The potential difference E thus created between metal and solution adds the electrical work $|qE|$ to the work of transporting an ion of charge q across the interface, always acting to oppose the further movement of ions in the same direction. Equilibrium is reached when the sum of the chemical and electrical work done in transporting ions across the interface in either direction is zero.

Two comments by Nernst which counter obvious arguments that might be raised against eq. (5) should be noted: i) because of the very large charge associated with metal ions (the large value of Faraday's constant, we might say), the quantity of ions moving across the interface to achieve equilibrium at open circuit is minute; ii) since in the nature of things P must always have a positive value, it seems that in contact with pure water ($p = 0$) all metals should be negatively charged to an infinite extent, which is not possible. We must therefore attribute to all metals the capacity of trace dissolution as ions.

Nernst is now able to write an expression for the emf of a concentration cell such as Cell 2 by addition of the predicted liquid junction potential and the two opposed electrode/electrolyte interfacial potentials, giving his equation (6)

$$E'_m - E''_m = 0.860\ \mathrm{T}\ \frac{2v}{u+v} \ln \frac{p_2}{p_1} \times 10^{-4}\ \mathrm{Volt} \qquad (6a)$$

which in modern terminology is

$$\mathrm{E} = \frac{\mathrm{RT}}{\mathcal{F}}\ 2t_- \ \ln \frac{c_2}{c_1} \qquad (6b)$$

Nernst shows that the emf of Cell 2, calculated from equation (6a) using the data of Hittorf (23) and Loeb and himself (24) on the transport number of Ag^+ in silver acetate and silver nitrate, is in agreement with Miesler's published data (25).

Turning to the similar case of concentration cells with anion-reversible electrodes such as

$$\mathrm{Hg,\ Hg_2Cl_2\ |\ HCl\ (c_1)\ \vdots\ HCl\ (c_2)\ |\ Hg,\ Hg_2Cl_2} \qquad \text{(Cell 3)}$$

Nernst gives the appropriate formula (his equation (11)) for the emf, namely

$$E'_n - E''_n = 0.860\ \mathrm{T}\ \frac{2u}{u+v} \ln \frac{p_1}{p_2} \times 10^{-4}\ \mathrm{Volt} \qquad (6c)$$

and shows the excellent agreement between calculated values and his own experimental data for Cell 3 and related cells in the table of data reproduced below, in which μ is the equivalent concentration.

Combination	*Elektromotorische Kraft bei 18°* beob.	ber.	
Hg, Hg_2Cl_2, HCl μ_1=0.105, HCl μ_2=0.0180, Hg_2Cl_2, Hg	0.0710	0.0717	Volt
Hg, Hg_2Br_2, HBr μ_1=0.126, HBr μ_2=0.0132, Hg_2Br_2, Hg	0.0932	0.0917	"
Hg, Hg_2Cl_2, LiCl μ_1=0.100, LiCl μ_2=0.010, Hg_2Cl_2, Hg	0.0354	0.0336	"
Hg, Hg_2Cl_2, KCl μ_1=0.125, KCl μ_2=0.0125, Hg_2Cl_2, Hg	0.0532	0.0542	"

Such data provided good verification of equation (6c). However, as Nernst acknowledges in paper (6), he was not the first to give the correct expression for the emf of a concentration cell. Helmholtz (26,27) had produced an equivalent expression by calculating in terms of the different vapor pressures of the two solutions the free energy change involved in equalizing the two concentrations, yielding for cells such as Cell 2 the expression

$$E'_m - E''_m = \frac{\pi_0 - \pi}{\pi} \cdot \frac{MS}{H_2O} \cdot 0.000933 \cdot \pi_0 V_0 \frac{v}{u+v} \ln \frac{c_2}{c_1} \times 10^{-8} \text{ Volt} \quad (6d)$$

where π_0 and π are the vapor pressures of pure water and a solution of unit concentration respectively, and V_0 is the volume of 1 gm water vapor at the prevailing temperature and pressure p_0.

Nernst shows that equation (6a) and (6d) are equivalent if the lowering of the vapor pressure by a binary electrolyte is twice as great as that produced by the same concentration of a non-electrolyte, and notes that it is *"nicht uninteressant"* that this is in agreement with Arrhenius's hypothesis of complete ionic dissociation. However, Nernst adds, Helmholtz's thermodynamic treatment gives no guide as to the location or locations of the potential differences which combine to produce the cell emf, whereas his summation is explicit and furthermore provides a mechanistic reason for the equivalence of electrical and chemical energy.

This careful attention to Helmholtz's work contrasts with silence from Nernst on the published contributions of Duhem and Gibbs. Duhem's 1886 book (28) on thermodynamic potential contains an exact expression for the emf of concentration cells with transference, although his argument is taken (with attributions) from Helmholtz. Nernst seems at the time of the publication of his papers (5-7) to have been unaware of Duhem's book. As for Gibbs, Ostwald translated his thermodynamic writings into German in 1892 and only thereafter did they become required reading for all students at Leipzig.

Three months after Nernst's paper (6) came paper (7), entitled *Die elektromotorische Wirksamkeit der Ionen*, the condensed published version of his *Habilitationsschrift*. Paper (7) is the consolidation of the approaches of papers (5) and (6), but also contains the extension of the earlier treatment to cells such as

$$Zn \mid ZnCl_2 \ (c) \mid Hg_2Cl_2, Hg \qquad \text{(Cell 4)}$$

These are simple in that they lack a liquid junction but complicated in that the two electrodes are made of different materials, so that the possibility of there being an additional interfacial potential difference across the metal|metal boundary must be considered. By combining the appropriate half-cell expressions obtained from eq. (5), Nernst obtains for the emf of Cell 4 the expression

$$E \quad = \quad \frac{p_0}{2} \ln \frac{P}{p} \quad + \quad p_0 \ln \frac{P'}{2p} \quad + \quad A \qquad (7a)$$

where the first term relates to zinc and the second to chloride and A is the (unknown) potential difference between the two metals. In modern nomenclature, any contribution from the A term (discussed below) may be regarded as subsumed under the standard emfs $E^{\ominus}$ in the following expression for Cell 4:

$$E \quad = \quad E^{\ominus}_{Hg_2Cl_2\,|\,Hg} - \frac{RT}{\mathcal{F}} \ln 2c \quad - \quad E^{\ominus}_{Zn^{2+}\,|\,Zn} - \frac{RT}{2\mathcal{F}} \ln c \qquad (7b)$$

Bearing in mind that the modern sign convention for electrode potentials relates to the potential of the solution with respect to the metal, whereas Nernst uses the old convention of the potential of the metal with respect to the solution, eqs. (7a) and (7b) are formally equivalent if P and $E^{\ominus}$ for each electrode are related by $P = \exp\,[-\mathcal{F}E^{\ominus}/RT]$.

Nernst presented fairly limited data in support of his expression for the potential difference across an electrode/ electrolyte interface, though he made more measurements of liquid junction potentials. Negbaur (29) made extensive measurements on concentration cells, using both Nernst's and Planck's liquid junction formulae; Nernst wrote a careful commentary on this work (30). Schöller (31) showed that sodium amalgam electrodes obeyed the appropriate Nernst equation, and Gordon (32) that silver electrodes immersed in various fused silver nitrate/potassium nitrate/sodium nitrate mixtures also did. Duane (33,34) and Tower (35) carried out more liquid junction work, including the case that the two solutions were at different temperatures.

In general, there was ready and rapid acceptance of the Nernst equations; by 1897, Moore (36) in his review of lead accumulators refers to "the well-known formula of Nernst for the E.M.F. of a cell". Bancroft's remarks (37) on the arbitrary nature of the dissolution pressure P are, however, telling, and Nernst himself had realized the difficulty, concluding paper (7) with the words: "When we finally succeed in measuring in absolute units the electrolytic dissolution pressures....we will have made an important advance by providing answers to questions that have engaged the attention of physicists since the times of Volta."

He reverted to the problem in 1894 in a paper on the seat of the emf of the Daniell cell (38). He dismissed the metal/metal contact as a source of emf, correctly noting that a transient flow of charge might result from first contact but not an enduring current: thus A

in equations such as (7a) could be set to zero. Moreover, calculated and experimental values of liquid junction potentials were always small, so this left only the electrode/electrolyte interfaces as the source of the emf, and hence the difference in the dissolution pressures of the two electrodes was crucial. Still the problem of *a priori* calculation of a single P value remained: it was uncertain whether it was a property of the electrode only or depended on the counter ion in solution, and its physical meaning was unclear.

By 1900, Nernst had shifted his stance further (39); he had abandoned the unsatisfactory P and essentially adopted the now-accepted view that interfacial potential differences between chemically dissimilar phases are not individually measurable, although the difference between any two such potentials is. He had therefore concluded that electrode potentials could only be measured relative to one another, and proposed the hydrogen electrode as the standard whose $E^{\ominus}$ value should be set at zero (39). He was led to these conclusions by noting that Helmholtz's hypothesis that the electrocapillary maximum (ecm) of a dropping mercury electrode identified the point of zero potential difference between mercury and solution is not necessarily true, since changing the ions in the solution may change the ecm. (This is of course so: in modern terms, however, the problem is not that specificially-adsorbed ions shift the ecm, but that even if they are absent so the interface indeed contains no ion layer, there remains the unmeasurable interfacial dipole potential.)

Gibbs was taking a similar view at the same time, writing in 1899: "Again, the consideration of the electric potential in the electrolyte, and especially the consideration of the difference of potential in electrolyte and electrode, involves the consideration of quantities of which we have no apparent means of physical measurement, while the difference of potential in "pieces of metal of the same kind attached to the electrodes" is exactly one of the things which we can and do measure." (40) However, the principle of what electrochemical potential differences could and could not be measured was not widely appreciated until much later when it was reformulated in terms of the electrochemical potential $\bar{\mu} = \mu + z\mathfrak{F}\phi$ by Guggenheim (41).

After his eponymous equation was established, Nernst, although gradually shifting his attention from 'wet' to 'dry' chemistry, did substantial further electrochemical work on electrolytic dissociation (42-48), electrolysis (49), capillary-active ions (50), cell resistance (51), electrostriction (52), hydrogen permeation (53,54), gas polarization (10), the stagnant layer at an electrode (now termed the *Nernst diffusion layer)* (55), and nervous electrical impulses (56,57). In the field of liquids and solutions, Nernst modified Wheatstone's bridge in order to measure the dielectric constants of liquids (58), and he proposed the theory of the solubility product (59) and generalized the distribution law (60,61).

After the publication of his *Habilitationsschrift*, Nernst was successful in obtaining a lectureship at Leipzig. In the summer of 1889, he took a temporary assistantship in Heidelberg, where a lecturer was required until Bunsen's successor was appointed. In 1890, he left Leipzig for good to take up a lectureship at Göttingen, in charge of a sub-department of the Physics laboratory under Riecke, and he stayed there in positions of increasing importance and influence for the next fifteen years. At Göttingen, and after 1905 in Berlin, under the patronage of Kaiser Wilhelm II and the powerful Friedrich Althoff, Permanent Secretary at the Prussian Ministry of Education, Nernst became a man of affairs, though not to the exclusion of his science (62). His heat theorem work, based on measurements of the heat capacities of solids at very low temperatures, earned him the Nobel prize in 1920.

Nernst was at heart a practical man rather than a theoretician, and his equation is an eminently practical one. Of Nernst's *Theoretical Chemistry* published in 1893, which became the standard text in physical chemistry until the twenties, Einstein in his affectionate obituary notice of Nernst could say no better than that it was "theoretically elementary" (63). But to Nernst's Leipzig work, Einstein accords a celebratory elegy: "Nernst ascended from Arrhenius, Ostwald and van't Hoff, the last of a dynasty that based its investigations on thermodynamics, osmotic pressure and ionic theory."

Literature Cited

1. Ettingshausen, A. v.; Nernst, W. Sitzungsber. Akad. Wiss. Wien 1886, 94, 560-610.
2. Ettingshausen, A. v.; Nernst, W. Ann. Phys. Chem. (Wied. Ann.) 1886, 29, 343-47.
3. Nernst, W. Ann. Phys. Chem. (Wied. Ann.) 1887, 31, 760-89.
4. Ettingshausen, A. v.; Nernst, W. Ann. Phys. Chem. (Wied. Ann.), 1888, 33, 474-92.
5. Nernst, W. Z. physik. Chem. 1888, 2, 613-37.
6. Nernst, W. Sitzungsber. preuss. Akad. Wiss. 1889, 83-95.
7. Nernst, W. Z. physik. Chem. 1889, 4, 129-81.
8. Bockris, J. O'M.; Reddy, A. K. N. Modern Electrochemistry; Macdonald & Co.: London, 1970; p 16, 23, 24.
9. Caspari, W. A. Z. physik. Chem. 1899, 30, 89-97.
10. Nernst, W.; Dolezalek, F. Z. Elektrochem. 1899-1900, 549-50.
11. Ostwald, W. Elektrochemie: Ihre Geschichte und Lehre; Veit & Comp.: Leipzig, 1896. Published in translation as Electrochemistry: History and Theory by Amerind Publishing Co. Pvt. Ltd., New Delhi, 1980 for the Smithsonian Institution and the National Science Foundation, Washington D.C.
12. Arrhenius, S. Mag.-Cirkular der B.A.C.f.E. 1887, 23.
13. Arrhenius, S. Z. physik. Chem. 1887, 1, 631-48.
14. van't Hoff, J. H. Z. physik. Chem. 1887, 1, 481-508.
15. Wiedemann, G. Ann. Phys. Chem. (Pogg. Ann.) 1858, 104, 162-70.
16. Long, J. H. Ann. Phys. Chem. (Wied. Ann.) 1880, 9, 613-641.
17. Lenz, R. Mem. Acad. Imp. Sciences Petersbourg 1882, 7th series, 30(8), 64 pp.
18. Nernst, W.; Pauli, R. Ann. Phys. Chem. (Wied. Ann.) 1892, 45, 353-59.
19. Planck, M. Ann. Phys. Chem. (Wied. Ann.) 1890, 40, 561-76.
20. Henderson, P. Z. physik. Chem. 1907, 59, 118-28.
21. Henderson, P. Z. physik. Chem. 1908, 63, 325-45.
22. Nernst, W. Z. Elektrochem. 1896, 52-54.

23. Hittorf, W. Ann. Phys. Chem. (Pogg. Ann.) 1853, 89, 177-211.
24. Loeb, M.; Nernst, W. Z. physik. Chem. 1888, 2, 948-63.
25. Miesler, J. Sitzungsber. Akad. Wiss. Wien (Math. Naturwiss.) 1887, 95, 642-45.
26. Helmholtz, H. Monatsber. Akad. Berlin, 1877, 713-26.
27. Helmholtz, H. Ann. Phys. Chem. (Wied. Ann.) 1878, 3, 201-
28. Duhem, P. Le Potentiel Thermodynamique et ses Applications; Hermann: Paris, 1886.
29. Negbaur, W. Ann. Phys. Chem. (Wied. Ann.) 1891, 44, 737-58.
30. Nernst, W. Ann. Phys. Chem. (Wied. Ann.) 1892, 45, 360-69.
31. Schöller, A. Z. Elektrochem. 1898, 5, 259-61.
32. Gordon, C. McC. Amer. Acad. Proc. 1898, 33, 59-68.
33. Duane, W. Sitzungsber. deutsch. Akad. Wiss. Berlin 1896, 39, 967-970.
34. Duane, W. Ann. Phys. Chem. (Wied. Ann.) 1898, 65, 374-402.
35. Tower, O. F. Z. physik. Chem. 1895, 18, 17-50.
36. Moore, B. E. Phys. Rev. 1897, 4, 353-74.
37. Bancroft, W. D. Phys. Rev. 1896, 3, 250-69.
38. Nernst, W. Z. Elektrotechnik Elektrochem. 1894, 7, 243-46.
39. Nernst, W. Z. Elektrochem. 1900, 19, 253-57.
40. Unpublished letter to W. D. Bancroft. See Gibbs, J. W. Collected Works; Longmans, Green and Co.: New York and London, 1928. Vol. 1, p 429.
41. Guggenheim, E. A. J. Phys. Chem. 1929, 33, 842-49.
42. Nernst, W. Chem. Zeitbl. 1890, 1, 65-68, 145-147.
43. Nernst. W. Z. physik. Chem. 1894, 13, 531-536.
44. Nernst, W. Z. physik. Chem. 1894, 14, 155-156.
45. Nernst, W.; von Wartenburg, H. Nachr. Ges. Wiss. Göttingen 1905, 64-74.
46. Nernst, W.; von Wartenburg, H. Z. physik. Chem. 1906, 56, 548-557.
47. Nernst, W. Z. physik. Chem. 1928, 135, 237-250.
48. Nernst, W. Sitzungsber. preuss. Akad. Wiss. 1928, 4-8.
49. Nernst, W. Ber, deutsch. Chem. Ges., 1897, 30, II, 1547-1563.
50. Nernst. W. Z. Elektrochem. 1897, 4 , 29-31.
51. Nernst, W.; Haagn, E. Z. Elektrochem. 1895-1896, 493-494.
52. Drude, P.; Nernst, W. Z. physik. Chem. 1894, 15, 79-85.
53. Nernst, W.; Lessing, A. Nachr. Ges. Wiss. Göttingen 1902, 146-159.
54. Tammann, G.; Nernst, W. Nachr. Ges. Wiss. Göttingen 1891, 202-212.
55. Nernst, W. Z. physik. Chem. 1904, 47, 52-55.
56. Nernst, W.; Barratt, J.O.W. Z. Elektrochem. 1904, 10, 664-668.
57. Nernst, W. Z. Elektrochem. 1908, 14, 545-549.
58. Nernst, W. Ann. Phys. Chem. (Wied. Ann.) 1896, 57, 209-214.
59. Nernst, W. Z. physik. Chem. 1901, 38, 487-500.
60. Nernst, W. Z. physik. Chem. 1891, 8, 110-139.
61. Nernst, W. Nachr. Ges. Wiss. Göttingen 1890, 401-416.
62. Mendelssohn, K. The World of Walther Nernst, Macmillan, London: 1973.
63. Einstein, A. Scientific Monthly, Feb. 1942.

RECEIVED August 3, 1988

Chapter 9

The Choice of the Hydrogen Electrode as the Base for the Electromotive Series

Alfred W. von Smolinski[1], Carl E. Moore[2], and Bruno Jaselskis[2]

[1]Department of Medicinal Chemistry and Pharmacognosy, University of Illinois at Chicago, Chicago, IL 60680
[2]Department of Chemistry, Loyola University of Chicago, Chicago, IL 60626

The electromotive series is a list of the elements in accordance with their electrode potentials. The measurement of what is commonly known as the "single electrode potential", the "half-reaction potential" or the "half-cell electromotive force" by means of a potentiometer requires a second electrode, a reference electrode, to complete the circuit. If the potential of the reference electrode is taken as zero, the measured E.M.F. will be equal to the potential of the unknown electrode on this scale. W. Ostwald prepared the first table of electrode potentials in 1887 with the dropping mercury electrode as a reference electrode. W. Nernst selected in 1889 the Normal Hydrogen Electrode as a reference electrode. G.N. Lewis and M. Randall published in 1923 their table of single electrode potentials with the Standard Hydrogen Electrode (SHE) as the reference electrode. The Commission of Electrochemistry of the I.U.P.A.C. meeting at Stockholm in 1953 defined the "electrode potential" of a half-cell with the SHE as the reference electrode.

The Status of the Hydrogen Electrode. Probably no area of electrochemistry is more greatly neglected in current texts than the history of the choice of the hydrogen electrode as the reference standard for electromotive force measurements. Since all tables of potentials of oxidation-reduction half-reactions are based on the half-cell reaction $\frac{1}{2}H_2 = H^+ + e^-$, it would seem that the selection of this reaction as the standard should warrant more attention. If the selection is treated at all, it is usually dismissed as an arbitrary choice, which it is, with no reference made to the people and events involved in establishing this fundamental reference point for the EMF scale. One possible exception may be noted (1). The referenced edition of this work is perhaps the best previously existing source on this topic. However, the subsequent edition omits the subject entirely.

0097–6156/89/0390–0127$06.00/0

Early History of the Electromotive Series. The roots of the choice of hydrogen as the standard for electromotive force measurement (2) may be traced to the decade of the 1790's and the discoveries of that period which were to change the whole state of science. Allessandro Volta, as a result of a series of experiments on what was later to be called galvanic electricity, published, as early as 1792, a list of substances (3,4) in an order such that "each is positive toward all which follow it and negative to all which precede." In a footnote to a report in Ostwald's Klassiker dated August 1796, Volta (4) reported as follows:

> In this table, which is not particularly different from that which Dr. Pfaff has made, zinc stands at the top; approximately in the middle are lead and tin, toward the end, silver; then finally graphite, carbon and copper sulfide. (Translated from German.)

Thus almost at the inception of electrochemistry as a science, there was a recognition of the intrinsic electrochemical properties of substances and an attempt to arrange these substances in an order consistent with observed properties. In 1798 Ritter (2) linked the electrical and chemical properties by showing that the same series was obtained when comparing the ability of the metals to separate other metals from their salt solutions.

These studies by Volta led to the discovery of the cell which now bears his name and which, for the first time, provided a source of electricity at low voltage and moderate current. This discovery opened the way for many other investigators, who immediately capitalized on it to do a wide variety of experiments.

In 1807 J.J. Berzelius (2), with Baron Hisinger as coauthor, published a paper based on the electrolysis of salt solutions. As a result of his extensive studies, he formulated an electrochemical series. He arranged 54 elements beginning with the negative elements oxygen, sulphur and selenium and concluding with the positive elements sodium and potassium. Berzelius' version of the electrochemical series played an important and orientating role in the direction of the chemical research which was to follow.

Almost one hundred years after Volta's report of the voltaic cell, electrochemistry had become an essentially quantitative science. In the 1830's Faraday had published his laws of electrolysis, and in the 1880's Nernst (5) had developed a mathematical treatment of cell potentials with respect to ion concentration.

The outpouring of electrochemical data that were reported after the discovery of the voltaic cell proved again and again, as Volta had shown, that there were intrinsic differences in what we would now call redox properties of substances and that these immutable constants of nature needed to be anchored to a suitable reference point so that data originating in different laboratories would be interchangeable.

In 1849 Beetz, in a paper entitled "The Electromotive Forces of Gases," (6) reported the use of the hydrogen electrode as a starting point for his EMF measurements. It should be noted that this paper by Beetz was published well before the birth of Nernst and Ostwald, who were to play determinative roles in the selection of the base of the EMF system. Thus the use of hydrogen as a reference antedates Nernst's proposal of its use by about fifty years.

The use of cadmium and zinc bars as reference electrodes (7) had become an established practice in the storage battery industry by the time that Nernst was concerning himself with the problem of a standard. Ostwald (8) also mentions that in earlier times amalgamated zinc in a concentrated zinc sulfate solution was used as a constant electrode and occasionally as a reference electrode but that its use posed some serious problems.

The need for a common reference point for the electromotive series was evident, and serious but ineffective attempts had been made to supply this need. There were philosophical as well as practical questions involved in the choice to be made. At this point in the history of electrochemistry, no recourse appears to have been made to committee selection of an appropriate reference electrode.

Modern Foundation of the Electromotive Series

Ostwald's Use of the Dropping Mercury Electrode. The modern foundation of the electromotive series is closely related to the effort of selecting a proper reference electrode for EMF measurements. At the turn of the century, W. Ostwald and W. Nernst did pioneering work in the field of reference electrodes, each approaching the problem from a different angle.

For a better insight into the disagreement on the choice of a reference point, it might be helpful to look a little more closely at the two principals. Ostwald (9) was a man of tremendous energy and enthusiasm. In 1883 he founded the *Zeitschrift fur physikalische Chemie* (10). Then in 1902, at a time when philosophy and science were not considered compatible by most scientists, he established the *Annalen der Naturphilosophie*. In his mid-thirties he succeeded to the chair of physical chemistry at Leipzig. He was in his prime when he started recruiting for his laboratory at Leipzig. Ostwald recognized Nernst as a young man of great promise and invited Nernst to join him as an assistant when he had completed his doctorate with Ettinghausen at Graz. Both Nernst and Arrhenius moved to Leipzig, where, under Ostwald's support and tutelage, each developed into the kind of forceful young scientist whose work leads to fame. Later, by age fifty, with thirty-four of his pupils now professors, Ostwald's interest had shifted. He resigned his position and devoted his time to philosophical problems.

Some gauge of the confusion that existed at this time may be obtained from the critical review (11) of the field of oxidation potentials prepared by Abegg, Luther, and Auerbach in 1911.

Ostwald antedated Nernst by several years in his proposal of a reference electrode for the electromotive series. He was, of course, aware of the early attempts to prepare reference electrodes. He specifically mentions that in earlier times amalgamated zinc in concentrated zinc sulfate solution had been used as a reference electrode, that this combination when used as a reference electrode (8) gave liquid junction potentials with other electrolytes which could not be calculated, and that the electrode also contaminated the experiments.

Ostwald had what appeared to be a very elegant concept. It involved the measurement of a single electrode potential. The method of measurement was in good accordance with his philosophical views and with the chemistry of the times, and it would, in his opinion, yield an absolute potential. An absolute potential was a sharp contrast to the relative potential obtained by referring a measured half-cell to another single electrode reaction arbitrarily set at zero. Ostwald's measurements of half-cell potentials could be directly related to heats of ionization (12). In his opinion, an absolute half-cell redox potential would allow the establishment of an electromotive series which would be analogous to the absolute temperature scale.

In 1887 Ostwald published a paper in which he showed that with the help of a dropping mercury electrode it is possible to measure the potential difference between a metal and a solution of electrolyte (13). He describes his use of the dropping mercury electrode and the Lippmann capillary electrometer, giving some experimental details and numerous tables of data. He clearly points out the low accuracy of the method based on the measurement of the surface tension of mercury, noting that it appears to be fundamentally sound but that near the maximum value of the surface tension there is only a very small change of surface tension for a relatively large change of electromotive force. This measurement problem inherent in the Ostwald electrode potential was one of the points attacked by Nernst.

Ostwald also discusses the background of the Null point method (14) in his text and quotes extensively from von Helmholtz, Wied. (1882), 16, 35. One von Helmholz quotation is the following:

> From this I conclude that if a rapidly dripping and otherwise insulated mass of mercury is in contact with an electrolyte through the dripping tip, then the mercury and the electrolyte cannot have different potentials. Had they different potentials - for example, if the mercury were to be positive - then each falling drop would form a double layer on its surface, which would take away +E from the mercury and render its positive potential smaller and smaller until it was the same as that of the liquid.
>
> Thus a dripping mass of mercury is an electrode with the help of which one can connect liquids with the electrometer without any change of potential.

In 1890 Ostwald introduced the normal calomel electrode (15) as a reference electrode of fixed potential in equilibrium with aqueous potassium chloride solution. Ostwald calibrated a normal calomel electrode against a dropping mercury electrode and obtained a mean value of 0.560 volt. He referred to this value as the absolute value of the electrode potential of the normal calomel electrode. Ostwald recommended the use of the normal calomel electrode as the null electrode, or the standard electrode, in the measurement of the potential difference at a metal-solution junction. He suggested that the electromotive potential measurement with the normal calomel electrode as reference electrode should be considered to be on an absolute scale.

A table of standard potentials, according to Ostwald, appears in Arrhenius' textbook (2) of electrochemistry. The potentials are as follows:

Potential Difference Between Metals and Their Salts in Normal Solution

	Volts		Volts
Magnesium	+1.22	Lead	-0.10
Zinc	+0.51	Hydrogen	-0.25
Aluminum	+0.22	Copper	-0.60
Cadmium	+0.19	Mercury	-0.99
Iron	+0.06	Silver	-1.01
Nickel	-0.02		

It is interesting to note that Arrhenius does not mention the hydrogen electrode in his textbook on electrochemistry which was published in 1902.

Nernst's Introduction of the Hydrogen Electrode. After two years at Leipzig, Nernst, using the theory of the galvanic cell as his thesis, completed a successful defense and was made a lecturer; this appointment opened the road to a professorship. He then left, as Mendelssohn (9) colorfully describes it, the large and benevolent shadow of Ostwald and proceeded to Göttingen by way of Heidelberg.

Nernst (10), whose father was a judge, had a broad range of interests. Fond of the theater, he wrote a play which was produced on the Berlin stage. At one time he even considered a career as an actor. As a child he developed a love for chemistry and was allowed to develop a basement laboratory. He finished the Gymnasium at the top of his class but with the ambition of becoming a poet, not a chemist. This strange, pudgy, little man projected a trusting and credulous image of innocent astonishment which never changed, even on the many occasions when he engaged in withering sarcasm. The following story is told in a biography of Nernst by Mendelssohn (9):

> "A story became current among his colleagues that one day God had decided to create a superman. He began his work on the brain and formed the most perfect and subtle mind, but then was unfortunately called away. The archangel Gabriel saw the unique brain and could not resist the temptation to shape the body, but unfortunately, due to his inexperience, he only succeeded in fashioning a rather unimpressive looking little man. Dissatisfied with his efforts, he left his work. Finally, the devil came along and saw this inanimate thing,

and he blew the breath of life into it. That was Walther Nernst."

Nernst rose rapidly up the academic ladder to the professorship. Professor Nernst was an excellent business man and became quite rich.

Nernst appears to have first introduced the normal hydrogen electrode, arbitrarily set at zero volts, in a lecture given before the German Chemical Society on May 24, 1897 (16). This lecture was published in the Berichte the same year. The following quotations are from pages 1556 and 1557 of this publication. Each is followed by an English translation; the numerical values of the chart are not repeated in the translation.

In der nachstehenden kleinen Tabelle-ihre Ferstellung und ihre Erweiterung für wichtigsten bekannten Ionen scheint eine Aufgabe von der grössten Bedeutung zu sein - sind einige Zahlenangaben zusammengestellt, an die ich einige Bemerkungen knüpfen möchte.

Zersetzungsspannungen für normale Concentrationen

E_1 *(Kationen)*		E_2 *(Anionen)*	
Ag^+	-0.78	I^-	0.52
Cu^{++}	-0.34	Br^-	0.94
H^+	0.0	$O^=$	1.08*
Pb^{++}	+0.17	Cl^-	1.31
Cd^{++}	+0.38	OH^-	1.68*
Zn^{++}	+0.74	$SO_4^=$	1.9
		HSO_4^-	2.6

In the following small table -- its ascertainment and its extension to the most important known ions seems to be a task of utmost importance -- some numerical values are compiled, to which I would like to tie some remarks.

Decomposition Potentials for Normal Concentrations

E_1 (cations) E_2 (anions)

Diese Zahlen beziehen sich auf Normalconcentration der Ionen; eine Verminderung der Concentration um eine Zehnerpotenz erhöht nach unseren früheren Betrachtungen die Werthe um 0.058 volt, (n = Zahl der Ladungen oder chemischer Werth des Ions). Die Lösungstension des Wasserstoffs ist null gesetzt; da wir immer Anode und Kathode haben, so kann zu allen obigen Zahlen ein beliebiges, aber gleiches additives Glied hinzugefügt werden, d.h. über seinen Werth durfen wir willkürlich verfügen.

These values refer to a normal concentration of ions; a tenfold reduction of the concentration increases, according to our former considerations, the values by 0.058 volts (n = number of charges or the chemical values of ion). The solubility tension of hydrogen has been set at zero; since we always have an anode and a cathode, we can add to all the above values some constant quantity -- i.e., a value which can be arbitrarily established.

Three years later at the Seventh Congress of the German Electrochemical Society, which met on August 6-8, 1900, in Zurich in the Chemiegebäude, Nernst again presented his choice of the hydrogen electrode as well as some criticism of Ostwald's calomel electrode (7). He gave this presentation at the second session on August 6 "at two hours past lunch". As an interesting sidelight, Van't Hoff chaired this session. Nernst questioned the validity of the theory of von Helmholtz which held that the polarized mercury electrode at its maximum surface tension has the same potential as the solution. He presented his own opposing view, which was to the effect that there is a surface layer of ions and that one could scarcely assume them to be without effect on the capillary voltage. In addition, he pointed out that the introduction of other ions into the layer would change the potential and deform and shift the maximum in an unknown way and that the use of absolute potentials was of no particular significance. He explained that for the last several years in his work and lectures he had used the normal hydrogen electrode, set at zero, as a reference point. He gave the following values based on hydrogen set at zero.

Elektrodenpotentiale

K	*+3.20*	*H*	*± 0*
Na	*+2.82*	*Cu*	*-0.320*
Ba	*+2.75*	*As*	*< -0.293*
Sr	*+2.54*	*Bi*	*< -0.391*
Ca	*+2.21*	*Sb*	*< -0.466*
Mg	*+1.85*	*Hg*	*-0.750*
Al	*+1.276(?)*	*Ag*	*-0.77*
Mn	*+1.075*	*Pd*	*< -0.789*
Zn	*+0.0770*	*Pt*	*< -0.863*
Cd	*+0.420*	*Au*	*< -1.079*
Fe	*+0.340*	*Fl*	*-1.96*
Tl	*+0.322*	*Cl*	*-1.415*
Co	*+0.232*	*O*	*-1.08*
Ni	*+0.228*	*Br*	*-0.993*
Sn	*< +0.192*	*J*	*-0.520*
Pb	*+0.148*		

Nernst then gave several advantages of the hydrogen electrode. We have translated them as follows:

1. Herr Wilsmore has shown that the hydrogen electrode is easy to prepare; if one bubbles hydrogen for just 15 minutes past a well-platinized platinum electrode, one can be sure of its potential to about 0.001 of a volt. The electrode, therefore, fulfills completely the conditions that are set for normal electrodes.

2. The hydrogen electrode is compatible with most acid or alkaline solution of known titer; therefore, it can often be added directly to a solution to be electrolyzed.

3. The potential difference of the hydrogen electrode vs. the calomel electrode was determined by Mr. Wilsmore through a set of calculations and measurements to be 0.283 volts, and thus the hydrogen electrode also becomes related to the calomel electrode, which quite properly is often used because of its constancy.

In addition, the hydrogen electrode has the following systematical advantages: The hydrogen electrode indicates the borderline between the metals that develop hydrogen and those that do not. If a metal has a positive potential against the hydrogen electrode, then it is, so to speak, less noble than hydrogen; the more noble metals always have negative potentials. Hydrogen is the reducing agent par excellence: Electrodes of higher potentials have stronger reducing power, and those of lower potential have weaker reducing power. Oxygen, the oxidizing agent par excellence, has against hydrogen a potential of 1.08 volts (force of the oxyhydrogen cell); electrodes that deviate from the hydrogen electrode by more than 1.08 volts possess a yet stronger oxidizing power than oxygen.

Nernst pursues his point further (7) by saying that it appears to him that electrode potentials derived in a chemical and electrochemical relationship through the introduction of hydrogen as the zero point have values which are immediately obvious. The calomel electrode, with its hypothetical potential of -0.560 volts, has no direct chemical relationship.

The following portion of the argument is given both in the original German and our English translation of it:

> *Mir scheint also, dass die an sich abstrakten Elektrodenpotentiale in chemischer und elektrochemischer Beziehung durch Einführung des Wasserstoffes als Nullpunkt der Zählung unmittelbar anschaulich werden, während die Kalomelelektrode mit ihrem hypothetischen Potential von -0.560 volt keinerlei direkte chemische Beziehungen in sich schliesst.*

> It seems to me that electrode potentials, which in themselves are rather abstract in the chemical and electrochemical sense, are becoming directly intuitive because of the introduction of hydrogen as the zero point of the measurement; whereas the calomel electrode, with its hypothetical potential of -0.560 volts, does not include any sort of direct chemical relationship.

In his lecture of August 6, 1900, before the Seventh Congress of the Electrochemical Society, Nernst gave credit and praise to N.T.M.

Wilsmore, who was working in his laboratory at that time. Wilsmore had not only measured some potentials but had also critically reviewed the literature and recalculated some of the old values. Nernst noted in his lecture that a paper by Wilsmore was soon to appear in the Zeitschrift fur Physikalische Chemie dealing with electrode potentials, and for that reason he would not give many details in this lecture.

The Confrontation - The Jointly Authored Paper (Wilsmore and Ostwald)

Norman Thomas Mortimer Wilsmore, an Australian, played an important role in this controversy. His excellent experimental work done in Nernst's laboratory and his detailed account of the Nernst concept of electrode potentials were immediately challenged by Ostwald, who was co-editor and founder of the Zeitscrift fur physikalische Chemie. Wilsmore was caught up in a controversy between two giants of his time, or perhaps any time. His two papers which followed out of Nernst's laboratory obviously reflected Nernst's thinking. One could speculate that Nernst was speaking to Ostwald through Wilsmore. On the other hand, Ostwald was possibly speaking to Nernst through Wilsmore with a vigor that may not have been acceptable to one of such high station. The second of the two papers (17) was a joint publication by Ostwald and Wilsmore, each giving opposing views. Again, it seems a little incongruous that the very famous Ostwald would publish a controversial paper with an investigator who was not very well known. Our speculations point to Wilsmore as an acceptable proxy for both Nernst and Ostwald.

In publishing with Nernst and openly challenging Ostwald in a joint publication, it would appear that Wilsmore was to have a brilliant future in electrochemistry. This was not to be the case.

While books have been written on Ostwald and Nernst, relatively little is known of Wilsmore. As a participant in this important controversy over the choice of hydrogen as a standard for EMF measurements as well as for his other good works, such as the discovery of ketene, he deserves better recognition. We are, therefore, including here a portion of his obituary notice, written by G. Tattersall (18), which gives a concise summary of his career.

> Norman Thomas Mortimer Wilsmore, who retired from the Chair of Chemistry in the University of Western Australia at the end of 1937, died at Claremont, W.A., on June 12th, after a brief illness. He was in his 73rd year.
>
> Wilsmore was born at Williamstown, Victoria, Australia, in 1868. After preliminary schooling, he entered Melbourne University in 1887. He took a science course and graduated in 1890. From this time to 1892 he was engaged in chemical research and consulting work under Professor (later Sir) David Orme Masson. In 1892 he obtained the degree of Master of Science and in 1907 was awarded the D.Sc.
>
> Leaving Australia in 1894, Wilsmore proceeded to University College, London, where for three years he was

engaged in research work under Professors William Ramsay and Norman Collie. In 1897 he went to Göttingen primarily to study physical chemistry under Nernst. Four years later he removed to Zurich for the purpose of making a further study of technical chemistry. Shortly after his arrival there he was appointed first assistant in the department of physical and electrochemistry in the Federal Polytechnic under Professor Richard Lorenz. In 1903 he returned to University College, London, as assistant in chemistry. Subsequently he was promoted to be Assistant-Professor. This post he held until the end of 1912, when he left England to take up the Australian appointment.

In 1911 the University of Western Australia was established by Parliament and in the following year applications were called for the original teaching appointments. Wilsmore was appointed to the Chair of Chemistry. Before he left England, the Chair of Physical Chemistry at Liverpool became vacant in consequence of Professor F.G. Donnan's removal from Liverpool to University College, London, as successor to Sir William Ramsay. The Liverpool Chair was offered to Wilsmore, but he was unable to accept it, as he had already accepted the Australian Chair.

After his removal to Western Australia Wilsmore found no opportunities for research. Departmental improvements advanced exceedingly slowly. From about 1920 until almost the time of his retirement, laboratory facilities were limited to those necessary for students; and in addition, the necessary literature and time for research were not available. The burden of teaching and administrative work alone was a serious hindrance to laboratory investigation, had that been otherwise possible.

Wilsmore's paper "Uber Elektroden-Potentiale" (19) is clear and uncomplicated. As indicated by Nernst at the August meeting of the Congress, Wilsmore addressed the question of the absolute potential. Our English translation of parts of the paper is as follows:

If we could measure the potential *E* for any one individual electrode, it would be very easy to arrive at the electrolytic potentials of all other electrodes. We have only to build up the units in which we combine the first with all others of the series; from the electromotive forces of these elements -- if need be, corrected to normal ion concentration -- we must then only subtract the E.P. of the first electrode in order to have all the others in absolute values. Unfortunately, up to now, we have in no case and in no way been able to do this except for the rather questionable calculation of the voltage between mercury and its salts from the Lippmann capillary phenomenon. In all galvanic cells, at least two electrodes are present. For this reason

> Nernst proposed the E.P. of hydrogen at atmospheric pressure as the null point and further proposed to relate all others to it. As a consequence, in the work that follows, this proposal is followed... .
>
> Fortunately, however, the E.P. of the calomel and other mercury electrodes has been established with definite accuracy with respect to hydrogen, and one can directly compare with it almost all metals in neutral solutions of their salts.

Wilsmore concedes, thus, that the hydrogen electrode has limited applicability but that it can be compared to the calomel electrode very successfully.

Ostwald made an immediate reply (8) to Wilsmore. He gave the practical and important advantages of the calomel cell, pointing out that it was very constant and reproducible, that these characteristics had been validated by others, that potassium and chloride ions had about the same mobility, and that the temperature coefficient (0.00007 volts/degree) was not large. Then he concluded that the experimental side was no less favorable than the theoretical side, pointing out again that the only route open to measure a single electrode potential was via a method based on the surface tension of mercury.

He wrote, "*Auf diesem Wege hat sich umittelbar ergeben, dass zwischen Quecksilber und normaler Chlorkaliumlösung, die mit Kalomel gesättigt ist, ein Potentialunterscheid von 0.56 volt besteht.*" A literal translation of the German is: In this way it results directly that the potential difference between mercury and a normal KCl solution saturated with calomel is 0.56 volts. Later in the article Ostwald pointed out that this value could deviate a few millivolts from the correct value. He arbitrarily chose to record number 0.560 volts for the standard.

Ostwald then discussed the hydrogen electrode, noting that some investigators had started using it as a reference electrode. He said that as far as he could see this method of calculation could be traced back to a remark by Nernst in which Nernst only said that for this choice one can arbitrarily assume the starting point since the potential arises between two electrodes. Ostwald said that in his opinion Wilsmore's attempts to popularize this electrode was not progress but a step backwards. He then contrasted the mobility of the hydrogen with the mobilities of potassium and chloride and concluded that the use of hydrogen leads to unnecessary error and is the worst possible choice while KCl gives the best case. He pointed further to the dependence of the hydrogen electrode on pressure, noting that the pressure dependence amounts to 1 mv. per 1/29 atmosphere. Ostwald continued by again repeating his claim that the calomel electrode is the best electrode now available and that the hydrogen is the most questionable.

Ostwald then conceded a point (8). He said that there are difficulties with the calomel electrode in its requiring an arbitrary assumption of a reference value so that, in that sense, both sides are even. He then claims ("*das Recht der ersten Besitzergreifung*") squatters rights. He said that his electrode was in use four or five years before the others came into use and, consequently, should be used.

In rebuttal, Wilsmore (17) and Ostwald published a joint paper presenting opposing positions. Wilsmore restated his purpose: namely, to develop a uniform table of redox potentials. He said that while there is an analogy between the redox potential scale and the temperature scale, redox potentials do not have the same theoretical importance as absolute temperatures. He fully agreed with Ostwald that the calomel electrode is a better starting point for many measurements than the hydrogen electrode, but he did not accept Ostwald's absolute value. He reaffirmed that the hydrogen electrode works well in acidic and basic solutions and has the distinct advantage of not introducing foreign ions. He corrected several errors that Ostwald had had the misfortune to make in his paper attacking Wilsmore and gave the corrected data. Then he quoted between fifty and sixty lines from Nernst's address (7) before the Seventh Congress. These lines contain the essentials of Nernst's argument.

Ostwald's rebuttal was relatively short and more philosophical in tone than that of Wilsmore. It conveyed a feeling that he knew his cause was lost but that it was being lost through a miscarriage of science rather than on sound logical grounds. He repeated that there are good reasons for not considering the choice of a null point as merely the difference between two potentials and cites Van't Hoff's lectures as supporting evidence. He pointed out that his method is in line with good chemical intuition. Once again he stated his position on absolute potentials by saying that the situation is somewhat analogous to the absolute temperature scale -- which had its critics as well as its problems of definition but that no one would question its utility. He closed on a very positive note by saying that we are in the fortunate position in this developing science to be able to introduce a rather plausible null point based on the surface tension maximum of mercury, that at a later date his value can be corrected by recalculation, if necessary, and that both camps have praised the calomel electrode. With the climactic statement that as a practical matter there does not seem to be any doubt about the choice, Ostwald ended his rebuttal (17).

The work of evaluating and organizing the old data as well as critically compiling the more recent material was undertaken by R. Abegg, Fr. Auerbach, and R. Luther. This task was supported by the Bunsen Society, with Nernst acting as a consultant. No further mention seems to have been made of the debate over absolute potentials, but the calomel cell was again mentioned rather prominently and in a rather strange way in Abegg, Auerbach, and Luther's instructions to contributors (20):

> *4...bei der Berechnung von Elektrodenpotentialen, gemäss den früheren Beschlüssen der Bunsen-Gesellschaft, die Werte entweder auf die Normal-Waserstoff-Elektrode = 0 zu beziehen und dann als* E_h *zu bezeichnen oder auf die Normal-Kalomel-Electrode = 0 (nicht = 0.56!) zu beziehen und dann als* E_c *zu bezeichnen; ....*

> ... in calculating electrode potentials, according to the earlier resolutions of the Bunsen-Gesellschaft, to refer the values either to the normal hydrogen electrode = 0 and then to designate as E_h or to refer to the normal calomel electrode = 0 (not = 0.56!) and then designate as E_c; ...

The instructions to contributors explicitly bar the use of the absolute value 0.56 volt value for calomel but permit setting the normal calomel electrode equal to zero.

E_h and E_c became a formal part of chemistry via a report by Nernst to the 5 June 1903 meeting of the Bunsen Society (21). He announced that these symbols along with others had been proposed by the Bunsen Society to, and approved by, the International Congress for Applied Chemistry, which met in Berlin.

Later Consideration of the Question

Nearly a century has passed since Ostwald formally introduced the use of absolute potentials in his Lehrbuch der Allgem. Chemie. Although the Nernst forces carried the day and established the normal hydrogen electrode as the basis of the redox scale, Ostwald's elegant concept of an absolute potential as a base for the system has attracted attention of prominent scientists from time to time since then. New concepts and new experimental approaches have been tested, but in no case does there seem to be a system developed likely to supplant hydrogen. The chemistry of the time led both Nernst and Ostwald to believe that they were dealing with systems much less complex than the experience of a century of research has proved.

The shades of time have been drawn, and one has very little chance of first-hand accounts of this important era in the development of electrochemistry. The authors were fortunate in this respect to have a letter from Professor Joel Hildebrand (22), who worked in Nernst's lab in 1907. Professor Hildebrand said that he could recall nothing of this controversy regarding the choice of the hydrogen electrode as the null point of the electromotive series. Thus the matter appears to have been settled by 1907, and research had gone on to other things.

The Normal Hydrogen Electrode (N.H.E.), as named by Nernst and used as a zero reference electrode, viz. a hydrogen electrode (1 atm H_2) in a normal solution of hydrogen ions, was a hypothetical electrode and no recipe accompanied the definition. No universally accepted operational definition developed before the introduction of activities rendered the concept of normal solutions of hydrogen ions, and hence the N.H.E., obsolete. The zero reference electrode was later redefined by G.N. Lewis (23) as the Standard Hydrogen Electrode (S.H.E.), thus leading to the standard hydrogen scale. By definition, this half-cell consists of a hydrogen electrode (unit fugacity of H_2) in a solution in which the hydrogen ion activity is unity. Unfortunately such a half-cell is also hypothetical, because we can only prepare a solution in which the mean ion activity is unity, not the individual ion activity. In 1923 G. N. Lewis and

M. Randall (23) published their table of single electrode potentials with the S.H.E. as the reference electrode.

Even though chemists have comfortably adopted the hydrogen electrode as the reference point for the electromotive series, the standard concept is still scrutinized from time to time. In 1939 Latimer, Pitzer, and Slansky (24) calculated the absolute value of the normal calomel electrode from the free energy of hydration of gaseous ions. They concluded, based on their work and that of Billitizer, Garrison, and Baur, that the old value of -0.56V was the best experimental value. This value (-0.56V) is the value proposed by W. Ostwald. In 1950 Patrick and Littler (25) proposed an experimental method for determining an absolute half-cell potential. The method was based on the immersion of selected metals in solutions of ions of the same metals and then following the potential build-up on the electrode surface. Some of the results appear promising but not definitive. The Commission of Electrochemistry of the I.U.P.A.C. Meeting at Stockholm in 1953 (See J.A. Christiansen and M. Pourbaix (26)) defined the "electrode potential" of a half-cell with the S.H.E. as the reference electrode.

The idea of the measurement of a single electrode potential in a manner independent of a reference electrode will probably continue to pose an intriguing question to the philosophically inclined chemist. The choice of hydrogen as the standard has posed problems for workers employing nonaqueous solvents, for H^+ associates strongly with almost every solvent. In 1947 Pleskov (27) suggested the use of rubidium, which would yield a very large singly charged cation as opposed to the very small singly charged H^+. Koepp, Wendt, and Strehlow remarked (27) that almost any other element would be better suited to this purpose than hydrogen. Bates suggested (27) that cesium would be even better than rubidium. Thus it does not seem unreasonable to expect that this subject will be re-examined from time to time in the light of new needs, new measurement technology, and, especially, the unifying influences of the philosophy of science.

Literature Cited

1. Bates, R. G. Electrometric pH Determinations - Theory and Practice; Wiley: New York, 1954.
2. Arrhenius, S. Textbook of Electrochemistry; Longmans: New York, 1902.
3. Moore, P. J. History of Chemistry; McGraw Hill: New York, 1939.
4. Volta, A. In Ostwald's Klassiker Der Exakten Wissenschaften, Galvanismus und Entdeckung Des Säulenapparates 1796-1800, Nr. 118; Engelmann: Leipzig, 1900.
5. Nernst, W. Z. phys. Chem. 1889, 4, 147.
6. Beetz, W. Pogg. Ann. 1849, 77, 493-511.
7. Nernst, W. Z. Elektrochem. 1900, 7, 253-256.
8. Ostwald, W. Z. phys. Chem. 1900, 35, 333-339.
9. Mendelssohn, K. The World of Walther Nernst; Univ. Pittsburgh: Pittsburgh, 1973; p 9.
10. Slosson, E. E. Major Prophets of Today; Books for Libraries: Freeport, NY, 1914.

11. Abegg, R.; Auerbach, F.; Luther, R. Messungen elektromotorischer Kräfte galvanischer Ketten; Abhandlungen der deutschen Bunsengesellschaft, No. 5; Knapp: Halle a.S., 1911.
12. Ostwald, W. Z. phys. Chem. 1893, 11, 521.
13. Ostwald, W. Z. phys. Chem. 1887, 1, 583-630.
14. Ostwald, W. Lehrbuch der Allg. Chemie; Engelmann: Leipzig, 1887.
15. Ives, D. J. G.; Janz, G. J., Eds. Reference Electrodes; Academic Press: New York, 1961; p 128.
16. Nernst, W. Ber. d.d. Chem. Ges. 1897, 30, 1547-1563.
17. Wilsmore, N. T. M.; Ostwald, W. Z. phys. Chem. 1901, 36, 91-98.
18. Tattersall, G. In The Australian Encyclopedia; VI, 337, 3rd ed.; The Grolier Society of Australia: Sydney, New South Wales, 1977.
19. Wilsmore, N. T. M. Z. phys. Chem. 1900, 35, 291-332.
20. Abegg, R.; Luther, R.; Auerbach, F. Z. f. Elektroch. 1906, 12, 500-501.
21. Nernst, W. Z. f. Elektroch. 1903, 9, 685-686.
22. Letter from Joel H. Hildebrand to C. E. Moore, dated Dec. 12, 1981.
23. Lewis, G. N.; Randall, M. Thermodynamics and the Free Energy of Chemical Substances; McGraw-Hill: New York, 1923; pp 404-405.
24. Latimer, W.M.; Pitzer, K.S.; Slansky, C. J. Chem. Phys. 1939, 7, 108.
25. Patrick, W. A.; Littler, C. L. J. Phys. and Colloid Chem. 1950, 54, 1016.
26. Christiansen, J. A.; Pourbaix, M. Conventions Concerning the Signs of Electromotive Forces and Electrode Potentials; Computes Rendus of the 17th Conference of the International Union of Pure and Applied Chemistry, held in Stockholm, 1953. Maison de Chimie: Paris, 1954; pp 82-84.
27. Koepp, H. M.; Wendt, H.; Strehlow. Z. f. Elektroch. 1960, 64, 483.

RECEIVED August 12, 1988

Chapter 10

Pursuit of the Elusive Single-Ion Activity

Roger G. Bates

Department of Chemistry, University of Florida, Gainesville, FL 32611

Although not accessible to direct measurement, the single electrode potential and the activities of individual ionic species nonetheless appear explicitly in many formulations of electrochemical metrology. Proposals for evaluating these quantities fall into two general categories. The first includes attempts to determine experimentally the potential of a single electrode through elimination or evaluation of phase-boundary potentials. The second depends on ionic solution theory for a guide to a reasonable separation of the mean activity coefficients of electrolytes into their ionic contributions. The result is a conventional basis for a scale of ionic activities which, despite its nonthermodynamic nature, satisfies the requirements of modern electroanalytical chemistry. Major developments of the past 75 years in these two directions are traced.

The absolute potentials of single electrodes have been a subject of interest since Ostwald, Nernst, and their contemporaries formulated the beginnings of modern electrochemistry in the nineteenth century. The twentieth century brought new methods of applying electrochemical measurements to analytical problems. For these, a knowledge of electrode potentials, or, alternatively, the activities of individual species of ions, could provide simplicity and accuracy not hitherto attainable. Nevertheless, ordinary thermodynamic procedures are incapable of measuring these quantities.

This paper reviews efforts to establish single ion activities for aqueous electrolytes. Nevertheless, a closely related problem, that of the energies of transfer of single ionic species from one solvent to another, has received much attention. Among the chief approaches on which these efforts are based are the following: choice of a reference electrode the potential of which may be independent of the solvent, such as Rb^+/Rb or the ferrocinium/ferrocene couple; assumption of the equality of the transfer energies of certain large ions such as tetraphenylarsonium and tetraphenylborate; and efforts to nullify the liquid-junction potential between ionic solutions in different solvents.

0097–6156/89/0390–0142$06.00/0

The past 75 years have seen the expenditure of considerable effort to solve the problem of individual ion activities in aqueous solutions. Two general approaches have been pursued. The first, experimental in nature, sought a means of nullifying or evaluating the space-charge energy barrier that accompanies the transfer of an ionic species into or out of solution. The second accepted the impossibility of defining exactly the single ion activity operationally and tried to find a conventional definition that appeared reasonable in relation to measured mean ionic activities and that would be consistent with modern electrolyte solution chemistry.

The concept of activity of a single ionic species is somewhat obscure and complex, as Noyes (1) has shown. Conway emphasizes that interactions between both like kinds and different kinds of ions are clearly involved, as are ion-solvent interactions (2,3).

In thermodynamic terms, taking cognizance of internationally accepted sign conventions, one can formulate the following expressions for the single electrode potential E, the cell potential difference V, the activity a, and the phase-boundary or liquid-junction potential ϕ_b. For the half reaction $M^{n+} + ne = M$,

$$E = E^0 + (RT/nF)\ \ln a_M \tag{1}$$

where a_M is the activity of the ion M^{n+}. To attempt a measurement of E, one must have recourse to a reference electrode in a cell such as

$$M|M^{n+}\ ||\ \text{reference} \tag{A}$$

where the double vertical line marks a phase boundary. The emf (V) is given by

$$V = E_{ref} - E \pm \phi_b = E_{ref} - E^0 - (RT/nF)\ \ln a_M \pm \phi_b \tag{2}$$

Thus,

$$-\ln a_M = \frac{V - E_{ref} + E^0 \pm \phi_b}{RT/nF} \tag{3}$$

Alternatively, one may elect to measure cells without transference. For simplicity, we consider solutions of the uniunivalent electrolyte MX in a cell composed of electrodes reversible to the cation M^+ and the anion X^-:

$$M|M^+,X^-|X \tag{B}$$

for which the cell reaction is

$$M + X = M^+ + X^- \tag{4}$$

and the emf is given by

$$V = E_r - E_l = E_r^0 - E_l^0 - (RT/F)\ln a_M a_X = E^{0\prime} - (2RT/F)\ln a_{MX} \tag{5}$$

where the subscripts r and l refer to the right and left electrodes, respectively. A cell of this type avoids the complication of the boundary potential but introduces the potential of a second electrode and the activity of a second ion. These relationships demonstrate 1) that the single ion activity and the potential ϕ_b at the boundary indicated by the double vertical line are interdependent, and 2) that only mean activities of neutral combinations of ions are accessible.

In the words of Harned (1924), "We are confronted with the interesting perplexity that it is not possible to compute liquid-junction potentials without a knowledge of individual ion activities, and it is not possible to determine individual ion activities without an exact knowledge of liquid-junction potentials." (4) The statement "This is a dilemma from which there is apparently no escape" is pertinent.

Experimental Approaches

Reference electrodes of mercury have been used by several investigators in an attempt to measure single electrode potentials. Stastny and Strafelda (5) concluded that the zero charge potential of such an electrode in contact with an infinitely dilute aqueous solution is -0.1901V referred to the standard hydrogen electrode. Hall (6) states that the potential drop across the double layer under these conditions is independent of solution composition when specific adsorption is absent. Daghetti and Trasatti (7,8) have used mercury reference electrodes to study the absolute potential of the fluoride ion-selective electrode and have compared their estimates of ion activities in NaF solutions with those provided by other methods. Their method is based on the assumption that the potential drop across the mercury|solution interface is independent of the electrolyte concentration once the diffuse layer effects are accounted for by the Gouy-Chapman theory.

Although the mercury reference electrode has given promising results, other electrodes, such as the common saturated calomel reference, are more convenient and have found extensive application. In such an arrangement the boundary is formed between two liquid phases of different composition, and the potential across this liquid junction must be evaluated theoretically. The Planck and Henderson formulas apply to boundaries of different structures, neither of which may be satisfactorily realized in practice. Although Morf et al (9,10) and Harper (11) have examined carefully the calculation and have made suggestions for improving its accuracy, most investigators have used the Henderson equation, together with limiting ionic mobilities, to arrive at reference electrode potentials, or have even neglected the boundary potential entirely (12).

In a series of papers, Hurlen (13,14) has reported "convenient" single ion activities derived from cells with transference. The liquid-junction potentials involved were estimated by the Henderson equation. In addition, Shatkay's ion-selective electrode measurements of the activities of Na^+ and Ca^{2+} ions in NaCl and $CaCl_2$ solutions, based on the Henderson equation, appear eminently reasonable in comparison with other estimates (15,16).

Pitzer and Brewer (17) have pointed out that the space-charge energy accompanying the transfer of ions out of a solution might, in principle, be minimized such that the net electrical charge of the solution could be measured or controlled. Perhaps the ideal electrochemical method would be realized if mass transfer at the boundary could be eliminated entirely. Indeed, in the late 19th century Kenrick (18) suggested the use of a "vertical stream method" in which an air gap exists between two half cells. This technique was developed by Randles (19) in 1956 and later by others (20-23). In the experimental arrangement, flowing streams of the two electrolytes are brought in such close proximity that potential measurements of the electrode assembly are possible with electrometers of high input impedance.

In Gomer and Tryson's treatment of this air-gap cell (24), the absolute half-cell potential is regarded as $V_{MS}-\phi_M$, where V_{MS} is the potential difference between a metal electrode M and the solution while ϕ_M is the work function of the metal in the solution. This quantity is equated to $V_{RS}-\phi_R$, where V_{RS} is the electrostatic potential between the reference electrode and air above the solution while ϕ_R is the work function of this same reference in air. From sensitive potential measurements, combined with photoelectric determinations, they concluded that the standard hydrogen half cell has an absolute potential of 4.73 ± 0.05V. Reiss and Heller (25) agree that absolute potentials cannot be obtained by measurement alone. A residual theoretical assessment, usually relating to the interfacial dipole layer, is always necessary. For the standard hydrogen electrode, they assign a value of 4.43V. Related procedures involving different estimates of interfacial potentials have given results in substantial agreement with these (26,27).

In the opinion of Pitzer (28), this experimental procedure, with improvements in sensitivity, may provide useful data in the future for the thermodynamic functions of single ionic species. That this goal may be achieved in the near future is suggested by the study of Farrell and McTigue (23). These investigators obtained a precision of ±0.1 mV in the potential differences between a mercury jet and $Pt|H_2$ and $AgCl|Ag$ electrodes in aqueous solutions of HCl.

A variety of other approaches to scales of absolute electrode potentials of an experimental nature have been suggested. Oppenheim (29) mentions the possibility, apparently not tried, of measuring the quadrupole radiation emitted by an electrode made to execute harmonic motion by mechanical or ultrasonic means. Goldberg and Frank, in a novel approach to liquid-junction potentials, noted that the time rise of the calculated potential is sensitive to the choice of single ion activity (30). Similarly, Leckey and Horne (31) have studied cell potentials as a function of time and have identified a complicated interdependence of electric transport and single ion activity coefficients which may reveal the magnitude of individual ionic properties. Cells involving a thermal gradient, the possible utility of which was suggested by Szabo in 1938 (32), have been studied extensively by Milazzo and his co-workers (33,34). In later work, Szabo and his associates (35) used a 3-electrode polarographic method, following a suggestion from the work of Janata et al (36,37). Potentials were referred to the half-wave potential of a ferrocinium/ferrocene reference electrode, and the cell contained no liquid junction. An accuracy of 5 mV was estimated. Elsemongy (38) proposes to obtain absolute electrode potentials for the hydrogen electrode by extrapolating standard potentials for $Pt;H_2|AgX;Ag$ cells as a function of the radius of the halide ion X^-. This procedure, however, does not appear to be generally applicable.

Assumptions and Conventions

In his classic study of liquid-junction potentials in concentration cells (1919), MacInnes (39) suggested that the activity coefficients of K^+ and Cl^- ions be assumed equal to the mean activity coefficient of KCl at all ionic strengths:

$$\gamma_K = \gamma_{Cl} = \gamma_{KCl} \tag{6}$$

This assumption seemed reasonable in view of the similarity of the sizes and limiting conductances of the two ions. It gave rise to the MacInnes convention, which enables the activity coefficients of other ionic species to be evaluated from the known mean activity coefficients of selected electrolytes. This process, called the "mean salt method", has been used extensively and with apparent success in studies of the seawater medium. Garrels and Thompson (40,41) used a glass electrode reversible to Na^+ to establish a reference point for the activity coefficient of that ion in seawater. The MacInnes convention was then applied to obtain data for the other ions present. The procedure proposed by Maronny and Valensi (42,43) for the determination of standard pH values utilizes Equation 6, where γ_{KCl} is the mean activity coefficient of KCl in each particular medium, rather than in aqueous KCl alone.

Nesbitt (44,45) has pointed out that ratios of the activity coefficients of ions of the same charge in mixtures can be obtained without ambiguity from mean activity coefficients of electrolytes with a common anion or cation. If HCl is one of the electrolytes, a pH measurement might provide a reference point for calculating the activity coefficient of a second cation as well as that of the anion involved. Equilibrium theory suggests that pH measurements of saturated solutions of a metal hydroxide or carbonate might also lead to the activity coefficient of the metal ion concerned (46). In these cases, a convention is necessary to provide numerical values of the pH.

The Debye-Hückel theory, which appeared in 1923, predicted that the activity coefficients of singly charged cations and anions would be equal at ionic strengths so low that departures from ideal behavior are caused solely by long-range electrostatic interactions. Guggenheim (47) proposed a formula applicable to mixed electrolytes, more general and complex than that of MacInnes, in which the numerical value of the single ion activity coefficient is weighted according to the contribution of that ion to the total ionic strength. For a single symmetrical electrolyte MX, Guggenheim's formula reduces to

$$\gamma_M = \gamma_X = \gamma_{MX} \tag{7}$$

The ionic interaction theories of Brønsted-Guggenheim (48) and Pitzer (49,50) have been conspicuously successful in accounting for the mean activity coefficients and other thermodynamic properties of electrolytes, singly and in mixtures of ionic solutes. They have proved especially useful in salt mixtures such as seawater (51,52). Unfortunately, specific parameters characteristic of single ions do not appear in the theory. For a single 1:1 electrolyte, the equations lead to equality of the activity coefficients of cation and anion, as in Equation 7.

The Debye-Hückel equation and its semi-empirical extensions have constituted the framework for a number of conventions. A useful version of this equation,

$$\ln \gamma_{\pm} = \frac{-A|z_+z_-|I^{\frac{1}{2}}}{1 + B\mathring{a}I^{\frac{1}{2}}} \tag{8}$$

represents rather satisfactorily the mean activity coefficients ($\gamma_{\pm}$) of many unassociated electrolytes up to ionic strengths (I) of 0.05 to 0.1 mol kg^{-1}. In this equation, A and B are constants dependent on the temperature, density, and dielectric constant of the medium, z_+ and z_- are the charges of the cation and anion, and $\mathring{a}$ is an adjustable parameter representing, in the original theory, the closest distance of approach of the cation and anion.

An empirical term linear in I is often added to the right of Equation 8 in order to account for the minimum in curves of $-\ln \gamma_{\pm}$ as a function of I, thus extending the applicability to higher concentrations:

$$\ln \gamma_{\pm} = \frac{-A|z_+z_-|I^{\frac{1}{2}}}{1 + B\mathring{a}I^{\frac{1}{2}}} + cI \tag{9}$$

where c is an adjustable parameter. As c is unknown, Equation 8 rather than Equation 9 has been used for estimating activity coefficients of ionic species. For this purpose, z_+z_- becomes z^2 while $\mathring{a}$ is regarded formally as the ionic diameter.

Kielland (53) examined critically varied data for unassociated electrolytes and recommended values of $\mathring{a}$ for a series of single cations and anions. This convenient procedure has found wide use. Tamamushi (54) has calculated activity coefficients for single ionic species by identifying $\mathring{a}$ in Equation 8 with the distance parameter in the Debye-Hückel-Onsager equations for electrolytic conductance. Neff (55) prefers to express ionic activity coefficients as a function of the sums of the activities of all species rather than the ionic strength.

In 1960, the "pH convention" was adopted by the International Union of Pure and Applied Chemistry, in response to the need for a consistent series of standards to define the pH scale (56). It was agreed that the activity coefficient of chloride ion in aqueous buffer solutions at ionic strengths no greater than 0.1 would be defined by

$$\ln \gamma_{Cl} = \frac{-AI^{\frac{1}{2}}}{1+1.5I^{\frac{1}{2}}} \tag{10}$$

at all temperatures. The activity coefficients of chloride ion at 25°C defined in this way are nearly the same as the mean activity coefficients of NaCl in its pure aqueous solutions. Although this convention is nonthermodynamic in nature, its adoption removed a source of ambiguity and placed pH measurements on a common basis. A proposal to extend this convention to ionic strengths higher than 0.1, as needed for the standardization of many ion-selective electrode measurements, proved impractical (57).

The MacInnes, Debye-Hückel, and pH conventions describe the ionic activity as a function only of ionic strength. It is, however, not reasonable to expect the chloride ion, for example, always to have the same activity coefficient at a fixed temperature and ionic strength, regardless of the nature of the counter cation. Bjerrum (58) showed in 1920 that the behavior of electrolytes, including the minima observed in plots of ln $\gamma_{\pm}$ *vs.* I, provides evidence for ion-solvent interactions. Hydration of the ions must be considered, and "single-parameter" conventions are inadequate from the standpoint of solution theory.

Bjerrum's two-parameter description of mean activity coefficients was refined and extended by Stokes and Robinson (59) and by Glueckauf (60). Both treatments, though differing in statistical detail, consisted essentially in expressing the parameter c in Equation 9 in terms of h, a hydration index for each specific electrolyte. In a later version of the hydration equation (61), Bates and Robinson replaced the Debye-Hückel electrostatic term, that is, the first term on the right of Equation 9, by the modification proposed by Glueckauf in 1969 (62), which is valid to higher ionic strengths.

In these treatments, the hydration index h was treated as a constant. It is, however, reasonable to expect h to decrease as the ionic strength increases. Indeed, Bates (1986) found support for a linear decrease with I from its limiting value (h^0) at zero ionic strength (63). His modification of the hydration equation for mean activity coefficients is as follows:

$$\log \gamma_{\pm} = -A'|z_+z_-|I_c^{1/2}\left[\frac{1+0.5B\mathring{a}I_c^{1/2}}{1+B\mathring{a}I_c^{1/2}}\right]^2 + \frac{0.018mr(r+h-\nu)}{\nu(\ln 10)(1+0.018mr)}$$

$$+ \frac{h-\nu}{\nu}\log(1+0.018mr) - \frac{h}{\nu}\log(1-0.018mh) \qquad (11)$$

where

$$h = h^0 - qI_c \qquad (12)$$

In Equation 11, log is the decadic logarithm, A'=0.4343A, ν is the number of ions from one molecule of the electrolyte, r is the ratio of the apparent molal volume of electrolyte to the volume of a mole of water, and I_c is the ionic strength in concentration units (mol dm^{-3}).

The three parameters $\mathring{a}$, h^0, and q were derived from mean activity coefficients by nonlinear least-squares procedures. The hydration indexes were found to decrease from 12 for Mg^{2+} and 11 for Ca^{2+} to 5.2, 3.8, and 2.5 for Li^+, Na^+, and K^+, respectively. Pan (64) has preferred to fix the ion-size parameter in the hydration equation at the sum of the crystallographic radii of the ions. If this is done, he has found that much larger values of h (about 21, 13, and 7 for the lithium, sodium, and potassium halides, respectively) are obtained from mean activity coefficients.

Mean activity coefficients can, in theory, be determined without ambiguity. For this reason, considerable attention has been directed to the use of solution theory as a guide to separating mean activities into their cationic and anionic components. The MacInnes assumption described above falls in this category. Frank (1963) appears to have been the first to apply the Bjerrum hydration approach to the calculation of single ion activities (65). From hydration numbers h_+ and h_- for the cation and anion, estimated in the manner of Bjerrum, he derived a correction term $\delta_\pm$ such that

$$\ln \gamma_+ = \ln \gamma_\pm + \delta_\pm \tag{13}$$

and

$$\ln \gamma_- = \ln \gamma_\pm - \delta_\pm \tag{14}$$

When the parameters of the Stokes-Robinson modification of the hydration equation for simple electrolytes were examined, Bates, Staples, and Robinson (66) noted that values of h for the alkali chlorides varied in nearly linear fashion with the reciprocal of the radii of the cations. The intercept for very large cations was near h=0. On the strength of this observation, together with the assumption that h values for cation and anion are additive, the convention that h=0 for chloride ion was advanced. From this point of departure, together with the Gibbs-Duhem equation, the following reasonable separation of $\gamma_\pm$ for a 1:1 electrolyte MX into its ionic contributions was derived:

$$\log \gamma_+ = \log \gamma_\pm + \frac{0.018}{\ln 10}(h_+ - h_-)m\phi \tag{15}$$

and

$$\log \gamma_- \quad \log \gamma_\pm - \frac{0.018}{\ln 10}(h_+ - h_-)m\phi \tag{16}$$

where m is molality and ϕ is the osmotic coefficient.

It has been pointed out (67) that the suggestion of hydration numbers near zero for the halide ions appears contrary to experimental evidence. Although the physical concept of a primary hydration number is reasonably clear, the precise nature of the hydration index is not nearly as well defined. Furthermore, the separation of ionic activity coefficients embodied in Equations 15 and 16 is rather insensitive to the choice of hydration indexes. For example, Bagg and Rechnitz's studies of cells with liquid junction (68) lead to a value of h=0.9 for chloride ion, instead of h=0. This difference produces a change of only about 0.015 in pM, the negative logarithm of the cation activity, for Na^+ in 2m NaCl and less than 0.001 for Ca^{2+} in 2m $CaCl_2$ (63).

Conclusion

What should one conclude from the foregoing summary of the past 75 years of research on individual ion activities? First, the inaccessibility of these quantities to thermodynamic measurement has been amply confirmed; second, that a start has been made toward the eventual establishment of absolute electrode potentials and phase-boundary potentials which may, in the future, lead to useful thermodynamic data for ionic species but which currently lacks the necessary sensitivity. Perhaps most striking is the recognition that conventional scales of ionic activity may fulfill many of the requirements of electrochemical measurement. A convention cannot be proved right or wrong, and its effectiveness depends upon its universal adoption.

Literature Cited

1. Noyes, R. M. J. Am. Chem. Soc. 1964, 86, 971.
2. Conway, B. E. J. Solution Chem. 1978, 7, 721.
3. Conway, B. E. In Activity Coefficients in Electrolyte Solutions; Pytkowicz, R. M., Ed.; CRC Press: Boca Raton, FL, 1979; Vol I, p. 95.
4. Harned, H. S. In A Treatise on Physical Chemistry; Taylor, H. S., Ed.; Van Nostrand Co.: New York, NY, 1924; Vol II, p. 782.
5. Stastny, M.; Strafelda, F. Coll. Czech. Chem. Commun. 1972, 37, 37.
6. Hall, D. G. J. Chem. Soc. Faraday Trans II 1978, 74, 405.
7. Daghetti, A.; Trasatti, S.; Trasatti, S. Inorg. Chim. Acta 1980, 40, X143.
8. Daghetti, A.; Trasatti, S. Can. J. Chem. 1981, 59, 1925.
9. Morf, W. E. Anal. Chem. 1979, 49, 810.
10. Dohner, R. E.; Wegmann, D.; Morf, W. E.; Simon, W. Anal. Chem. 1986, 58, 2585.
11. Harper, H. W. J. Phys. Chem. 1985, 89, 1659.
12. Schwabe, K.; Kalm, H.; Queck, Chr. Z. Physik. Chem. (Leipzig) 1974, 255, 1149.
13. Hurlen, T. Acta Chem. Scand. 1979, A33, 631, 637.
14. Hurlen, T. Acta Chem. Scand. 1981, A35, 457, 587.
15. Shatkay, A.; Lerman, A. Anal. Chem. 1969, 41, 514.
16. Shatkay, A. Electrochim. Acta 1970, 15, 1759.
17. Pitzer, K.; Brewer, L. Thermodynamics; McGraw-Hill Book Co.: New York, NY, 1961; p. 310.
18. Kenrick, F. B. Z. Physik. Chem. 1896, 19, 625.
19. Randles, J. E. B. Trans. Faraday Soc. 1956, 52, 1573.
20. Rybkin, Yu. F.; Shevchenko, N. F.; Aleksandrov, V. V. Zh. Anal. Khim. 1965, 20, 26.
21. Case, B.; Parsons, R. Trans. Faraday Soc. 1967, 63, 1224.
22. Rabinovich, V. A.; Alekseeva, T. E.; Voronina, L. Elektrokhimiya 1973 (Eng. translation), 9, 1354.
23. Farrell, J. R.; McTigue, P. J. Electroanal. Chem. 1982, 139, 37.
24. Gomer, R.; Tryson, G. J. Chem. Phys. 1977, 66, 4413.
25. Reiss, H.; Heller, A. J. Phys. Chem. 1985, 89, 4207.
26. Gurevich, Yu. Ya.; Pleskov, Yu. V. Elektrokimiya 1982, 18, 1477.
27. Trasatti, S. Pure Appl. Chem. 1986, 58, 955.
28. Pitzer, K. S. In Activity Coefficients in Electrolyte Solutions; Pytkowicz, R. M., Ed.; CRC Press: Boca Raton, Fl, 1979; Vol I, p. 157.
29. Oppenheim, I. J. Phys. Chem. 1964, 68, 2959.
30. Goldberg, R. N.; Frank, H. S. J. Phys. Chem. 1972, 76, 1758.
31. Leckey, J. H.; Horne, F. H. J. Phys. Chem. 1981, 85, 2504.
32. Szabo, Z. G. Z. Physik. Chem. 1938, A181, 169.
33. Milazzo, G.; Bonciocat, N.; Borda, M. Electrochim. Acta 1976, 21, 349.

34. Bonciocat, N. Electrochim. Acta 1977, 22, 1047.
35. Szabo, Z. G.; Barcza, L.; Ladanyi, L.; Ruff, I. J. Electroanal. Chem. 1976, 71, 241.
36. Janata, J.; Jansen, G. J. Chem. Soc. Faraday Trans I 1972, 68, 1656.
37. Janata, J.; Holtby-Brown, R. D. J. Electroanal. Chem. 1973, 44, 137.
38. Elsemongy, M. M. Thermochim. Acta 1984, 80, 239.
39. MacInnes, D. A. J. Am. Chem. Soc. 1919, 41, 1086.
40. Garrels, R. M.; Thompson, M. E. Am. J. Sci. 1962, 260, 57.
41. Garrels, R. M. In Glass Electrodes for Hydrogen and Other Cations; Eisenman, G., Ed.; Marcel Dekker, New York, NY, 1967; p. 344.
42. Maronny, G.; Valensi, G. J. Chim. Phys. 1952, 49, C91.
43. Maronny, G. Electrochim. Acta 1960, 2, 326.
44. Nesbitt, H. W. Chem. Geol. 1980, 29, 107.
45. Nesbitt, H. W. Chem. Geol. 1981, 32, 207.
46. Hostetler, P. B. Am. J. Sci. 1963, 261, 238.
47. Guggenheim, E. A. J. Phys. Chem. 1930, 34, 1758.
48. Brønsted, J. N. J. Am. Chem. Soc. 1922, 44, 877; 1923, 45, 2898.
49. Pitzer, K. S. J. Phys. Chem. 1973, 77, 268.
50. Pitzer, K. S..; Mayorga, G. J. Phys. Chem. 1973, 77, 2300.
51. Millero, F. J. In Activity Coefficients in Electrolyte Solutions; Pytkowicz, R. M., Ed.; CRC Press: Boca Raton, FL, 1979; Vol II, p. 63.
52. Whitfield, M. In Activity Coefficients in Electrolyte Solutions; Pytkowicz, R. M., Ed.; CRC Press: Boca Raton, FL, 1979; Vol II, p. 153.
53. Kielland, J. J. Am. Chem. Soc. 1937, 59, 1675.
54. Tamamushi, R. Bull. Chem. Soc. Japan 1974, 47, 1921.
55. Neff, G. W. Anal. Chem. 1970, 42, 1579.
56. Bates, R. G.; Guggenheim, E. A. Pure Appl. Chem. 1960, 1, 163.
57. Bates, R. G.; Alfenaar, M. In Ion-Selective Electrodes; Durst, R. A., Ed.; U.S. Government Printing Office: Washington, DC, 1969; p. 191.
58. Bjerrum, N. Z. Anorg. Chem. 1920, 109, 275.
59. Stokes, R. H.; Robinson, R. A. J. Am. Chem. Soc. 1948, 70, 1870; J. Solution Chem. 1973, 2, 173.
60. Glueckauf, E. Trans. Faraday Soc. 1955, 51, 1235.
61. Bates, R. G.; Robinson, R. A. In Ion-Selective Electrodes; Pungor, E., Ed.; Akadémiai Kiadó, Budapest, 1978; p. 3.
62. Glueckauf, E. Proc. Roy. Soc. (London) 1969, A310, 449.
63. Bates, R. G. Anal. Chem. 1986, 58, 2939.
64. Pan, C.-F. J. Phys. Chem. 1985, 89, 2777.
65. Frank, H. S. J. Phys. Chem. 1963, 67, 1554.
66. Bates, R. G.; Staples, B. R.; Robinson, R. A. Anal. Chem. 1970, 42, 867.
67. Gennero de Chialvo, M. R.; Chialvo, A. C. Electrochim. Acta 1987, 32, 331.
68. Bagg, J.; Rechnitz, G. A. Anal. Chem. 1973, 45, 271.

RECEIVED August 3, 1988

Chapter 11

Historical Development of the Understanding of Charge-Transfer Processes in Electrochemistry

B. E. Conway

Chemistry Department, University of Ottawa, Ottawa, Ontario K1N 6N5, Canada

The true nature of electrolytic processes in electrochemistry took many years to be understood. An historical outline of the development of these ideas from the pre-Faraday period until the present time is given. One of the matters of outstanding importance for chemistry and electrochemistry was the eventual realization that electricity itself is "atomic" in nature, with the electron as the natural unit of electric charge. Not until this concept was established experimentally, and understood in its theoretical ramifications, was it possible for the microscopic basis of electrolytic processes to be established, and developed more quantitatively with the correct qualitative basis. The final and correct perception of the nature of these processes provided one of the important bases for recognition of the electrical nature of matter itself and the foundations of physical chemistry.

In this paper, an historical outline is given of the principal developments in electrochemistry concerning the mechanism and phenomenology of charge transfer in electrolytic processes. In tracing early contributions in this topic, it will be necessary to examine first the historical evolution of ideas about electricity and electric charge.

Concepts of Electricity, Charge and the Electron

Historically, the understanding of charge-transfer processes in electrochemistry required a long period of gestation. It depended on and, in fact, led in part to, recognition of several complementary factors: the nature of electric charge itself; the association of electric charges with atomic and molecular species in chemistry, giving rise to ions; the solvation and movement of such ions in liquids and some solids, giving rise to conductivity; the discharge or formation of such ions at metal interfaces, characterizing the process of electrolysis and perception of the role of electrons as

0097–6156/89/0390–0152$06.00/0

a delocalized system of mobile charges in metals with an energy distribution represented by an energy "band structure". The understanding of the nature of electric charge and, in particular, the atomic entity of electricity, the electron, played a principal role in the foundations of physical and electrochemistry in the last century, and in recognition of the electrical nature of matter. The latter aspect was not fully understood until the earlier years of this century through the development of electronic theories of valency and atomic and molecular structure, chemical bonding and spectroscopy, and electronic theories of metals. In these developments, especially in the last century, electrochemistry played a major role, especially through Faraday's investigations on exchange of charge in processes at metal interfaces that we now call electrolysis, a term, like many others in electrochemistry, that was first used by Faraday himself.

It must first be pointed out that the idea and phenomenology of electric charge originated in the 18th century work on the physics of electrostatics, well before any of the principal developments of electrochemistry historically associated with the names of Luigi Galvani, Alessandro Volta, Sir Humphry Davy and Michael Faraday. The phenomenology of generation, storage and transfer of electricity as electric charge by use of electrostatic machines (the Wimshurst machine) and other devices (the Leyden jar and the Electrophorus) was well developed by the middle of the 18th century. (One of the most impressive collections of historically significant examples of such machines is to be found in the Museum of the History of Science in Florence, Italy). However, the "nature of electricity" remained for many years later, through Faraday's time, essentially a mystery and a source of active controversy(1). Electricity was regarded, somewhat like the Aether, as an imponderable fluid and its intimate connection with the structure and properties of matter itself was not appreciated until later in the mid 19th century. In fact, the connection between electrostatic and electrochemical, as well as electromagnetic, manifestations of electricity took a very long time to crystallize; indeed, for a lengthy period, two "kinds" of electricity continued to be referred to: electrostatic and current or Galvanic electricity.

The stimulation of the frog's leg nerve in the historical experiment of Galvani, with electricity generated from the contact of two dissimilar metals, marked the point of origin of electrochemistry and its divergence from electrostatics. At first, and indeed for some length of time, the so-called electrostatic (or "common") and "Galvanic" or "animal" manifestation of electricity were thought to correspond to two distinct types of electricity. Again, it was a relatively long time between the first demonstration of the Galvanic effect (1786) and its controlled utilization in Volta's pile (1800) for production of a steady voltage and a current of electricity. During that period, much controversy raged about the supposed differences of the "two kinds" of electricities, and later about electromagnetically induced currents of electricity(1).

It could be said, therefore, that until the basic nature of electricity was recognized, the electrochemistry of charge-transfer processes could not be understood and taken further. However, historically, this would be an incorrect perception as the study of the nature and phenomenology of charge-transfer processes in electrolysis and in conductivity of salt and acid solutions, was itself precisely one of the main bases of the ultimate understanding of the electrical nature of matter and of electricity itself. The second and complementary basis was, of course, the discovery and characterization of the electron and its charge(e)-to-mass(m) ratio by J.J. Thomson(2), and its charge in experiments by Townsend(3) and by Millikan(4). These experiments demonstrated the ubiquitous involvement of a negatively charged particle, termed the "electron" by Johnstone Stoney(5-6), in electrically stimulated ionization of gases. Electron photoemission from metals was another key discovery.

Interestingly enough, for chemistry, one of the most thorough accounts of static electrical phenomena was that published by Joseph Priestley, "The History and Present State of Electricity", London, 1767, in two volumes. It was an article in Encyclopaedia Britannica on the history of electricity by Tytler, copiously illustrated by diagrams of electrostatic machines, based in part on Priestley's volumes, that first stimulated Faraday to construct his first scientific instruments.

Generation of static electricity was based primarily on the use of friction machines and the effect became later widely known as "triboelectricity". In the mid-18th century, it was not, of course, recognized that static electricity originated from a mechanical separation of electronic charges (i.e. ionization) at suitable interfaces, classically amber, or glass and resins. However, it is interesting that in one of the earliest lectures given by Faraday he took up a "two-fluid" theory of Major Eeles (1771) and described electricity as a compound body capable of being divided by friction or excitation into two "Portions or Powers" one of which, when separated, always attracts itself to the rubber; the other to the excited Electric. It was also recognized that the "separated" electricities would tend spontaneously to reunite, but the same electricities to repel one another. Faraday was also influenced, at an early age, by a book by Jane Marcet, "Conversations on Chemistry" (2 volumes, London, 1809), written for an audience newly created by Sir Humphry Davy's Royal Institution lectures. In these two volumes, Faraday was introduced to electrochemistry and, according to the author, electrochemical effects, like the origin of static electricity, seemed to require two "electrical fluids", i.e. states we now recognize as positive and negative charge associated with excess or deficiency of electron density.

At this point a quotation from Mrs. Marcet's book must be given as it is of much historical significance for chemistry and electrochemistry in relation to the role of electrical effects:

> "Mr. Davy...whose important discoveries have opened such improved views on chemistry, has suggested an hypothesis which may throw great light upon that science. He supposes that there are two kinds of electricity with one or other of which all bodies are united. These we distinguish by the names of positive and negative electricity; those bodies are disposed to combine which possess opposite electricities, as they are brought together by the attraction which these electricities have for each other. But

whether this hypothesis be altogether founded on truth or not, it is impossible to question the great influence of electricity in chemical combinations."

It is clear from this paragraph that a close connection between chemistry and electricity was already perceived in Davy's time through his work on applications of electrolysis, including the preparation of several of the most electropositive and electronegative elements for the first time. However, it remained for Faraday to quantify the relations between passage of current (charge) and chemical change as expressed in his two laws of electrolytic action:

"....the chemical power of a current of electricity is in direct proportion to the absolute quantity of electricity which passes; and

"....the equivalent weights of bodies are simply those quantities of them which contain equal quantities of electricity"

Thus were the essentials of chemical combination first enunciated!

Faraday's own commentary on the consequences of the laws of electrolysis is of interest:

"The harmony which this theory of the definite evolution and the equivalent definite action of electricity introduces into the associated theories of definite proportions and electro-chemical affinity, is very great. According to it, the equivalent weights of bodies are simply those quantities of them which contain equal quantities of electricity, or have naturally equal electric powers; it being the ELECTRICITY which determines the equivalent number, because it determines the combining force. Or, if we adopt the atomic theory or phraseology, then the atoms of bodies which are equivalents to each other in their ordinary chemical action, have equal quantities of electricity naturally associated with them."

This is an interesting statement, historically, not only for electrochemistry but chemistry in general, for it indicates that Faraday almost, but not quite, reached the vital and important conclusion for chemistry and physics that electricity itself had an "atomic" nature, the electron, as we now know it. Despite his own conclusion cited above, Faraday tended to adhere to older ideas of electricity as a fluid. The "atomic" view of electricity seems not to have been clearly stated, however, until it was expressed in von Helmholtz's Faraday Memorial Lecture (1) in 1881, although a similar idea was advanced by G. Johnstone Stoney at the British Association for Advancement of Science in 1874(5) but not published until 1881. In his later paper of 1891(6), the name "electron" was proposed for this unit of charge, but the latter was not associated with a particular, physically significant, "particle" until Thomson's classic experiment(2) (1897) in which the charge to mass ratio, e/m for the electron "particle" was definitively measured. (Shuster(7) had previously made an electromagnetic deflection experiment similar to Thomson's but found values about 500 times too large).

Probably the first to recognize the possibility that atoms, as ions, might be associated with a definite fundamental unit of electric charge was Clerk Maxwell in 1873(8). With some reservation, he stated the following with reference to the electrical properties of electrolytes:

> "If we....assume that the molecules of the ions within the electrolyte are actually charged with definite quantities of electricity, positive and negative, so the electrolytic current is simply a current of convection, we find that this tempting hypothesis leads us into very difficult ground. Suppose we leap over this difficulty by simply arresting the constant value for the molecular charge, and that we call this....one molecule of electricity."

One of the difficulties in understanding the true mechanism of the process of electrolysis at the time that Faraday enunciated his laws of electrolysis was the absence of the idea of spontaneous electrolytic dissociation, postulated much later by Clausius and by Arrhenius. In fact, Faraday believed that the electric force at (between) electrodes split up molecules in the electrolyte, giving rise to conductivity. This idea was connected with Freiherr von Grotthus's theory of a series of dissociations and recombinations of charged species in the conductance of aqueous solutions.

We now recognize(9) that two significant physico-chemical concepts that had not been developed at that time, gave rise to these conceptual difficulties: one was that it was not realized that so-called salts were already ionic in the pure solid state and the other was that it was not understood that a large solvation (hydration) energy would accompany dissolution of ions of a salt in a polar medium such as water, and largely compensate for the attractive energy amongst ions of unlike sign of charge in a crystal lattice. In fact, an elementary theory of ionic hydration, in terms of dielectric polarization, was not available until Born's treatment in 1920(10) or, in more molecular terms, until Bernal and Fowler's classic paper in 1933 (see also ref. 10).

We thus see that a surprizingly long time elapsed between the formulation of the Faraday Laws (1834) and a proper understanding of their significance for chemistry and electrochemistry. Also, until an electron theory of metals was available, no proper microscopic description of charge-transfer in electrolytic processes at electrode interfaces was conceivable. Also, the theory of ionic dissociation in solution in relation to electrolytic conductivity was only developed slowly and accepted with difficulty and much contro-- versy through the works of Arrhenius, van't Hoff and Ostwald (the "Ionists"--see the paper in this symposium by K. J. Laidler); in fact, it was not until the early 20th century that "complete dissociation" of so-called "strong" electrolytes became accepted through the works of Lewis, Bjerrum, and Debye and Hückel.

Treatments of Charge-transfer in Electrode Processes

Historically, in Nicholson and Carlisle's demonstration (April, 1800) of water electrolysis, producing hydrogen and oxygen, the cathodic hydrogen evolution reaction, with proton discharge as its initial step, was the first electrolytic decomposition reaction to be effected following development of Volta's pile, utilizing electricity generated by electrochemical action at dissimilar metals wetted by an electrolyte solution. Nicholson and Carlisle's experiment resulted from a personal communication to them from the then President of the Royal Society, Sir Joseph Banks, to whom Volta had earlier communicated a letter for publication which appeared in the Philosophical Transactions, 1800, and was read on 26 June, later that year. For electrochemistry, Nicholson and Carlisle's discovery of cathodic hydrogen evolution in electrolysis of water was as monumental as Volta's demonstration of the electrochemical generation of electricity described in that letter. The principle was immediately utilized by Sir Humphry Davy who applied it to molten salts producing, for the first time, the very reactive alkali metals.

From an historical point of view, it seems likely(11) that electrolysis of water was already achieved earlier in an experiment with a large electrostatic generator by Paets van Troostwijk and Deimann in 1789 but separation of hydrogen and oxygen at different poles was not noted. Later, Faraday himself examined the chemical effects of electrostatic electricity and showed them to be the same as those of Galvanic electricity. At that time, it was not possible, from electrostatic machines, to produce a continuous and strong enough current to conduct a significant study of electrochemical reactions as became possible immediately in the 19th century with Volta's pile.

As was indicated earlier in this article, a detailed microscopic understanding of the phenomena underlying electrolysis took a very long time to mature and depended on postulation, proof and acceptance of such concepts as spontaneous ionization in solution, ionic hydration, free ion mobility in solution and the free-electron theory of metals.

From a phenomenological point of view, the first quantitative representation of electrolysis as a kinetic process arose from the work of Tafel in 1905 at Erlangen(12) on cathodic H_2 evolution. Tafel had been active also in the field of electro-organic chemistry, an area that had been already extensively investigated in earlier years of the 19th century, e.g. by Kolbe, but more from the preparative than the kinetic-mechanistic directions.

Tafel observed that an extra driving voltage, the overpotential η, was required to cause electrolysis to proceed at appreciable net rates expressed in terms of current-density, i. His empirical equation representing this behavior

$$\eta = a + b \log i \qquad (1)$$

is fundamental in the kinetics of electrode processes. In Equation 1 a and b are empirical constants: a represents the overpotential at unit current-density (1 A cm^{-2}) while b represents the slope of the increase of overvoltage with logarithmically increasing current-density, $d\eta/d \log i$.

The theoretical significance of the logarithmic increase of η with current-density was not understood until much later in the present century in terms of activation ideas in chemical kinetics, starting with Arrhenius in 1889. However, the proper representation of this behavior in electrode kinetics was not formulated until the independent works of Butler (15) in England in 1924 and of Volmer (14) in Germany around 1930 (see later), some 20 years after Tafel's empirical relation for electrode-kinetic behavior.

Some authors have attributed this delay to the great influence exercised by the electrochemical thermodynamic work of Walther Nernst, commencing in 1889 and continuing onwards for many years. This work emphasized information that can be obtained from measurements on electrochemical cells at equilibrium, with virtually no reference to the kinetic-mechanistic aspects of the processes involved. In fact, it was not until 1924 that Butler (15) gave a kinetic derivation of the Nernst equation, based on the potential-dependence of rates of the backward and forward directions of the process at equilibrium, and 1936 that this approach was developed further by Butler (13) for the reversible H_2/H^+ electrode. The components of the energy of the overall process, corresponding to the Gibbs energy of the electrode reaction that determines the Nernst reversible potential, were also treated in some detail in several papers by Butler, e.g. refs. 13 and 15.

The Energetics of Electrochemical Charge-transfer Processes

It was mentioned in the preceding section that the (standard) Gibbs energy, ΔG^o_R, for an overall process of charge transfer at an electrode, can be considered in terms of component energy contributions. Thus, for a metal ion (M^{z+}) deposition process

$$M^{z+}_{aq} + ze^-(\text{in M}) \overset{\Delta G^o_R}{\rightleftharpoons} M_{crystal} \tag{2}$$

the essential component energies can be visualized through the following Born-Haber cycle of processes that are equivalent to the overall process:

$$\begin{array}{ccccc} M^{z+}_{gas} & + & ze^-(\text{in gas}) & \xrightarrow{-I_{M/M^{z+}}} & M_{gas} \\ \uparrow -G^o_{hyd.M^{z+}} & & \uparrow +z\Phi_M & & \downarrow G^o_{lattice} \\ M^{z+}_{aq} & + & ze^-(\text{in metal}) & \overset{\Delta G^o_R}{\rightleftharpoons} & M_{crystal} \end{array} \tag{3}$$

where G^o_{hyd} is the Gibbs energy of hydration of the M^{z+} ion (a large negative quantity), $I_{M/M^{z+}}$ is the ionization energy of M atoms, Φ_M is the electronic work function of the metal M and $G^o_{lattice}$ is the Gibbs energy for lattice formation of bulk M. ΔG^o_R can therefore be represented by

$$\Delta G^o_R = -G^o_{hyd.M^{z+}} + z\Phi_M - I_{M/M^{z+}} + G^o_{lattice} \tag{4}$$

Equation 4 applies to the hypothetical single-electrode interface reaction (2). In an overall, experimentally accessible, cell reaction, it is to be noted that the work functions of the metals involved as the electrode materials cancel out when the metal/metal contact in the external circuit is taken into account.

An important development in the 1930's was the representation of the energetic course of electrochemical reactions, especially the step (I) of hydrated-proton discharge in the cathodic hydrogen evolution reaction, as a function of the course of the reaction along the "reaction coordinate" from reagents to products.

In the hydrogen evolution reaction, the steps are

$$H_3O^+ + e^-(M) + M \rightarrow MH_{ads} + H_2O \qquad \text{I}$$

followed by either

$$MH_{ads} + H_3O^+ + e^-(M) \rightarrow H_2 + M + H_2O \qquad \text{II}$$

or

$$2MH_{ads} \rightarrow 2M + H_2\uparrow, \qquad \text{III}$$

the latter process being attributed to Tafel as a rate-controlling step at some metals, e.g. Pt. In steps I, II, or III, the immediate product of discharge is regarded as the H atom chemisorbed at a site on the metal electrode surface, as first recognized by Horiuti and Polanyi in 1935 and treated theoretically by Butler in 1936 (13).

The course of the proton-discharge step in electrolytic H_2 evolution was first considered in a semi-quantitative way by Gurney and Fowler(16). The interaction of the proton with water, as in H_3O^+, was represented as a function of distance from the O atom by a potential-energy function, L. In the course of neutralization of H_3O^+ by an electron transferred from the metal electrode, the energy Φ has to be supplied (see below) and, at the same time, the proton charge in H_3O^+ loses its hydration energy, leaving a free H atom in a repulsive relation to the conjugate water molecule left as the product of the discharge process. In Gurney's representation, the possible binding or adsorption of the discharged H atom at the metal was not taken into account (cf. ref. 13).

In electrode kinetics, as empirically represented by Tafel's equation, a basic feature is the potential-dependence of the reaction rate (current-density). This effect arises in Gurney's representation in a fundamental and general way: as the electric potential V, of the electrode metal is changed by ΔV relative to that of the solution (in practice, measured relative to the potential of a reference electrode at open-circuit), the effective value of the electron work function Φ of the metal is changed according to

$$\Phi_V = \Phi_{V=0} \pm e(\Delta V) \qquad (5)$$

This has the effect of raising or lowering the energy level of the initial state of the reaction [here $H_3O^+ + e^-(M)$] through change of the energy level of the sub-system of electrons in the metal relative to their hypothetical zero kinetic-energy level in vacuum. That is, the "electron affinity" of the metal can be regarded as being changed by changes of applied potential according to Equation 5.

Apart from representing the course of an electrochemical charge-transfer process in terms of a reaction energy-profile diagram, Gurney brought into consideration two other fundamental ideas involved in the treatment of electrode kinetics(16):

a) introduction of wave-mechanical tunneling of the electrons involved in the process; and

b) consideration of the distribution of energy levels of electrons in the metal (i.e. in modern terms, the band structure in relation to the Fermi level from which Φ is measured relative to vacuum) and of unoccupied and occupied orbital states in the reagent(s), here H_3O^+.

These two factors formed the general basis of all succeeding treatments of electron transfer in electrode processes. Another general condition arises indirectly from (a) in Gurney's treatment, that is, that the electron transfer takes place as a radiationless transition. Thus, the e transfer takes place from an occupied energy level in the metal, at or near the Fermi level, to an unoccupied level in the reagent (usually the lowest unoccupied molecular orbital, LUMO, in the case of a cathodic process) at the same energy as the level in the metal from which the electron originates. For an anodic process, the opposite considerations apply.

The energy-level distribution factor is now recognized as a fundamental factor in the quantum-mechanical representation of electron-transfer rates both in heterogeneous redox reactions(7) at electrode surfaces and in homogeneous ones in bulk solution, as well as at semi-conductors(18).

One factor of importance for electrode-kinetics was, however, omitted in Gurney's treatment: the role of adsorption of the discharged H atom at the metal electrode surface. Neglect of this factor led Gurney's treatment to predict activation energies for proton discharge that were far too high and independent of the electrode metal. This adsorption energy factor, A, was taken into account in the 1936 paper of Butler(13) who showed how it could account for substantial differences of activation energy for proton discharge at Ni compared with Hg. In fact, in this paper, can be traced the birth of the important field of electrocatalysis. The H atom adsorption energy, A, is also a distance-variable in the potential-energy diagram representation of proton discharge and must, like L, be represented as a function of distance from the metal surface by a potential-energy function, e.g. a Morse function for the quasi-diatomic molecule MH, representing the adsorbed state of H at M.

The role of the H adsorption energy factor in the kinetics of steps I, II and III of the H_2 evolution reaction was taken by Parsons(19) in 1958 in a quasi-thermodynamic way, using kinetic equations at the reversible potential. In that paper he showed how the exchange-current density parameter, i_o, characterizing the kinetic reversibility of the reaction, depended, in a volcano-like plot, on the standard Gibbs energy of chemisorption, ΔG^o_H, of the H at the electrode metal through the equilibrium coverage, θ_H, by H and the corresponding free-site fraction $1-\theta_H$. Maximum activity arises when $\Delta G^o_H = 0$.

An interesting development from Gurney's paper(16) and from a paper of Bell(20) published soon after, was the application of quantum-mechanical tunneling concepts to the transfer of the proton in the H_2 evolution reaction by a tunneling mechanism, as treated in a paper by Bawn and Ogden(21). Further treatments of this kind were made later by Conway(22), Conway and Salomon(23), Christov(24), and Bockris and Matthews(25). However, despite attempts to detect proton transfer by the tunneling mechanism by means of electrochemical experiments at low temperatures(23) (e.g. down to 180 K), no clear experimental basis for such behavior could be found. Attempts have also been made to detect such effects in proton-transfer-controlled homogeneous reactions, e.g. processes in which enolization is a rate-controlling step; unfortunately, in most cases, indications of proton-tunneling in the kinetics have been ambiguous.

Kinetic Representation of Electrode Processes

Following the seminal paper of Butler (37) in 1924 on the kinetic basis of Nernst equilibrium potentials, an electrochemical rate equation was written by Erdey-Gruz and Volmer (14), for a net current-density i, in terms of components of i for the forward and backward directions of the process. They recognized that only some fraction (denoted by α or β) of the electrical energy change ηF associated with change of electrode potential, would exponentially modify the current, giving a potential-dependent rate-equation of the form:

$$i = zF\vec{k}c_R \exp[\beta\eta F/RT] - zF\overleftarrow{k}c_0 \exp-[\alpha\eta F/RT] \tag{6}$$

where $\vec{k}$ and $\overleftarrow{k}$ are electrochemical rate constants for the forward and backward directions of the process, containing exponential terms in the reversible potential, and β and α are charge-transfer factors that must be taken as ca. 0.5 in order to represent correctly the experimental value of $d\eta/d \ln i$, as was recognized by Gurney but apparently almost as an afterthought by Fowler (16). Equation 6 refers to a process that is first order in each direction with an electron number n=1.

In order that Equation 6 gives rise to the equilibrium potential when zero net current flow takes place (i=0), it is necessary that $\alpha = 1-\beta$. Also, in Equation 6, the overpotential η is defined as the difference between the actual electrode potential, E, required to pass a density of current i A cm ^{2} and the reversible potential, E_{rev}, when i=0 and η=0; i.e. $\eta = E - E_{rev}$.

Equation 6 is referred to as the Butler-Volmer equation. Normally, for significant overpotentials, either one or the other of the two terms is dominant, so that the current-density exponentially increases with η, i.e. ln i is proportional to $\beta\eta F/RT$ in the case, for example, of appreciable positive η values. Here the significance of Tafel's b coefficient (Equation 1) is seen: $b = d\eta/d \ln i = RT/\beta F$ for a simple, single-electron charge-transfer process.

Equation 6 represents the experimentally realizable possibility, in electrode kinetics, of changing the rate of a reaction over many magnitudes simply by turning the knob of a potential-controller.

For more complex processes, e.g. in the case of the desorption steps II or III in the H_2 evolution reaction, b can take limiting values of $RT/(1+\beta)F$ or $RT/2F$, depending on the mechanism. The evaluation and interpretation of the Tafel b coefficient has played an important role in elucidation of mechanisms of multi-step electrode processes involving coupled charge-transfer and surface-chemical steps, and many cases have been worked out (26).

Extension of the Gurney-Butler treatments of the kinetics of electrochemical charge-transfer was made in terms of the "transition-state" theory of Eyring, Glasstone and Laidler in a paper (27) by these authors. This provided the formal basis for representation of electrode reaction rates in terms of the standard Gibbs energy of activation and a transmission coefficient κ usually taken, at least until recently, with the value of unity:

$$i(V) = zF\frac{kT}{h}.\ C_R \exp[-\Delta G^{o\ddagger}/RT].\ \exp[\alpha\Delta\phi F/RT] \quad (7)$$

where C_R is a local reactant concentration at the electrode surface and k, T and h have their usual significance. i is the current-density for a z-electron transfer process proceeding at a metal/solution potential of $\Delta\phi$. Correspondingly Equation 7 can be expressed in terms of exponentials in $-\Delta H^{o\ddagger}/RT$ and $\Delta S^{o\ddagger}/R$.

The parameter α (or β) in Equation 7 was clearly recognized in the Eyring, Glasstone and Laidler paper(27) as representing the potential at the transition state as a fraction of the overall metal-solution potential difference. When the transition state is "symmetrically situated", then α or β is 0.5.

There is an interesting relation between the effect of changing electrode potential on the rate of an electrode process and the result of changing acid or base strength (pK) on the rate constant, k, of homogeneous proton-transfer processes, as expressed by the Brønsted relation: $\Delta \ln k = \alpha \Delta \ln K$, where α is Brønsted's coefficient, *ca*. 0.5. Thus, the value of α or β in the Butler-Volmer equation expresses the fraction (α, $\beta \cong 0.5$) of the applied change of electrical energy, ηF, that is effective in modulating the Gibbs energy of activation of the reaction. The analogy of the electrochemical α or β to the Brønsted coefficient for linear free-energy relations in proton transfer was already recognized by Brønsted himself with Ross-Kane(32), and in papers by Frumkin(33) and the present author(34). The Tafel relation, especially for electrochemical proton transfer at Hg, in fact, provides one of the best examples of a linear free-energy relation over a wide range of changes of energy corresponding to *ca*. 9 decades of change of rate constant.

In recent work it has been found(35) that β is dependent on temperature, an effect that can be explained formally in terms of the electrode potential, or interphasial field, affecting not only the *energy* of activation, through change of the electron's energy, but also the *entropy* of activation(36) through changes of the interphasial environment of the reacting particle, e.g. H_3O^+.

This historical account of the development of ideas and experiments on charge transfer in electrochemistry should not conclude without reference to the Faraday Discussion(28) in 1947, held at the University of Manchester. This discussion marked an important turning point in electrode kinetics towards more modern and quantitative analyses of electrode process mechanisms and utilization of relatively new (for that time) techniques, e.g. a.c. impedance studies in the papers by Randles (37) and by Ershler (38). It also brought together many European electrochemists, following the war years, during which little scientific intercourse had taken place on fundamental aspects of electrochemistry.

Electron Transfer in Inorganic Ionic Redox Reactions

Much of the earlier work on charge-transfer processes was directed to studies of the water electrolysis processes which involve "atom transfer" as well as electron transfer, and to some electro-organic reactions. Around the mid-1950's, a more generalized approach was made to the problem of electron transfer, separately by Hush(29) and by Marcus(17,30), who considered the conditions applying to electron transfer between the conjugate ions of a redox couple, both at an electrode and homogeneously. A central concept in this approach(30) was the "reorganization energy", λ, associated with change of the solvated state of one ion or the other of the redox couple when electron transfer takes place. The activation energy in this approach is determined, in part, by one-quarter of the quantity λ which itself is related to the change of dielectric polarization energy of the ion upon being oxidized or reduced. In its original form, this theory was based on non-specific, long-range dielectric polarization effects, contrary to the types of approach that had been made for the proton-discharge process where short-range, hydration-shell energy changes had been the basis of earlier activation energy calculations(31). The Marcus type theory has had a great influence on the treatment of mechanisms and rates of inorganic inner and outer-shell redox reactions at electrodes and the relations of their kinetics to those of corresponding processes conducted homogeneously.

Acknowledgment

The author makes grateful acknowledgment to Alcan International and to the Natural Sciences and Engineering Research Council for appointment to the Research Chair in Electrochemistry at the University of Ottawa, and for support of this and related work.

Literature Cited

1. Williams, L. Pearce, Faraday, Michael A biography, Chapman and Hall: London 1965.
2. Thomson, J.J. Phil. Mag. 1897; 44, 293.
3. See Wilson, H.A. Proc. Cambridge Phil. Soc. 1897; 9, 244.
4. Millikan, R.A. e.g. see " The Electron ", 2nd edn., Univ. of Chicago Press: Chicago, 1924.
5. Stoney, G.Johnstone Phil. Mag. 1881, 11, 381.
6. Stoney, G.Johnstone Sci. Trans. Roy. Dublin Soc. 1891, [2] 4, 583.
7. Shuster, A. The Progress of Physics, Cambridge Univ. Press: Cambridge, 1911, p.56.
8. Maxwell, J.Clerk A Treatise on Electricity and Magnetism, Clarendon Press: Oxford, 1873, sec. 259, 260.
9. Conway, B.E. Electricity and Chemistry, Dow Staff Research Lecture, 1966, Univ. of Ottawa, publ. Univ. of Ottawa Press: Ottawa, 1967.
10. Born, M Zeit. Physik, 1920, 1, 45; see also Bernal, J.D.; Fowler, R.H. J. Chem. Phys., 1933, 1, 515.
11. Leicester, H.M. The Historical Background of Chemistry, Dover Publications: New York, 1956.

12. Tafel, J. Zeit. Phys. Chem. 1905, 50, 641.
13. Butler, J.A.V. Proc. Roy. Soc. 1936, A157, 423.
14. Erdey-Gruz, J.: Volmer, M. Zeit. Phys. Chem. 1930, 150, 203; 1932, 162, 53.
15. Butler, J.A.V. Trans. Faraday Soc. 1924, 19, 729 and 734.
16. Gurney, R.W. Proc. Roy. Soc. London, 1931, A134, 137; Fowler, R.H. Trans. Faraday Soc. 1932, 28, 368.
17. Marcus, R.A. Ann. Rev. Phys. Chem. 1964, 15, 155; and Biochim. Biophys. Acta 1985, 811, 265.
18. Gerischer, H. Zeit. Phys. Chem. 1960, 26, 223; 1960, 26, 325.
19. Parsons, R. Trans. Faraday Soc. 1958, 54, 1053.
20. Bell, R.P. Proc. Roy. Soc. London 1933, A139, 466; and 1936, A154, 414.
21. Bawn, C.E.H.; Ogden, G. Trans. Faraday Soc. 1934, 30, 432.
22. Conway, B.E. Can. J. Chem. 1959, 37, 178.
23. Conway, B.E.; Salomon, M. J.Chem. Phys. 1964, 41, 3169 and Discussions Faraday Soc. 1965, 39, 223.
24. Christov, St. G. Zeit. Elektrochem. 1958, 62, 567.
25. Bockris, J.O'M.; Matthews, D.B. Proc. Roy. Soc. London, 1966, A292, 479; and J. Chem. Phys. 1968, 24, 298.
26. E.g. see Bockris, J.O'M. Chapter 4 in Modern Aspects of Electrochemistry, Vol. 1, Ed. Bockris, J.O'M. Butterworths: London, 1954, and J. Chem. Phys. 1956, 24, 817.
27. Eyring, H., Glasstone, S.; Laidler, K.J. J. Chem. Phys. 1939, 7, 1053.
28. Discussions of the Faraday Society, 1947, 1, various papers.
29. Hush, N.S. J. Chem. Phys. 1958, 28, 962.
30. Marcus, R.A. J. Chem. Phys. 1956, 24, 966.
31. E.g. Parsons, R.; Bockris, J.O'M. Trans. Faraday Soc. 1951, 47, 914.
32. Brønsted, J.N.; Kane, N.L. Ross J. Amer. Soc. 1931, 53, 3624.
33. Frumkin, A.N. Zeit. Phys. Chem. 1932, 160, 116.
34. Conway, B.E.; Salomon, M. J. Chem. Educ. 1967, 44, 554.
35. Conway, B.E. Chapter 2 in Modern Aspects of Electrochemistry, Vol. 16, Eds. Conway, B.E., Bockris, J.O'M.; White, R.E. Plenum Publ. Corp.: New York, 1986.
36. Agar, J.N. Discussion Faraday Soc. 1947, 1, 84.
37. Randles, J.E.B. Discussion Faraday Soc. 1947, 1, 11.
38. Ershler, B.V. Discussion Faraday Soc. 1947, 1, 197.

RECEIVED September 12, 1988

Chapter 12

William Lash Miller (1866–1940)

W. A. E. McBryde

Department of Chemistry, University of Waterloo, Waterloo, Ontario N2L 3G1, Canada

Physical chemistry was still in its infancy when Miller earned his second German Ph.D. under Ostwald in 1892. His first Ph.D. in 1890 had been in organic chemistry. His thesis work at Leipzig incorporated a validation of the theoretical ideas of Willard Gibbs by demonstrating that the electrode potentials of several fusible metals at their melting points were the same whether the metals were solid or liquid. Miller went on to become one of the leading exponents of Gibbs' chemical thermodynamics in North America; and electrochemistry -- theoretical and applied -- under his leadership became a major research interest at the University of Toronto throughout and even beyond his own lifetime. In his heyday he was undoubtedly the Canadian chemist best known in the United States, and held high office in many chemical and scientific societies in that country as well as in Canada. This paper will examine parts of his career especially pertinent to pure and applied electrochemistry.

The Department of Chemistry at the University of Toronto is housed in a handsome building known as the Lash Miller Chemical Laboratories, opened just twenty-five years ago. The person after whom this edifice was named is William Lash Miller, the fourth professor or head-of-department of chemistry to serve during the first 95 years in which that university -- originally known as King's College, Toronto -- was in operation. It would be no exaggeration to state that for a time Miller was the most distinguished, the most colorful, and the best-known chemist in Canada. The occasion of this great North American Chemical Congress in Toronto is a suitable time in which to illuminate some aspects of his career; and this particular symposium affords an opportunity to dwell on one part of that career -- his achievements in electrochemistry.

0097–6156/89/0390–0165$06.00/0

Academic Career

Miller might be described as a quintessential Canadian of his time. His great-great grandfather settled as a loyalist in Niagara, British North America, in the early 1780s. His descendants gradually moved inland into what became Canada West, and ultimately Ontario; several of them made careers in law. Lash Miller's grandfather served as a lawyer in Dundas, and then as a judge in Galt; his father was a prominent Toronto lawyer. When young Miller entered the University of Toronto in 1883 he may well have had in mind following his forebears' vocation. He arrived from high school with honors in classics, and continued the study of classics for two more years. However, in his first year he also received instruction in chemistry from William H. Pike, who had come as Toronto's second professor of chemistry from Merton College, Oxford, and whose credentials included a Ph.D. from Göttingen. Pike's lectures appear to have taken Miller's fancy, for in succeeding years his academic program veered to the sciences and his academic standing rose significantly. In his graduating year Miller stood at the top of his class in Natural Science, and he received a gold medal from the chemistry department.

Pike was an enthusiastic advocate of the merits of chemical education in Germany at that time, and he encouraged Miller to undertake further studies in that country. Following the prevailing practice Miller spent some time in each of three universities: with Hofmann in Berlin, with Victor Meyer in Göttingen, and with Claisen in Munich. He submitted a dissertation at Munich on certain derivatives of acetone oxalic esters, thereby earning a Ph.D. in 1890. That summer he went to Leipzig, where he worked in Ostwald's laboratory; subsequently he spent the summer of 1892 also with Ostwald, and incorporated the work he did there into a second dissertation for a second Ph.D. From Ostwald, Miller quickly gained insight into the still comparatively young subject of physical chemistry, and especially into the language and method of Gibbsian thermodynamics. No other aspect of chemistry was ever to be so closely linked with Miller's name as this last-mentioned one. The investigation described in his second thesis was a demonstration that the emf of an electrochemical cell containing a fusible metal electrode (mercury, tin, or lead) was independent of whether the metal was in the liquid or solid state. Given the equality of the chemical potential in either phase of the electrode substance, the importance of the Gibbs free energy in fixing the electrode potential was confirmed. This was Miller's first exploit in electrochemistry, a topic with which he and many of his students became closely identified.

University of Toronto. Miller was appointed a Fellow in chemistry at the University of Toronto in the autumn of 1890, Demonstrator the following year, and Lecturer in 1894. In 1898 Pike was afflicted by a severe throat ailment and was given a year's leave of absence; in the following year he submitted his resignation. However, no action was taken to fill the chair until 1900, and consequently Miller was in charge of the chemistry department for two years, Pike having returned to England. Then occurred one of those strange obliquities of judgment of which examples litter the pages of history. The vacant professorship was advertised and attracted a number of excellent candidates (including one who later won a Nobel prize, and two others who were subsequently knighted and became Fellows of the Royal Society). Miller was in the running with Pike's strong endorsement. In the event, the position was awarded to one W.R. Lang, a recently-appointed lecturer at the University of Glasgow, Miller's junior by three years, and with a very modest record of research accomplishment. It was a bitter snub for Miller, who had served ten years as a staff member in the department, and who was by this time gaining recognition as a productive scholar. He very nearly left Toronto and, indeed, went so far as to lease a house in Ithaca, N.Y. for a year (although he never occupied it) on the promise of an appointment at Cornell University. In the end he decided to stay on and fight. His situation was made at least more palatable by his immediate appointment as Associate Professor (then a relatively rare title at Toronto) of Physical Chemistry.

Relations between Lang and Miller at no stage ever became cordial, but Miller's title, which was changed to Professor of Physical Chemistry in 1908, gained for him a measure of independence, especially in budgetary and research matters. Tensions, which were acute during the first year or two of Lang's professorship, gradually eased, especially through the intervention of W.H. Ellis -- the Professor of Applied Chemistry in the School of Practical Science (which later became the Faculty of Applied Science and Engineering in the University of Toronto). Ellis had been assistant to Henry Croft, the first professor of chemistry in the university; he was an older man of sterling qualities, and appears to have served as an effective peace-maker among the chemists. He also appears to have facilitated the creation of a sub-department of electrochemistry in the university, and its installation in a new Chemistry and Mining Building built in 1905. (This housed the Department of Applied Chemistry, while the Department of Chemistry occupied the Chemical Laboratory across the street.) An account of activities in that unit will be given presently. but first it will be appropriate to complete an outline of Miller's academic career at Toronto.

When the first World War broke out in 1914, Professor Lang, who had had a long record of service with militia unit in Britain and then in Canada, enlisted for active duty. On this account Miller inherited much of the day-to-day running of the chemistry department, consulting with Lang as necessary on important matters. After the war Lang was appointed to a professorship of Military Studies in the University of Toronto, and Miller finally became head of the department in 1921. He served in that capacity until 1937, retiring at age 70 with the title Professor Emeritus. He died a few days before his 74th birthday.

Electrochemistry and Thermodynamics.

Miller's researches were varied, and not always easy to identify because he frequently omitted his own name as an author of papers describing work done by his students. Given the particular emphasis of this symposium we shall give prominence in this account to electrochemical investigations conducted under his supervision or at his instigation. As early as 1903 he began with T.R. Rosebrugh (a professor of Electrical Engineering) the evaluation of certain functions of e^{-x} required for the quantitative study of diffusion and similar phenomena. Later (1910) the same two men published an important mathematical treatment of diffusion and chemical action at electrodes. This collaboration is suggestive of the interdisciplinary character of the sub-department of electrochemistry to which reference has already been made. Undergraduate instruction in electrochemistry was given to chemistry and chemical engineering students together for decades, and graduates of either department engaged in postgraduate research in the unit. Papers published from the electrochemical laboratory bear titles suggestive of a wide range of interests and activities.

As well as the topics already mentioned, work of a fundamental character is suggested by the numerical evaluation of infinite series and integrals which arise in certain problems of linear heat flow, electrochemical diffusion, etc. (1931), or the theory of the direct method of determining transport numbers (1908). There were studies of electrodeposition from aqueous solutions, including polarization and concentration changes at electrodes (1924). There were papers dealing with more practical matters, such as the formation of a badly-conducting film on copper anodes working in cyanide solutions (1915), or electrodeposition of copper or nickel on aluminum (1922). There were papers dealing with electrical conductivity of various refractories (1922, 1925), multiple electrode systems and throw in electroplating baths (1923), current-voltage relationships in carbon arcs (1923-24), and so on. It was quite an active school of research. Two of Miller's students in this field remained as staff members: J.T. Burt-Gerrans and A.R. Gordon. The latter earned an enviable reputation for the excellence of researches done in his laboratory on conductivity, transference numbers, diffusion coefficients, and other thermodynamic properties of solutions.

Lest the preceding information create the impression that Miller's researches were restricted solely to electrochemistry, it may be appropriate to mention briefly the considerable amount of other work in which he and his students were engaged over the years. This included quantitative studies of the kinetics of a number of familiar reactions in solution, and several investigations of vapor pressures of binary systems as a basis for evaluating partial free energies in solution. This latter work was extensively continued in the department by John B. Ferguson, a former student who was appointed to the teaching staff in 1921. And, finally, some allusion must be made to Miller's search, throughout a period of more than twenty years, for the identity of the nutrient substances essential for the growth and reproduction of micro-organisms. This almost obsessive undertaking to determine the nature of Wildier's 'bios' became the quest for Miller's personal scientific 'holy grail'. In retrospect, it must be acknowledged to have yielded very little of consequence. It was an attempt too far in advance of the sorts of separative and analytical techniques that are commonplace today to have permitted a successful outcome, and represented an investment of time and effort too costly in proportion to the meagre and probably now inconsequential results that it yielded.

Miller's fascination with Gibbsian thermodynamics, to which reference was made previously, was a cornerstone not only for his own approach to chemistry but also for much of the teaching in the

Toronto chemistry department. Miller himself taught this in each of three years to the students specializing in chemistry. He did not lecture, but rather conducted seminars based on Socratic inquiry -- probing, cajoling, bullying, always making his point by relentless logic. He taught direct from Gibbs, and there is the story told dramatically by Wilder Bancroft of the time when Miller personally bought up the publisher's entire residual stock of Gibbs' volume on thermodynamics to ensure continuity of supply for his students. He retained Gibbs' Greek symbols for the thermodynamic functions in all his teaching and writing; of these, only the use of μ has survived to the present day.

Miller's first major independent application of Gibbs' theories came with his published explanation of the unexpected observation, made by one of his students, that the addition of salt to aqueous alcohol raised the partial vapor pressure of the alcohol. He was able to explain by thermodynamic reasoning how this phenomenon came about in a paper published in 1897, "On the Second Differential Coefficient of Gibbs' Function ζ" (1). This title provides a clue to a seldom-realized difference between Miller's treatment of the thermodynamics of binary systems and that developed by G. N. Lewis, which ultimately came to be adopted worldwide. In Miller's system, of which an account was published in 1925 (2), the critical quantity from which partial free energies of components were calculated was the symmetrical second derivative of the Gibbs free energy of the system, symbolically represented by μ_{12}. This method had a certain mathematical elegance, and was perhaps easier to grasp than that based on a standard state and an activity coefficient, as devised by Lewis, but the latter proved to be the more practically useful (3).

Professional Service

In surveying Miller's career certain things stand out as highlights, and warrant mention as we summarize his achievements. He was in his time undoubtedly the Canadian chemist best known in the United States. He had encountered and acquired some mastery of the new sub-discipline of physical chemistry at a time when this was not yet well understood or developed in North America. He was fortunate in having found a close friend and personal ally in his fellow student at Leipzig, Wilder Bancroft, who provided Miller with scope for self-development as a key editor of the former's new Journal of Physical Chemistry. He was endowed with a brilliant mind and boundless energy and enthusiasm, and these qualities earned for him a high level of respect and many honors.

He took an active and often leading role in scientific associations. He was a charter member and later chairman of the Canadian Section of the Society of Chemical Industry, the first professional association for chemists in Canada, formed in 1902. Early in this century he was persuaded to join the American Chemical Society, and it is of some interest that he served as secretary of the local committee for the 36th Annual Meeting of that body, held in Toronto in 1907. Later he served on the Executive Committee of the Division of Physical and Inorganic Chemistry, and in 1926 was made the first Canadian honorary member of that society. He was an active participant in the American Electrochemical Society, and served as its

President in 1912; at that body's meeting in Toronto in 1929 he was made an honorary member. He was a member of the Washington Academy of Science, and of the American Association for the Advancement of Science, serving as Vice-President of the latter in 1921. Three years later he was Vice-President of the British Association for the Advancement of Science. Later, in 1934, he served as President of the Royal Society of Canada. In 1923 he was chosen as "one of the seven greatest chemists of the world", who were invited to give addresses on the occasion of the opening of the new Sterling Chemistry Laboratory at Yale University. His address was later published as Reference 2 of this paper. Even when allowance is made for the hyperbole in which such occasions are invested, the inclusion of Miller in this group was no trifling accolade. In 1935, two years before his retirement, he was invested with a national honor as a Commander of the Order of the British Empire (C.B.E.)

Miller was a fascinating person, and he acquired both disciples and detractors. A forthcoming history of the Chemistry Department at the University of Toronto, written by the author of this paper, will provide an opportunity for a more complete account of his life, his strengths and his weaknesses.

Literature Cited

1. Miller, W. L. J. Phys. Chem. 1897, 2, 633-642.
2. Miller, W. L. Chem. Rev. 1925, 1, 293-344.
3. LeRoy, D. J. *In* Dictionary of Scientific Biography; Gillespie, C. C., Ed.; Scribner's: New York, 1974; Vol. 9, pp. 393-95.

RECEIVED August 9, 1988

ORGANIC AND BIOCHEMICAL ELECTROCHEMISTRY

Chapter 13

History of Organic Electrosynthesis

Manuel M. Baizer[1]

Department of Chemistry, University of California, Santa Barbara, CA 93106

Very many types of organic synthesis can now be carried out electrochemically, often more advantageously than by other means. Organic electrosynthesis, always multi-disciplinary, now includes photoelectrochemistry, electro-catalysis, bioelectrochemistry and others. Some of the scientific and technological developments which led to the present status of the field will be cited: the use of potential control, the invention of the potentiostat, the combination of electrochemistry with spectroscopy, the use of ion-selective membranes; the large-scale production of sorbitol, adiponitrile, and dimethyl sebacate.

Although organic electrosynthesis is only a small part of electrochemistry, recounting its history chronologically and in detail would take more time than is allotted to this presentation and in any event would serve no useful purpose. The author will therefore pick out the important milestones on the way to the relative sophistication we enjoy today in designing and carrying out organic electrosyntheses. These landmarks are usually associated with scientists' names so this will be an opportunity for those not previously acquainted with this field to learn at least the names of those responsible for major advances.

The first experiment in organic electrochemistry was said to have been the transformation of wine into vinegar. This was of course not a synthesis but a practice, electrochemical or not, that still seems to be followed by the French wine industry in exporting moderately priced products to the U.S. Nor was Faraday's observation of the formation of a gaseous hydrocarbon from acetic acid a synthesis. He was too preoccupied with other activities to pursue this finding and develop the Kolbe Reaction.

Hermann Kolbe, however, did go much further and developed a reaction which is one of the widely used in organic electrosynthesis, namely the oxidation at platinum of carboxylates to hydrocarbons (oxidative decarboxylation):

$$2RCOO^- - 2e^- \rightarrow RR + 2CO_2$$

[1]Deceased

0097-6156/89/0390-0172$06.00/0

Several variants have become established. The Hofer-Moest Reaction uses carbon rather than platinum anodes; the intermediate radical loses another electron and becomes a carbocation which reacts with nucleophiles $(Nu)^-$ purposely added or adventitiously present (e.g., OH^- from water):

$$RCOO^- - 2e^- \rightarrow R^+ + CO_2$$
$$R^+ + Nu^- \rightarrow RNu$$

A second variant is the Crum Brown-Walker oxidation of monocarboxylate monoesters to diesters:

$$2\,^-OCO(CH_2)_nCOOR - 2e^- \rightarrow ROCO(CH_2)_{2n}COOR + 2CO_2$$

This process is commercial for the preparation of sebacates from adipates because, in spite of the loss of a substantial portion of the molecular weight as CO_2, it is more economical than the alternatives.

Fritz Haber showed that for selective syntheses it is more important to control the electrode voltage than to select a given electrode material. E.g., nitrobenzene can, depending on pH, be reduced to a variety of products including those of condensation and rearrangement but, with certain limitation, can be reduced largely to one specific product by controlling the cathode potential. This type of control was not automated until the 1940's when Hickling (Liverpool) designed and demonstrated the first potentiostat. This development provided one of the most important tools of organic electrosynthesis.

Of inestimable value also has been the invention of polarography by Nobel Laureate J. Heyrovsky (first at Charles University in Prague then at the Institute which now bears his name). It became possible to detect, sometimes quantitatively, a mixture of metallic ions or several functional groups in one molecule using the dropping mercury electrode and regularly increasing the cathode voltage in the negative direction. One of his students, Petr Zuman (now at Clarkson University) used the polarography of organic compounds to clarify the reactions occurring in the course of the reduction. Kolthoff introduced the techniques in the U.S. and extended it; he and Lingane are the authors of a two-volume book on the subject.

S. Wawzonek (University of Iowa) concentrated on the polarography of organic compounds. He used as solvents organic compounds (e.g., dioxane, alcohol) diluted with a little water - therefore these were not aprotic systems - and usually helped elucidate the polographic phenomena by means of preparative-scale experiments.

At one time polarography was the most widely used analytical technique in the world. Variants, particularly of wave-form, have been invented. However the technique has been supplemented and in many cases replaced by cyclic voltammetry which gives reduction (and oxidation potential) and, at the same time, information on the rapidity (reversibility) of the electron-transfers, i.e., on the kinetics of the reactions. In this area seminal work was done by Nichols and Shain both then at the University of Wisconsin.

In addition to doing much original research on organic electrosynthesis Fr. Fichter (Basel) produced the first authoritative book on the entire subject ("Organische Elektrochemie", Steinkopff, Dresden and Leipzig, 1942). This has served as a complete guide to the literature until the time it was published. Since 1973 there have been several comprehensive but not encyclopedic books on the subject. A bibliography of Electroorganic Syntheses 1801-1975 was published in 1980 (The Electrochemical Society Inc.). It had been a life-long project of Sherlock Swann, Jr. (University of Illinois) and was brought to completion by Richard Alkire (University of Illinois). Four volumes (XI-XIV, 1978-80) of the Encyclopedia of Electrochemistry of the Elements (Marcel Dekker, Inc.) under the general editorship of A. J. Bard were edited by H. Lund and are the Organic Section.

With some exceptions there was a hiatus of activity in this area from about 1940 to 1955. In the U.S. Swann, mentioned above, kept the field alive and in Japan Kiichiro Sugino. (They were acquainted with each other.) Swann, from a background in metallurgy, emphasized electrode (usually cathode) material, shape, method of preparation, etc. Sugino, an organic chemist, was interested in the synthetic results, less so with mechanisms of reactions except speculatively. However, there was one outstanding development: the commercialization of the electrochemical process for reducing glucose to sorbitol (+ mannitol) which grew out of the work of H. J. Creighton. This was practiced by the Atlas Powder Co. 1937-1948 and replaced by a catalytic process.

A new dimension was added to at least the understanding of organic electrosyntheses by combining electrolysis with spectroscopy to yield information on the nature of the first-formed intermediate and the kinetics of the follow-up reaction(s). One early development was the characterization of nitro-aromatic anion radicals by electron-spin-resonance (e. s. r.) spectroscopy due to D. H. Geske and his associates. This field has expanded enormously and now includes internal reflectance, resonance Raman, specular reflectance and transmission spectroscopies. This range of techniques is now commonly used in elucidating mechanisms: A. Bewick and M. Fleischmann (University of Southampton, England), A. J. Bard (Austin, Texas), D. H. Evans (University of Delaware), Vernon Parker (Trondheim), J.-M. Saveant (University of Paris).

The field of electrochemical polymerization, after a shaky start, began to develop on a rational basis about 1960 due to the efforts of J. W. Breitenbach (University of Vienna) and B. L. Funt (now at Simon Fraser University in Vancouver). To the best of my knowledge none of these processes has been commercialized; further research is being overshadowed by the frantic activity in the field of conducting polymers (A. MacDiarmid, University of Pennsylvania, A. Heeger and F. Wudl, University of California at Santa Barbara, A. Diaz, I.B.M., San Jose, CA).

The entire field of organic electrosynthesis received its biggest stimulus in about 1965 after Nalco's tetralkylleads and Monsanto's adiponitrile process were commercialized. In the former case, non-aqueous media were used successfully on a large scale; in the latter a relatively inexpensive product was made on a multi-tonnage basis thus finally upsetting the old dictum that electrochemical methodology could be considered for the production only of speciality chemicals, medicinals, etc. This upsurge was nowhere more noticeable than in Japan which now has probably the most extensive programs in the world. Outstanding results have been achieved by Y. Ban (Hokkaido), T. Osa (Tohoku University in Sendai), T. Shono (Kyoto), S. Torii (Okayama), T. Nonaka (Graduate School, Tokyo Institute of Technology), I. Nishiguchi (Municipal Technical Institute, Osaka) and their co-workers.

Although its history encompasses only the last fifteen years insofar as its relevance to organic electrosynthesis is concerned, the study and development of chemically modified electrodes (C.M.E.'s) has stimulated a great deal of work which may eventually lead to unique syntheses. L. L. Miller (now at the University of Minnesota) attached covalently to oxidized surfaces of graphite a chiral moiety which, it was hoped, would induce chirality during a cathodic reduction. The results were very modest but encouraging.

What does this brief review of the highlights of the history of organic electrosynthesis tell us about the state of its health? It is alive and well and living in all industrial countries in the world. If has progressed on the basis of a wealth of ideas that this methodology has inspired. It has been accompanied (necessarily) by the development of instrumentation to test, modify, and refine these ideas. It has spawned new sub-areas whose impact upon the field cannot yet be assessed. A review paper written in 2001 covering the period from 1950 will undoubtedly recount more exciting developments than were produced in the century and a half 1801-1950.

RECEIVED August 3, 1988

Chapter 14

Anodic Electroorganic Chemistry and Natural Products

James M. Bobbitt

Department of Chemistry, University of Connecticut, Storrs, CT 06268

Possible and actual relationships between electroorganic chemistry and natural product chemistry will be discussed. Attempts to carry out biogenetic type reactions at an electrode surface will be summarized. Some specific examples of unique electrochemistry that have been discovered in natural materials or materials similar to natural materials will be described.

Without doubt, the historical driving force behind organic chemistry has been the study of natural materials. Living tissue has managed to assemble compounds containing combinations of functionality and stereochemistry that would be far beyond the imagination of mere mortal chemists. The unique reactions and rearrangements that have been discovered in natural product work have been explored synthetically and mechanistically to form the backbone of our science. Thus, it is scarcely surprising that the study of natural products should yield some unique electroorganic chemistry. The more surprising aspect is that the yield has been so small.

Preparative electroorganic chemistry and natural product chemistry impinge on one another in several ways. First, electroorganic reactions have been used to synthesize natural materials of many types. Second, knowledge of the biogenetic reactions that are used in nature to make compounds has been used as a guide in the electrochemical synthesis of natural compounds. Finally, advantage has been taken of the unique polyfunctionality and stereochemistry that exist in many natural materials to learn about new electrochemical reactions.

There are numerous examples, both old and new, of electrosynthetic steps that have been used to prepare natural materials. However, a discussion of this topic would simply be organic chemistry and would show no special relationship to natural

0097–6156/89/0390–00176$06.00/0

materials; it will not be considered in this chapter. Numerous references are available (1-7).

The last two relationships between electroorganic chemistry and natural products, biogenetic reasoning and new chemistry, will be considered in more detail. Emphasis will be placed on natural materials from plants, namely the so-called secondary plant metabolites (alkaloids, steroids, pigments, terpenes, *etc*.). Furthermore, we will be concerned only with those reactions from which products have been isolated and identified. The electrochemistry of such primary metabolites as purines, pyrimidines, flavins, porphyrins, *etc*. has been dealt with by Dryhurst (9), but the work has been centered more on mechanistic analysis than preparative experiments.

Electrochemistry and Biogenesis of Natural Products

Nature uses an array of enzymes to carry out a large number of organic reactions. In a sense, a redox enzyme is a surface which provides or accepts electrons, much like an electrode surface (Fig. 1). The most striking similarity between the two is that electrons are transferred *one at a time* in both systems. This has been pointed out by a number of workers, most recently by Guengerich and MacDonald (8) in a study of cytochrome P-450 systems. The extent to which an electrode actually functions as surface will be discussed in the latter part of this paper. The biggest difference between the two systems is that an enzyme surface has a more sophisticated structure which holds substrate molecules in certain ways or conformations so that unique reactions may take place. A second major difference is that an enzyme is part of a system to transfer electrons between a substrate and a secondary redox reagent such as air or peroxide (in oxidations). In an electrochemical system, the electrons go into the circuit.

The ultimate goal would be to coat an electrode with some substance which would make it function like an enzyme. Many such coated electrode systems have been prepared, but few have been used in preparative reactions (10). Recently, however, Scheffold and his coworkers have attached vitamin B-12 to graphite felt and used it to preparatively couple ethyl iodide with acrylonitrile by a reductive process. (11, 12). Coche and Moutet coated felt electrodes with a redox polymer containing palladium and used it to catalytically reduce nitrobenzene and various double bonds (12a).

There are, however, certain types of natural reactions that occur in many situations, that are probably brought about by similar enzymes, and that may profit from an electrochemical study. Three of these that we have worked on for many years have been phenol oxidation, oxidative decarboxylation, and indole oxidation. The coupling of aromatic ethers electrochemically is unlikely to be a natural reaction, but the results have been so interesting and so analogous to phenol coupling, that this topic will be discussed briefly also.

Phenol Oxidation. Phenol oxidation is a major reaction in the elaboration of many types of natural materials, ranging from

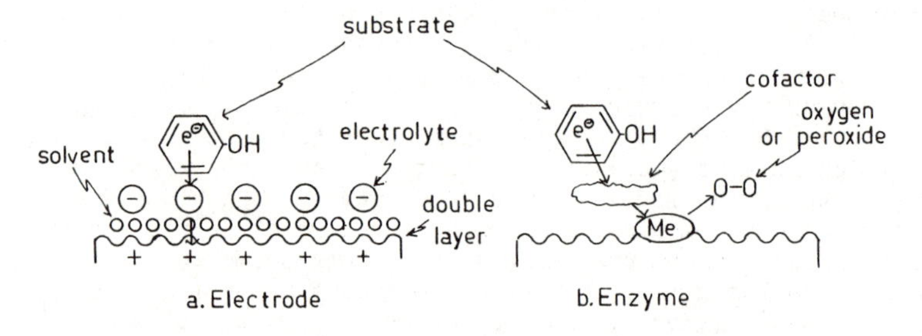

Figure 1. (a) Electrooxidation and (b) Enzyme Oxidation Comparison.

alkaloids to pigments to antibiotics, *etc.* (13). The preparative oxidation of phenols using many reagents has been extensively studied, but there is always a problem of reaction control and overoxidation. It was for this reason that we began our studies on phenols, although we were not the first to do so. Fichter and his students published a series of papers on the electrooxidation of simple phenols between 1915 and 1930 (14). In general, two classes of reactions and three types of products were formed. The classes were one- and two-electron reactions, and the products were quinones such as 4 (from a two-electron reaction in water, through the diphenol, 3), carbon-oxygen dimers such as 5, and carbon-carbon dimers such as 6. The dimers were thought to arise from a one-electron reaction and coupling of the radical forms shown as 2a, 2b, and 2c in Fig. 2; isomers other than 5 and 6 were also formed. Yields, if measured at all, were low. The reactions were carried out without potential control (which became practicable only with the design of a superior three-electrode potentiostat in the 1940's).

The electrooxidation of vanillin, 7, to dehydrovanillin, 8, in 65% yield was reported in 1964 by Vermillian and Pearl (15). They combined a thorough voltammetric study with preparative experiments and were able to clarify the mechanisms of phenol and phenoxide oxidation. Specifically, they pointed out that the oxidation of phenols in neutral solution was a two-electron oxidation and that oxidation of the phenoxide ion (in base) produced a one-electron oxidation and coupling reactions. This work prompted us to apply the method to the synthesis of some isoquinoline alkaloids.

A large number of complex isoquinoline alkaloids are formed in nature by the oxidative coupling of relatively simple phenols (16, 17). For example (Fig. 3), phenolic coupling of the two phenol groups in coclaurine, 9, gives rise to a large number of compounds known as the "bisbenzylisoquinoline alkaloids." Two examples are duaricine, 10, which contains one carbon-oxygen linkage, and the useful anesthetic, tubocurarine, 11, which contains two such linkages.

We were able to show that electrooxidation in basic solutions was an effective method for the intermolecular coupling of various phenolic tetrahydroisoquinolines to give natural materials or materials containing the same functional groups as natural materials. Furthermore, we were able to show that the ratio of carbon-carbon products to carbon-oxygen products was influenced by the electrolytic conditions. Oxidation in aqueous systems with platinum electrodes yielded predominately carbon-oxygen products, and oxidation on carbon felt in acetonitrile gave mainly carbon-carbon products. This work has been reviewed (18), and the synthesis of duaricine (in 8 % yield, the dotted line shows the bond that was formed electrolytically) has been published (19).

Intramolecular coupling was more difficult using our methods although we were able to prepare the N-nor-N-carbethoxy derivatives of isosalutaridine, 12, and kreysiginone, 13, in yields of 18% and 36% respectively (20). Unfortunately, isosalutaridine is the *para-para* coupled isomer and not the *ortho-para* isomer used in the synthesis of morphine by oxidative coupling. Kreysiginone is only

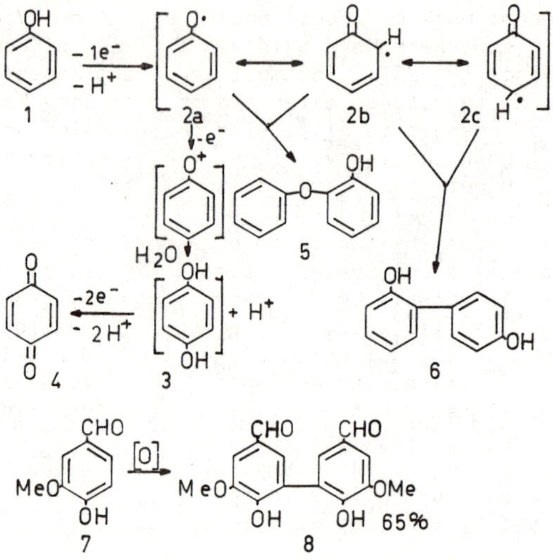

Figure 2. Phenol Oxidations

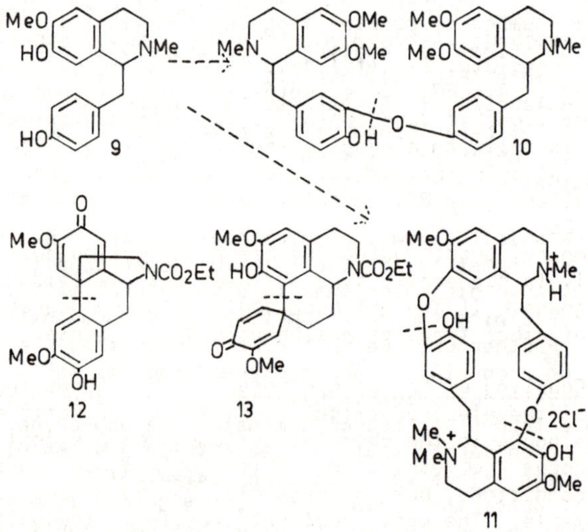

Figure 3. Phenol Coupling in Isoquinoline Alkaloid Biosynthesis

one of the two possible isomers of 13. These isomers differ in the relation of the methoxy group on the dienone ring and the C-1; no preference was noted for one isomer over the other in the electro-oxidation.

In some recent work on anodic phenol oxidation, Yamamura and Iguchi and their coworkers have oxidized a number of complex phenols in methanol (21-23). The most important aspect of their work, in reference to natural materials is that they have extended the reactivity of phenols to allyl side chains which are often encountered in lignan chemistry. Thus, the oxidation of (E)-isoeugenol, 14, in methanol, gave 15 and 16, among many other products (21) as shown in Fig. 4.

Compound 15 shows the formation of a bond between the terminal ends of a conjugated system activated by the oxidation of a phenol, and 16 shows a bond between a phenoxy group and such an end position. Ferulic acid, 17, undergoes oxidation to give a product analogous to 15, but which closes to form a dilactone, 18 (22) rather than reacting with methanol. These types of bonds are found in natural lignan substances. In lignan formation, the carbon-carbon coupling reactions are probably one-electron reactions analogous to the various alkaloid reactions discussed above. In the electrooxidations in methanol, however, the carbocation formed by the loss of a second electron is captured by methanol.

Two-electron reactions also probably play a role in some biogenetic reactions. Dihydroferulic acid (dihydro 17) gave 19 on oxidation (22), a result reminiscent of the work of Scott a number of years ago (25). Finally, the synthesis of the naturally occurring compound, asatone, 22, from the electrooxidation of 20 through intermediate, 21, (followed by a Diels-Alder reaction) was reported (23). Although this synthesis does not involve a phenolic coupling reaction, it does contain a two-electron oxidation of a phenol and is quite an elegant piece of chemistry. The oxidation of 23 to a cationic intermediate, 24, followed by reaction with an activated alkene gave 25, a useful material for natural product synthesis (26).

In summary, anodic oxidation has been applied quite successfully to biogenetic type syntheses of alkaloids and, to some extent, lignans. It has not been applied in any degree to the synthesis of other types of polyphenols and pigments.

Oxidative Coupling of Aromatic Ethers. Much interesting work has evolved from the electrochemical oxidation of aromatic methyl ethers, analogous to the phenols mentioned above except that the phenolic groups are methylated. This requires much more rigorous oxidation conditions, but the products, having no phenol groups, are more stable. The most prominent example is the oxidation of laudanosine, 26, to flavantine, 27, in yields approaching quantitative by Miller (27, 28) and Tobinaga (29) and their coworkers (Fig. 5). Similar work has also been done by Tobinaga to yield 28 (29), by Kametani to yield 29 (30), and by Sainsbury to obtain 30 (31). The theory of such reactions as well as a number of other similar examples has been studied by Ronlan and Parker (32) and is well described by Torii (5).

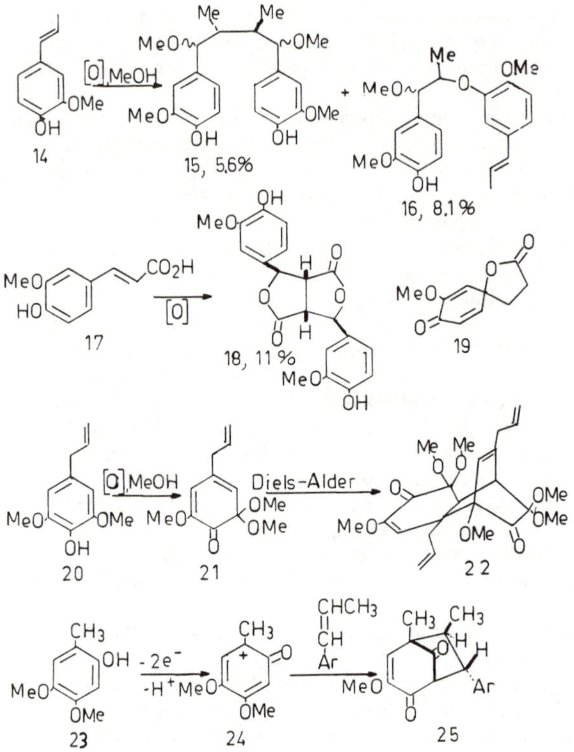

Figure 4. Oxidations of Some Vinyl and Allyl Phenols

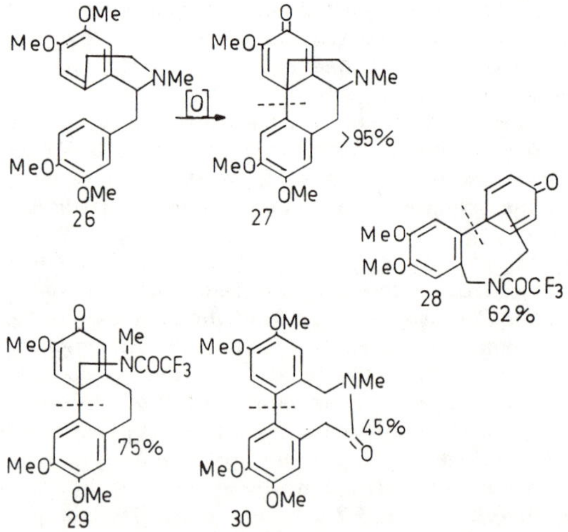

Figure 5. Oxidatively Coupled Products from Aromatic Ethers

Oxidative Decarboxylations. Oxidative decarboxylation, as in the Kolbe reaction, is one of the oldest of all electrochemical oxidations. Whether such reactions parallel any reactions that occur in nature is still in question. Actually, the reaction can take two courses: loss of an electron from a carboxylate and decarboxylation to form a radical which dimerizes or reacts with another radical, or loss of an electron followed by decarboxylation and loss of a second electron to form a carbocation (The Hofer-Moest reaction, 33). The carbocation may then be neutralized by reaction with nucleophile or another source of electrons.

We became interested in oxidative decarboxylations in alkaloid chemistry in an attempt to understand a primary step in the biosynthesis of isoquinoline and indole alkaloids. These two large classes of plant bases are derived from tyrosine, 31, and tryptophane, 35, respectively (17). The first step in the formation of many alkaloids is an oxidation in the aromatic ring to yield such compounds as 32 and 36 where the oxygen functions may be hydroxyl or methoxyl. Little is known about such oxidations, and there seems to be no appropriate electrochemical information about them. The next step is a ring closure as in Fig. 6. The closure can be visualized to take place through a pyruvic acid derivative as written or through a simple aldehyde which would not yield a 1-carboxyl group. The original carboxyl group in tyrosine or tryptophane (in parentheses in Fig. 6) is lost in most cases before ring closure, but may be carried through the reaction. Closure with a pyruvic acid would yield 1,2,3,4-tetrahydroisoquinoline-1,3-dicarboxylic acid, 33, from 32, and a similar compound, 38 from 37. Ring closures with both pyruvic acids and aldehydes take place easily under "physiological conditions" (ambient temperature, near neutral pH in water). It is probable that simple alkaloids where R is methyl are formed through the pyruvic acid pathway, but for the more complex alkaloids where R is benzyl or a terpene derivative, an aldehyde is involved (17). Although the various carboxylic acids can usually be decarboxylated by chemical oxidation, the conditions are vigorous and certainly not "physiological." We postulated that since phenols and indoles are isoelectronic and are both oxidized with ease, that the carboxy groups in both the 1 and 3 positions would fall off if the compounds were oxidized. In other words, the phenol or indole portion would act as an "electrophore" and trigger the loss of carbon dioxide. Since mild oxidations are quite common in nature, such decarboxylations could be considered "physiological".

In the isoquinoline series, we were able to demonstrate clearly that a carboxyl group in the 1-position was lost quantitatively at the oxidation potential of a phenol in the ring (34). The products were 3,4-dihydroisoquinolines such as 39 which were reduced to tetrahydroisoquinolines for isolation. When no phenol group was present, the oxidation potential was much higher, but the decarboxylation did take place. In a sense, the reaction is a Hofer-Moest reaction where the cation is stabilized by the electron pair on the neighboring nitrogen. This does not explain the relationship between the oxidation potential of the phenol and the decarboxylation, but it has been established that phenylacetic

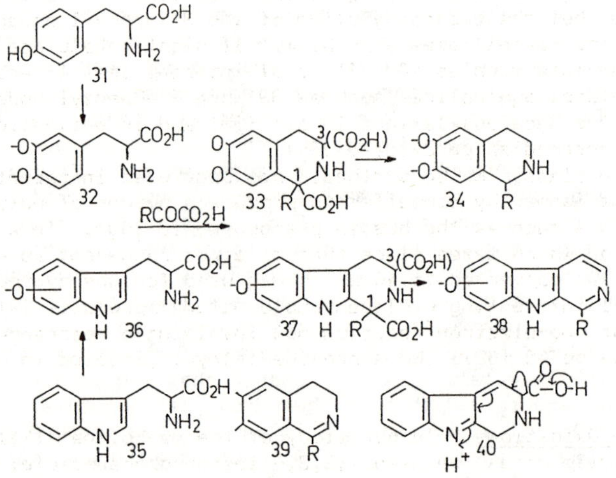

Figure 6. Biosynthesis of Isoquinoline and Indole Alkaloids

acids in which the phenyl ring is electron rich are prone to undergo a Hofer-Moest reaction (also called a pseudo Kolbe reaction) (35,36). No experiments were done on compounds containing 3-carboxyl groups.

In the indole series, it was clearly demonstrated that carboxyl groups in the 1- and 3- positions were lost at the oxidation potentials of the indole nucleus (37). When only one carboxyl group was present in the 1-position, a 3,4-dihydrocarbazole could be isolated, analogous to the isoquinoline series. When two carboxyl groups were present, or when only one was present in the 3-position, the products were always the carbazoles such as 38. Thus, even when only a single carboxyl group was present, on carbon 1, a further oxidation took place to give a completely aromatic product. The decarboxylation of the 1-carboxyl group is analogous to the reactions in the isoquinoline series, but the decarboxylation of the 3-carboxyl group in the carboline series poses a problem. It might take place through an intermediate such as 40. It is of interest that 1-carboxy-tetrahydroisoquinolines such as 33 when R = benzyl undergo oxidative decarboxylations in air (38) and in enzymatic oxidations with horseradish peroxidase (39).

The classic Kolbe synthesis has been used in recent years in an elegant manner by Schaeffer to prepare a number of natural materials such as the beetle pheromone, 28 (40). In a two-electron reaction which takes place through a rearrangement as shown in Fig. 7, 42 was converted to 43 which was used for the synthesis of several interesting cyclopentanoid natural products (41). A similar two-electron reaction not involving a rearrangement, led from 44 to 45 (42). No stereochemistry is implied in 44 and 45.

Indole Oxidations. In our study of the carboline alkaloid carboxylic acids, we used 1,2,3,4-tetrahydrocarbazole, 46, as a model compound to determine the oxidation potential of an indole nucleus. We found that the electrooxidation of 46 gave a good yield of a dimer which, after the assignment of an incorrect structure (37), was correctly assigned as 47 (43). Oxidation of the anion of 46 led to the formation of two more dimers having structures 49 and 50 (44), as shown in Fig. 8. Both 49 Both 49 and 50 have a restricted rotation around the bond marked with a dotted line.

The meso and dl forms of 49 were isolated with no stereochemical preference. Their structures were assigned through the study of the dynamic nuclear magnetic resonance spectra of the two compounds at various temperatures. The bond between the two moieties is only partially restricted at room temperature for one isomer (the dl), but is completely restricted at low temperature for both. The dimers are an interesting example of a situation where the normally favored conformation, anti, is forbidden by steric effects. Rotation in 50 was completely restricted at room temperature, but the rotational isomers could not be separated; the C-13 N.M.R. spectrum of the mixture has 48 peaks. It is of interest that the bonding found in 47 (dotted line) has been observed in the alkaloid cimiciphytine, 48 (45) and that the bonding found in 49is present in the alkaloid chimonanthine, 51 (46).

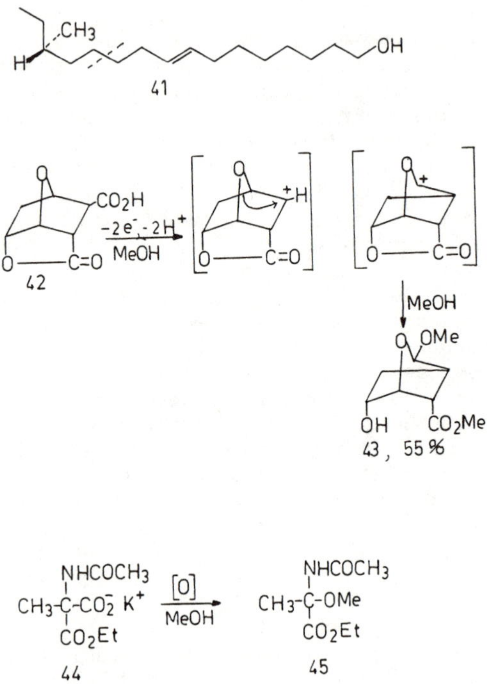

CH_3 OH
41

CO_2H $-2e^-, -2H^+$ MeOH O C=O 42 → [+H O C=O] [O + O C=O]
↓ MeOH
OMe OH CO_2Me
43, 55%

$NHCOCH_3$ $CH_3-C-CO_2^-$ K^+ CO_2Et 44 $\xrightarrow[MeOH]{[O]}$ $NHCOCH_3$ $CH_3-C-OMe$ CO_2Et 45

Figure 7. Decarboxylation Reactions In Natural Products

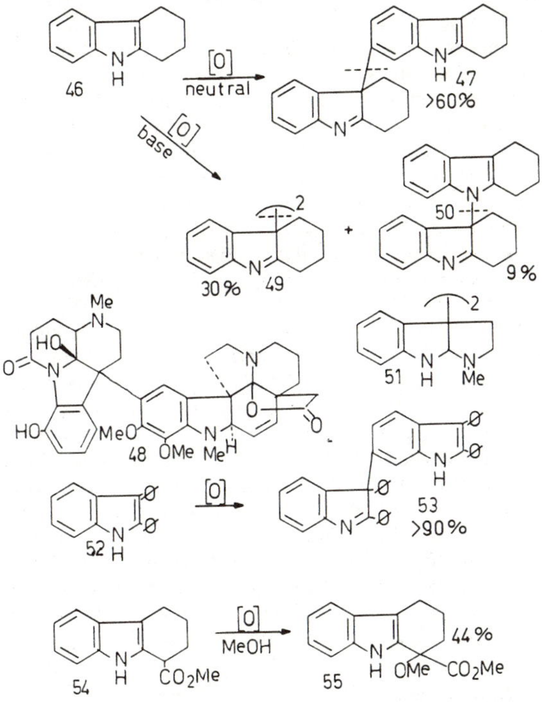

Figure 8. Oxidations of Tetrahydrocarbazoles and Indoles

The oxidation of 2,3-diphenylindole, 52, had been found to give a dimer also (47) which was initially assigned the incorrect structure. This was corrected to 53 which is analogous to 47 (48). The electrooxidation of indole compounds was reviewed (49) but the structure for 47 is incorrect in that review. In related work, we have demonstrated that 1-carbomethoxy-1,2,3,4-tetrahydrocarbazole, 54, can be electrochemically oxidized to give a 1-methoxy derivative, 55 (50).

A detailed mechanistic study of the neutral dimerization of 46 was made (51) before we knew about the products formed in base, but no simple explanation of the data was reasonable. We now believe that, in at least a qualitative sense, the reactions shown in Fig. 9 will explain the product data. The difference between oxidation under neutral and acid conditions should be the first intermediate formed. Under neutral conditions, the cation radical, formulated as two (of several) contributing forms, 56a and 56b, should be formed. In base, the first intermediate would be a radical, again formulated as two structures, 57a and 57b. Thus, the difference in products probably lies in the reactions of a cation radical vs. a neutral radical. The reactions of the radical are easiest to explain, because the two logical products of a coupling reaction are the observed 49 (from two 57a's) and 50 (from 57a plus 57b). Any dimerization reaction between cation radicals, 56a, to give a symmetrical product such as 49, however, would require that two positive charges approach one another in the transition state. Therefore, the more separated radical cation form, 56b, would provide the easiest dimerization reaction, as observed in the formation of 47 (from 56a plus 56b). It should be noted that the mechanistic study (51) does not support the formation of 47 from 56 as shown. It is not certain, however, that the same mechanism would prevail in the micro scale reactions used in the mechanistic studies and in the macro, preparative reactions.

Unique Electrochemistry From Natural Products

Unique electrochemisty can arise from natural product work in a number of ways. One is simply the oxidation (or reduction) of complex natural materials. A second might be chemistry that has been observed in the course of electrochemical work on materials similar to natural compounds or work along biogenetic lines as discussed in the previous section.

Oxidation of Complex Natural Materials. While it it is certainly true that much of the early electrochemistry of natural materials was carried out on simple materials such as alcohol or acetic acid, more complex substrates would be more likely to produce unpredicted chemistry. Our first interest in this type of work was aroused by the paper of Allen and Powell (52) in which the behavior of a number of complex indole alkaloids was examined voltammetrically. Although products were not isolated, such complex molecules should produce some new and unique chemistry. Unfortunately, the work has never been followed up. We did confirm that yohimbine gives products on controlled potential oxidation, but they were not isolated (53).

An almost perfect example of this type of work is shown in the remarkable rearrangement observed in a glycyrrhizin triterpene, 58 to 59 (54). The reaction is triggered by the proximity of an oxidizable unsaturated ketone and a hydrogen in the molecule. A possible route for the reaction is shown in Fig. 10, and R may be an acetyl group, a hydrogen, or even a sugar moiety. The reaction is similar to one observed by Miller in aliphatic ketones (55). In another reaction of this general category, Mori has shown that some beta-lactams such as 60, when oxidized in methanol, gave various methyoxylated derivatives as shown in Fig. 10 (56). Several steroid hydrocarbons have been acetoxylated and methoxylated by oxidations in acetic acid or methanol (57). For example, androstane, 61, gave acetoxlylation in positions 6, 7, and 12 in the ratio shown in 62.

Several other reactions that have been described in other parts of this paper would also fall into this class. Dryhurst has carried out oxidations and reductions of many primary metabolites (9), and other similar work is summarized by Torii (5). It is unfortunate that so little work has actually been done on readily available complex natural materials. However this work is not easily justified or funded since one can only predict that "something interesting will probably happen."

Unique Surface Chemistry Observed From Natural Product Studies. One might profitably conjecture on just what type of surface chemistry might be observed on an electrode. When organic chemists think of electrochemistry, they automatically equate electrode surface reactions with those stereospecific surface reactions which occur on catalytic surfaces used in hydrogenation. Unfortunately, the relationship is not that simple because an electrode surface has a complex layer structure (see Fig. 1a) consisting of electrolyte ions and solvent. Thus, one is dealing with a coated surface, and the extent and thickness of that coating depends upon the electrostatic charge of the surface with respect to the electrolyte solution (58). This electrostatic charge in turn, depends primarily upon the electrode material and is well understood with respect to mercury because one can measure the electrocapillary properties of that liquid with respect to potential. However, the relationship is poorly understood with respect to preparative reactions on such solid electrodes as platinum or carbon. Thus, on solid electrodes, one rarely knows much about the coating or "cleanness" of the surface. The cleanness and the adsorptive properties of the molecules being oxidized will determine whether the reaction takes place near the surface and can be influenced by it or not. Few examples have been reported where the surface played a strong role in product composition.

A second type of reaction may take place on an electrode surface. One normally thinks of an electrochemical reaction as taking place by one-electron shifts, regardless of the total number of electrons exchanged. However, on an electrode surface, there is

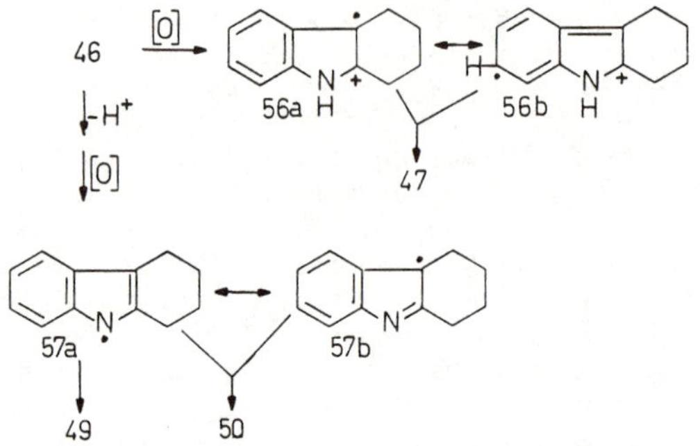

Figure 9. Mechanism of Tetrahydrocarbazole Dimerization

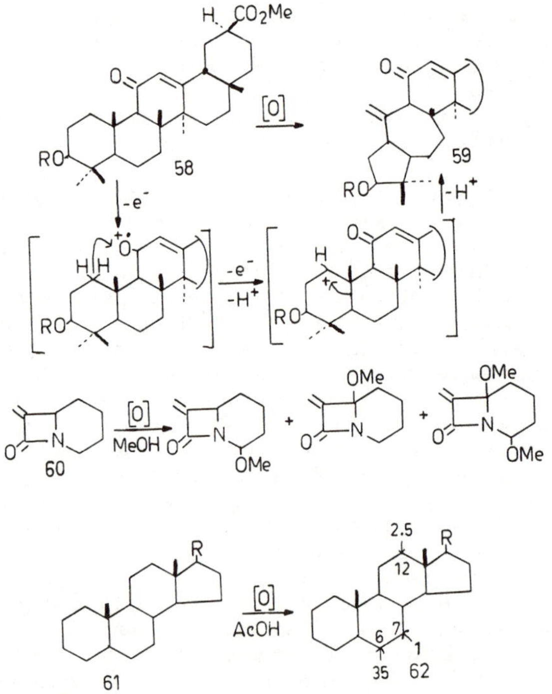

Figure 10. Oxidation of Some Natural Products

no reason to say that electrons may not come out of different parts of the molecule in a concerted reaction with some shuffle of electrons between the two sites. This would be possible if the molecule could be considered to touch the electrode at more than one position. Such a possibility as shown schematically in Fig. 11.

We have observed three reactions in our work on isoquinoline alkaloids which may illustrate the roles described above for a surface influence.

The first of these reactions is the oxidative dimerization of 1,2-dimethyl-6-methoxy-7-hydroxy-1,2,3,4-tetrahydroisoquinoline, 63 to 64 (59) as shown in Fig. 12. The dimerization provides a carbon-carbon dimer in good yield with completely restricted rotation around the new bond. This means that there can be three possible pairs of enantiomers. Although all three were obtained by chemical oxidation with potassium ferricyanide, only one of the three was obtained on electrooxidation. We rationalized this result by proposing that the isoquinoline molecule is adsorbed to the surface in a "tilted manner", see Fig. 12, and that bond formation from this configuration would lead to the isomer which was formed. Adsorption of similar molecules was, in fact, found to take place on carbon electrodes (60). Although no other rationale has been forthcoming for this result, there are two problems that have not been resolved to our satisfaction. The first is the positive proof for the structure of the product, although convincing optical data has been given in its favor (61). No suitable crystals of 64 could be obtained for X-ray study. The second is that we have never been able to devise an electrochemical experiment in which the preference for only one isomer _is lost_. It would seem logical that some potential or type of coating would destroy the tight adsorption needed to produce the results, but such data have never been obtained (62). It is of interest that the dimerization of 46 to give the symmetrical dimer, 49, also produces a product with restricted rotation having three possible pairs of enantiomers (at very low temperature), but that _no_ stereochemical preference is shown. This may be due to the highly oxygenated and therefore very electron rich nature of 63 which may make it adsorb tightly to a positively charged anode.

We suggest that the fragmentation of the 1-(p-hydroxybenzyl)-6,7-dimethoxy-1,2,3,4-tetrahydroisoquinoline, 65, takes place by concerted reaction on the electrode surface, Fig. 13 (19). Compound 66 was isolated in low yield and the quinoneimide, 67, was trapped by starting material, 65, as a 1,2-di(p-hydroxybenzyl)-tetrahydroisoquinoline. The reaction does not take place when the nitrogen is acylated or when the hydroxy group is methylated. It is of interest that the same reaction takes place in enzyme oxidations of similar materials (63) and in mass spectral fragmentations. Such reactions have not been noted in chemical oxidation, although it is questionable whether they have been looked for carefully.

Another possible concerted reaction was observed in a study of the dimerization of simple phenolic tetrahydroisoquinolines (64). When the anion of N-methyl-6-methoxy-7-hydroxy-1,2,3,4-

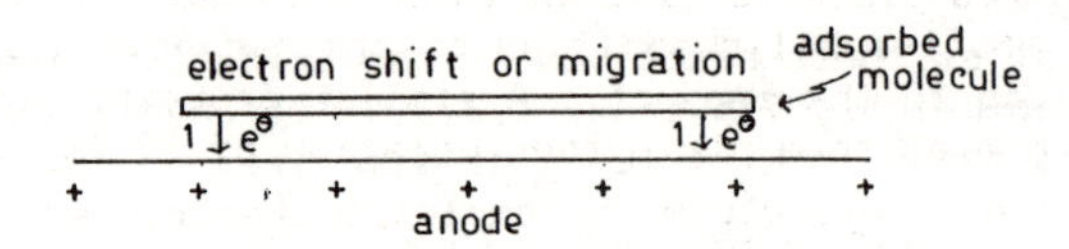

Figure 11. Schematic Concerted Two-electron Oxidation

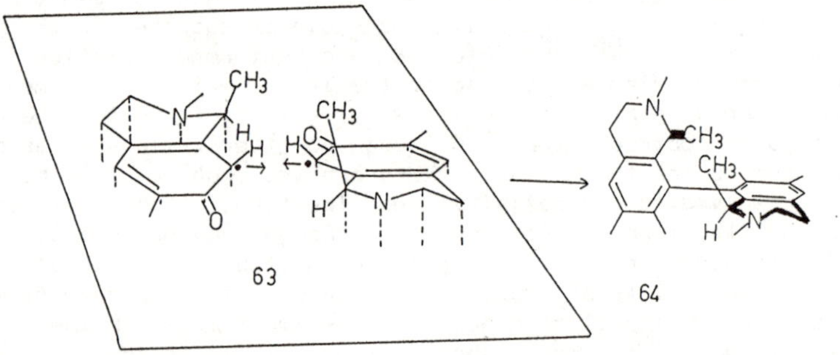

Figure 12. Surface Effects in a Phenolic Isoquinoline Dimerization

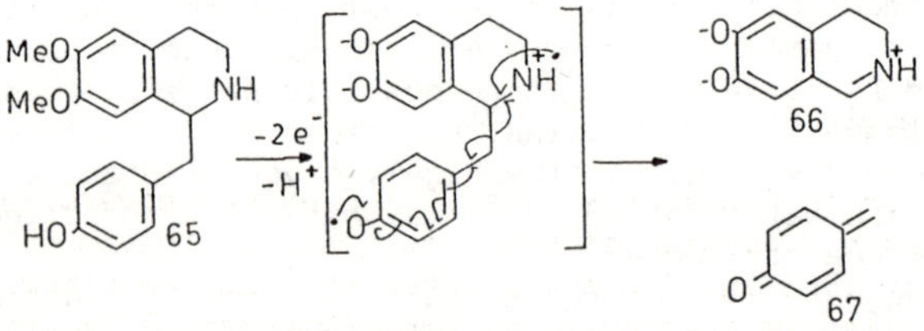

Figure 13. Oxidative Fragmentation of 1-(p-Hydroxybenzyl)-1,2,3,4-tetra-hydroisoquinoline

tetrahydroisoquinoline, 68, was oxidized, a carbon-carbon dimer, 69 was isolated in 85 % yield. When the compound with the hydroxy and methoxy groups reversed, 70, was oxidized, only the carbon-oxygen dimer, 71, was isolated in about 50 % yield with no carbon-carbon dimer. When the nitrogen in 70 was non basic, as in its hydrochloride salt or when acylated, the product isolated was a carbon-carbon dimer corresponding to 69. We suggest that a concerted loss of two electrons, one from the oxygen and one from the N of 70 takes place to give an intermediate as shown in figure 14) Nucleophilic addition of another molecule of phenol as shown would lead to the product. A similar argument has been suggested by Miller to explain the formation of 27 (65).

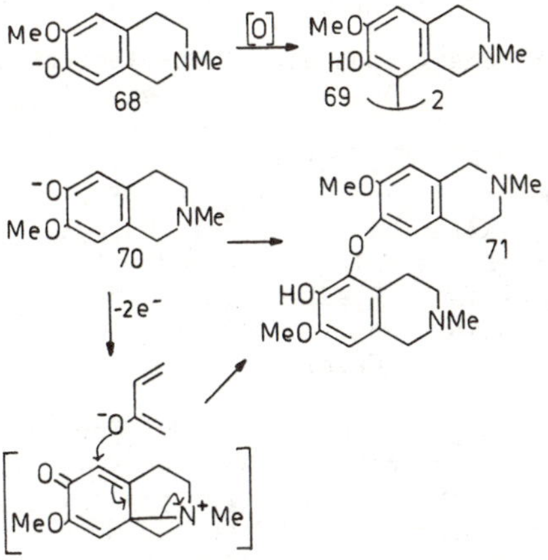

Figure 14. A Concerted Electrooxidation Leading to a Carbon-oxygen Bond

Acknowledgment

I am indebted to Profesor S. Torii of Okayama University in Japan for the superb Symposium of Electroorganic Synthesis that he organized in 1986 and for the proceedings that have been so copiously cited in this chapter. Furthermore, I am indebted to Professor A. J. Fry of Wesleyan University for some pertinent references and especially to my competent and delightful students who have done all of the work cited from our laboratories. Most of our work was supported by various grants from the National Science Foundation and the National Institutes of Health.

Literature Cited

1. Baizer, M. M.; Lund, H., Eds. Organic Electrochemistry; 2nd Ed. Marcel Dekker, Inc.: New York, 1983.
2. Weinberg, N. L. Technique of Electroorganic Synthesis; John Wiley & Sons: New York, 1975.
3. Swann, Jr., S.; Alkire, R. C. Bibliography of Electro-Organic Synthesis, 1801-1975; The Electrochemical Society: Princeton, N. J., 1979.

4. Yoshida, K. Electro-Oxidation in Organic Chemistry; The Role of Cation Radicals as Synthetic Intermediates; John Wiley & Sons: New York , 1983.
5. Torii, S. Electroorganic Syntheses, Methods and Applications, Part 1: Oxidations, Kodansha: Tokyo, 1985.
6. Torii, S. Recent Advances in Electroorganic Synthesis-Proceedings of the 1st International Symposium on Electroorganic Synthesis, Oct. 1986; Elsevier, Kodansha: Tokyo, 1987.
7. Shono, T. Tetrahedron 1984, 40, 811.
8. Guengerich, F. P.; MacDonald, T. L. Acc. Chem. Res. 1984, 17, 9.
9. Dryhurst, G. Electrochemistry of Biological Molecules; Academic Press: New York, 1977.
10. Faulkner, L. R. Chem. Eng. News 1984, 69, No. 9, 28.
11. Ruhe, A.; Walder, L.; Scheffold, R. Helv. Chem. Acta 1985, 68, 1301.
12. Scheffold, R.; Ruhe, A.; Walder, L. ref. 6, p. 325.
12a. Coche, L.; Moutet, J.-C. J. Am. Chem. Soc. 1987, 109, 6887.
13. Taylor, W. I.; Battersby, A. R. Eds. Oxidative Coupling of Phenols; Marcel Dekker, Inc.: New York, 1967.
14. Fichter, F. Organische Elektrochemie; Theodor Steinkopff: Dresden and Leipzig, 1942; pp 109-115.
15. Vermillian, Jr. F. J.; Pearl, I. A. J. Electrochem. Soc. 1964, 111, 1392.
16. Shamma, M. The Isoquinoline Alkaloids, Chemistry and Pharmacognosy; Academic Press: New York, 1972.
17. Herbert, R. B. in The Chemistry and Biology of Isoquinoline Alkaloids; Phillipson, J. D.; Roberts, M. F.; Zenk, M. H., Eds.; Springer-Verlag: Berlin, 1985, p 213.
18. Bobbitt, J. M. Heterocycles 1973, 1, 181.
19. Bobbitt, J. M.; Hallcher, R. C. Chem. Comm. 1971, 543.
20. Bobbitt, J. M.; Noguchi, I.; Ware, R. S.; Chiong, K. Ng; Huang, S. J. J. Org. Chem. 1975, 40, 2924.
21. Nishiyama, A.; Eto, H.; Yukimasa, T.; Iguchi, M.; Yamamura, S. Chem. Pharm. Bull. 1983, 31, 2834.
22. Nishiyama, A.; Eto, H.; Yukimasa, T.; Iguchi, M.; Yamamura, S. Chem. Pharm. Bull. 1983, 31, 2845.
23. Nishiyama, A.; Eto, H.; Yukimasa, T.; Iguchi, M.; Yamamura, S. Chem. Pharm. Bull. 1983, 31, 2820.
24. Iguchi, M.; Nishiyama, A.; Ito, H.; Suzuki, E.; Miyabe, Y.; Kato, Y. ref. 6, p. 53.
25. Scott, A. I.; Dodson, P. A.; McCapra, F.; Meyers, M. B.; J. Am. Chem. Soc. 1963, 85, 3702.
26. Shizuri, Y.; Suyama, K.; Yamamura, S. J. Chem. Soc., Chem. Commun. 1986, 63.
27. Miller, L. L.; Stermitz, F. R.; Falck, J. R. J. Am. Chem. Soc. 1973, 95, 2651.
28. Falck, J. R.; Miller, L. L.; Stermitz, F. R. Tetrahedron 1974, 30, 931.
29. Tobinaga, S. Bioorg. Chem. 1975, 4, 110.
30. Kametani, T.; Shishido, K.; Takano, S. J. Heterocyclic Chem. 1975, 12, 305.
31. Sainsbury, M. Tetrahedron 1980, 36, 3327.

32. Palmquist, U.; Nilsson, A.; Parker, V. D.; Ronlan, A. J. Am. Chem. Soc. 1976, 98, 2571.
33. Hofer, J.; Moest, M. Ann. 1902, 323, 284.
34. Bobbitt, J. M.; Cheng, T. J. Org. Chem. 1976, 41, 1978.
35. Coleman, J. P.; Utley, J. H. P.; Weedon, B. C. L. Chem. Commun. 1971, 438.
36. Coleman, J. P.; Eberson, L. Chem. Commun. 1971, 1300.
37. Bobbitt, J. M.; Willis, J. P. J. Org. Chem. 1980, 45, 443.
38. Bobbitt, J. M.; Kulkarni, C. L.; Wiriyachitra, P. Heterocycles 1976, 4, 1645.
39. Coutts, I. G. C.; Hamblin, M. R.; Tinley, E. J.; Bobbitt, J. M. J. Chem. Soc., Perkin I 1979, 2744.
40. Jensen, U.; Schaeffer, H. J. Chem. Ber. 1981, 114, 292.
41. Schaeffer, H. J. Ref. 1f, p. 3.
42. Iwasaki, T.; Horikawa, H.; Matsumoto, K.; Miyoshi, M. Bull. Chem. Soc. Japan 1979, 52, 826.
43. Bobbitt, J. M.; Scola, P. M.; Kulkarni, C. L.; DeNicola, Jr., A. J.; Chou, T.-t. Heterocycles 1986, 24, 669.
44. Bobbitt, J. M.; Chou, T.-t.; Leipert, T. K. Heterocycles 1986, 24, 687.
45. Lakshmikantham, M. V.; Mitchell, M. J.; Cava, M. P. Heterocycles 1978, 9, 1009.
46. Cordell, G. A. Introduction to Alkaloids, A Biogenetic Approach; John Wiley & Sons, Inc.: New York 1981, p. 590.
47. Cheek, G. T.; Nelson, R. F. J. Org. Chem. 1978, 43, 1230.
48. Cheek, G. T.; Nelson, R. F.; Bobbitt, J. M.; Scola, P. M. J. Org. Chem. 1987, 52, 5277.
49. Bobbitt, J. M.; Kulkarni, C. L.; Willis, J. P. Heterocycles 1981, 15, 495.
50. Rusling, J. F.; Scheer, B. J.; Owlia, A.; Chou, T.-t.; Bobbitt, J. M. J. Electroanal. Chem. 1984, 178, 129.
51. Kulkarni, C. L.; Scheer, B. J.; Rusling, J. F. J. Electroanal. Chem. 1982, 140, 57.
52. Allen, M. J.; Powell, V. J. J. Electrochem. Soc. 1958, 105, 541.
53. Bobbitt, J. M.; Scola, P. M. unpublished results.
54. Yoshikawa, M.; Kitagawa, I. ref. 6, p 97.
55. Becker, J. Y.; Byrd, L. R.; Miller, L. L.; So, Y.-H. J. Am. Chem. Soc. 1975, 97, 853.
56. Mori, M. ref. 6, p 89.
57. Hembrock, A.; Schaefer, H. J. ref. 6, p 121.
58. Fry, A. J. Synthetic Organic Electrochemistry; Harper & Row: New York, 1972, pp. 48-54.
59. Bobbitt, J. M.; Noguchi, I.; Yagi, H.; Weisgraber, K. H.; J. Org. Chem. 1976, 41, 845.
60. Braun, R. D.; Stock, J. T. Anal. Chem. Acta 1973, 65, 177.
61. Lyle, G. G. J. Org. Chem. 1976, 41, 850.
62. Podolney, S.: M. S. thesis, University of Connecticut, 1979.
63. Inubushi, Y.; Aoyagi, Y.; Matsuo, M. Tetrahedron Letters 1969, 2363.
64. Bobbitt, J. M.; Yagi, H.; Shibuya, S.; Stock, J. T. J. Org. Chem. 1971, 36, 3006.
65. Miller, L. L.; Stermitz, F. R.; Becker, J. Y.; Ramachandran, V. J. Am. Chem. Soc. 1975, 97, 2922.

RECEIVED August 3, 1988

Chapter 15

Oxygen Electrode

History, Design, and Applications

Mary Virginia Orna[1]

Laboratory of Molecular Biophysics, National Institute of Environmental Health Sciences, Research Triangle Park, NC 27709

The measurement of dissolved oxygen levels in aqueous solutions is an important index in medicine, biochemistry, molecular biology, and environmental, sanitation and industrial chemistry. This paper reviews the development of oxygen electrodes which work well under a variety of conditions.

The literature on ion-sensitive electrodes is vast and is growing at an enormous rate, so much so that the development of gas sensors and selective bioelectrode systems comprises only a small portion of this fast-developing field (1). However, the importance of gaseous oxygen and its measurement in numerous biologically-related systems has led to a large body of literature devoted to this very specific application. For example, in my laboratory, we have applied the use of the oxygen electrode to structure-activity correlations of known and potential carcinogens by measuring the kinetic parameters of enzymes reacting with these chemicals (2). This paper addresses the sub-field of oxygen sensors with respect to their early history, principles of measurement, their current design, and applications of the modern generation of these sensors.

There is no doubt that Priestley discovered oxygen in August of 1774, although a work by Klaproth, as cited by Duckworth (3), indicates that an oxygen-like component of the atmosphere was recognized by the Chinese as early as the eighth century. The confirmation of Priestley's discovery was related to the consumption of oxygen by a biological system. The oxygen-biology link has driven the development of methods for measuring oxygen production and consumption. Two distinct methods of measurement have arisen based on quite different principles, *viz.*, manometry and electrochemistry.

The Warburg manometer, the quantitative culmination of the "blood-gas manometer" of Barcroft and Haldane, has been a standard

[1]Current address: Department of Chemistry, College of New Rochelle, New Rochelle, NY 10801

0097–6156/89/0390–0196$06.00/0

piece of apparatus in biology and biochemistry laboratories for many years. The Warburg manometer is based upon measurement of the pressure or volume changes resulting from the evolution or consumption of oxygen by a system, and it has certain advantages over the electrochemical technique in that: (1) it lends itself to multiple simultaneous runs, (2) is insensitive to product accumulation, and (3) is able to cope with high endogenous levels of oxygen (4). However, the sensitivity of the method is fairly low, and the necessary apparatus is complex, fragile and expensive. In addition, the sample flasks are relatively large and cumbersome compared to the micro-chambers that are now available for mini-oxygen electrodes, a technical advance that allows for rapid, accurate determination of oxygen even in severely sample-limited conditions. The three most serious limitations of manometric techniques are: (1) the inability to follow rapid changes in the gas phase; (2) the relatively long time required for equilibration and measurement; and (3) the inability to probe different regions of the same biological system (5).

Because of the limitations listed above, manometry as the method of choice for measurement of oxygen tension has long since been superseded by the use of the oxygen electrode. The electrochemical technique is rapid and sensitive, provides continuous information about oxygen tension changes rather than just average values, lends itself to addition of reagents and substrates during a run, and can be used in a variety of settings *in vitro*, *in vivo* and in the field. Furthermore, the apparatus is simple, relatively inexpensive, and can easily be interfaced with digital data acquisition devices for purposes of data analysis.

History of Oxygen Measurement by Electrochemical Techniques

The fundamental reaction that allows electrochemical measurement of oxygen tension is the 4-electron reduction of O_2

$$O_2(g) + 4H^+ + 4e^- \longrightarrow 2H_2O \qquad (1)$$

at a cathode maintained at a fixed potential sufficiently negative to reduce all of the oxygen that diffuses to its surface, essentially a polarographic technique. Although such an arrangement seems relatively simple, it was over 150 years in development from the first description of the phenomenon of polarization.

Recognition of electrode problems, particularly the problem of electrode polarization (6), links the history of electrochemistry and bioelectric phenomena. In 1801, Gautherot described the existence of a residual current of brief duration that remained in charged silver and platinum wires even after removal of the source of charge. Later, Oersted recognized that this so-called "secondary current" could produce muscular contractions in frogs, but it was not until 1826 that the term "polarization" was applied to the phenomenon by de la Rive. During the course of the remainder of the 19th century, numerous workers wrestled with this phenomenon, particularly since it was a problem in dealing with the physiological effects of electricity. Becquerel, Peltier and Matteuci attributed polarization to the deposition of thin films of gas or foreign materials on the electrodes. The concerted work of de la Rive, Fechner, Schroeder, Ohm, Lenz and Beets finally gave rise to the concept of what is now known as concentration polarization. Further developments in the field of electrochemistry as applied to biological systems were inextricably entwined with the interpretation and, often, the avoidance, of polarization effects.

Around the turn of the century, the electrochemical reduction of oxygen was discovered by Salomon (7) and von Danneel (8) working in Nernst's laboratory, and shortly thereafter by F. Cottrell (9) and Grassi (10). It was later observed that when oxygen was reduced at a platinum cathode, a current-voltage curve with a plateau current proportional to oxygen concentration could be obtained (5). However, 25 years were to elapse before Jaroslav Heyrovsky (11) clarified and developed (12) the basic principles and instrumentation of polarographic analysis, a feat for which he was awarded the Nobel prize in 1959 (13). Another 20 years passed before Heyrovsky's work was translated into a reliable tool for oxygen analysis by Davies and Brink, who were the first to utilize a platinum cathode in tissue studies (14), and yet another 14 years went by before Clark's membrane-covered platinum electrode (15) enabled routine measurement of oxygen tension in a variety of sample types. Despite the fact that the measurement of oxygen tension is essentially a polarographic technique, the use of the dropping mercury electrode never became popular for biochemical measurements, although some instances of its use can be found in the literature (16). The platinum electrode has always been the electrode of choice for biochemical systems. An excellent historical review of the development of this so-called "oxygen cathode" is given by Davies (17).

Basic Principles

Measurement of oxygen availability in tissues or fluids, *i.e.*, the oxygen tension of a system when expressed as pressure of O_2, is influenced by many factors that render absolute measurements extremely difficult. Attention must be paid to such variables as the diffusion coefficient of oxygen in different tissues, the degree of stirring and the possibility of poisoning the electrode surface by a variety of mechanisms (18). The different designs of these electrodes are all attempts to overcome these limitations, but for this very reason are not universally applicable.

Basically, the oxygen cathode consists of a small area of platinum or gold that is polarized to approximately 0.6 V with respect to a reference anode, usually a calomel or a silver-silver chloride half cell; the accompanying circuitry is designed to measure the current passing through this cell. The reaction taking place at the cathode has been proposed to follow several different courses, *viz.*, according to Laitinen and Kolthoff (19), a two-electron reduction,

$$2H^+ + O_2 + 2e^- \longrightarrow H_2O_2 \tag{2}$$

or, according to Davies and Brink (14), a two-step process,

$$2H_2O + O_2 + 2e^- \longrightarrow H_2O_2 + 2OH^- \quad (3)$$

$$H_2O_2 + 2e^- \longrightarrow 2OH^-. \quad (4)$$

While the precise stoichiometry of the electrode reaction is unimportant, it is very important that the reaction be consistent for the duration of any one experiment.

When a potential is initially applied to the oxygen cathode, the concentration of oxygen in its immediate vicinity decreases rapidly, since the rate of reaction at the electrode is much more rapid than the rate of diffusion through the medium. As soon as this concentration gradient has developed, the rate of oxygen reduction is diffusion-dependent, and in this particular situation, Fick's diffusion law can be applied. If we express the quantity of oxygen passing through the diffusion layer in unit time as $d[O_2]/dt$, then

$$d[O_2]/dt = -D\ (dc/dx) \quad (5)$$

where D is the diffusion coefficient of oxygen in aqueous solution and (dc/dt) is the concentration gradient in the diffusion layer. Since the measured current, i, is proportional to the amount of oxygen reaching and being reduced at the cathode, we can rewrite Equation 5 as

$$i = kD\ (dc/dx). \quad (6)$$

However, the magnitude of the concentration gradient is proportional to the amount of oxygen in the medium; therefore, the following generalization can be made (20):

$$i = k'D[O_2]. \quad (7)$$

This equation is the fundamental basis for oxygen electrode measurements since the reduction current is directly proportional to oxygen concentration (strictly speaking, activity), which means, in practical applications, the amount of oxygen reaching the cathode. This equation assumes that the diffusion coefficient, D, is constant for a particular experiment and that the oxygen is transported to the electrode by the process of diffusion alone.

If the applied voltage to the oxygen cathode is varied and the corresponding current produced measured, it is possible to construct current-voltage curves similar to those displayed in Figure 1 (21, 22). The current-voltage relationship varies with the type of electrode employed and the nature of the solution. A normal working applied voltage is that giving the least variation in current with voltage, *i.e.*, a voltage in the plateau region of the curve where the current is limited by diffusion alone. For most systems, an applied voltage of -0.6 V fulfills this condition.

The Design and Modification of Oxygen Electrodes

Numerous designs for the oxygen electrode have been described in the literature. The variety in designs arises from the many applications

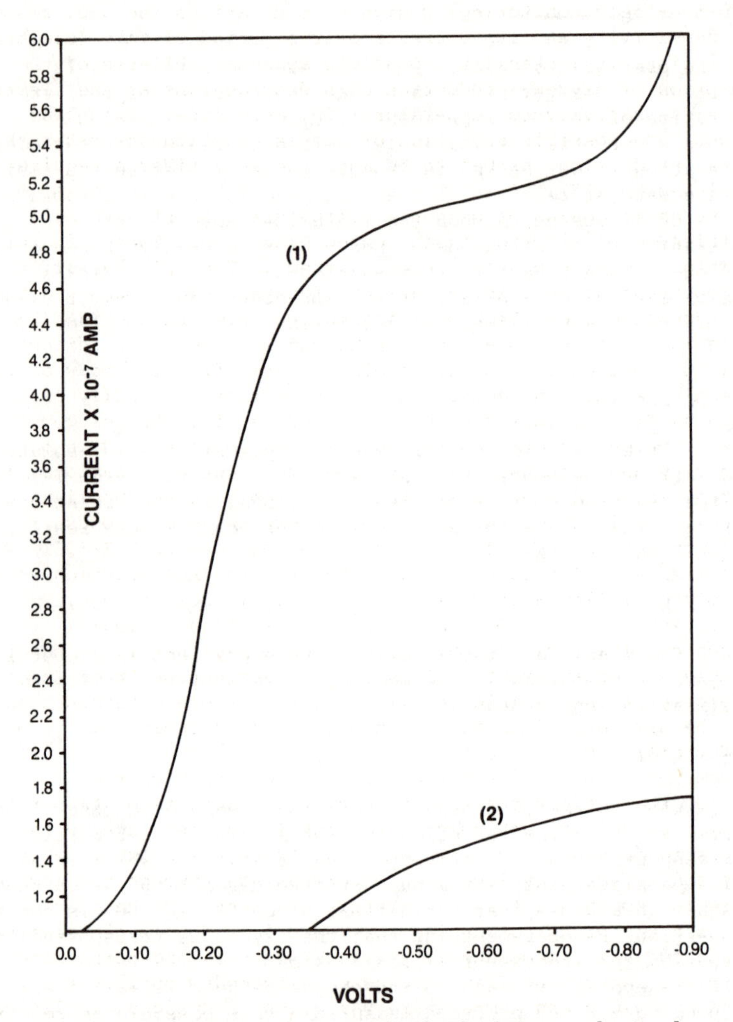

Figure 1. Voltammograms for a typical Clark-type electrode indicating the nature of the plateau at (1) 95.0% oxygen and (2) air saturation.

to which the basic electrode has been put, from homogeneous, controlled parameter experiments to heterogeneous biological media and in-the-field monitoring. Design considerations include: response time; sensitivity and its decrease over a period of time (electrode aging); linearity; stability; possible secondary effects of the consumption of oxygen; production of hydroxide ions at the electrode surface; sensitivity to temperature; pH; mechanical mixing and movement; and possible reduction of sample constituents other than oxygen. In addition, probes to be inserted into tissues require special design (23).

The basic design of oxygen electrodes falls into two main categories, open electrodes and coated electrodes. These, in turn, fall into several subcategories as listed in Table I. By far, the most important advance in the design of oxygen electrodes was the Clark-type electrode (Figure 2) introduced by Clark in 1956 (15). Clark's idea was to isolate both platinum cathode and reference anode, both bathed in a suitable electrolyte, from the body of the solution by a thin, nonconducting membrane that is readily permeable to oxygen (24,25). This design became and remains the design of choice in oxygen electrodes, and continuous improvements in both its specificity and size have made it ideal for most applications. The electrode is now commercially available from several suppliers and the ancillary instrumentation is both inexpensive and robust.

In my laboratory, we have used a Teflon covered YSI-5331 platinum/silver electrode with a half-saturated KCl solution as electrolyte (Yellow Springs Instrument Co., Yellow Springs, OH) inserted into a 1.8 mL incubation chamber (Gilson Medical Electronics, Inc., Middleton, WI) in such a way that it is possible to complete a measurement in less than five minutes. Under these conditions, it is possible to make rapid, multiple determinations under the same experimental conditions, thereby enhancing reproducibility of the data.

Mathieu (26) has given a complete description of the variables that affect the electrode system - effects of temperature on oxygen solubility and porosity of the Teflon membrane, effect of atmospheric pressure and of water salinity on the solubility of oxygen and effect of velocity of water passage through the membrane - and how to compensate for them. Ferris and Kunz (27) have summarized the principal design considerations for oxygen-sensing electrodes, *viz.*, membrane materials, thickness and performance, electrolyte nature and film thickness, electrode configuration, external pressure and temperature variation, and signal amplification requirements, and Ultman and co-workers (28) developed a spherical model that relates the sensitivity of the Clark electrode to cathode radius and electrolyte layer permeability and thickness for two different data sets, one obtained from polypropylene-covered and the other obtained from Teflon-covered Clark electrodes. The material from which the membrane is constructed has been the subject of much study (29-32) with respect to permeability, selectivity and hydrophobicity.

Reports of numerous modifications of the Clark electrode have appeared in the literature since it was first devised, mainly because the traditional Clark cell must be miniaturized for many biochemical and biological applications. However, miniaturization introduces a

Table I. Types of Oxygen Electrodes

Type	Characteristics
A. Open Electrodes	
1. Stationary	Small current; rapid aging Requires little ancillary apparatus Sensitive to external vibrations
2. Moving	Relatively large currents Rapid establishment of steady-state current
a. Rotating	High sensitivity; relatively small response time Decreased sensitivity to external vibrations Very thin diffusion layer
b. Vibrating	Diffusion current independent of vibration frequency above a minimum frequency Relatively small response time
3. Recessed	Utilizes pulsed polarizing voltage Improved stability; insensitivity to convection Possible to obtain absolute O_2 measurements
B. Coated Electrodes	
1. Platinum/Polymer Coat	Decreased sensitivity; increased response time Reduction in aging; decreased movement artifact
2. Clark	Robust; very stable; readily changeable membrane Insulation of electrode circuit from sample Anode & cathode behind same membrane

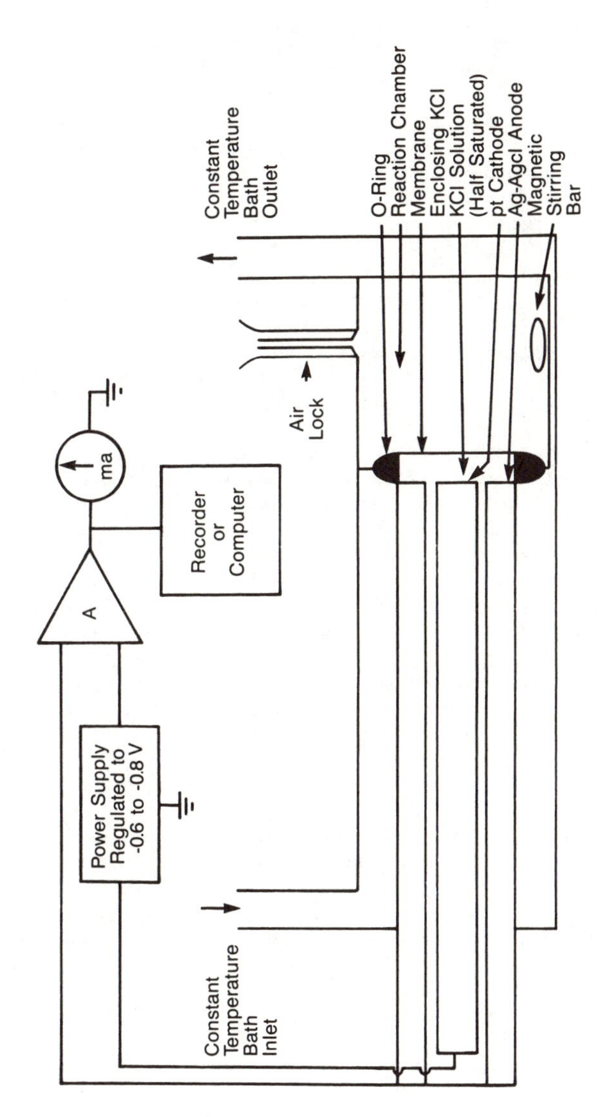

Figure 2. Schematic drawing of a Clark-type oxygen electrode.

number of additional problems, including different electrochemical behavior of microelectrodes and sensor lifetime limitations (33). Fatt (34) has described a modified Clark electrode that gives 99% of final response in 1.24 s for a 155 mmHg change of $p[O_2]$ in the gas phase and also an ultra-micro electrode which shows only a 2% difference in response between a vigorously stirred and an unstirred solution (35). Friese (36) describes a new type of thermo-sterilization resistant Clark electrode in which the platinum cathode is insulated with glass; the electrode compartment between anode and cathode is made of a porous sintered glass, and the electrode body is ground flat and covered with two layers of Teflon. This modification allows measurements to be made in systems where there is a great deal of mechanical and thermal stress.

Other workers have reported improvements in the ancillary circuitry surrounding the electrode, in particular, stable, sensitive amplification (37), and multichannel recording permitting automatic reading and recording of oxygen consumption from as many as six cells simultaneously (38). Modifications for simultaneous spectrophotometric and polarographic measurements have also been described (39,40). Details for the construction of an oxygen polarograph designed for undergraduate use have also appeared in the literature (41).

Mancy and co-workers (42) have described a galvanic cell consisting of a silver-lead couple for determination of oxygen concentrations in fluids. This cell has been used successfully as a completely portable instrument for oxygen monitoring in natural waters and wastes. Mackereth has found long term instability of sensitivity in this and similar devices, as well as a problem with the small current output (on the order of tens of microamperes). He has described an improved galvanic cell (43) which has a current output roughly ten times that of the conventional diffusion-controlled electrode. It is most advantageous for measurements where there is a high degree of pollution; however, it is necessary to continuously renew the membrane surface by techniques such as air bubbling (44).

A recent improvement in sample cell size has been offered by Yellow Springs Instrument Co. in their product literature. The YSI 5356 Micro Oxygen Chamber provides capability of handling samples as small as 600 microliters in either batch mode or in a flow-through mode.

As applications in the use of oxygen electrodes continue to multiply, design modifications and adaptations of existing apparatus can be expected to keep pace.

Performance Characteristics and Behavior

Although the direct relationship between current output and oxygen concentration (activity) as given in Equation 7 holds for most systems under diffusion-controlled conditions, it is admittedly an oversimplification of the situation. Using a first order, one-dimensional two-layer model, and applying Fick's laws of diffusion, Mancy *et al.* (42) derived expressions for the transient current obtained on stepping the dissolved oxygen cell from open circuit to a potential where diffusion-controlled reduction of oxygen takes place at the cathode. They have shown that if one considers the fact that the diffusion coefficients of the electrolyte thin film and the membrane are different, and that the permeability coefficient of the membrane is a non-negligible parameter, the steady state current of the system is given by

$$i = nFAP_m C/b \tag{8}$$

where i = steady state current
n = number of electrons transferred
F = the Faraday
A = cathode area
P_m = membrane permeability coefficient
C = oxygen concentration of the test solution
b = membrane thickness.

This equation is based upon the assumption that only transport in the membrane is important and that the electrolyte layer between the membrane and the cathode is infinitely thin.

Koudelka (33), in treating the time interval between t=0 and attainment of the steady state, has pointed out that between 3 and 20 seconds after the potential step is applied, the sensor follows the Cottrell equation

$$i_t = nFA(D/\pi t)^{1/2}\ C \tag{9}$$

where i_t = initial current
D = membrane diffusion coefficient
t = time.

If this is the case, a plot of i_t vs. $1/(t)^{1/2}$ is linear and $i_t(t)^{1/2}/C$ is a constant.

In practice, the behavior of amperometric oxygen sensors is considerably more complicated than implied by the simple diffusional model. For example, the time for a potentiostatic current to reach a steady value is often much longer than theory would predict, and the value of the steady state current finally achieved depends both on the thickness of the electrolyte layer and on the type of electrolyte in the cell. Hale and Hitchman (45) have examined examples of more complicated behavior, taking into account the transient current variation at switch-on, the effect of electrolyte layer thickness, and the pH of the electrolyte in the steady state situation.

Additional studies on the behavior and performance characteristics of the oxygen electrode include: a three-dimensional mathematical modeling (46); the effect of the accumulation of cathodic reaction products on probe linearity; signal overshooting; tailing and hysteresis of transient characteristics for step changes in the concentration of oxygen for Clark-type probes (47); error produced by finite response time in rate determinations of oxygen consumption (48); and variations in responses of oxygen electrodes to variables in construction and use (49).

Obviously, this list of performance characteristics is not exhaustive, nor is the associated bibliography. It is merely the intent of this section of the paper to make the reader aware of the simplifying assumptions governing routine use of the oxygen electrode system and some attempts to address these problems.

Calibration

When it is necessary to obtain absolute values for oxygen uptake, the oxygen electrode must be calibrated. The simplest way to do this is to equilibrate the solution under study with air and make use of the solubility of oxygen in conjunction with its known partial pressure in air. The difficulty with this method is that it is limited to aqueous solutions and it is difficult to take such variables as pressure, temperature and ionic strength into account. In addition, any absolute measure of oxygen concentration must include corrections for the fact that activity, not concentration, of oxygen is the parameter being measured, and that the presence of both electrolytes and nonelectrolytes influences the solubility and the activity coefficient of oxygen in aqueous solutions.

A number of calibration methods that address these problems are in the literature. The classic method of Winkler (50) is extremely tedious for routine use and is no longer used. The method that took its place, that of Estabrook and Mackler (51), measures the oxygen consumed in the stoichiometric oxidation of a specified amount of the reduced form of nicotinamide adenine dinucleotide (NADH) as catalyzed by lysed mitochondria. The main drawback of this method is that it requires the preparation of mitochondria, a procedure which could take up to four hours.

In the past two decades, several methods for calibration of the oxygen electrode, using readily available chemicals and relatively simple reactions, have been proposed. These will be described in the order in which they first appeared in the literature.

The method of Robinson and Cooper (52) utilizes the reduction of N-methylphenazonium methosulfate (PMS) by NADH according to the following scheme:

$$PMS + NADH + H^{+} \longrightarrow PMSH_2 + NAD^{+} \quad (10)$$

$$PMSH_2 + O_2 \longrightarrow PMS + H_2O_2 \quad (11)$$

$$H_2O_2 \xrightarrow{\text{catalase}} H_2O + 1/2\ O_2 \quad (12)$$

giving a net reaction of

$$NADH + H^{+} + 1/2\ O_2 \longrightarrow NAD^{+} + H_2O \quad (13)$$

This method has the advantage of directly determining the activity of oxygen, which is the species monitored, under the same physical conditions as the experiment, thus eliminating errors that can be as high as 25%.

Billiar and co-workers (53) have proposed a calibration method based upon the stoichiometry of the xanthine oxidase reaction:

$$\text{Xanthine} + H_2O + O_2 \xrightarrow{\text{xanthine oxidase}} \text{uric acid} + H_2O_2 \quad (14)$$

The difficulty with this reaction is the possibility of contamination of xanthine oxidase with catalase or the catalytic oxidation of the product hydrogen peroxide to oxygen by trace metals present in the system, thus leading to spuriously low recorded amounts of oxygen consumption. The group reported that they did not observe any of the mechanisms of hydrogen peroxide breakdown operative in the system, but the method leaves other experimenters vulnerable to these problems unless care is taken to check for such contamination.

Holtzman's method (54) circumvents the problems associated with the Billiar method by including catalase in the hypoxanthine-xanthine oxidase reaction chain according to the following scheme:

$$\text{Hypoxanthine} + O_2 \xrightarrow{\text{xanthine oxidase}} \text{uric acid} + 2H_2O_2 \quad (15)$$

$$2H_2O_2 \xrightarrow{\text{catalase}} 2H_2O + O_2 \quad (16)$$

with a net reaction of

$$\text{Hypoxanthine} + O_2 \longrightarrow \text{uric acid} + 2\,H_2O \quad (17)$$

This method is highly advantageous in that it gives highly reproducible results, the reagents are readily available, inexpensive and stable for long durations. In addition, possible metal ion catalysis of hydrogen peroxide decomposition would not interfere with the overall stoichiometry of the system.

Misra and Fridovich (55) have proposed a method of calibration that entirely avoids the use of enzymatic reactions. Their chemical system takes advantage of the fact that phenylhydrazine is rapidly oxidized to phenyldiimide by ferricyanide ion, and that the phenyldiimide formed reacts very rapidly with oxygen to produce phenol and nitrogen gas. The authors were able to demonstrate that the accumulated reaction products do not alter the sensitivity of the method.

Finally, Wise and Naylor (56) reported a calibration procedure that involves the auto-oxidation of duroquinol at pH 9.0 to produce hydrogen peroxide, which subsequently disproportionates to form water and molecular oxygen. The regeneration of oxygen presents the same problem as in the method of Billiar, but the authors claim that fast electrode response times at higher temperatures allow the method to give good agreement with calculated values while at the same time allowing for highly accurate determination of the starting material, duroquinol, since it has a strong absorbance between 260 and 270 nm. This method has been shown to provide accurate values over the temperature range 5-45 $^{\circ}$C.

While the choice of calibration method depends upon the reagents available and the specific application of each worker, it must be emphasized that calibration should not be overlooked since the monitor in itself is relative and must be related to actual oxygen concentrations (activities) in order to render any but the most qualitative oxygen consumption measurements meaningful.

Applications

Applications of the oxygen electrode are many and varied, and to cite all of the references related to these applications would comprise a volume in itself. In addition to the many review articles already cited (4,5,13,17,18,20,21,26), there exist several additional reviews including general references on polarography (57-60). In addition, the Scientific Division of Yellow Springs Instrument Co. has published a booklet, "YSI Oxygen Electrode References," (61) which is available free of charge from the company. Although it is three years old, this is the most comprehensive bibliography available. It is divided into ten parts and covers Bacteria, Yeasts and Molds, Biochemical Assays and Techniques, Cell Fractions, Intact Organisms, Pathology and Toxicology, Pharmacological Studies, Radiation Research, Special Methods and Applications, Tissues and Cells and a list of Reviews. It is briefly annotated, which is a great advantage, but the list is limited to those applications using the YSI Model 53 (now 5300) Biological Oxygen Monitor or some modification thereof.

In addition to the applications covered in (61), this author has found that kinetics research was done with the oxygen monitor as early as 1938 (62), and reaction mechanisms studies (63) are also susceptible of measurement. Stoichiometry (64-67), monitoring the levels of other gases (68-70), pollution control (71) and biochemical assays (72-76) are also areas of use.

It is hoped that this brief review will help to situate the oxygen electrode among other electroanalytical methods as an important tool in biochemical research.

Literature Cited

1. Arnold, M.A.; Meyerhoff, M.E. Anal. Chem. 1984, 56, 20R-48R.
2. Orna, M.V.; Mason, R.P. In preparation.
3. Duckworth, C.W. Chemical News, 1886, 53, 250.
4. Trudgill, P.W. In CRC Handbook of Methods for Oxygen Radical Research; Greenwald, R.A., Ed.; CRC Press, Inc.: Boca Raton, FL, 1985; pp. 329-42.
5. Lessler, M.A.; Brierley, G.P. In Methods of Biochemical Analysis; Glick, D., Ed.; Wiley-Interscience: New York, 1969; Vol. 17, pp. 1-29.
6. Feder, W. Ann. N.Y. Acad. Sci. 1968, 148, 3-4.
7. Salomon, E. Z. Physik. Chem. 1897, 24, 55.
8. von Danneel, H. Z. Elektrochem. 1897-98, 4, 227.
9. Cottrell, F.G. Z. Physik. Chem. 1903, 42, 385.
10. Grassi, U. Z. Physik. Chem. 1903, 44, 460.
11. Heyrovsky, J. Trans. Far. Soc. 1923-24, 19, 785.
12. Heyrovsky, J.; Shikata, M. Rec. Trav. Chim. 1925, 44, 496-8.
13. Lubbers, D.W. Int. Anaesthes. Clinics 1966, 4, 103-27.
14. Davies, P.W.; Brink, Jr., F. Rev. Sci. Instr. 1942, 13, 524.
15. Clark, L.C. Trans. Soc. for Art. Int. Organs 1956, 2, 41-57.
16. Petering, H.G.; Daniels, F. J. Am. Chem. Soc. 1938, 60, 2796-2802.
17. Davies, P.W. In Physical Techniques in Biological Research; Nastuk, W.H., Ed.; Academic Press: New York, 1962; Vol. 4, pp. 137-79.
18. Silver, I.A. Int. Anaesthes. Clinics 1966, 4, 135-53.
19. Laitinen, H.A.; Kolthoff, I.M. J. Phys. Chem. 1941, 45, 1061.

20. Beechey, R.B.; Ribbons, D.W. In Methods in Microbiology; Norris, J.R.; Ribbons, D.W., Eds.; Academic Press, New York, 1972; Vol. 68, pp. 25-53.
21. Connelly, C.M. Fed. Proc. 1957, 16, 681-84.
22. Hagihara, B. Biochim. Biophys. Acta 1961, 46, 134-42.
23. Connelly, C.M.; Bronk, D.W.; Brink, Jr., F. Rev. Sci. Instr. 1953, 24, 683-95.
24. Clark, Jr., L.C.; Wolf, R.; Granger, D.; Taylor, Z. J. Appl. Physiol. 1953, 6, 189-93.
25. Clark, Jr., L.C.; Lyons, C. Ann. N.Y. Acad. Sci. 1962, 102, 29-45.
26. Mathieu, J.L. Eau et Industrie 1980, 48, 99-112.
27. Ferris, C.D.; Kunz, D.N. Biomed. Sci. Instrum. 1981, 17, 103-8.
28. Ultman, J.S.; Firouztale, E.; Skerpon, M.J. J. Electroanal. Chem. 1981, 127, 59-66.
29. Severinghaus, J.W.; Bradley, A.F. J. Appl. Physiol. 1958, 13, 515-20.
30. Refojo, M.F.; Yasuda, H. J. Appl. Poly. Sci. 1965, 9, 2425-35.
31. Rattner, B.D.; Miller, I.F. J. Biomed. Mat. Res. 1973, 7, 353-67.
32. Michaels, A. J. Poly. Sci. 1961, 50, 393-412.
33. Koudelka, M. Sensors and Actuators 1986, 9, 249-58.
34. Fatt, I. J. Appl. Physiol. 1964, 19, 550-53.
35. Fatt, I. J. Appl. Physiol. 1964, 19, 326-29.
36. Friese, P. J. Electroanal. Chem. Interfacial Electrochem. 1980, 106, 409-12.
37. Carr, L.J.; Hiebert, R.D.; Currie, W.D.;Gregg, C.T. Anal. Biochem. 1971, 41, 492-502.
38. Carr, L.J.; Larkins, J.H.; Gregg, C.T. Anal. Biochem. 1971, 41, 503-509.
39. Ribbons, D.W.; Hewett, A.J.W.; Smith, F.A. In Biotechnology and Bioengineering; Gaden, Jr., E.L., Ed.; John Wiley and Sons, New York, 1968; Vol. X, pp. 238-42.
40. Trudgill, P.W. Anal. Biochem. 1974, 58, 183-89.
41. Reed, K.C. Anal. Biochem. 1972, 50, 206-212.
42. Mancy, K.H.; Okun, D.A.; Reilley, C.N. J. Electroanal. Chem. 1962, 4, 65-92.
43. Mackereth, F.J.H. J. Sci. Instr. 1964, 41, 38-41.
44. Fujimoto, E.; Iwahori, K. Environ. Technol. Lett. 1983, 4, 397-404.
45. Hale, J.M.; Hitchman, M.L. J. Electroanal. Chem. 1980, 107, 281-94.
46. Lemke, K. Internationales Wissenschaftliches Kolloquium - Technische Hochschule Ilmenau 1985, 30, 165-8.
47. Linek, B.; Sinkule, J.; Vacek, V. J. Electroanal. Chem. 1985, 187, 1-30.
48. Ito, S.; Yamamoto, T. Anal. Biochem. 1982, 124, 440-45.
49. Carey, F.G.; Teal, J.M. J. Appl. Physiol. 1965, 20, 1074-77.
50. Winkler, L.W. Z. Anal. Chem. 1914, 53, 665.
51. Estabrook, R.W.; Mackler, B. J. Biol. Chem. 1957, 229, 1091-1103.
52. Robinson, J.; Cooper, J.M. Anal. Biochem. 1970, 33, 390-99.

53. Billiar, R.B.; Knappenberger, M.; Little, B. Anal. Biochem. 1970, 36, 101-4.
54. Holtzman, J.L. Anal. Chem. 1976, 48, 229-30.
55. Misra, H.P.; Fridovich, I. Anal. Biochem. 1976, 70, 632-34.
56. Wise, R.R.; Naylor, A.W. Anal. Biochem. 1985, 146, 260-64.
57. Estabrook, R.W. Meth. Enzymol. 1967, 10, 41-57.
58. Fatt, I. The Polarographic Oxygen Sensor: Its Theory of Operation and Its Application in Biology, Medicine and Technology; CRC Press: Cleveland, 1976, pp. 1-278.
59. Christian, G.D. Adv. Biomed. Eng. Med. Phys. 1971, 4, 95-161.
60. Lessler, M.A. In Meth. Biochem. Anal.; Glick, D., Ed.; Wiley-Interscience: New York, 1982; Vol. 28, pp. 175-99.
61. Yellow Springs Instrument Co., Inc.; YSI Oxygen Electrode References; Yellow Springs, OH, 1985.
62. Blinks, L.R.; Skow, R.K. Proc. Nat. Acad. Sci. U.S. 1938, 24, 420-27.
63. Chappell, J.B. Biochem. J. 1964, 90, 225-37.
64. Saniewski, M.; Plich, H.; Millikan, D.F. Trans. Mo. Acad. Sci. 1973, 7-8, 154-59.
65. Schwartz, M. Biochim. Biophys. Acta 1966, 131, 548-58.
66. Peterson, J.A.; McKenna, E.J.; Estabrook, R.W.; Coon, M.J. Arch. Biochem. Biophys. 1969, 131, 245-52.
67. Spiller, H.; Bookjans, G.; Boger, P. Z. Naturforsch. 1976, 31C, 565-8.
68. Kuchnicki, T.C.; Campbell, N.E.R. Anal. Biochem. 1985, 149, 111-16.
69. Kuchnicki, T.C.; Campbell, N.E.R. Anal. Biochem. 1983, 131, 34-41.
70. Srinivasan, V.S.; Tarcy, G.P. Anal. Chem. 1981, 53, 926-29.
71. Strand, S.E.; Carlson, D.A. J. Water Pollution Control Fed. 1984, 56, 464-67.
72. Chance, B.; Williams, G.R. J. Biol. Chem. 1955, 217, 383-93.
73. Chance, B.; Williams, G.R. Nature 1955, 175, 1120-21.
74. Cheng, F.S.; Christian, G.D. Clin. Chim. Acta 1979, 91, 295-301.
75. Del Rio, L.A.; Gomez Ortega, M.; Leal Lopez, A.; Lopez Gorge, J. Anal. Biochem. 1977, 80, 409-15.
76. Pacakova, V.; Stulik, K.; Brabcova, D.; Barthova, J. Anal. Chim. Acta 1984, 159, 71-79.

RECEIVED August 9, 1988

Chapter 16

Bioelectrochemistry—Before and After

M. J. Allen

Biophysical Laboratories, Department of Chemistry, Virginia Commonwealth University, Richmond, VA 23284 and Department of Physiology, University of Nottingham, United Kingdom

"Bioelectrochemistry" was first used as a descriptive term by the writer (1) in the early 1960s' to categorize his studies on the electrochemical behavior of biological systems. Although the "term" referred to is of recent origin, the subject matter had its birth in GALVANI's laboratory toward the end of the eighteenth century, and was continued by others in a somewhat empirical fashion to the middle of this century. At this time investigations were initiated with the objective of quantifying the electrochemical response of microbial systems to metabolisable substrates. The successful outcome of these studies led to further investigations on erythrocyte behavior. isolated organs, plant tissues and in more recent years to the electronic properties of plant and mammalian membranes.

As far as can be ascertained Luigi Galvani was probably the initiator of electrochemical studies on biological systems. In 1786 he began a series of experiments in which he used a device made of copper and iron to induce muscular contractions in a frog by bringing one metal into contact with a nerve and the other with a muscle. He published his report in 1791, De veribus electricitatis in motu musculari commentarius, in which he attributed his results as due to the electricity generated by the frog's tissues. This assumption was incorrect and was undoubtedly due to his observations of the electric fish which does generate a current.

Volta deduced from Galvani's publication that the current he observed was due to the two dissimilar metals in contact with the frog's tissues. This led to investigations which demonstrated that electrochemical reactions occurred between two dissimilar metals in an appropriate electrolyte.

0097–6156/89/0390–0211$06.75/0

Probably the first application of redox indicator as a method of studying biological phenomena can be attributed to the pioneer efforts of Paul Ehrlich. In 1885 he published reports describing his studies on the different abilities of animal organs to reduce dyes as correlated with the oxygen requirements of the organs (2). Ehrlich's findings initiated an era of tremendous progress which ultimately resulted in the development of staining techniques for microorganisms. This development too led to the period in which numerous investigators studied the redox potentials of various biological and biochemical systems by use of the indicator dyes or by direct measurement with electrodes, on metabolizing animal and microbial cell systems.

Insofar as cellular systems are concerned it is believed that Potter (3) should receive credit for being the first to associate change in electrochemical potential with metabolic activity. He observed that the electrode potential of culture medium containing viable bacteria was more negative than that of a comparable sterile medium. These findings were confirmed and extended by investigators such as Cannan, Clark, Cohen, (4) and others in the 1920's who demonstrated that the potential of the medium decreases or becomes more strongly reducing as bacterial growth proceeds towards the stationary or terminal phase of the growth cycle. Thereafter a large number of investigations were performed on the potentials developed by numerous types of multiplying unicellular systems, as well as on that observed by interaction of pure biochemical systems. This was splendidly reviewed by Hewitt in 1950 (5).

It wasn't until the late 50's that a renewed interest occurred in developing techniques for *in vivo* measurements of electrochemical phenomena (other than in areas related to ion transport through membranes). Increased refinement and the sensitivity necessary for biological studies also evolved.

The interest in the electrochemistry associated with microbial metabolism came about as a result of observations made by Yudkin (6) related to the potentials developed by these systems. He came to the conclusion that the observed potentials were strictly fortuitous and they were not directly related to the activity of the bacteria. Rather, they represented the potentials developed by the secretion products of microbial metabolism.

In view of the demonstrated inadequacy of these potential measurement studies were undertaken related to the determination of another electrochemical parameter, namely, the capacity factor of the system to do work, which appeared to offer a distinct advantage over measurements of potential.

The more recent studies on the electronic properties were stimulated in 1941 by Szent Gyorgyi who suggested that biological systems could be investigated in terms of semiconduction mechanisms. He indicated the possibility that energy in living systems might be transferred by conduction bands (7). Application of this concept to the study of substances of biological origin e.g., proteins has yielded interesting and useful information. The fact that a vast majority of these biochemical substances were studied in a dehydrated state, made it extremely difficult to extrapolate the results to their behavior in a biological system.

Due to the presence of water as part of the organization of the cell in this type of system, we not only have electronic, but also ionic conduction. However, though we were not dealing with a truly solid state, but rather a fluid state, the concepts of solid state physics could be applied to a biological system.

Activity of Microorganisms: Coulokinetic Technique

In the experiments to be described washed *Escherichia coli* suspended in a pH 6.7 0.1 M phosphate buffer was used. Detailed information regarding growth of these bacteria and preparation for use as well as studies on growing systems have been described elsewhere as have the electrochemical cell used in many of the studies and the details of the methods of obtaining the data (8-21).

Early in these studies it was suspected that the currents developed by metabolishing bacterial systems might be due to one of several factors. The generated current could be, in part, the result of the oxidation of a microbial secretion product, or products; the direct transfer of electrons from microorganisms to the electrode surface; or both. The substrate itself was not considered to be responsible for current production, as in the absence of the bacteria essentially no current was produced. In order to define the source of the electrons available to the electrode, experiments were performed in which the bacteria were prevented from coming in contact with the electrode (22). This forestallment was accomplished by enclosing the anode in a dialysis membrane.

The data from the experiments performed with membraned, as compared with non-membraned, anodes are shown in Figure 1. It can be seen from the current *versus* time recordings obtained with membraned and non-membraned anode systems that the current and resultant coulombic outputs are a result of electron transfer through contact of the microorganisms with the electrode surface. Therefore, it appears that this coulokinetic technique permits the direct measure of some biological activities of the microorganisms. It can be assumed that there are electrons present on the wall surface of the microbial cell. During the course of active metabolism, there is a continuous transfer of electrons from the cytoplasm to the cytoplasmic membrane and the cell wall surface. It is also probable that some of the terminal oxidative processes occur predominantly at the cytoplasmic membrane membrane surface. In either case those electrons not captured by the electrode serve to reduce the hydrogen ions, formed at the same time as the electron release, to hydrogen gas. In all probability, the microorganism makes contact with the electrode by the combined process of electrophoretic migration and the motility of the organism itself. It is of interest that the organisms do not produce a significant current (curve 5, Figure 1). Therefore, the coulombic outputs obtained with a useable substrate were due completely to exogenous metabolism.

As the electron transfer appeared to occur *via* microorganism electrode contact, it was suspected that greater transfer potentiality could be achieved if the number of cells in the

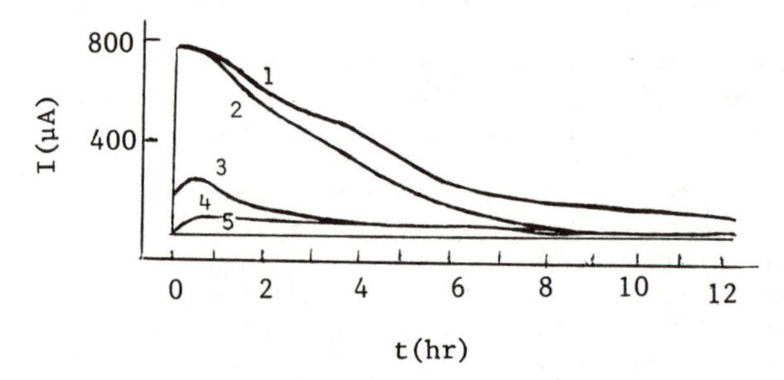

Figure 1. Current-time plots for various membraned and non-membraned anode systems, (1) buffered suspension of *E. coli* with glucose substrate, (2) as 1 with formate substrate, (3) (inside membrane) buffered formate solution, (outside membrane) buffered *E. coli* suspension with formate substrate, (4) as 3 with glucose substrate, and (5) buffered *E. coli* suspension only. (Reproduced with permission from Ref. 22. Copyright 1966 Pergamon Press.)

solution of substrate was increased. Use of twice the number of cells resulted in approximately a two-fold increase in coulombic output. On this basis, it was anticipated that with agitation of the anolyte, the microorganisms might bombard the electrode surface at a greater rate, and thus possibly increase the current output. The reverse effect was observed, even under relatively mild conditions. Accordingly, it can be concluded that agitation removes the cells or prevents them from coming into intimate contact with the electrode. This finding was of great importance, for it indicated that not only must the microorganism make contact with the electrode, but also that the cells must remain at the electrode for some period in order to adhere to its surface. This assumption was supported by the fact that the electrode surface was coated with bacteria when examined at the termination of an experiment.

A current decrease, as a result of agitation, also eliminated the possibility of the generated current being due to an oxidizable nondialysable polymeric metabolite. If this were the case, by analogy to the behavior of organic and inorganic depolarizers when stirred, an increase in current would be observed (23).

To summarize, it can be said that the current produced by the metabolising microorganisms is a result of their making intimate contact with, and transferring potentially available electrons to, the electrode surface. Therefore, the observed parameters i.e., current and coulombic output, are a direct indication of the activity of the biological cell.

The metabolic studies involved investigations related to the coulombic outputs obtained from various prime and intermediate substrates common to the glycolytic and monophosphate pathways i.e., glucose, fructose, gluconolactone, pyruvate, and formate (24). According to the available information, glucose was metabolised by *E. coli* predominantly via the glycolytic pathway (25). On this basis, one would expect to obtain the equivalent of two pyruvates for every one glucose molecule. However, if a portion of the glucose formed dihydroxyacetone phosphate from fructose-1,6-diphosphate, which in turn formed glycerol, only that portion of the phosphorylated fructose which formed glyceraldehyde-3-phosphate could be further converted into pyruvate. Furthermore, pyruvate can be reduced to lactate. Under the experimental conditions used we found that lactate, as well as glycerol, was electrochemically inert. These factors would contribute to the possibility of observing electrochemically fewer than two pyruvates. It was expected that fructose could enter the glycolytic pathway *via* fructose-6-phosphate, and that it would then be treated by the microorganisms in a manner similar to that of glucose. Alternatively gluconolactone, by the nature of the shunt pathways would yield the equivalent of two pyruvates, or a greater coulombic output than observed with glucose.

The results shown in Table 1 indicate that the original premise was reasonable in that the coulombic outputs of glucose and fructose were essentially the same as that resulting when pyruvate or formate was used as substrates. It is interesting to note that the observed coulombic output for gluconolactone was approximately twice that of pyruvate.

Table I. Coulombic Outputs Obtained from the Metabolism to Various Substrates by E. coli at pH 6.7

Substrate	Coulombic Output (coulombs)
Glucose	12.6
Fructose	14.4
Gluconolactone	22.0
Pyruvate	11.5
Formate	11.5

In order to verify the behavior of gluconolactone it was felt necessary to examine an intermediate in the monophosphate shunt pathway. The only substrate available was L-arabinose which although not a normal intermediate, can be isomerized and phosphorylated by the bacteria to yield ribulose-5-phosphate (26), one of the intermediates in the shunt pathway. The coulombic output using arabinose yielded 20.8 coulombs as compared to 22.0 coulombs for gluconolactone. This added credence to the assumption that gluconolactone was indeed being metabolized predominatly via the shunt pathway (24).

The addition of a metabolisable substrate to a suspension of bacteria will give a linear rise in current with time. The slope of this straight line differs with the nature of the substrate and is probably dependent on the rate of transport to the site of enzymatic activity. Arabinose, however, appeared to give a curve for the initial current rise with time which was concave upwards, indicating that the measured rate of oxidation was increasing with time. This suggested the possibility that induction phenomena related to the synthesis of arabinoisomerase, D-ribulokinase, or both of these enzymes were being demonstrated.

In order to further evaluate the electrical characteristics of enzyme induction, a system was studied (27) which had been intensively investigated and was clearly defined, namely the β-galactosidase system. As is well known, the production of β-galactosidase can be induced in *E. coli* cells by growth in lactose, (28) and repressed in glucose media (29). A comparison of the initial I *versus* t curves observed for induced and uninduced *E. coli* cells demonstrated a distinct difference with lactose as an oxidisable substrate and essentially no difference towards glucose as substrate since each system will have the requisite enzymes necessary for the metabolism of glucose as shown in Figure 2.

In conclusion, it is quite certain now that the measurement of potentials of metabolising microbial systems can only suggest that one system is more strongly reducing than another and that this potential is in all probability related to the products secreted by the bacteria. The coulokinetic technique, however, appeared to offer a more satisfactory alternative to the study of metabolic behavior of microorganisms, and to future studies of animal cells, because it measured the capacity factor of available free energy at the site of metabolic activity. It has been demonstrated that this electrochemical technique can serve as another tool for the

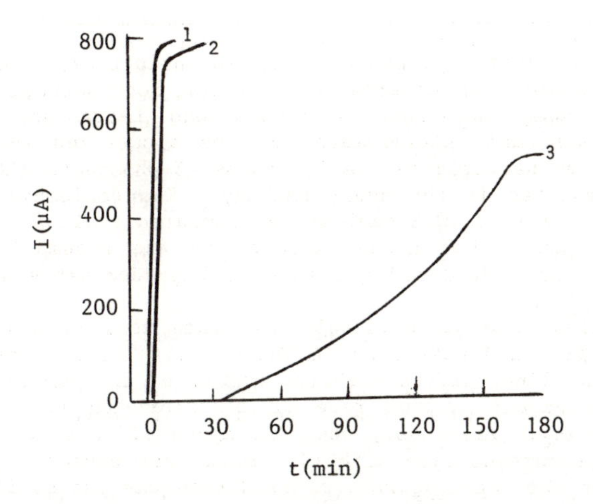

Figure 2. Response of lactose grown E. coli to glucose and lactose (1) and glucose grown E. coli to glucose (2) and lactose (3). (Reproduced with permission from Ref. 27. Copyright 1967 North Holland Publishing Co.)

investigation of the pathways taken by a biological system in the metabolism of substrates under both normal and abnormal physiological conditions. Its usefulness has also been shown in the study of induction phenomena.

Extension of the studies with microbial systems demonstrated that under conditions of externally imposed potentiostatic control, the rate of substrate utilization by a suspension of microorganisms could be controlled, and the growth behavior of an inoculum in a suitable medium altered to give a more rapid growth induction followed by increased yields of microorganisms (30).

In the cultured growth of aquatic biosystems i.e., bacteria, algae, etc., the organisms pollute their own aquasphere by the processes of living and dying. This pollution process undoubtedly disturbs the normal redox systems operating within the biosystem. As a result there will be a diminished response to the normal reproductive stimuli, rate of growth, and maintenance of a normal *status quo* in the fully grown organism. Therefore, it is reasonable to envisage the possible advantages of imposed potentiostasis of appropriate magnitude on the metabolic behavior of useful biosystems. In this manner the growth of these systems can be enhanced, yielding more of the desired product for the same quantity of *food* consumed.

Electrochemical Studies on Plant Photosystems

The investigation of the plant photosystems was a natural extension of earlier studies with the objective of determining the possibility of obtaining a direct conversion of radiant energy into some form of an electrical output. With this information the determination of the effects of various inhibitors and potentiators on the charge transfer processes occurring within these photosystems could be established (31).

As our interests were primarily with the photosystems, the thylakoid membrane components of the chloroplasts were chosen for the investigations to be described. The principal function of the thylakoid membrane system is to use light energy to decompose water. Free oxygen is evolved and the electrons are transported via a complex carrier system to form useable reducing equivalents ($NADPH_2$). Some of the free energy lost during this transport is conserved in the form of ATP (32). The methods for isolation of the chloroplasts and thylakoids and the electrochemical cell used in many of the studies have been described previously (31).

The working electrode and counter electrode compartments of the electrochemical cell contained 15 cm^3 and 35 cm^3 respectively of 0.1M KH_2PO_4-K_2HPO_4 pH 6.8 buffer. To the working electrode compartment we added 0.2 cm^3 of a thylakoid suspension equivalent to 200 μg of chlorophyll. After equilibration in the dark to the circulating water temperature (20°C) the potential U_R of the platinum working electrode *versus* the S.C.E. was determined and this null potential (U_R) was imposed upon the cell with a potentiostat (U_C). The working electrode compartment was exposed to light using a cut-off filter (WRATTEN No. 29) which transmits light predominantly greater than 635 nm the most effective region for activation of photosystems I and II. A typical I versus t curve for one light-dark period of 10 and 30 min. respectively is shown in Figure 3.

Coulombic outputs (millicoulombs) obtained by integration of the areas under the I versus t curves of consecutive light periods for one representative experiment are given in Table II.

Table II. Coulombic Outputs from a Thylakoid Suspension on Consecutive Light Pulses

Light Pulse (10 min)	mC/Pulse	% of Pulse I
1	9.73	--
2	8.16	83.9
3	6.56	67.4

Although it has been found that there is considerable variation in coulombic outputs of the thylakoids from day to day depending on seasonal conditions e.g., 8.5 to 11.65 mC in the summer to 5.5 to 7.5 mC during less favorable periods, the percentage coulombic output for pulses 2 and 3 as compared with that for pulse 1 remains essentially constant (82-86% for 2 and 67-73% for 3).

The compounds o-phenanthrolene (1 μMol.) (33) and 2-chloro-4-cyanoisopropylamino-6-ethyl-amino-1,3,5-triazine (34) (CCET) are known to inhibit electron flow from water to the reducing side of photosystem II. Addition of I μMol o-phenanthrolene and 6.4 μMol CCET to thylakoid suspensions resulted in a decrease in charge 41.0 and 0.5% respectively. Their precise sites of action are not known and are probably not identical. When the thylakoids were exposed to 55° for 5 minutes their ability to split water is selectively destroyed leaving the remaining electron transfer system intact (35). In this instance essentially no activity was obtained.

To further define the site, or sites, in the thylakoid photosystems which contribute the electrons to the electrode, we investigated the effect of polylysine on an illuminated thylakoid suspension under null potential conditions. Polylysines are effective inhibitors of photosystem I activity (36). Polylysine (0.3 mg MW 2600) was added to 15 cm^3 buffered thylakoid suspension immediately after the first light pulse. The coulombic output from the second light pulse was 35% of pulse 1 compared with 83% from an untreated suspension.

These observations with the inhibitory agents implied that the prime source of electrons was the decomposition of water and that electrons ultimately detected by the electrode came from the reducing side of photosystem I.

The general procedure followed in a blind test of the validity of the method involved exposure of a buffered thylakoid suspension (15 cm^3 0.1 M KH_2PO_4-K_2HPO_4, pH 6.8 + 0.2 cm^3 thylakoid concentrate equivalent to 200 μg chlorophyll) to two successive light pulses of 1 minute each separated by a 15 minute dark period. In instances where a chemical compound was evaluated, 10 mm^3 of a (0.03 μmol)

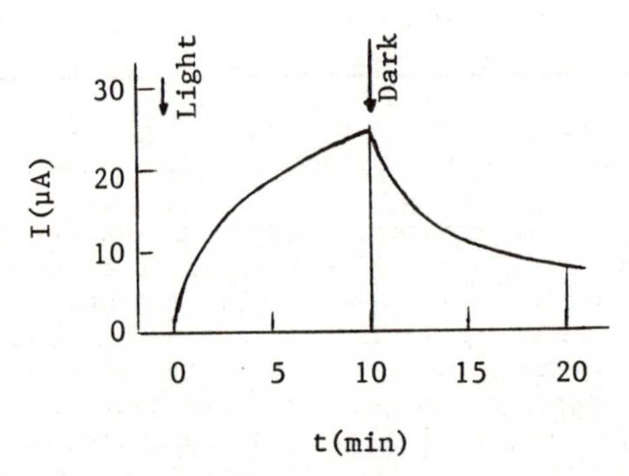

Figure 3. Current response of irradiated thylakoid suspension. (Reproduced with permission from Ref. 31. Copyright 1974 Birkhauser Verlag.)

solution was added immediately after the first light pulse. The results are shown in Table III.

Table III: Blind Study of Agents Inhibiting Electron Flow in Plan Photosystems

	Coulombic Output % Pulse 1	Corrected for % Decrease of Control	% Inactivation	Activity Fraction CCET
Control	95.4	--	--	--
CCET	29.1	80.5	69.5	1.0
BS6	87.1	91.2	8.8	0.1
BS7	66.9	70.1	29.9	0.4
BS8	21.1	22.1	77.9	1.1
BS9	79.1	82.9	17.1	0.2
BS10	86.9	91.0	9.0	0.1

The electrochemical results obtained were in agreement with those obtained by evaluation of total plant behavior under greenhouse conditions.

During the course of earlier studies it was found that imposing more positive or less positive conditions by potentiostatic means prior to illumination under null potential conditions resulted in the measurement of significant alterations of the parameters. Similar abnormal effects were noted when the thylakoid system was examined under oxygen-saturated and anaerobic conditions (31). These results suggested that under circumstances wherein more positive conditions exist, the photosensitizers might in their more oxidised state--being hyperactivated receptors--increase the rate of water decomposition, resulting in concurrent increase in electron release. Conditions less positive than the null state result in an opposite effect.

Results of investigations using catalase as a radical scavenger, suggested the formation of •OH (ads) as a product of the oxidation of water by the irradiated thylakoids. This conclusion was verified by irradiation of a thylakoids suspension containing allyl alcohol. At the end of the irradiation period glycerol was detected by gas chromatography (37).

Polarography of Red Blood Cell Suspensions

The purpose of this study was to determine if any electrochemical parameters could be related to metabolic activity, to correlate these parameters with the age of stored RBC in acid citrate dextrose medium (ACD), and finally, to correlate the behavior of normal and diseased state cells.

A Tast-polarographic system was used (Radiometer PO4 Polariter and a E65 mercury drop electrode) with a 1 s drop time with a 0.5 s drop life. The collection and preparation of the RBC for a Tast-polarography as well as the experimental details and findings have been reported (38). A summary of this work is presented here.

Results of initial voltammetric studies on washed suspensions of intact human RBC in a pH 7.4 isotonic phosphate buffer suggested that at a dropping mercury electrode (D.M.E.) there was an interaction between the electrode and the sulfhydryl groups associated with the membrane. This conclusion was based on the findings that the observed wave ($U_{1/2}$ = -0.220 V *versus* S.C.E.) was not present when using a rotating platinum or glassy carbon electrode, or with the D.M.E. after exposure of the RBC to a sulfhydryl blocking agent (i.e., N-ethyl-maleimide). The presence of reduced sulfhydryl groups in RBC is essential for the maintenance of the integrity of the RBC membrane and the numerous enzyme systems associated with the cell, as well as the integrity of the haemoglobin molecule, and the normal concentrations of methaemoglobin.

Reduction of nicotinamide-adenine-dinucleotide (NAD) to NADH and of NAD-phosphate (NADP) to NADPH occurs as a result of the metabolism of glucose *via* the glycolytic and pentose phosphate pathways. The availability of NADH and NADPH is responsible for maintaining the sulfhydryl groups in their reduced form. On this basis I investigated the effect of the metabolism of glucose by the RBC under sterile aerobic conditions for 18 h at 37°C. A comparison of the polarographic waves obtained before and after incubation in this medium with those obtained in the absence of glucose is shown in Figure 4.

The difference between the polarographic limiting currents before and after incubation is given the symbol ΔId and represents the change in the quantity of sulfhydryl groups detected by the measured polarographic limiting currents.

From a comparison of the two sets of curves the control shows a small negative value of ΔId after incubation, whereas the RBC in the presence of glucose shows a significant positive value of ΔId.

The RBC performs its many functions optimally at pH 7.4. Therefore, it was not unexpected to find that a negative value of ΔI_d was obtained when washed RBC were incubated in a pH 7.0 phosphate buffered glucose medium. Similar behavior was noted with a haemolysed sample of RBC in a pH 7.4 buffered glucose solution. Its non-haemolysed counterpart gave the usual positive value of ΔId. Although there is generally no significant deterioration of individual enzymes as a result of haemolysis, there is a destruction of the *enzymatic transmission network*. As a result, the normal sequence of metabolic events cannot occur and a negative value of ΔId might be expected.

The association of the +ΔId with active metabolism was further confirmed by experiments in which the glucose substrate was replaced by fructose (as fructose-6-phosphate) and gluconolactone (as 6-phosphogluconolactone) which are intermediates in the glycolytic and pentose phosphate pathways respectively. Pyruvate was also incorporated into this series to determine what effect this intermediate, which precedes the formation of the terminal product of metabolism, namely lactate, would have on the ΔId value. It was expected that pyruvate, because of its position in the metabolic scheme, would show either no effect or manifest an adverse effect on the metabolic behavior of the RBC. The adverse

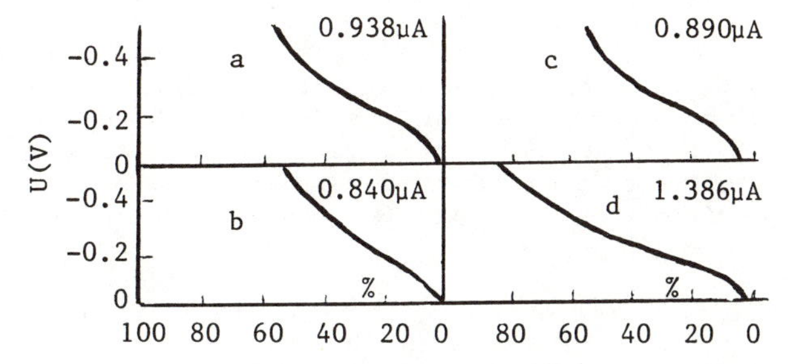

Figure 4. Comparison between washed RBC from single source donor, (a) in pH 7.4 buffer, (b) after incubation 18 hr 37°C, (c) in buffer containing 0.04 M glucose, and (d) after incubation. (Reproduced with permission from Ref. 38. Copyright 1971 Coll. Czech. Chem. Commun.)

effect might occur owing to the absence of sufficient NADH to accomplish the reduction of pyruvate to lactate. Normally the production of NADH accompanies the oxidation of glyceraldehyde-3-phosphate to 1,3-phosphoglyceric acid obtained as a result of the metabolism of a substrate such as glucose by either the glycolytic or the pentose phosphate pathway. As a consequence of the NADH deficiency, the sulfhydryl groups could conceivably serve as electron donors to accomplish the requisite reduction. The percentage change in ΔId of RBC after incubation with glucose, fructose, gluconolactone, and pyruvate were found to be 188, 158, 123, and 79% respectively of a substrateless control. The $U_{\frac{1}{2}}$ values obtained from these polarographic waves were within the range mentioned earlier. This suggested that regardless of the substrate used, a particular sulfhydryl species was apparently being quantitated in all instances. The low value obtained with pyruvate indicated that this substrate does have an adverse effect on the RBC.

The metabolic activity of red blood cells from stored whole blood was evaluated and found to deteriorate rapidly for the first 10 days and then more slowly for the balance of the normally accepted useful life of the stored pack. Application of the technique to studies on patients with various metabolic disorders also indicated significant differences from the norm as shown in Table IV.

Table IV. Correlation of RBC Activity in Normal and Pathological State

Condition	No. of Patients	ΔId(μA)
Normal	18	+0.315 to +0.422
Folate deficiency	10	+0.012 to +0.139
Muscular dystrophy	4 (Age 6-9 yr.)	+0.345 to +0.370
	6 (Age 11-14 yr.)	+0.244 to +0.316
	5 (Age 20-23 yr.)	+0.156 to +0.203

Isolated Heart Studies

These investigations were carried out in the early days of heart transplantation in humans. It was then thought that one of many problems to be solved was the determination that a metabolically viable organ was available for a successful, long-term transplant.

Rabbit hearts were used in these studies. After rapid removal from the animal and appropriate preparation, the organ was transferred to a modifier Lindburgh perfusion pump (39). The pump modification consisted primarily of adapting the organ chamber to permit introduction of a plantinum-saturated calomel (S.C.E.), or any other suitable electrode system for the measurement of the potential of the efflux medium, and the various electrodes required to obtain electrocardiographic and frequency of ventricular beat data. In addition, a heat exchanger was inserted as part of the ascending connection from the medium reservoir to the organ chamber to permit maintaining predetermined perfusion temperatures.

All normal hearts used demonstrated ventricular contraction prior to introduction into the pump. As soon as perfusion started with the NCTC 135-new born calf serum medium, it immediately induced strong rapid ventricular contractions (130 to 160 beats per minute).

The perfusion fluid was introduced into the aorta at 37°C with pressures of 120/80 mm Hg and a pulse rate of 60 per minute. Prior to passage through the heart the medium was aerated with a mixture of 95% air and 5% CO_2 in the course of pump operation. The purpose of the CO_2 was to maintain the pH of the medium essentially constant at 7.4. Relevant potential _versus_ time data for the efflux perfusion medium were obtained with a suitable recorder. Electrocardiographic data and frequency of heart beat were obtained with a A103 amplifier and A107A cardiotachometer (Lexington Instruments, Corp., Waltham, Mass.).

Plots of the potential _versus_ time recordings demonstrated an abrupt change in potential from approximately -122 mV to -69 mV _versus_ S.C.E. starting at about 230 to 360 min and continue to beat for an additional 270-380 min. The characteristics of the ECG QRS complex changes significantly within 30-60 min after the rapid potential decrease. In the absence of an organ the medium demonstrated a potential of approximately -50 mV _versus_ S.C.E. for the same period. Therefore, the initial higher negative potential of the perfusing organ suggested that the heart was oxidizing various components of the medium e.g., glucose, which resulted in electron contribution to the platinum anode of the Pt-S.C.E. electrode system. The abrupt change in potential suggested that at some point in the life of the perfused organ there is a drastic change in the metabolic activity. At this point only 12% of the total quantity of glucose initially present in the medium had been used. As the medium is quite complex it may be that one or more critical constituents were completely used up, which limited the normal metabolic behavior of the organ, or that the enzyme systems required to accomplish the oxidative processes have deteriorated.

In order to determine what effect a damaged heart would have on the observed efflux potential, experiments were performed using hearts which had been subjected to anoxic conditions for 10 minutes at 25°C prior to perfusion. The potential _versus_ time recordings in these instances demonstrated an abrupt potential change from about -130 mV to -100 mV _versus_ S.C.E. at 150-170 minutes and was followed within a matter of 5 to 10 minutes by a very rapid deterioration in amplitude and heart beat frequency.

Electronic Properties of Biological Membranes

A. Plant Systems. A system of null potential voltammetry was developed for investigating the effects of biocides and surfactants on the transport properties of leaf epidermal membranes. These membranes were obtained by a procedure which was found suitable for a wide variety of plant species. A residual translucent membrane composed of epidermal cells and the associated cuticular surface was exposed by gently rubbing the abaxial surface of the leaf with a moistened cotton-tipped applicator until all of the green material was removed. In these studies advantage was taken of the electrical activity of irradiated plant thylakoids by using it as an indicator (40). As a result of this work it was decided to attempt an approach which might bring to light some of the electronic properties of "normal" and abnormal membrane systems under various conditions of physical and chemical stress.

In the initial studies with plant membranes, Arrhenius plots of conductance vs 1/T at 20-50°C demonstrated discontinuities which were consistent with a change in the activities of transport proteins and the lipid constituents (41). The behavior of these membranes toward various biocides also brought to light many interesting phenomena, especially those related to the action of Paraquat and Diquat, two extremely active bipyridyl herbicides (42). The effects of pressure on these membranes resulted in potentiation in charge transfer which was attributed to a compression of intermolecular distances and energy barriers. An equivalent circuit was also developed to describe the observations (43). Investigation of the effects of low frequency electronic excitation on the discharge related behavior which varied with the frequency duration of exposure and plant species (44). The results also suggested that the differential responses to positive and negative going voltage pulses might produce genetically related behavioral patterns from leaf membranes of different plant varieties. The detailed performance of experiments has been described (45). The results obtained are shown in Table V.

Table V. Charge-Energy Relations

Plant	∫Q/μc +:-	∫E/μJ +:-	E/Q*
Ia	738:1933 1:2.62	120:202 1:1.68	0.64
Ib	541:1060 1:1.96	89:124 1:1.39	0.70
Ic	646:802 1:1.24	117:93 1:0.79	0.63
Id	692:2508 1:3.62	121:295 1:2.44	0.67
IIa	3386:2582 1:0.76	421:494 1:1.17	1.54
IIb	2033:2153 1:1.06	246:417 1:1.69	1.59
IIc	1169:2243 1:1.92	140:460 1:3.29	1.71
IIIa	661:1440 1:3.55	102:197 1:2.93	0.86
IIIb	665:2365 1:3.55	97:285 1:2.93	0.82
IIIc	581:1016 1:1.74	80:114 1:1.42	0.81

I a B. napus (Pridaora); b (Victor); c (Elvira); d b. rapa (Just right);
II a V. faba (Longpod); b (Aq. claudia); c (Imp. white windsor);
III a P. vulgaris (Canadian wonder); b(Masterpiece); c (Panamanian)
* E/Q from negative ratio values. (Reproduced with permission from Ref. 44.

It is quite apparent from a comparison of ∫Q and ∫E + :- ratios for each of the varieties within each species and within a particular genus, as is the case with Brassica, that significant differences will be noted in relating one variety to another. These differences at the moment, can only be attributed to the genetics of the variety. However, it is likely that these genetic differences between varieties are reflected in differences in protein structure and lipid composition. This may more specifically account for the dissimilar results obtained between varieties within each genus.

The E/Q factor obtained from the negative portion of the E and Q ratios indicates a distinct categorisation with average values of 0.66, 1.61, and 0.83 respectively for each genus studied. This suggests that this factor is related to the similar genetic foundation which exists within each of the genus studied. Additional studies on a number of other types of plant systems appear to verify this assumption.

It is of interest to note that it was found possible by the method described above to differentiate the behavior of an F-1 variety as compared to its parent varieties and similarly their responses to low temperature (46).

B. Mammalian Systems. Investigations of the electronic properties of plant systems served as an excellent entree into the examination of mammalian membranes using the techniques already developed. Because of the writer's long-time interest in muscle physiology, this tissue was chosen for exploration with the objective of obtaining more intimate information regarding muscle behavior.

In these studies, the diaphragm muscle tissue used was obtained from mice. The isolated tissue was equilibrated to Hank's solution, a defined nutrient media, for a minimum of 6 hours at 4°C. It was found that this tissue retained its viability for at least 24 hours stored in this fashion.

A section of the membrane was plated on a platinum electrode previously treated first with 50% aqua regia and then with nitric acid, both maintained at 50°C. The "plated" electrode was mounted on a temperature block consisting of a thermoelectric refrigeration unit in combination with a pulsitile heating unit. This unit is controlled by a Eurotherm Model 818 control temperature indicator, which has a range of -15 to +50°C and is programmable for a sequence of temperatures, charge and dwell periods. The mounted electrode is then enclosed by a Perspex compartment containing a 3 mm diameter opening for exposure of the membrane, and a channel for admission of the reference electrode bridge containing Hank's solution. A small platinum counter electrode is added and is covered by another Perspex plate. The total system is then bolted together as shown in Figure 5.

After equilibrium to temperature and potential vs. S.C.E. (10 min), voltage clamp conditions are imposed with a potentiostat and the muscle membrane is exposed to potential ramps of +300 or -300 mV for each membrane sample from a diaphragm obtained from the same animal. Each sample is exposed to a total of six 5 s ramps intercepted by 5 m recovery periods. The average current and voltage passing between the membrane 'plated' electrode and the counter

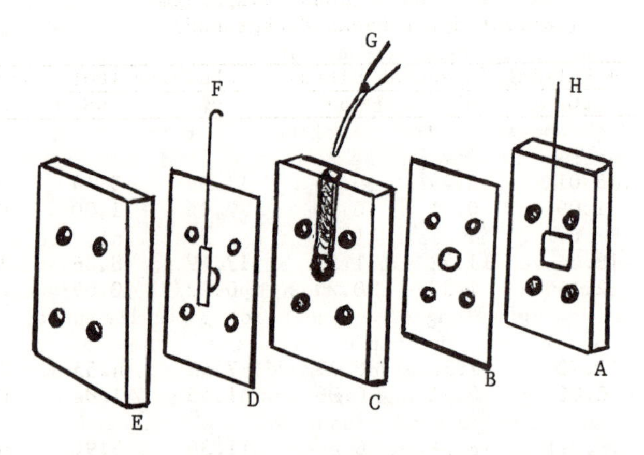

Figure 5. Cell for electronic studies, (A) thermo-electric temperature block 5x7.5 cm, (B) silicone rubber gasket with centrally located 0.3 cm diam. window, (C) Perspex compartment 5x7.5x0.45 cm with 0.3 cm diam. window and 0.3 cm channel for introduction of reference electrode bridge-G, (D) silicone rubber gasket with 0.3 cm diam. window, (E) Perspex cover plate, (F) Pt counter-electrode, and (H) membrane plated Pt electrode.

electrode is used to obtain the various electrical parameters which represent the electro-kinetic behavior (EKB) of the system.

Initially, a series of Arrhenius plots of conductance versus 1/T were obtained which demonstrated transitions at about 15°C and 30°C. These coincided with the change in lipidity of the membrane in the first instance and at the higher temperature, the activation of calcium ATP-ase.

Examination of two different strains of mice, namely the CDI and ICR strains, was made to determine if the genetic differences would become apparent. Both strains were male animals of 35 - 38g in weight. The results obtained are presented in Table VI.

Table VI. EKB of Mouse Diaphragm
(Corrected for Hanks Background)

	+ Pulsing (Electron Sites)			- Pulsing (Hole Sites)		
	μC	μS	μJ	μC	μS	μJ
CDI Strain						
20°	15.01	8.79	0.96	11.54	5.14	0.88
S.D.	1.09	0.51	0.10	2.08	1.00	0.17
37°	19.24	11.22	1.27	17.49	8.36	1.29
S.D.	2.11	1.22	0.20	0.92	0.09	0.43
ICR Strain						
20°	11.72	7.21	0.71	7.49	4.55	0.44
S.D.	0.82	0.31	0.09	1.45	1.02	0.16
37°	13.92	8.74	0.84	11.36	5.94	0.77
S.D.	1.23	1.02	0.14	0.80	0.07	0.32

Normally a preponderance of negative charges would be expected due to the fact that the sarcolemma membrane consists of a superficially located "basement membrane" and a plasma membrane which contains an abundance of negatively charged free or protein bound mucopolysaccharide material with specific affinity for calcium. Negatively charged phosphorylated proteins superficially located in the sarcolemma membrane also act as binding sites for calcium.

It became apparent that it would be of interest to examine a number of the "Group A" calcium antagonists. Compounds in this group inhibit 90 to 100% of the slow inward Ca^{2+} current without concomitant influence on transmembrane Na^{+} conductivity (47). To determine if electronic difference could be observed, the compounds of Group A were divided into two subgroups - those soluble in Hanks or Hanks containing 0.03% ethanol (Group A1, Table VII) and the other soluble only in Hanks containing 1.0% ethanol (Group A2, Table VII). The reason for this division was the observations which suggested that the use of a 1% ethanol had a synergistic effect on the action of these compounds in the experimental system. The structures of the compounds examined together with the results obtained are described in the table on the following page.

Group A1 Calcium Antagonists

Diltiazem

Verapamil

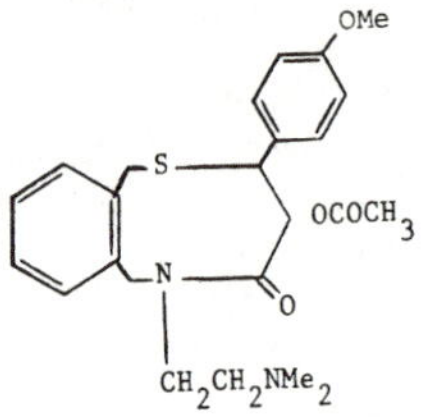

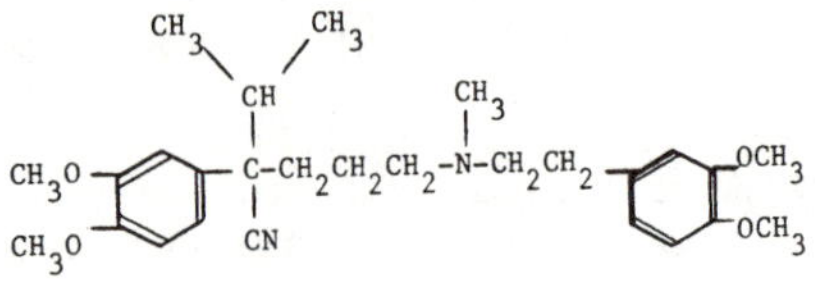

Bepridil

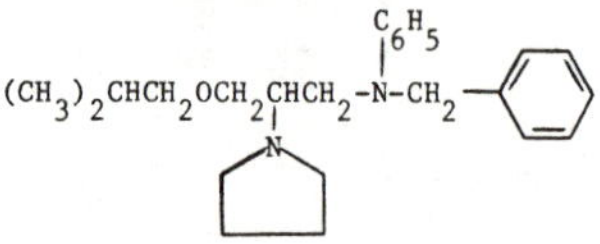

Table VII: Effects of Group A1 Calcium Antagonists EKB-MD + Antagonist vs. EKB-MD Control (CDI Strain)

	Relative Behavior					
	+ Pulsing			- Pulsing		
	μC	μS	μJ	μC	μS	μJ
All Controls at Unity	1.000	1.000	1.000	1.000	1.000	1.000
10^{-5}M Diltiazem in Hanks						
20°	1.225	1.232	1.306	1.154	1.159	1.167
37°	0.498	0.563	0.434	0.632	0.714	0.553
10^{-5}M Verapamil in Hanks -.03% Ethanol						
20°	1.785	1.254	1.513	0.573	0.633	0.516
37°	1.403	1.099	1.000	0.204	0.193	0.198
10^{-5}M Bepridil in Hanks						
20°	1.055	1.170	1.000	1.342	1.423	1.455
37°	0.776	0.842	0.724	1.502	1.240	1.746

Group A2 Calcium Antagonists

Nifedipine

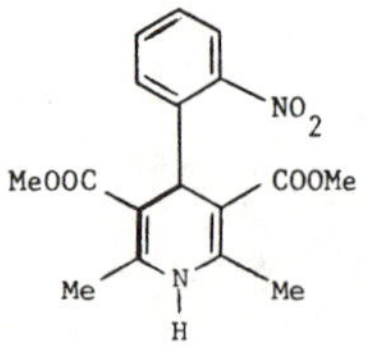

Nimodipine

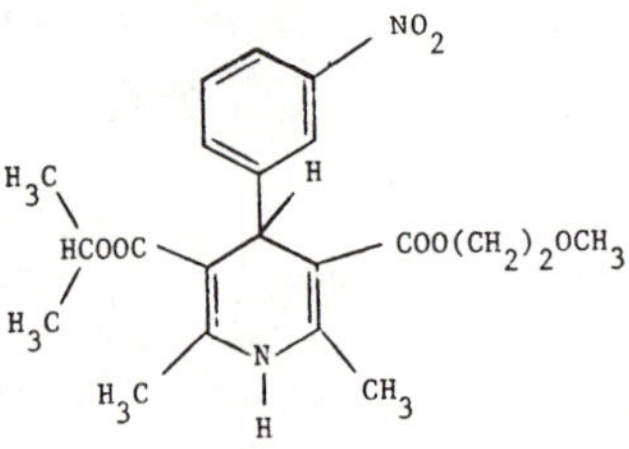

Nitrendipine

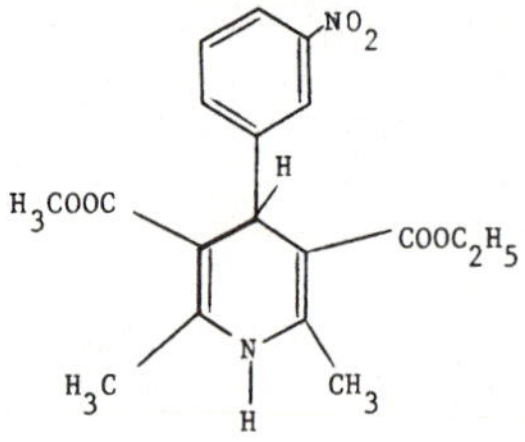

Table VII Continued: Effects of Group A2 Calcium Antagonists
EKB-MD + Antagonist vs. EKB-MD Control
(CDI Strain)

	Relative Behavior					
	+ Pulsing			- Pulsing		
	μC	μS	μJ	μC	μS	μJ
All Controls at Unity	1.000	1.000	1.000	1.000	1.000	1.000
10^{-5}M Nifedipine in Hanks - 1.0% Ethanol						
20°	3.224	2.572	3.883	0.589	0.712	0.970
37°	1.546	1.469	1.622	0.690	0.848	0.963
10^{-5}M Nimodipine in Hanks - 1.0% Ethanol						
20°	0.736	1.016	0.604	0.220	0.687	0.866
37°	1.197	1.140	1.243	0.225	0.616	0.369
10^{-5}M Nitrendipine in Hanks - 1.0% Ethanol						
20°	0.611	0.654	0.571	0.796	0.880	1.088
37°	0.576	0.845	1.027	0.061	0.031	0.891

The behavioral differences noted are undoubtedly due, in part, to the dissimilar chemical structures of the antagonists within each of the groups described. These results suggest, as one possibility, that these agents have varied affinities for the binding sites within the superficial basement membrane and the internal plasma membrane.

One also cannot overlook the fact that these substances may have other physiological effects which contribute to the overall electrical behavior of the membrane. Continuing studies in this area may give a more defined explanation for the differences noted.

Conclusion

More than two decades ago some studies on the electrochemical behavior of microbial systems using some rather novel approaches were presented (48). The author ended his paper with a quotation by Winston Churchill:

> "This is not the end. It is not even the beginning of the end. But it is perhaps the end of the beginning."

Since then the investigations of microbial systems were continued and extended to the electrochemical and electronic behavior of plant and mammalian tissues. Therefore, now one can say that we are really in the midst of it all. Undoubtedly life might sometimes appear simpler if research activities were restricted to isolated biochemical systems incorporated into artificial membranes. Unfortunately it is almost impossible to extrapolate these findings to that which might exist in a living system. Therefore, studies must be done by those who persevere, on the viable organism.

Acknowledgments

The author wishes to thank Gwen Geffert, Deborah Donati, and Jeanne Roberts for their kind assistance in the preparation of this manuscript. The supply of the calcium antagonists by Miles Laboratories and McNeil Pharmaceuticals is greatly appreciated.

Literature Cited

1. Allen, M. J. Bioelectrochemistry; Encyclopedia of Electrochemistry; Hempel, C. A., Ed.; Reinhold Publishing Corp.: New York, 1964; p 102.
2. Ehrlich, P. Das Sauerstoff-Bedurfmis des Organimus. Ein farben-analytisches Studium Cent. Wissensch. 23, Berlin, 113, 1885).
3. Potter, M. C., Proc. R. Soc., Ser. B 84, London, 260, 1911.
4. Cannan, R. K.; Cohen, B.; Clark, W. M. U.S. Public Health Pre. Suppl. 55 1926.
5. Hewitt, L. F. Oxidation-Reduction Potentials in Bacteriology and Biochemistry; Williams and Wilkens Co.: Baltimore, 1950.
6. Yudkin, J. Biochem J. 1935, 29, 1130.
7. Szent-Gyorgyi, A. Nature 1941, 148, London 1941, 157.
8. Allen, M. J.; Bowen, R. J.; Nicholson, N.; Vasta, B. M. Electrochim. Acta 1963, 8, 991.
9. Allen, M. J.; Januszeksi, R. L. Ibid. 1964, 9, 1423.
10. Allen, M. J. Ibid. 1964, 9, 1429.
11. Allen, M. J. Ibid. 1966, 11, 1.
12. Allen, M. J. Ibid. 1966, 11, 15.
13. Allen, M. J. Ibid. 1966, 11, 1503.
14. Allen, M. J. Ibid. 1967, 12, 563.
15. Allen, M. J. Ibid. 1967, 12, 569.
16. Allen, M. J. Bacteriol. Rev. 1966, 30, 80.
17. Charnecki, A.; Allen, M. J. Curr. Mod. Biol. 1968, 1, 325.
18. Allen, M. J.; Charnecki, A; Ibid. 1968, 2, 215.
19. Allen, M. J.; Bellanti, J. A.; Jackson, A. L. Ibid. 1968, 2, 329.
20. Allen, M. J. Electrochim. Acta 1970, 15, 1565.
21. Allen, M. J.; Charnecki, A. Ibid. 1971, 16 1055.
22. Allen, M. J. Ibid. 1966, 11, 7.
23. Allen, M. J. Organic Electrode Processes; Chapman and Hall: London, 1958; pp 15-16.
24. Allen, M. J. Curr. Mod. Biol. 1967, 1, 116.
25. Wang, C. H. et. al., J. Bacteriol. 1958, 76, 207.
26. Linn, R; Cohen, S. J. Biol. Chem. 1966, 241 4304.
27. Allen, M. J. Curr. Mod. Biol. 1967, 1, 177.
28. Monad, J.; Cohen-Bazire, G.; Cohen, M. Biochim. Biophy. Acta 1951, 7 585.
29. Cohn, M.; Horibata, J. J. Bacteriol. 1959, 78, 624.
30. Allen, M. J; Dahloff, J. A. Bioelectrochem. Bioenerg. 1975, 2, 198.

31. Allen, M. J.; Curtiss, J. A.; Kerr, M. W. Bioelectrochem. Bioenerg. 1974, 1, 408.
32. Hill, R.; Bendall, F. Nature 1960, 186, London 1960, 136.
33. Gaffron, H. J. Gen. Physiol. 1945, 28, 269.
34. Kerr, M. W. unpublished results.
35. Hinkson, J. W.; Vernon, L. P. Plant Physiol. 1959, 34, 268.
36. Brand, J., et. al., J. Biol. Chem. 1972, 247, 2814.
37. Allen, M. J.; Dahlhoff, J. Bioelectrochem. Bioenerg. 1975, 2, 177.
38. Allen, M. J. Coll. Czech. Chem. Commun. 1971, 36, 658.
39. Lindbergh, C. A. Exp. Med. 1935, 62, 409.
40. Allen, M. J. Experientia 1980, 36, 1268.
41. Allen, M. J. J. Exp. Botany 1981, 32, 855.
42. Allen, M. J. J. Bioelectricity 1981, 1, 161.
43. Allen, M. J. Studia Biophysica 1982, 90 19.
44. Allen, M. J.; Davies, E. C. Studia Biophysica 1983, 94, 153.
45. Allen, M. J. Charge and Field Effects in Biosystem; Allen, M. J.; Usherwood, P. N. R., Eds.; Abacus Press: G.B., 1984; pp 369-375.
46. Allen, M. J. Water and Ions in Biological Systems; Pullman, A.; Vasilescu, V.; Packer, L., Eds.; Soc. for Medical Sciences: Bucharest, 1985; pp 109-129.
47. Fleckenstein, A. Calcium Antagonism in Heart and Smooth Muscle; Wiley Interscience: New York, 1983; pp 36-41.
48. Allen, M. J. Bacteriological Rev. 1966, 30, 80.

RECEIVED August 3, 1988

ELECTROANALYTICAL CHEMISTRY

Chapter 17

Development of Electrochemical Instrumentation

H. Gunasingham

Department of Chemistry, National University of Singapore, Kent Ridge, Singapore 0511

The development of instrumentation for electrochemistry is separated into three periods; the early history, the electronic age and the computer age. Progress in each period is seen in terms of the impact of technology as well as specific innovations brought about by individuals.

The Early History

The early history of electrochemical instrumentation could be said to have begun shortly after the discovery of electricity by Galvani in 1780 (1). The first distinct record of an electrochemical instrumental technique was probably the use of electrolytic deposition as a qualitative analytical technique by Cruikshank in 1800 (2). In the following years there was much activity on the study and use of electrode potential for identifying a wide range of metals.

Quantitative electrochemical analysis was not introduced, however, until 1864, when electrogravimetry was developed by Wolcott Gibbs (3) and C. Luckow (4). The technique was further studied by a number of workers including Alexander Classen who wrote the first book on the subject (1). By 1900, electrogravimetry was accepted as a standard analytical technique and was routinely used to determine many metals.

Initially, the major limitation of electrogravimetry was the inability to resolve several metals in a single solution. However, it was soon realized that by controlling the potential, metals could be selectively deposited. By the 1920's, controlled potential electrolysis was widely used as an analytical tool. It also found use in preparative work for both organic and inorganic chemicals.

Although it is very accurate, electrogravimetry is a cumbersome technique from a practical standpoint. The alternative approach in analytical determinations is to measure the amount of electricity. However, despite the fact that Faraday's Law was well known

0097–6156/89/0390–0236$06.00/0

coulometric analytical techniques came long after the development of electrogravimetry. An early coulometric method was developed in 1917 by Grower (5). However, coulometry became accepted as a routine analytical technique only in the 1940's, largely because of the work of Somogy (6) Lingane (7) and others.

Electrolysis at controlled current was known as early as 1879 (8). However, its practical application to analytical and physical chemistry dates only from the 1950's (8).

During the early history of electrochemical instrumentation, much of the work was directed more towards understanding the chemistry involved, rather than the development of instrumentation. By the 1920's, however, scientists confronted by the limitations of existing techniques, particularly the fact that they were manual and required considerable skill to operate, sought to exploit the newly available electromechanical technology. One of the first improvements to arise was instruments that could automatically record an experiment. Potentiometric titrations were one of the techniques that benefitted. Although the concept of electrometric titrations was exploited as early as 1893 by Behrend (9) the first instruments to enable the automatic recording of titration curves were electromechanical systems developed in the early 1920's.

Another significant advance that exploited the available electromechanical techonology was the invention of the polarograph, by Heyrovsky and Shikata (10), enabling the automatic recording of polarographic current potential curves. More than any other development, it was the invention of the polarograph and the rapid acceptance of polarography and voltammetry that provided the impetus to modern electroanalytical chemistry.

The theory and application of the polarographic technique was developed over the next decade, largely due to the work of Heyrovsky and co-workers (11). Among refinements to the basic technique were improvements in sensitivity and in recording, including derivative and a.c. oscillopolarographic techniques.

The Electronic Age

The electronic age was marked by the advent of the vacuum tube. This enabled the construction of highly accurate and versatile automatic instruments. Perhaps the first impact was on potentiometric methods, especially to pH measurements using the glass electrode. Although the glass electrode was known as early as 1909 it was only after the development of vacuum tube voltmeters around 1935 that it was used for practical pH measurements (8).

The vacuum tube also heralded the development of the first reliable potentiometric autotitrators which became commercially available aroung 1940. In the recording of titration curves the rate of the addition of the of titrant has to be synchronized with the travel of the recorder chart. Furthermore, in order to minimize end point overshoot, the titration rate must be slowed down as the "end point" is approached (8). Such problems are more readily handled by electronic systems than by electromechanical ones.

The impact of electronics on instrumentation for voltammetry and polarography can be traced along two lines; the generation and control of the perturbing potential waveform and the method of current measurement. In both areas, major developments arose because of specific advances in electronics technology. Table 1 outlines the major developments in electroanalytical techniques along with the concurrent advances in electronics, signal generation and current measurement.

Developments in electrochemical instrumentation originate with Heyrovsky's invention of the polarograph in 1922. The electromechanical nature of the polarograph reflected the state of technology at the time of its development. Following enormous enthusiasm and widespread application during the 30's and 40's, by the early 1950's the trace inorganic analytical uses of the technique were losing out to competing instrumental methods such as atomic absorption spectrophotometry which were more sensitive (12). This resulted in the start of a dormant period for electroanalytical chemistry which lasted until the advent of integrated circuit instrumentation in the late 1960's.

Classical dc polarography suffers from two major defects. First, sensitivity is limited due to contributions from background charging current and second, the electrode itself is mechanically unstable. A further problem is linked to the use of the original, two electrode technique. Due to the ohmic drop between the micro-cathode and the large anode, measurements in poorly conducting, organic, or aprotic solutions were impractical. Although Hickling's invention of the three electrode potentiostat in 1942 (13) as a means of obviating this problem was well known, most commercially available instruments up to the early 1950's employed the two electrode mode. As a result, the area of electrochemistry in non-aqueous solvents was largely neglected.

The advent of low cost operational amplifiers in the late 1950's radically changed this situation, enabling modification of the basic dc polarographic technique to overcome much of the inherent limitations. Perhaps one of the most important consequences was in the construction of simpler and more reliable potentiostats. Figure 1 compares the operational amplifier based circuit for three-electrode potentiostatic control with a two electrode circuit. In the two electrode mode, the effective potential of the working electrode depends upon the electrochemical resistance of the cell system, an effect that is serious when this resistance is not small.

In a potentiostatic system, the current flows between the auxiliary and the working electrodes while the third, or reference, electrode operates under essentially zero current conditions and controls the potential of the working electrode through the feedback arrangement of the operational amplifier circuit.

With continued improvement in the performance of operational amplifiers, operational characteristics of potentiostats have also advanced. The original vacuum tube op-amps were surpassed by the introduction of solid state devices, first discrete transistor based devices and then by integrated circuits. Increasing the level of integration of electronic circuits has resulted in the

Table 1

The Evolution of Instrumentation for Electrochemistry

DATE	TECHNOLOGY	VOLTAGE CONTROL	CURRENT MONITORING
1990	Advanced Software	Cybernetic Instruments with Inferential Capabilities	
	Automated Analysis of Voltammetric Data		
1980	VLSI Technology (microcomputer-based instruments)	Integrated Intelligent Control	
	On-Line Minicomputers	Software Waveform/DAC	ADC/Software Control
1970	Hybrid Analog-Digital Circuits	Logic Driven Pulse Rechniques	Sampled Current
1960	Integrated Circuit	Integrator	i/E Conversion
	Solid State Transistor	Improved Speed and Accuracy	
	Vacuum Tube	Operational Amplifier-Based Instrumentation	
1950			
	Servomechanism	Servomechanism	Galvanometer
1940			
1930			
	Electromechanical	Motor-Driven Potentiometer	Galvanometer
1920			

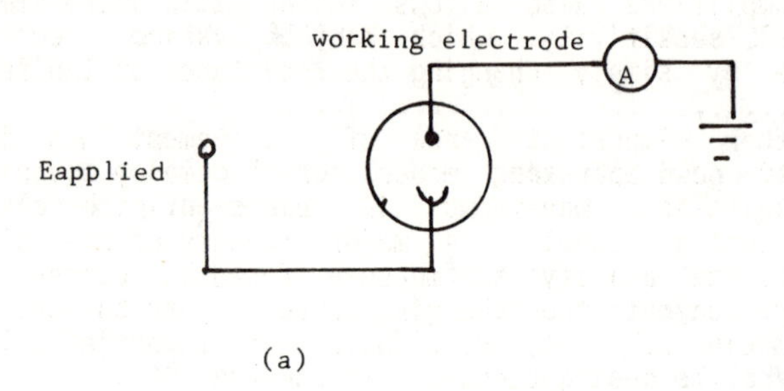

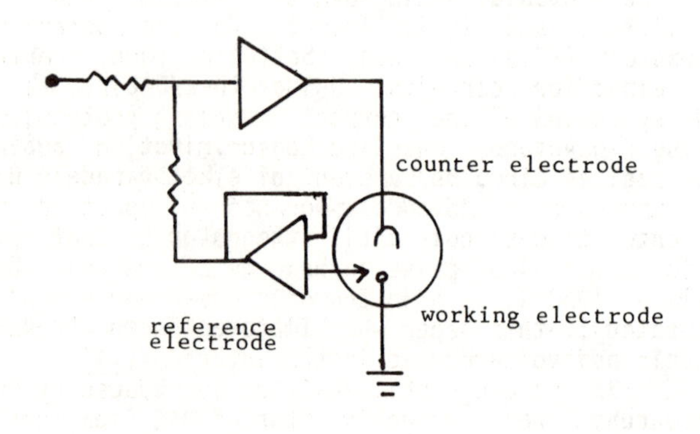

Figure 1. a) Two electrode and b) three electrode potential control circuits.

development of potentiostats with improved speed of response and accuracy.

In addition to control of potential, operational amplifiers can also be used for current measurement by means of current to voltage conversion. The high stability and noise immunity of operational amplifiers also allows for dramatic improvements in the analytical sensitivity, which can be varied over a wide dynamic range by simply changing the resistance of the feedback circuit.

The other important area of improvement was in the development of new operating modes, which employ a variety of potential excitation waveforms in place of the classical d.c.linear potential input. A major feature of most of these variants is the ability to measure faradaic current while discriminating against the charging current contribution. Pulse techniques, including Tast, d.c. pulse, differential pulse and square wave were the most successful.

The circuits that were developed to support these new techniques were hybrid analog-digital systems that employed digital oscillators and logic circuitry for the control of field effect transistor (FET) switches that, in turn, control the operational amplifier circuits. Using these circuits it was possible to synchronise the current sampling protocol with the applied pulse waveform, enabling discrimination against the charging current in the measurement of small faradaic currents. While these instrumental advances soon had an impact on homebuilt instrumentation, it was not until the 1960's that commercial instrumentation employing pulse techniques became available (14). Until the late 1960's, such instruments were inaccessible to the average laboratory; then began what Flato calls the "renaissance" in polarographic and voltammetric instrumentation (14).

The PAR 174 polarographic analyser introduced by Princeton Applied Research, was arguably the first low cost, high performance instrument to properly exploit the new electronics technology (14). It is no exaggeration to say that the PAR 174 was a key factor in the popularisation of polarography and voltammetry as a routine technique in the 1970's. A feature of this instrument is the availability of normal pulse and differential pulse modes. After nearly two decades, this instrument is still being sold virtually unchanged. The major operating mode of the PAR 174, differential pulse polarography (DPP) has had a major impact on the use of electrochemical methods in trace analysis. The technique developed from the work of Barker on square wave polarography in the 1950's (15). In differential pulse polarography (DPP) the applied waveform is a symmetrical pulse superimposed on a linear potential ramp. The current is sampled at the beginning and end of each pulse and the difference is recorded. The differential current when plotted against potential yields a peak shaped response.

By screening out the bulk of the background charging current contribution, the DPP mode allows much higher sensivities and lower detection limits to be achieved compared with the DC mode. The technique has found wide application; in trace metal analysis, especially in anodic stripping voltammetry (ASV), as well as for the analysis of low levels of organics, biological materials and toxic pollutant species.

More recently, a new generation of pulse techniques has been developed. The pioneering work in this area has been done by Robert and Janet Osteryoung and their co-workers. These include reverse pulse, differential normal pulse and square wave techniques (16,17).

Miniaturisation

Since the advent of the vacuum tube, microelectronics technology has been continually driven by the need to increase the performance/cost ratio. The way this has been done is through integration and miniaturisation of electronic circuits. It is evident that production technology for integrated circuits has advanced faster than perhaps any other manufacturing industry.

Electrochemical instrumentation has also benefitted from this new technology in several ways. One example is the recent interest in the use of micron sized electrodes, either singly or in arrays. Integrated circuit manufacturing technology such as photolithography and vapor deposition have been employed for making such electrodes.

One interesting point to note is that the instrumentation used has gone full circle and reverted to the original two-electrode mode. This is because the faradaic and capacitive currents of microelectrodes are much smaller than those of macro electrodes. Generally, analytical currents in the picoamp to nanoamp range are measured. Also due to the much lower charging currents, high speed cyclic voltammetry with scan rates in excess of 1000V/sec can also be employed (18).

Microelectrodes are also opening up a new world for electrochemists to explore systems where the space and time dimensions are dramatically reduced. Such systems enable the probing of micro-environments, such as in-vivo monitoring of neurotransmitters in human body fluids. Although microelectrodes are still at the research stage, a wide spectrum of practical applications are envisaged.

A further aspect of miniaturisation is the possibility of integration of the electrochemical sensor system with the measurement electronics. This has already been accomplished to a limited extent by Hill and co-workers who have integrated an entire amperometric measurement system, including the microprocessor, into a holder of the size of a pen (19). The device has been used in conjunction with disposable enzyme electrodes for glucose measurements.

At a higher level of miniaturisation, work has been directed at constructing miniature electrochemical cells for biosensor applications (20). Chemical field effect transistors (CHEMFETs) and ion selective field effect translators (ISFETs) have received a high level of research interest for almost a decade (21) but with little tangible results in terms of commercially available systems. Some recent research has been directed at integration of the sensor and control electronics into a discrete, disposable device (21). Although further development is still required, the development of implantable, low cost, disposable electrochemical sensors is almost certainly attainable within the next decade.

The Computer Age

The advent of the Computer for instrumental applications has led to the third major period of development for electrochemical instrumentation.

Voltammetry is perhaps one of the earliest techniques to be computerized. Pioneering work was carried out by Smith (22), Osteryoung (23) and Perone (24) in the early 1970's, using minicomputers for generating the voltage waveform and for recording current response. Figure 2 illustrates a typical arrangement for computer controlled electrochemistry. The interface, comprising one or more digital to analog (D/A) converters, an analog to digital (A/D) converter, and timing and synchronisation circuits, provides the two way link between the digital computer and the analog potentiostat.

The applied potential waveform is readily constructed using software and translated into the analog waveform input signal via the D/A converter. For the current measurement, digitized data is obtained by A/D conversion which translates the voltage level from the current to voltage converter into digital values. Again, complex current sampling protocols can be devised using software. The major advantage of computer control is the ability to construct entirely flexible input waveforms and current sampling routines with minimal hardware modification requirements.

Perhaps the greatest impact of the computer in regard to potential generation and current monitoring is in pulse techniques. The computer has contributed to the development of novel pulse waveforms that would be difficult to implement using dedicated electronic circuits. Of particular interest is the ability to finely adjust pulse frequencies and waveshapes to suit specific experimental conditions. For example, the pulse width can be optimised to ensure discrimination against charging current while maximising the sampling frequency. Robert and Janet Osteryoung have made some of the major contributions to the development of computer aided pulse voltammetric techniques (16,17). Osteryoung square wave voltammetry is a notable example of a technique that is practically feasible only because of the computer. The waveform consists of a large amplitude square wave superimposed on a staircase waveform. The technique affords a much higher sensitivity for both reversible and irreversible systems. It is also faster than any other pulse mode (17).

One of the advantages of the computer is the ability to perform electrochemical experiments having short time scales. Previously,the use of storage oscilloscopes was the only means of capturing fast data. For a permanent record the tracing on the oscilloscope had to be photographed and the points read manually from the photograph. Using fairly inexpensive A/D converters having acquisition times of the order of 10 microseconds, computers can be used to record fast experiments in the millisecond scale. For example, fast cyclic voltammograms with scan rates in excess of 1 volt/sec can be obtained, while with the

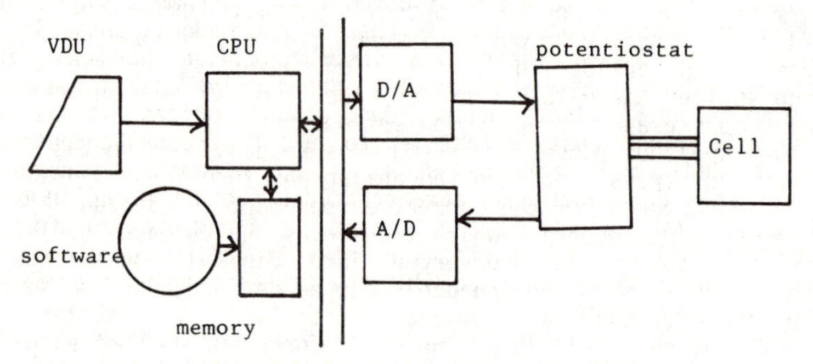

Figure 2. Computer-controlled electrochemistry.

use of special interfacing hardware, such as direct memory access, even faster experiments can be monitored.

One early example of this type of application of computers to short time scale data acquisition experiments was reported by Lauer and co-workers (25,26). The work showed the enormous advantages of the computer route over the earlier tedious, less accurate oscilloscope procedure.

Micro-Computer Based Systems

The trend in the 1980's has shifted towards the development of microcomputer controlled voltammetry. Several reports have appeared in the literature where an inexpensive personal computer has been interfaced to a potentiostat (27). The most significant advantage of the microcomputer is that it has made it possible for researchers to build their own instrumentation hardware, thus dispelling much of the mystery previously associated with minicomputers (28). Also, with decreasing prices it is now considerably more cost effective to use a personal computer in place of dedicated hardware such as potentiometric recorders. This fact has been admirably demonstrated by Bond's group in their development of microcomputer based instrumentation for ac voltammetry (29). Of particular interest is the replacement of hardware functions by software with the tradeoff being flexibility and cost against speed.

Also with the decline in the cost of electronic components it is now possible to devise even more sophisticated hardware. For example, the use of multi (micro) processor systems has been proposed for improving the real time response of microcomputers. In multiprocessor systems, various tasks can be distributed among several processors. Thus, data acquisition is handled by one processor while the processing of the data is handled by a another and the experimental control by a third.

In the commercial area, work has been concerned with integration of the microcomputer as a component part of the instrument. Already, a few "smart" instruments have appeared in the market. Instruments such as the PARC 273 (Princeton Applied Research Corporation) and BAS 100 (Bioanalytial Systems Inc.) enable the chemist to perform a wide range of techniques using the same instrument. Smart instruments provide the user with the flexibility of software control. However, the software is transparent to the user as experimental procedures are usually invoked using front panel function keys.

Computer-Aided Interpretation of Electrochemical Data

In general, electrochemical measurements seek to analyse the time-dependent response of the electrochemical cell to an externally applied, perturbing potential waveform. In this context, a major area of activity in computer-aided voltammetry has been the development of software that enables the analysis of voltammetric data to provide analytical, mechanistic and thermodynamic information on the measured electrode process.

At the rudimentary level, one of the advantages of the computer comes from the ability to handle experimental data in digital form. The benefits are increased accuracy, the ability to store vast amounts of data and the ease of transmission. The computer can also present the results to the experimenter in the form of graphs, charts or tables. However, perhaps the most important benefit is the ability to transform and analyse the computer-stored data after completion of the experiment. Various techniques are also available for enhancing the quality of electrochemical data including digital smoothing, background subtraction, and signal transformations.

The application of Fourier transform techniques to the study of electroanalytical chemistry would not hve been feasible without the computer. Smith (30) has been responsible for most of the pioneering work in this area. The Fast Fourier Transform (FFT) can be used both for data acquisition and for data treatment. For example, the FFT approach in ac impedance measurements has the advantage that a range of frequencies can be superimposed during a single potential scan.

Transformation of data from the time to the frequency domain enables noise to be more easily eliminated. The FFT approach also has an advantage over digital smoothing techniques in that the latter introduces distortions in the smoothed signal (31). Another example of a useful transform technique is the work of Soong and Maloy who employed the Riemann Liouville transform to discriminate the faradaic current form the capacitive and adsorptive currents in polarographic measurements (32). The use of this transform technique affords a sensitivity that compares favourably with conventional pulse or differential pulse polarographic techniques.

Cybernetic Electrochemistry

Perhaps the most striking capability endowed by the computer is the facility to carry out a number of techniques from a single workstation. The Osteryoung group demonstrated this feature early in the development of pulse techniques (16). Recently, Faulkner and co-workers have described the design of a "cybernetic potentiostat" which enables the user to bring to bear a wide range of electrochemical techniques on a single problem (33,34). The electrochemist is now able to explore different experimental strategies without having to make any physical changes to the instrumentation. Moreover, experimental data is automatically transformed and plotted in a useful form. For example, Faulkner and He (34) have reported that up to 50 experiments can be performed in one hour with their cybernetic potentiostat.

The advantage of this facility in maximizing the amount of information that can be extracted from several experiments is readily apparent. Less obvious is the profound impact on productivity and on the way the electrochemist actually goes about his work. In Faulkner's cybernetic potentiostat the operator still plays the vital role of directing the experimental work and finally deriving conclusions based on the results. The next stage will be systems that have inferential and decision making capabilities in closed loop interaction with experimental control functions.

The key to decision making systems is the availability of software that can correctly interpret electrochemical data. Data interpretation methods invariably rely on a model that can be used as a basis for comparison. Work in this area has been developing steadily. For example, Olsen and Evans have used simulated theoretical cyclic voltammograms to predict mechanisms from experimental results (35); Harrison and Small have also reported on the use of a library of mechanisms (36).

Unfortunately, real data from electrochemical experiments are usually confused by noise and experimental artifacts. Although, signal processing techniques, such as Fourier transformation and digital filtering, can be used to improve the quality of the data, there is still a need for techniques that can extract useful information where there is inherent data uncertainty which cannot be removed. Pattern recognition techniques are powerful tools for this purpose (34,37,40). They have been applied to classify electrochemical data on the basis of electrode mechanism (37,38) and chemical structure (39,40).

Even with the use of pattern recognition, a single experiment (or even several experiments based on a single technique) is often insufficient. As remarked by He and Faulkner (34), the future of cybernetic electrochemistry will depend on the ability to combine the evidence derived from a number of experiments using different electrochemical techniques. Undoubtedly, use will be made of methodologies drawn from the artificial intelligence area. Of particular interest are expert systems that employ knowledge based search techniques.

A start has already been made in this direction. Eklund and Faulkner have designed a system that allows the use of a combination of evidence to derive conclusions on the 'n' value of an electrode process, the stability of an electrogenerated product and the existence of adsorption at an electrode (34,41).

Generally, the approach to problem solving using expert systems may be defined as a search of a problem space. The key to such intelligent systems will be the availability of diagnostics that can be used to guide the search towards a satisfactory solution. Therdteppitak and Maloy have developed a model for a cybernetic instrument that utilises a computed function serving as a general diagnostic criteria. As an example of its application, the function was successfully used to distinguish between diffusion control and kinetic control (42,43). Often, however, algorithmic approaches to deriving a solution are impractical in cases where the number of possible solutions is large or where an appropriate diagnostic is unavailable or insufficient. In such cases, heuristic techniques can be applied. These are simply rule of thumb approaches based on an expert's experience in the problem domain.

Gunasingham has recently addressed this issue (44). Figure 3 shows a schematic of a general design for a cybernetic instrument where the control software is separated into plan, execute, interpret and refine phases. In this view, heuristic methods are used in conjunction with algorithmic, deterministic

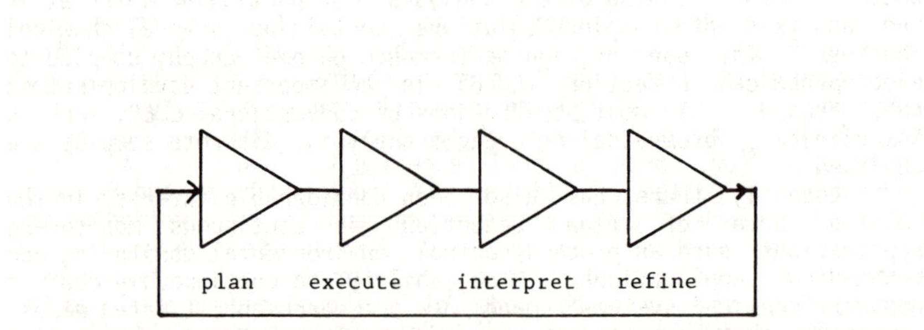

Figure 3. General design for a cybernetic potentiostat.

procedures in the interpretation of experimental data. The development of expert systems that combine heuristic and deterministic techniques will require the use of special programming languages. Gunasingham and Wong described the use of Prolog for this purpose (44,45). One of the advantages of Prolog is its declarative style based on first order predicate calculus. As a consequence, heuristics can be added to or deleted from the system without worrying about the order in which they are to be used. This approach is directed to experimental systems that have a learning capability.

Automation of Analytical Processes

One of the most active areas of electrochemical research and development in the last decade has been the use of electrochemical detectors for flowing stream analysis. Of particular interest is the analysis of complex solutions containing several chemical species. The use of high performance chromatography coupled to electrochemical detection (LCEC) is an important development in this regard. In most applications of conventional LCEC, such as in clinical, biomedical or organic analysis, discrete samples are analysed.

Recently there has also been considerable interest in the related area of flow detection in continuous monitoring applications such as process control, environmental monitoring and bio-medical applications. The ability to continuously monitor complex chemical systems demands new and more sophisticated modes, involving the combination of a number of analytical schemes in a single system. Such systems invariably require the use of microcomputer control for accurate timing and sequencing of operations. Also, the computer enables autonomous operation which is usually necessary in real applications. A good example of hybrid systems made possible by the computer is the work of Bond and co-workers in combining ultraviolet absorption and electrochemical detection for the monitoring of heavy metals (46). Another approach to monitoring heavy metals is through the use of continuous monitoring anodic stripping voltammetry (ASV). In anodic stripping voltammetry,the plating and stripping steps can be optimised by the judicious choice of medium (47). Periodic cleaning with a nitric acid wash is needed to limit the contamination of the electrodes. Techniques such as acid digestion and membrane separation can also be used to minimise or eliminate the effects of impurities.

Another example of a continuous electrochemical monitoring system is an extracorporeal analyzer system that has been developed for continuously monitoring human blood glucose levels. Figure 4 shows a schematic diagram of the monitoring system, which includes a dialysis chamber, an enzyme reactor and an electrochemical detector. A modified stopped flow technique is employed. A small sample of blood (around 10 ul) is sampled through a double lumen catheter and loaded into the dialyser. The glucose dialyser is then injected through an enzyme reactor and the generated hydrogen peroxide monitored at a platinum electrode. In this system the microcomputer performs a number of

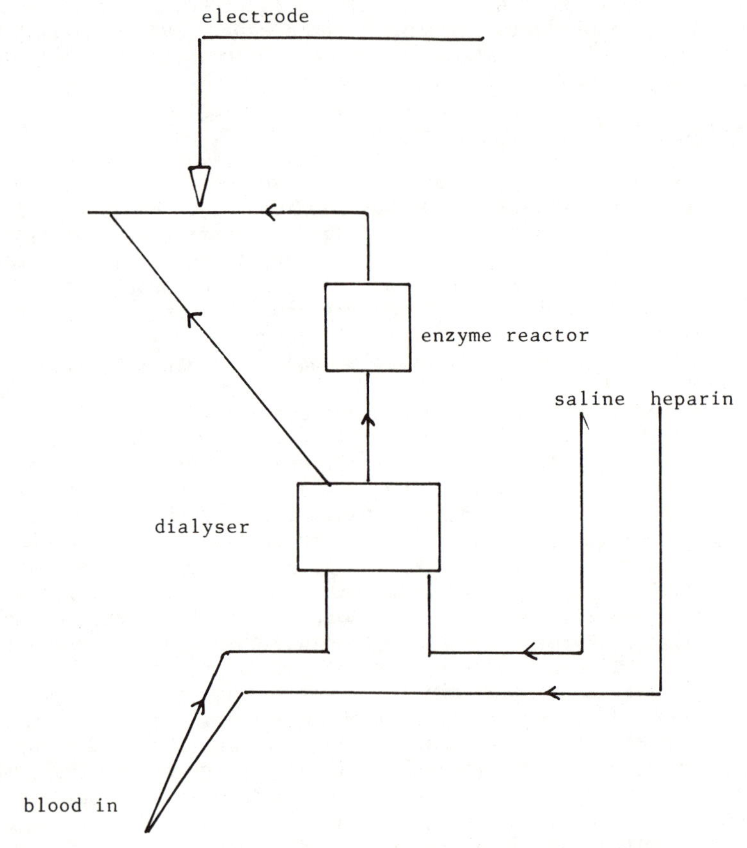

Figure 4. Extra-corporeal monitor for blood glucose.

functions including timing and control of the sampling and analysis sequence, data acquisition, background subtraction, auto calibration, and recording and display of results (48).

Conclusions

The development of electrochemical instrumentation has been shown to be an evolutionary process, progressing as a result of the impact of emerging technologies such as electronics and the advent of the microcomputer. Three distinct periods in this evolution have been identified: the early history and electromechanical systems, the electronic age and finally the computer era. In each period, progress has been punctuated by specific, occasionally revolutionary, contributions by individuals. Often, a single discovery, such as that of polarography has resulted in the transformation of previously accepted practices and techniques. This, of course, is a general feature in the evolution of all branches of science.

Electrochemical instrumentation has seen tremendous developments during its 200-year history. These advances have been fuelled in part by the involvement of electrochemical technology in some of the key problem areas of society, including public health, energy generation and conservation and environmental pollution.

The development and application of electrochemical instrumentation as evidenced by the volume of published research, the emergence of new journals and level of commercial systems development is clearly a rapidly expanding field. While the advances in microelectronics and solid state sensor devices have far outstripped the development of comparable chemical sensor devices, there is strong evidence that recent developments in electrochemical sensors and associated instrumentation may bridge the gap. Current research and development trends are towards a greater involvement and integration of computers, cybernetics, microelectronics and miniaturisation of measurement devices. With an increasing level of demand by industry for the practical demonstration of these technologies, the future indeed looks very bright.

Literature Cited

1. Szabadvary, F. History of Analytical Chemistry, Pergamon Press, 1966, Oxford, p309.
2. Cruikshank, W. Ann. Physik, 1801, 7, 105.
3. Gibbs, W. Z. Anal. Chem. 1864, 3, 334.
4. Luckow, C., Z. Anal. Chem. 1869, 8, 1.
5 Grower, G.G., Proc. Ann. Soc. testing Materials, 1917, 17, 129.
6. Szebelledy, L.; Somogyi, Z. Z. Anal. Chem., 1938, 112, 313.
7. Lingane, J.J., J. Amer. Chem. Soc., 1945, 67, 1917.
8. Delahay, P., New Instrumental Methods in Electrochemistry, Interscience Publishers Inc., New York, 1966.

9. Behrend, R., Z. Physik Chem., 1893, 11, 460.
10. Heyrovsky, J.; Shikata, M., Rec. trav. Chim., 1925, 44, 496.
11. Heyrovsky, J.; Zuman, P. Practical Polarography; Academic Press., N.Y., 1968.
12. Borman, S.A., Anal. Chem., 1982, 54, 698A.
13. Hickling, A., Trans. Faraday Soc., 1942, 38, 27.
14. Flato, J.B., Anal. Chem., 1972, 44, 75A.
15. Barker, G.C. and Gardner, A.W., Z. Anal. Chem., 1960, 173, 79.
16. Borman, S., Anal. Chem., 1982, 698A.
17. Osteryoung, J.; Osteryoung, R.A., Anal. Chem., 1985, 57, 101A.
18. Pons, S.; Fleischmann, M., Anal. Chem., 1987, 59, 139A.
19. Higgins, I.J.; McCann, J.M.; Davis, G.; Hill, H.A.O.; Zwanziger, R.; Triedl, B.L.; Birkett, N.N.; Plotkin, E.V. European Patent Application, EP127958, 1984.
20. Miyahara, Y.; Matsu, F.; Monizuri, T; Matsuoka, H.; Karube, I.; Suzuki, S., Proc. Symp. Chemical Sensors, Fukuoka, 1983, p501.
21. Ko, W.H.; Fung, C.D.; Yu, D.; Xu, Y.H., Proc. Symp. Chemical Sensors, Fukuoka, 1983, p496.
22. Smith, D.E., in "Computers in Chemistry and Instrumentation" Vol. 2, (Editors: Mattson, J.S., Mark, H.B., and MacDonald, H.C.), Marcel Dekker, New York, 1972, p369.
23. Osteryoung, R.A., in ref (22), p353.
24. Perone, S.P., in ref (23), p423.
25. Lauer, G.; Abel R.; Anson, F.C., Anal. Chem., 1967, 39, 765.
26. Lauer, G., Osteryoung, R.A., Anal. Chem., 1966, 70, 2857.
27. Gunasingham, H.; Ang, K.P., J. Chem. Educ., 1985, 62, 610.
28. Anderson, J.E.; Anal Chim. Acta, 1985, 169, 309.
29. Bond, A.M.,Modern Polarographic Methods in Analytical Chemistry, Marcel Dekker, New York, 1980.
30. Smith, D.E., Anal. Chem., 1976, 48, 221A.
31. Smith, D.E., ibid, 517A.
32. Soong, F.C. and Meloy, J.T., J. Electroanal. Chem., 1983, 153, 29.
33. He., P., Avery, J.P. and Faulkner, L.R., Anal. Chem., 1982, 54, 1313A.
34. He, P.; Faulkner, L.R., J. Chem. Inf. Comput. Sci., 1985, 25, 275.
35. Olsen, B.A.; Evans, D.H., J. Am. Chem. Soc., 1981, 103, 839.
36. Harrison, J.A. and Small, C.E., Electrochim. Acta, 1980, 25, 447.
37. DePalma, R.A.; Perone, S.P., Anal. Chem., 1979, 51, 839.
38. Rusling, J.F., Trends Anal. Chem., 1984, 3, 91.
39. Schachterle, S.D. and Perone, S.P., Anal. Chem., 1981, 53, 1672.

40. Dyers, W.A.; Freiser, B.S.; Perone, S.P., Anal. Chem., 1983, 55, 620.
41. Eklund, J.A.; Faulkner, L.R., Electroanalytical Symposium, BAS Press, 1985, West Lafayette, p225.
42. Therdteppitak, A.; Maloy, J.T., Anal. Chem., 1984, 56, 2592.
43. Therdteppitak, A.; Maloy, J.T., Electroanalytical Symposium, BAS Press, 1985, West Lafayette, p217.
44. Gunasingham, H., J. Chem. Inf. Comput. Sci., 1986, 26, 130.
45. Gunasingham, H.; Wong, M.L., J. Chem. Inf. Comput. Sci., 1988, In press.
46. Bond, A.M.; Wallace, G.G., Anal. Chem., 1984, 56, 2085.
47. Gunasingham, H.; Ang, K.P.; Ngo, C.C.; Thiak, P.C.; Fleet, B., J. Electroanal. Chem., 1985, 186, 51.
48. Gunasingham, H.; Teo, P.Y.T.; Tan, C.B.; Ang, K.P.; Tay, B.T.; Lim, S.H.; Thai, A.C.; Aw, T.C., Anal. Chim. Acta, submitted.

RECEIVED August 18, 1988

Chapter 18

Development of the pH Meter

Bruno Jaselskis[1], Carl E. Moore[1], and Alfred von Smolinski[2]

[1]Department of Chemistry, Loyola University of Chicago, Chicago, IL 60626
[2]Department of Medicinal Chemistry and Pharmacognosy, University of Illinois at Chicago, Chicago, IL 60680

The history of the measurement of hydrogen-ion concentration parallels not only progress in understanding of the acid-base concept but also the development of suitable instrumentation for H^+ indication. This brief survey explores the early beginnings, the development of the potentiometer, standard cells, and null point devices. These led to the vacuum tube voltmeter and the design of the first commercial pH meter.

Early Beginnings

Dr. Arnold O. Beckman, Gold Medalist of the American Institute of Chemists, 1987, in his address "Instruments and Progress in Chemistry" (1) stated:

> We all tend to take for granted the benefits and conveniences we get from things with which we are familiar. We easily forget and often do not know what life was like before we had them. I think it is well to maintain some perspective between the present and the past.

The development of the measurement of the hydrogen ion has not been an isolated phenomenon of science. It has closely paralleled the development of our understanding of acids and bases and the availability of instrumentation suitable to the measurement task. It presents, perhaps, a classic case of man's ingenuity in achieving a desired goal.

Prior to Arrhenius (1883), our understanding of acids and bases was rather minimal, and at the best, acids were defined in terms of their sour taste and their behavior in reactions with other substances.

0097–6156/89/0390–0254$06.00/0

The non-metal oxide reactions with water led Antoine Laurent Lavoisier (1713-1794) to associate the *acidity principle* with elemental oxygen, but Sir Humphry Davy, in the beginning of the XIX century, was the first to suggest that the acidifying principle was owed to hydrogen – as in the case of hydrochloric acid.

Confusion prevailed at this time, for there were many hydrogen compounds that were not acids. This prompted Sir Humphry Davy to state (2):

> ...acidity does not depend on any particular elementary substance, but upon peculiar combinations of various substances.

A few years later Justus Liebig (1838) defined acids in terms of replaceable hydrogen (3):

> Säueren sind hiernach gewisse Wasserstoff-Verbindungen, in denen der Wasserstoff vertreten werden kann durch Metalle.

> The acids are certain hydrogen compounds in which hydrogen can be replaced by metals.

Almost 50 years after Liebig's definition of acids Arrhenius presented his theory as his doctoral thesis, entitled "Reserches sur la Conductibilite Galvanique du Electrolytes," before the Swedish Academy of Science on June 6,1883. In this thesis he gave data pertaining to the dissociation of salts and acids in dilute solutions into "electrically charged ions." The thesis was accepted with a passing grade, but the oral defense was rated with a superior performance.

Arrhenius summarized his recollections of his early work as a student in the Sullivan lecture series presented in New Haven in 1912 (4):

> ...I came to my professor, Cleve, whom I admire very much, and I said: "I have a new theory of electrical conductivity as a cause of chemical reactions." He said: "That is very interesting" and then said, "Goodbye."

Arrhenius' research was not forgotten, even though many of his well recognized contemporaries (Clausius, Thomson, Wiedemann, von Helmholtz...) did not think much of his theory of electrolytic dissociation. Ostwald was an exception; he realized the importance of this work and promoted the concept. As described by the following quote from Ostwald, acids were recognized in the process of dissociation to produce protons (H*) and bases to yield hydroxyl (OH) ions (5):

> Eine Säure enthält neben dem Anion Wasserstoffion, H*, eine Basis neben dem Kation Hydroxylion, OH';...

> An acid contains in addition to the anion the hydrogen ion, H* ; and a base in addition to the cation contains the hydroxyl ion, OH';...

Before a quantity can be truly assessed, one must be able to identify it. Then it can be measured quantitatively as accurately as the means of the *State of the Art* permit. Lord Kelvin in his "Popular Lectures and Addresses" lecture on Standard Units of Measurement, 1883, expressed his views on measurements as follows (1):

> I often say that when you can measure what you are speaking about and express it in numbers you know something about it;...

With a satisfactory definition of acids and bases in place, the measurement of hydrogen-ion concentration became a reality. At first, it could be inferred from the neutralization reactions using colored indicators and then from the potentiometric methods.

Prior to the development of the glass electrode and pH meter, the color indicator application found a wide use for the direct determination of hydrogen-ion concentration. William Mansfield Clark (De Lamur Professor Emeritus of Physiological Chemistry at Johns Hopkins University) spent most of his lifetime investigating organic substances suitable for acid-base indicators. He never accepted the pH meter. In his classic book *Oxidation-Reduction Potential of Organic Systems* he stated (6):

> The trust placed in the beautifully engineered instruments now available and in the pH numbers written on the labels of standard solutions together with the sloppy conditions one often observes leads one to say that too often users are mere meter readers who know little about the instruments or the principles of pH measurements.

Limitations of the acid-base indicators for pH measurement instigated investigation of new and alternative methods. Rapid advances in the measurement of the electromotive force, resistance, and current set the stage for measurement of the concentration of the hydrogen ion by potentiometric methods. To a great extent, successful potentiometric measurement depended on the development of the basic elements of potentiometric circuitry — i. e., null point detectors, high sensitivity galvanometers, standard potential sources, and electrodes whose potential varied as a function of the hydrogen-ion concentration. Early workers recognized that it is very important in the measurement of an unknown cell potential that the cell not be polarized.

Development of the Potentiometer—the Classical Age

In 1826 J. C. Poggendorff, who may be considered the founder of modern potentiometry, reported a lamp mirror device for improving magnetic deviation measurement – probably the forerunner of the lamp scale galvanometer. On August 5, 1841, he presented a paper before the Royal Prussian Academy in Berlin: "Methode zur quantitativen Bestimmung der elektromotorischen Kraft inconstanter galvanischen Kette" (7) (Methods for Quantitative Determination of the Electromotive Force of Non-Constant Galvanic Cells). This paper laid out the basic principles of modern potentiometry using a compensation method; it showed that unknown potentials could be measured with great accuracy.

Poggendorff realized that the accuracy of the measurement depended:

(1) von der Genauigkeit und Sorgfalt, mit welcher die Compensation bewerkstellig worden; sie erfordert einen sehr emfindlichen Multiplicator und möglichste Ausschliessung der Polarisation auf dem angegebenen Wege;
(2) von der Constanz und genauen Kenntniss der elektromotorischen Kraft **k'**;
(3) von der richtigen Bestimmung der Wiederstände **r** und **r'**

(1) on the accuracy and care with which the compensation is carried out; which requires a very sensitive multiplicator and the greatest possible exclusion of polarization in the way which has been indicated;
(2) on the constancy and accurate knowledge of the electromotive force **k**;
(3) on the correct determination of the resistances **r** and **r'**.

The Poggendorf cell arrangement is shown in Figure 1. In this arrangement the wire **a** of the constant and non-constant cells provides resistance **r'** of the constant cell, while the wire **b** connecting platinum electrode of the constant cell to the zinc electrode of the non-constant cell provides the resistance **r**. When the ratio of **r'/r** is brought into correct ratio, then the current in wire **c** is negligible, and the meter shows zero current. In this arrangement the meter serves as the null point indicator, and the unknown cell potential is compensated against the known cell potential. The unknown cell potential **k''** is then related to the constant cell potential **k'** by the equation $k'' = k' \, r/(r + r')$. This procedure became known as the Poggendorff compensation method.

Shortly after Poggendorff's publication, Professor Charles Wheatstone, whose contributions to the development of instrumentation for the study of problems associated with telecommunications are classics, developed a bridge circuit. This circuit was described in the Bakerian Lecture of June 15, 1843: "An Account of Several New Instruments and Processes for Determining the Constants of Voltaic Circuit" (8). In this lecture Wheatstone described not only his devices but also those of Professor Jacobi of St. Petersburgh. He foresaw a great use of his rheostat and the principle of summation of electromotive forces in a voltaic circuit. To this effect he stated:

> But the use of new instruments is not limited to this special object; they will, I trust, be found of great assistance in all inquiries relating to the laws of electrical currents, and to various and daily increasing practical applications of this wonderful agent...The principle of my process is as follows: In two circuits, producing equal rheometric effects, the sum of the electromotive forces divided by the sum of the resistances is a constant quantity, i.e., $E/R = (n/n)\ E/R$. Knowing therefore the proportion of the resistances in two circuits producing the same effect, we are able immediately to infer that of the electromotive forces.

Standard Cells and Early Null Point Devices

Standardization of the Wheatstone's bridge or of Poggendorff's circuit required the availability of a reliable standard cell. Latimer Clark addressed this question and described the development of a standard cell in a paper communicated by Sir William Thomson in 1873, "On a Standard Voltaic Battery" (9):

> ...Practically electricians have been compelled to define electromotive forces by comparison with those of the Grove's or Daniell's cell, the copper and zinc cell, or other electromotive sources;... not one has been hitherto found which could be relied upon to give a definite electromotive force: however pure the materials, and however skilful the manipulation, differences varying from four to five percent upwards constantly occur without any assignable cause; and different observers using different materials of course meet with still larger discrepancies....
>
> In this paper Clark described the battery composed of mercury-mercurous sulfate, zinc and saturated zinc sulfate (known as the Clark cell).

The instrument used for this purpose was designed by Clark in 1859 and is shown in Figure 2. In this arrangement the current of the working battery, WB, is adjusted with a rheostat, R, to be equal to that of the standard cell, SC, as indicated by the galvanometer. Once this has been achieved, the unknown cell, UC, is connected to the slide wire, and the potential is estimated by moving the contact **i** along the slide wire **b,b'** until the galvanometers shows no current flow.

Twenty years later Edward Weston also addressed the question of a primary standard of EMF. The following is reported in *The Electrical Engineer* (10) and *The Electrician*, April 12, 1893 (11). The reporter in the publication stated:

> The standards of electromotive force available up to the present time did not come up to the ideal which Mr. Weston had set before him, even the best of these, the Clark cell, being subject to a variation of 0.077 per degree centigrade.

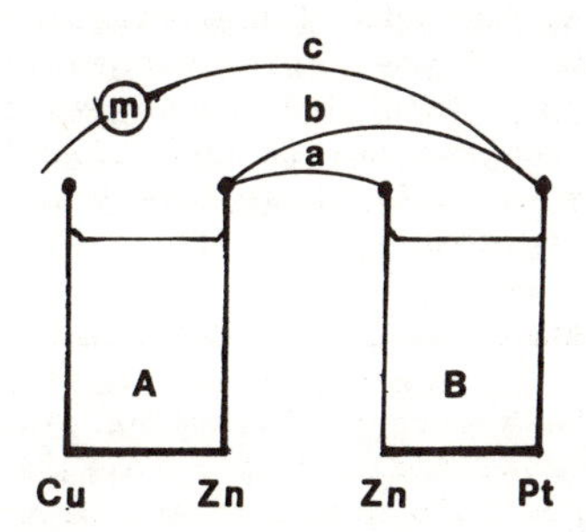

Figure 1. Poggendorff's potentiometer. The electrodes of the constant cell, B, are connected by wires **a** and **b** of variable resistance to the non-constant cell, A, and wire **c** with a meter in the series connects platinum and copper electrodes. (Diagram reproduced from Ref. 7, 1841)

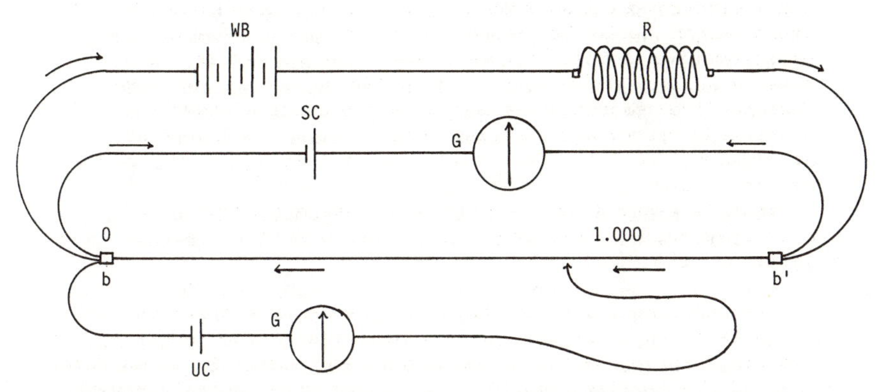

Figure 2. Clark's instrument for measuring unknown cell potentials. (Diagram reproduced from Ref. 9 1874).

Thus Weston constructed a cell practically without temperature error and free from local action via the use of cadmium salts.

By the end of the XIX century, potentiometric or Wheatstone circuits and standard EMF sources were in place; also, null detectors such as galvanometers were being developed for the measurement of smaller and smaller currents or smaller potential differences. A satisfactory null-point detector was an almost insurmountable problem. Galvanometers gave poor sensitivity and reproducibility in these early bridge circuits when they involved high resistances, and they required a considerable amount of time to stabilize in carrying out the measurements. The most satisfactory null detection device appears to have been the quadrant electrometer. However, its use required great skill and patience. William Mansfield Clark provided us with a good description of the quadrant electrometer (12):

> In the form useful for the purpose at hand a very light vane of aluminium is suspended by an extremely fine thread, preferably of quartz, which is metallized on the surface in order to conduct charges to the vane. The vane is surrounded by a flat, cylindrical metal box cut into quadrants. Two opposite quadrants are connected to one terminal and the remaining quadrants to another terminal. If now the vane or needle be charged from one terminal of a high-voltage battery the other terminal of which is grounded, and a difference of potential be established between the two sets of quadrants, the needle will be deflected by the electrostatic forces imposed and induced....

The Lippmann-Ostwald capillary electrometers also gave satisfactory results and provided accuracy a little better than 0.001 V.

Null point detection, also, could be achieved via the use of a sensitive galvanometer, such as the Deprez-d'Arsonval, in conjunction with high cell resistances (100 to 1000 megohms). The measurement was achieved by allowing the cell to be measured to charge a precision microfarad condenser for one half hour. Then the capacitor was discharged through the galvanometer. An improved design of this arrangement was described years later by Jones and Kaplan (13).

Thus, by the end of the nineteenth century the electromotive force from high resistance sources could be measured with a relatively high precision. The introduction of the glass membrane electrode for the hydrogen-ion concentration measurement presented a challenge and a test of instrument capabilities. The State of the Art (1905) for the measurement of membrane potentials had been described by Max Cremer in the paper entitled (14):

"Über die Ursache der elektromotorischen Eigenschaften der Gewebe, zugleich ein Beitrag zur Lehre von den polyphasischen

Elektrolytketten" (About the Causes of the Electromotive Properties of Tissues, as well as a Contribution toward the Study of Polyphase Electrolytic Cells).

In this paper Cremer also stated that he was the first to measure the glass electrode potential at room temperature:

> ...kann ich wohl sagen, dass ich mit Hilfe desselben zum ersten Male die elektromotorische Kraft von Glassketten bei gewöhnlicher Temperatur galvanometrisch gemessen habe.

> ...I can certainly say that I have measured for the first time, with the help of these, the electromotive force of the glass cell at room temperature.

Development of the Vacuum Tube Voltmeter-Potentiometer

Thus, in the beginning of the 20th century means were at hand to measure the very small currents, high resistances, and potential differences that were obtained with the glass electrode. The methods described by Cremer were used by Haber and Klemensiewicz (15) and others. Although the use of the glass electrode for the measurement of hydrogen-ion concentration was established, the tedious nature of its use precluded a wide application until the late 1920s. Usually the measurements were carried out using quadrant electrometers. A number of applications of the glass electrode were described in papers by Hughes (16) and Brown (17). A schematic diagram of the arrangement used by Brown is shown in Figure 3.

In the rapidly developing field of telecommunications, it became necessary to improve the strength and quality of the telegraphic signal. Improvements came with the invention of vacuum tubes. In 1904 the vacuum diode was reported, by J.A. Fleming, and in 1906 De Forest introduced the first triode, which he called "the Audion." The multitude of possibilities of the triode was not grasped by the inventor or his contemporaries until years later. In his paper entitled "The Audion, Its Action and Some Recent Applications," which was read before the Franklin Institute of Philadelphia, January 25, 1920, De Forest stated (18):

> "The Audion," suggestive of sound, prompts the consideration of an analogy in the realm of sight—the microscope.... But when the first steps were taken in the work which eventually resulted in the Audion of today, I no more foresaw the future possibilities than did the ancient who first observed magnification through a drop of water. In 1900, while experimenting with an electrolytic detector for wireless signals, it was my luck to be working by the light of Welsbach burner. That light dimmed and brightened again as my little spark transmitter was operated.... I had become convinced that in gases enveloping an incandescent electrode resided latent forces which could be utilized in a detector of Hertzian oscillations....

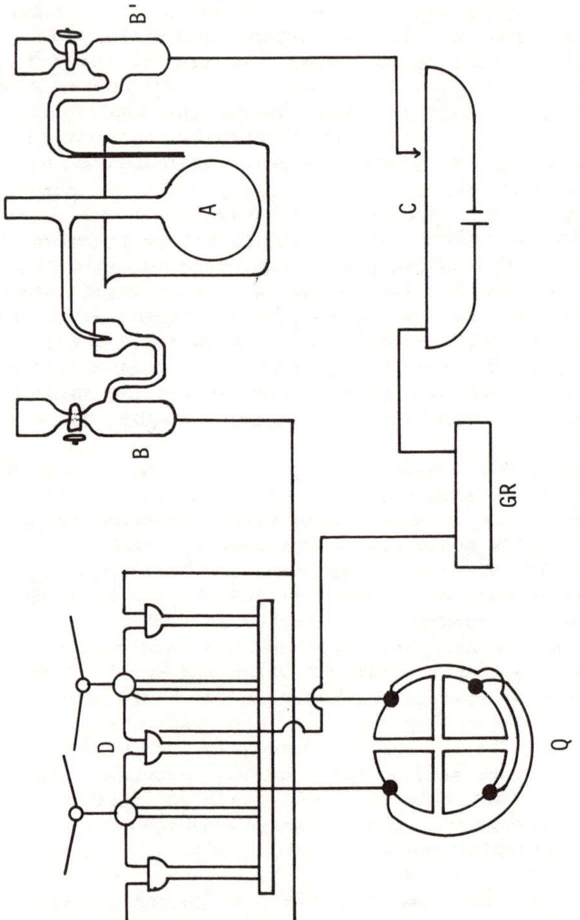

Figure 3. Brown's potentiometer for measurement of hydrogen ion concentration using glass electrode. B and B' - calomel electrodes; A - glass electrode; D - high voltage battery; C - potentiometer; GR - ground; Q - quadrant electrometer. (Reproduced with permission from Ref.17 Copyright 1925 American Institute of Physics.)

Thus the first commercial Audion — which appeared in 1906 — was, therefore, no accident or sudden inspiration. However, it rested on a very perceptive observation.

The first application of the triode to hydrogen-ion chemistry appears to have been by Kenneth H. Goode, who carried out research leading to the M.S. degree at the University of Chicago. He defended his thesis, entitled "Continuous Reading Electrotitration Apparatus," in August of 1924. In this thesis Goode indicated that he originated this use of the vacuum tube and gave the following reasons for doing the research (19):

> In order to avoid the use of the telescope and scale, which do not lend themselves to convenient use either for classroom demonstrations or for routine titration work, it occurred to the writer to make use of the amplifying characteristics of the vacuum tube to magnify the currents to such a value that they could be easily measured with a portable milliammeter.

His research, moreover, was published two years before the defense of his Master's thesis, and with the same title as his thesis (20). The research was so much in the forefront of the science of the time that neither his thesis nor the published paper had references to previous work. He measured the potentials developed between a calomel and a hydrogen electrode during acid-base titrations. His continuous reading apparatus gave a readout expressed either in volts or Sorensen units — i.e., pH units.
In the published paper Goode stated:

> The simplest possible apparatus for electrometric titration, from a theoretical standpoint, would be a sensitive voltmeter connected between a calomel electrode and a hydrogen electrode. In practice, however, an ordinary voltmeter cannot be used, because the instrument would consume current enough from the cell to discharge the hydrogen electrode, and render it inoperative. For this reason all types of apparatus hitherto in use have depended upon balancing the unknown e.m.f. of the cell being used against variable known e.m.f. produced by a potentiometer system, the balance being determined by a "null-point" galvanometer. With this type of apparatus...there is considerable uncertainty in the readings when the potential of the cell is rapidly changing.

The thrust of Goode's contribution, as he recognized it at that time, was the continuous readout. However, he made the point that this is an ideal voltmeter which draws virtually no current during measurement. For this purpose he designed the two circuits shown in Figures 4 and 5.

In the first circuit the steady plate current is balanced by an equal and opposite current produced by a dry cell acting through about 3000 ohms resistance. In the second diagram of

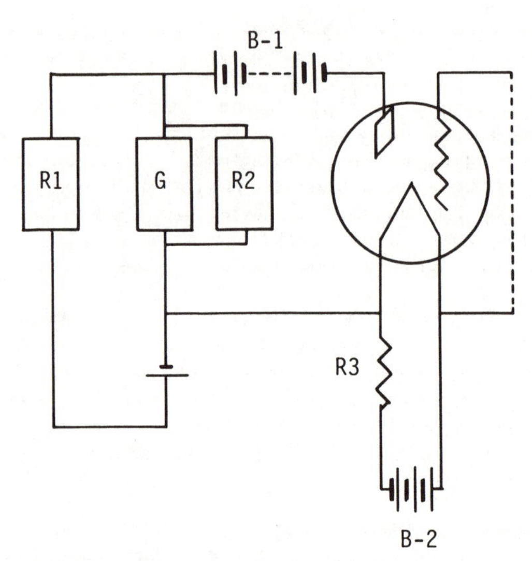

Figure 4. Diagram for the continuous reading electrotitration apparatus- original design. R1, R2 and R3 are 3000, 25 and 0.6 ohm resistors; B1 and B 2 are 22.5 and 6 volt batteries; G- galvanometer. (Reproduced from Ref.20 Copyright 1922 American Chemical Society).

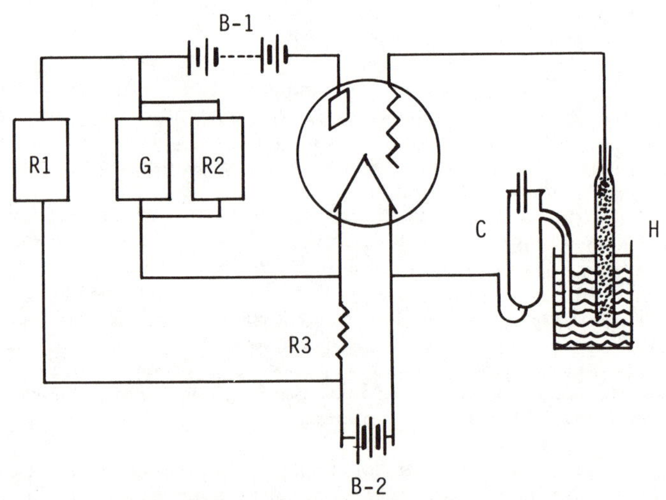

Figure 5. Diagram for the continuous reading electrotitration apparatus- improved version. R1, R 2 and R3 are 1800, 25 and 0.6 ohm resistors; B-1 and B-2 are 22.5 and 6 volt batteries; H - hydrogen electrode; C - calomel electrode and G- galvanometer. (Reproduced from Ref. 20 Copyright 1922 American Chemical Society).

the same paper, Goode obtained a more stable zero point by replacing the dry cell by the potential difference at the terminals of the filament resistance acting through an adjustable 1800 ohm resistance.

During the titration the plate current, $I_p - I_o$, values were read from a galvanometer which could be calibrated either in volts or in Sorensen pH units. In an ordinary titration the hydrogen electrode was connected to the grid and the calomel to the filament. Potential values -5.8 to - 6.7 volts were obtained in the range where the grid current is entirely negligible. The pH could be determined to a precision of 0.1 pH unit.

Goode's pioneering work set in motion a new era for the major development and application of instruments of similar design for continuous redox, precipitation (21), and conductimetric titrations (22). Treadwell (1925) described an instrument to be used for conductimetric titrations. In his paper he stated (22):

> Die Elektronröhre stellt unter diesen Umständen ein Potentiometer dar, das praktisch ohne Stromverbrauch arbeitet, indem das Gitter lediglich durch seine elektrostatische Wirkung den Anodenstrom beeinflusst.

> The electron tube under these conditions acts as a potentiometer, which draws practically no current, in that the grid is influenced solely by the action of the anode current.

In remarks on Goode's design, Treadwell stated (22):

> Durch eine sinnreiche Schaltung kompensiert Goode den constanten Anodenstrom, so dass er den ganzen Skalenbereich seines Galvanometers für die Messungen zur Verfügung hat.

> Through an ingenious hookup, Goode compensated the constant anode current, so that he had the entire scale of his galvanometer for measurements.

Other authors also acknowledged Goode's original contribution. Williams and Whitenack (23), in a paper describing precipitation titrations, stated in their summary:

> It has all the advantages of electro-titration; in addition, it gives a continuous automatic record of the progress of a reaction since the electron tube may be regarded as a direct-reading, sensitive voltmeter which draws no current from the titration cell...

A direct reading hydrogen meter was also described by Pope and Gowlett (24). In this design Goode's instrument was improved by eliminating the high voltage "B" battery and by measuring plate current with a millivoltmeter calibrated directly in terms of pH units. After the publication of Goode's paper, more than seven years would pass before the application of vacuum tube potentiometer was made to the glass electrode.

Determination of pH with a Vacuum Tube Potentiometer and Glass Electrode

In 1928 two independent research groups – Partridge at New York University and Elder and Wright at the University of Illinois – applied the vacuum tube potentiometric circuit to the measurement of the hydrogen-ion concentration via the use of a glass electrode. H. M. Partridge submitted a paper to the *Journal of the American Chemical Society* on March 24, 1928, entitled "A Vacuum Tube Potentiometer for Rapid E.M.F. Measurements" (25). This paper, published on January 8, 1929, was also presented on September 13, 1928, at the Fall Meeting of the American Chemical Society held in Swampscott, Massachusetts.

On October 24, 1928, L.W. Elder and W.H. Wright submitted a communication to the National Academy of Science entitled "pH Measurement with the Glass Electrode and Vacuum Tube Potentiometer" (26). This communication appeared shortly after the submission in 1928 with a footnote:

> After most of this work had been completed (August 12th) the authors learned that H.M. Partridge has carried out similar measurements of glass cell potentials with a vacuum tube potentiometer, at New York University.

In 1928 Walter H. Wright, a young radio experimenter undergraduate student at the University of Illinois, took a course, Advanced Instruments, under Lucius Elder. As a result of the experience in this course, Wright applied his knowledge of radio and wireless telegraphy to a B.S degree thesis project which dealt with the construction of a device for pH measurement using a vacuum tube potentiometer and a glass electrode. Both the thesis (27) and paper were entitled "pH Measurement with Glass Electrode and the Vacuum Tube Potentiometer."

In the following communication Elder and Wright gave their reasons for doing the study and also the rationale for the design of the instrument (26):

> Although it has been shown that solutions of oxidizing agents and of materials which poison or are catalytically decomposed by platinum i.e., solutions in which the ordinary hydrogen or quinhydrone electrodes are inapplicable, may be titrated by a number of electrometric methods, only two of these methods involve electrodes of sufficient reversibility or reproducibility to serve in pH measurement. Although a potentiometer involving a high capacity condenser and ballistic galvanometer has been described as serviceable for rough measurements of potential in cells of high internal resistance, it occurred to one of us that a vacuum tube potentiometer might be preferable.

To achieve this goal the authors described the circuit shown in Figure 6. The method of measurement was based on Morton's design, in which an unknown e.m.f. is measured by compensation with a calibrated potentiometer in the grid circuit (28). In a later publication (27) Elder described in greater detail the pH measurement with the glass electrode and compared the results of the new vacuum tube instrument to the Lindemann quadrant electrometer.

Mr. Wright in a letter of November 7, 1987, written to one of us (Letter from Walter H. Wright to Carl E. Moore, Loyola University Archives), gave reasons for his work:

> However, the glass electrode required a quadrant electrometer to avoid polarizing the glass electrode such as when ordinary Wheatstone bridge type instruments were used. Unfortunately, low humidity is essential for accurate readings with quadrant electrometers, particularly during a humid summer at U. of Ill. Goode's proposal (1921) to use a 3-electrode vacuum tube failed to recognize the basic space charge and ion or electron flow factors of the triode....

H.M. Partridge in his paper (25) first outlined his purpose and goals and then described the instrumental design requirements as follows (some of these requirements are quoted):

(a) it is a direct reading, no current device;
(b) it is independent of the characteristics of the valves and of their power supply;
(c) the "balance" condition is definite and very stable;...
(g) when used with a cell containing properly prepared electrodes, P_H values may be rapidly determined, and are reproducible to 0.02 Sorensen units.

The schematic diagram of this circuit is shown in Figure 7. The circuit used a tetrode and triode in which the grid potential of the tetrode was kept negative with respect to the filament by 2 to 3 volts.

The author also stated:

> The device described herewith is somewhat more convenient than most vacuum tube potentiometers for chemical measurements in that valves do not have to be calibrated and it is not essential that their characteristics remain constant over long periods. It is considerably more rapid than the average potentiometer in that only one adjustment is necessary for the measurement.

Shortly after the Elder and Wright and Partridge publications, further improvements in the measurement of the potentials using the glass electrode were reported by Stacie (29) and Goodhue et al. (30) at Iowa State College. Goodhue et

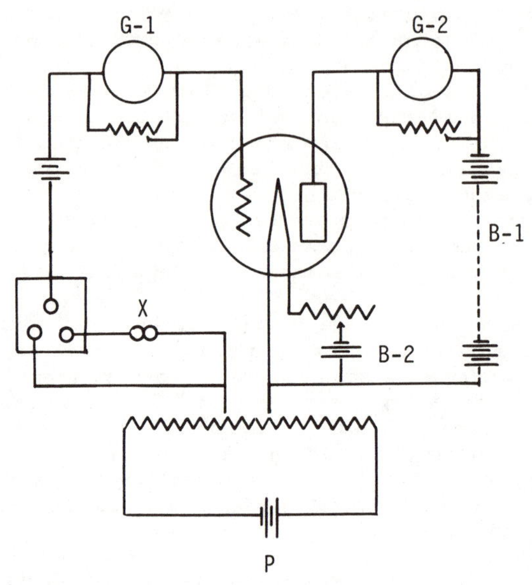

Figure 6. Vacuum tube potentiometer for pH measurement with the glass electrode. X- cell with glass and AgCl-Cl electrodes; P- potentiometer, G-1 and G-2 - high sensitivity galvanometers, B-1 and B-2 - 22.5 and 6 volt batteries. (Reproduced from Ref. 28 Copyright 1929 American Chemical Society).

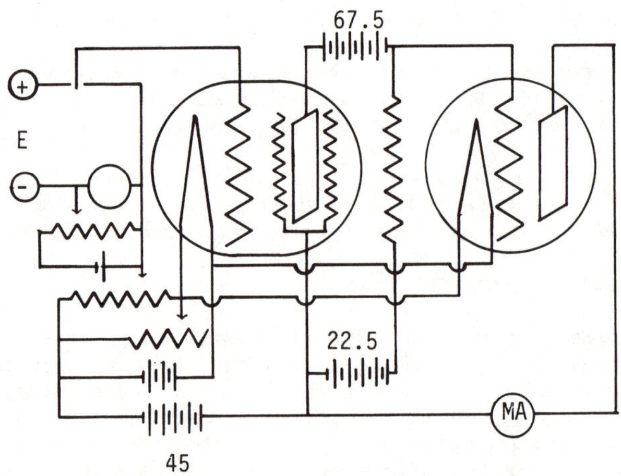

Figure 7. Diagram of a vacuum tube potentiometer for rapid E.M.F. measurement using tetrode and triode. (Reproduced from Ref. 25 Copyright 1929 American Chemical Society)

al in their paper "A Simple, Inexpensive Electron Tube Potentiometer for Use with the Glass Electrode" described some improvements made in the previous circuits. These improvements arose because (1) the tube characteristics changed with a change of resistance in the grid circuit, and (2) the tube required more current in the grid circuit to keep it in equilibrium than was furnished by the glass electrode. This problem was solved by putting a large resistance in the grid circuit, thereby changing the characteristics to the point that the addition of more resistance caused very little change.

The next major step in the measurement of pH was by A.O. Beckman. In his recollections he discussed the reasons for the development and production of the self-contained portable pH instrument that was to become a major factor in the mass production of modern laboratory instrumentation (1):

> Twenty five years later, in 1934, I experienced the type of luck in which preparedness played a role. I was then an assistant professor of Chemistry at California Institute of Technology. One of my University of Illinois classmates, Dr. Glen Joseph, happened to stop by one day for a chat. He told me of a serious problem he was having in making pectin and other by-products from lemon juice....He was forced to use a glass electrode, because it was not affected by the SO_2. He used a thin-wall glass bulb about an inch in diameter, very fragile and easily broken. ...Furthermore, because...of the extremely high resistance of the glass bulb, forced him to use the highest sensitivity galvanometer available to serve as a null point in the Wheatstone bridge....

Thus, Beckman suggested the use of a vacuum tube voltmeter which could be used with smaller and more sturdy glass electrodes than previously possible. As a result of these discussions, he built a two-stage, directly coupled vacuum tube voltmeter for Dr. Joseph – which "luckily" worked. In a few months his friend asked him for another instrument for his personal use. This was the starting point for further consideration (1):

> The thought occurred to me – if he needed two of these instruments in his modest laboratory, perhaps other chemists might have a need for such an instrument.

In 1935 Beckman set forth to produce a portable and rugged pH meter costing $195 – more than one month's salary for an analytical chemist. In this pH meter the amplifier consisted of two vacuum tubes (a tetrode and triode) and a milliammeter placed in the plate circuit of the triode to serve as a balance indicator. Strange as it may seem, the introduction of this instrument was not received with enthusiasm. At the American Chemical Society meeting in San Francisco in the late summer of 1935, the leading analytical chemists were rather dubious and suggested that he approach the leading chemical apparatus dealers. Most dealers were pessimistic, and one of the most optimistic – Arthur H. Thomas Company in Philadelphia – estimated that as many as 500 or 600 instruments might be sold over a ten-year period. Instead of letting the project drop, Beckman continued to make pH meters on his own, and in ten years several hundred thousand had been made without an end to the demand.

After the introduction of the first commercial pH meter, many instrument manufacturers recognized the great market potential and started production of their own versions of the pH meter. Garman and Droz published two papers on pH meters (32,33). One of the designs used a 110 V power supply while the other design used simple batteries and a low current consumption. The latter was sold by the Leitz Company as the Electrotitrator (33). Then Leeds and Northrup introduced their universal pH indicator, No. 7663, based on the Cherry design (34). Coleman marketed a pH meter, Model 20, having a two-stage amplifier – a potentiometric instrument. Thus a variety of pH meters came into the marketplace. These included direct reading, line-operated and negative feedback meters.

Vincent Dole, professor emeritus at a symposium held at the Rockefeller Institute in 1985, commented on what he termed the "Classical Age" (1):

> The Classical Age, the end of which I witnessed briefly in the early 1940's, featured personal skill, precise techniques, and style...With the instrumentation of the Beckman pH meter in 1940, every laboratory could estimate the pH of solution almost as easily as measuring its temperature.... In retrospect, the introduction of the pH meter signaled a new era in biochemistry.

So the new era was born, and continuous improvements were introduced. More sophisticated models incorporated all the frills and features of modern instrumentation–integrated circuits, microprocessor chips, automatic calibration, and computer readout. Thus, today, in the late 1980s, an inexperienced technician can measure the pH with a relatively high precision without understanding the instrumental design or even the concept of pH. We have come a long way.

Literature Cited

1. Beckman, A. O. The Hexagon., Fall 1987, 41.
2. Davy, Sir H. Phil. Trans., 1815, 105, 219.
3. Liebig, J. Ann. Pharm. 1838, 26, 178.
4. Arrhenius S. J. Am. Chem. Soc., 1912, 34, 353.
5. Ostwald, W. Die Wissentschaflichen Grundlagen der Analytischen Chemie, elementar dargestellt Verlag von Wilhelm Engelmann: Leipzig, 1894; p 69
6. Clark, W. M. Oxidation Reduction Potential of Organic Systems; Williams and Wilkins Co.: Baltimore, 1960; p 294.

7. Poggendorff, J. C. LIV. Ann. der Phys. Chem, 1841, 10, 161.
8. Wheatstone, C. XIII. Phil. Trans, 1843, 133, 303.
9. Clark, L. Phil. Trans. 1874, 164, 1.
10. (Unsigned). The Electrical Engineer, XIV April 12, 1893, 355.
11. (Unsigned). The Electrician, 1893, 30, 741.
12. Clark, M. W. The Determination of Hydrogen Ions, 2nd ed.; The Williams and Wilkins Co.: Baltimore, 1927; pp 214, 215.
13. Jones, G.; Kaplan, B. B. J. Am. Chem. Soc., 1922, 44, 2685.
14. Cremer, M. F. Z. für Biologie, 1909, 47, 562.
15. Haber, F.; Klemensiewicz Z. Physik. Chem., 1909, 67, 385.
16. Hughes, W S. J. Am. Chem. Soc., 1922, 44, 2860.
17. Brown, W. E. L. J. Sci. Instrum., 1925, 2, 12.
18. DeForest, L. Father of Radio: The Autobiography of Lee DeForest; Wilcox and Follett Co.: Chicago, 1950.
19. Goode, K. H. M.S. Dissertation, The University of Chicago, 1924.
20. Goode, K. H. J. Am. Chem. Soc., 1922, 44, 26.
21. Calhane, D. F.; Cushing, R. E. Ind. and Eng. Chem, 1923, 15, 1118.
22. Treadwell, W. D. Helv. Chim. Acta, 1925, 8, 89.
23. Williams, W.; Whitenack, T. A. J. Phys. Chem., 1927, 31, 519.
24. Pope, C. G.; Gowlett, F. W. J. Sci. Instrum., 1927, 4, 380.
25. Partridge, H. M. J. Am. Chem. Soc., 1929, 51, 1.
26. Elder, L. W.; Wright, W. H. Proc. Natl. Acad. of Sci., 1928, 14, 938.
27. Wright, W. H. B.S. Thesis, University of Illinois, 1929.
28. Elder, L. W., Jr. J. Am. Chem. Soc., 1929, 51, 3266.
29. Morton, C. Trans. Faraday Soc., 1928. 24, 14.
30. Stadie, W. C. Biol. Chem., 1929, 83, 477.
31. Goodhue, L.D.; Schwarte, L.H.; Fulmer, E. I. Iowa State College J. of Sci., 1932, 7, 10.
32. Garman, R. L.; Droz, M. E. Ind. Eng. Chem., Anal. Ed., 1935, 2, 341.
33. Garman, R. L.; Droz, M. E. Ibid, 1939, 11, 398.
34. Cherry, R. H. Trans. Electrochem. Soc., 1937, 72, 3.

RECEIVED August 18, 1988

Chapter 19

Development of the Glass Electrode

Carl E. Moore[1], Bruno Jaselskis[1], and Alfred von Smolinski[2]

[1]Department of Chemistry, Loyola University of Chicago, Chicago, IL 60626
[2]Department of Medicinal Chemistry and Pharmacognosy, University of Illinois at Chicago, Chicago, IL 60680

The development of the glass electrode as a practical working device has been an evolutionary process. The paper describes the electrode's developmental stages which involve experiments on the conductivity of glass, the use of glass as an electrolyte in voltaic cells, the discovery of the response of glass to the hydrogen ion, the improvements in electrode response through control of glass composition and fabrication, and the development of devices for measuring voltages in circuits containing high resistance components.

The glass electrode is the premier indicating electrode in the field of analytical chemistry when breadth of use and the quality of selectivity are the only considerations. Its practical importance puts it high on the list of historical topics which need to be included in the teaching of the culture of chemistry, but it must be noted that the history of the glass electrode is of interest from several points of view other than that of the immense importance of the electrode itself.

Reflected in the records that embody the development of the glass electrode—in a way rarely equaled—are the tenacious spirit of scientific inquiry and the important part frequently played by intersecting technologies in technological developments. The history of this electrode emphasizes the critical roles occasionally played by unsung heroes and also, by means of its very rich literature, describes in great detail the development of our most widely used indicating electrode system.

Even a rudimentary knowledge of State of the Art electronics of our time—the 1980s—makes it abundantly clear that most of the difficulties encountered in the early developmental work on the glass electrode are directly traceable to the fact that

0097–6156/89/0390–0272$06.00/0

workers of that period lacked a means for permitting accurate and reproducible measurements of the voltages of cells in which glass—i.e., a very high resistance—was a circuit element. The problem of measuring potentials in circuits containing very high resistances was not resolved in a practical way until there was an intersection of the electrochemical and communications technologies. This technological crossover came about via the utilization by the electrochemists of the development of the **Audion**, the first triode, by Lee de Forest in 1906. A major thrust of the research at the time of de Forest was to produce devices that would improve the strength of the wireless-telegraphic signal.

Early Work

The study of electrical properties of materials started systematically with the development of the Electric Machine and the Leyden Jar. These developments allowed the researcher to produce and store the electrical charge on command and by this means to become able to exert a significant amount of control over the electrical experiment.

In 1761 Benjamin Franklin (1) reported that some of the work of Lord Charles Cavendish, the father of the famous Henry Cavendish, demonstrated that glass heated to 400°F became a conductor of electricity. About a century later (1857) remarks by H. Buff (2) make it seem likely that the electrical conductivity of hot glasses was a well-known phenomenon. Buff pointed out that everyone knows that different kinds of glasses differ significantly in insulating ability, that potash glasses are usually the best insulators, and that sodium glasses are less effective insulators. He mentions a very conductive *Zuckerglas* whose analysis shows it to be an almost potassium-free sodium glass. We have not been able to find other references to *Zuckerglas*. Both the Corning Museum and the Jena Glass Works have been unable to supply further information on *Zuckerglas*. We suggest the possibility that the term refers to a cheap glass used in the manufacture of containers for storing candies.

Soon after the publication of the voltaic cell in 1800, the academician Ritter (~1802) of Munich prepared a Zn | Glass | Cu sandwich and noted that glass acted like a moist conductor. Buff (1857) also reported experiments with such arrangements as:

Carbon Powder | Glass | Liquid Zinc Amalgam

Manganese Dioxide | Glass | Liquid Zinc Amalgam

He found that these arrangements would charge a condenser. He concluded that the glass acted as water would act under the same circumstances. The early experimenters frequently used well or spring waters for their experiments. Even the rain water collected in cisterns contained significant amounts of electrolytes.

In 1875 Professor Sir William Thomson (3)—later to become Lord Kelvin—carried out experiments on flint glass sandwiched between copper and zinc plates. He noted that the conductivity increased with an increase in temperature and that there were signs of chemical reaction on both the metal plates.

von Helmholtz, Giese, and Meyer

In 1888 W. Giese, (4) a former assistant of von Helmholtz, reported on a flask containing sulfuric acid which he had wrapped in tin foil. The arrangement was as follows:

Tin Foil | Glass | H_2SO_4 | Pt

In this paper he described an American Virginian glass, widely distributed in the marketplace, which showed such a high conductivity that it behaved like a metal.

On April 5, 1881, von Helmholtz (5) gave the Faraday Lecture at The Royal Institution. The following quote from this English language lecture will give some insights into the state of electrical experimentation in the 1880s and von Helmholtz's recognition of the importance of the high resistance problem (see Figure 1).

> I show you, therefore, this little Daniell's cell...constructed by my former assistant, Dr. Giese, in which a solution of sulphate of copper with a platinum wire, **a**, as an electrode, is enclosed in a bulb of glass hermetically sealed. This is surrounded by a second cavity, sealed in the same way, which contains a solution of zinc sulphate and some amalgam of zinc, to which a second platinum wire, **b**, enters through the glass. The tubes **c** and **d** have served to introduce the liquids, and have been sealed afterwards. It is, therefore, like a Daniell's cell, in which the porous septum has been replaced by a thin stratum of glass. Externally all is symmetrical at the two poles; there is nothing in contact with the air but a closed surface of glass, through which two wires of platinum penetrate. The whole charges the electrometer exactly like a Daniell's cell of very great resistance, and this it would not do if the septum of glass did not behave like an electrolyte; for a metallic conductor would completely destroy the action of the cell by its polarization.
>
> All these facts show that electrolytic conduction is not at all limited to solutions of acids or salts. It will, however, be rather a difficult problem to find out how far the electrolytic conduction is extended, and I am not yet prepared to give a positive answer. What I intended to remind you of was only that the faculty to be decomposed by electric motion is not necessarily connected with a small resistance to the current. It is easier for us to study the cases of small resistance, but the illustration which they give us about the connection of electric and chemical force is not at all limited to the acid and saline solutions usually employed.

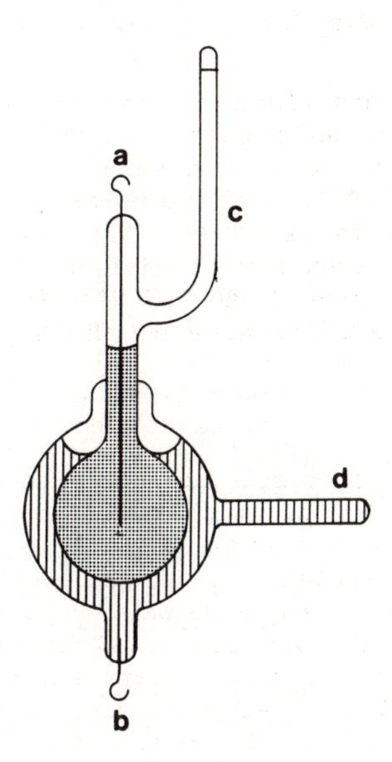

Figure 1. Dr. Giese's Daniell Cell.

In 1890 (6) Georg Meyer, who had worked in Warburg's laboratory in Freiburg, gave an inaugural dissertation at Heidelberg on the topic of the electromotive force developed between glass and amalgams. In this address he described in great detail the preparation of the glass surfaces that he used. Meyer pointed out that in some of his experiments he obtained puzzling results. He noted that in the manufacture of the test tubes which he was using as cells the end closures resulted in alkali-deficient outer-end surfaces, but alkali-rich (normal composition) inner-end surfaces. Flame tests showed the outer-end surface to be alkali-deficient but the inner-end surface to be alkali-rich.

He proposed a model that involved an alkali solution via a film of water (*Wasser Haut*) on both the inner and outer glass surfaces. (Translation from the German)

> These phenomena can be explained through the assumption that the outer glass wall of the test tube was made alkali-deficient when it was exposed to the direct action of the flame during manufacture and thus was not covered by an electromotively active water layer and that, on the other hand, the inner wall, which consists of alkali-rich glass, adsorbs a water layer on its surface. We would observe the electromotive force of
> Hg | Glass | Alkali Solution | Hg since the water on the surface of the glass forms an alkali solution.

Meyer used this model, which he supported with several suitably chosen experiments, to explain the lack of electromotive activity of dry glasses. This explanation was a giant step forward in the evolution of the glass electrode and undoubtedly influenced the thinking of Cremer. Many authors give Cremer full credit for the discovery of the glass electrode.

Cremer's Key Observation

In 1906, at the Institute of Physiology at Munich, Max Cremer published a paper titled "On the Origin of the Electromotive Properties of Tissues, at the Same Time a Contribution to the Study of Polyphasic Electrolytic Cells" (7). This paper is considered by many authors as the announcement of the discovery of the glass electrode.

He pointed out that Warburg considered glass to be a solution of silicic acid containing sodium silicate as the dissolved electrolyte and that, according to Warburg, sodium could be replaced by lithium when a strong current was pushed through hot glass. Cremer referred to the "remarkable electromotive forces of glasses" and said that if the Warburg model is correct, then this would constitute the ideal case of the semipermeable membrane. His discovery—that appreciable voltages were produced when acid was placed on one side and base on the other side of a glass membrane—was the key observation that caught the attention of other researchers.

He described his experiment as follows: (Translation from the German)

> Thus I obtained in the first experiment, when I prepared the following cell:
> Physiological sodium chloride solution — 1/10N sulfuric acid - glass - physiological sodium chloride solution — a difference of potential of 230 millivolts, when conduction was done with physiological saline solution, in the sense that the current goes from the acid into the glass. I used a flask which gave only an insignificant potential difference between the inner and outer wall and now prepared the following cell with it: Zinc in zinc sulfate - 0.6 percent sodium chloride solution - the same plus about 0.01 N sulfuric acid - glass wall - 0.6 percent sodium chloride solution - zinc in zinc sulfate.
>
> If the glass wall had not been there, this cell, like the first, would have shown no current. I was not a little surprised when I was able to discern an electromotive force of around 190 millivolts (current from the acid into the glass). I removed the acid again and replaced it by physiological sodium chloride solution after washing the flask. The potential difference was minimal.
>
> ...When I added dilute sulfuric acid to the 0.6 percent sodium chloride solution on the outside of the flask and sodium hydroxide to the 0.6 percent sodium chloride on the inside, the electromotive force of the combination increased to 0.55 volts when I used a physiological sodium chloride solution as the acting conductor.

Haber and Klemensiewicz

In January of 1909, Haber and Klemensiewicz(8) presented a paper before the Karlsruhe Chemical Society which was published later in the year in the *Zeitschrift für physikalische Chemie* under the title "On Electrical Forces at Phase Boundaries." It is important to note that even though Haber is remembered as a physical chemist, in this study of phase boundary potentials he also had an interest in physiology. He felt that phase boundary phenomena might help explain the action of muscles. In this work he studied the systems glass | water, benzene | water, toluene | water, and metaxylene | water.

Haber developed a model and a theory that explained the forces developed by acids and bases at the boundaries of electrolytically conducting aqueous phases. Haber's "conceptualization" for the glass | water interface envisioned, as reported in the thesis of Hans Schiller, (9) was

1. that the water-wetted surface layer of glass swelled,
2. that the swollen surface layer was a solid water phase (Haber refers his readers to the publications of Otto Schott and Fritz Forster for further information on the swelling process),

3. that the H^+ and OH^- were of constant concentration and immobile in the glass, and
4. that this kind of solid water phase must behave as a metallic hydrogen electrode.

Haber and Klemensiewicz reported the following: (Translation from the German)

> In all cases the test gave confirmation of a theoretical law whose main characteristic is that the boundary potential changes, as does the potential at a reversible hydrogen or oxygen electrode, on both sides of the neutral point, with the logarithym of the acidity, respectively alkalinity. In the case of soft glass, the observation of the interface potentials is so easy that one can base a titrimetric procedure on this voltage.

An interesting insight into this research is contained in a letter from Klemensiewicz to Malcolm Dole published in the *Journal of Chemical Education* in 1980 (10). We quote from the letter:

> I came to Karlsruhe in November 1908....I was then 22. Haber proposed to me to explore the glass electrode, the interest for this topic having been suggested to him by the earlier work of Cremer. They tried already in Karlsruhe some experiments before I came, but without success. I have been handed over the respective apparatus, consisting of a piece of broken glass —cylinder about 3 mm thick, with a tin-foil sticked around. I saw at once that such an element could never work, being short-circuited all over the moist glass surface. Although I didn't know at this stage that glass-balloons have been used by Cremer (and previously by Giese), I blew the thin bulb which remains until today the classical form of the glass electrode. I also installed a quadrant electrometer with the use of which I was well acquainted. With this arrangement I got at once positive results especially as I discovered at the very beginning the aid of steaming and soaking the glass, quite independently, of course. I applied steaming at first as a method of cleaning, being in this case reluctant to use either chromic acid or organic liquids. I also learned in the first days the need of avoiding drying out of the glass-bulb and the superiority of soft over hard glass. Finally, I chose the kind of diagram for plotting the bilogarythmic curve as the most adequate way of presentation. When Haber went to see me in the laboratory after two days, I was able to show him a very good curve in HCl-KOH, with an efficiency of about 0.5V. He would not believe first that it was possible to get these results in such a short time and I had to let him look at the reading telescope that he could plot the points by himself. The experiment proceeded smoothly, so he gave himself an outbreak of enthusiasm, leaped, embraced and praised me in his cheerful manner.

Haber and Klemensiewicz also made an important contribution to the knowledge of the effect of glass composition and pretreatment on glass electrode response.

The Decade of the 1920s

World War I disrupted the work of the great laboratories of central Europe and brought fundamental research to a halt so that the next landmark paper did not appear until 1920, when Freundlich and Rona (11) confirmed the observation of Haber and Klemensiewicz that the glass electrode acted as a completely reversible electrode against the hydrogen and hydroxyl ions and that it was completely insensitive to capillary active substances such as organic dyes. The decade 1920-1930 was a particularly important and fertile period in the development of the glass electrode into a practical tool for the chemist. A listing of seminal papers of this period is given in Table I.

Horovitz, Schiller, and Others of the Period

The mechanism of the process occurring at the glass | solution interface continued to intrigue researchers. In a further study of this problem, Horovitz (12) and Schiller (13), out of Horovitz's laboratory, extended Haber's theory. In this extension they considered

1. that glass is a solid electrolyte in which the Na ion preferentially affects conduction and must act as a metallic sodium electrode,
2. that glass also contains a series of other metal ions so that one cannot treat the glass as a simple electrode function but must consider it a mixed electrode in which the equilibrium condition must be fulfilled for all ionic species,
3. that the glass surface will exchange ions with the solution, and
4. that in aqueous solution hydrogen ions are taken up by ion exchange, causing the glass to act as a hydrogen electrode.

Schiller pointed out in his dissertation that this hydrogen-ion sensing electrode is the most important of the electrodes generated by ion-exchange.

Hughes (14) attempted to establish the limits of the glass electrode by posing the following questions for experimental answers:

1. What is the exact character of the curve representing the relationship between glass-surface potential and hydrogen-ion as measured by the hydrogen electrode?
2. Can the potential of a glass surface be used as a measure of hydrogen-ion concentration in cases where the hydrogen electrode cannot be used, as in the presence of oxidizing agents?

Table I. Important Glass Electrode Papers Appearing in the Period 1920-1930

Year	Author(s)	Title
1920	Freundlich and Rona	On the Relations between the Electrokinetic Potential Break and the Electrical Phase Boundary Force
1921	Freundlich	On Concentrations and Potentials Developed at Boundary Surfaces
1922	Goode	A Continuous Reading Electrotitration Apparatus
1922	Hughes	The Potential Difference between Glass and Electrolytes in Contact with Glass
1923	Horovitz	Ion Exchange on Dielectrics I. The Electrode Functions of Glasses
1924	Schiller	On the Electromotive Properties of Glasses
1924	Trümpler	On a New Way of Determining the Potentials of the Alkalimetals
1924	Brown	The Measurement of Hydrogen Ion Concentrations with Glass Electrodes
1924	Kerridge	The Use of Glass Electrodes
1924	von Steiger	On the Determination of the Hydrogen Ions of Acid Concentrations of Solutions with The Help of the Glass Electrode
1926	Michaelis	The Permeability of Membranes
1928	Elder and Wright	pH Measurement with the Glass Electrode and Vacuum Tube Potentiometer
1930	MacInnes and Dole	The Behavior of Glass Electrodes of Different Compositions

3. What are the conditions under which the glass-surface potential is not a reliable measure of hydrogen-ion concentration, and what other ions influence this potential?

Brown (15) reworked Haber's theory of potential formation at the glass electrode. He noted that the kind of glass used was of great importance and that the electrode was very promising from the point of view of the biologist. He was among the first to mention the problem of shielding and insulation when dealing with high resistance circuits.

Michaelis (16) mentioned that the glass membranes acted like his membranes of dried collodion. This observation points toward the current concept of membrane electrode theory.

Kerridge (17) concerned herself with measurement of the voltage of the cell. She determined the relative utility of the Dolazalek, Compton and Lindemann electrometers. She preferred the Lindemann instrument because of its portability.

From the very earliest studies on the electrical behavior of glass, it had been apparent that the properties of glass were very dependent on composition. However, nearly two centuries were to pass before a significant systematic study was undertaken on the effect of chemical composition on glass electrode response. MacInnes and Dole (18) reported on January 8, 1930—in what should be considered one of the landmark papers on the glass electrode—a systematic study of glass composition versus electrode response to the hydrogen ion. They formulated a number of glasses and measured (1) the potential existing in the diaphragm of the glass to be tested, (2) the resistance of the electrode, (3) the deviation from the theoretical in a buffer at pH 8, and (4) the deviation in an approximately 0.1 N sodium hydroxide solution of pH 12.75. They concluded that the best hydrogen-ion response was obtained from a glass of the composition SiO_2, 70%; Na_2O, 22%; and CaO, 6%.

The response of Thuringer glass to alkali and alkaline earth metals prompted Trümpler (19) to propose the glass electrode as a means of determining normal potentials via the use of amalgams. He found a value of 2.72 V for sodium.

An acceptable model for the mechanism of the behavior of the glass electrode was now in place. The type of glass used, the nature of the pretreatment, the other ions in the solution, and the thickness of the glass bulb were all recognized as experimental parameters requiring control. However, the formidable problems in measurement circuitry associated with the high resistance of glass still remained to be solved, and, as is so often the case in chemistry, the answer to the problem—the measurement of voltages in circuits of high resistance—was already in the literature albeit hidden in a new and poorly understood field of physics.

The Vacuum Tube Is Applied to Acid-Base Chemistry

A significant portion of the science effort during the middle and late 1800s was directed toward the problems involved in improving telegraphic communications. As a consequence of the intense interest and the resulting progress in the field of communications technology, an environment was developed that was suitable for the creation of the vacuum tube. Sir John Ambrose Fleming paved the way for the next major step forward by providing this invention—the diode—in 1904. Fleming used the diode for rectifying the alternating current generated in an antenna by incoming radio waves.

The triode version of the vacuum tube was soon developed by Lee de Forest, who filed for patent coverage on October 26, 1906. An excerpt (20) from de Forest's description of the invention follows:

> It now occurred to me that the third, or control, electrode could be located more efficiently between the plate and the filament. Obviously, this third electrode so located should not be a solid plate. Consequently, I supplied McCandless with a small plate of platinum, perforated by a great number of small holes. This arrangement performed much better than anything preceding it, but in order to simplify and cheapen the construction I decided that the interposed third electrode would be better in the form of a grid, a simple piece of wire bent back and forth, located as close to the filament as possible.
>
> I now possessed the first three-electrode vacuum tube—the **Audion**, granddaddy of all the vast progeny of electronic tubes that have come into existence since.

According to the autobiography, de Forest had trouble making suitable glass envelopes for his experimental vacuum tube. McCandless, an independent manufacturer of what in today's language would be called light bulbs, had a shop near de Forest's laboratory. When de Forest turned to him for help, he not only prepared the glass envelopes successfully but also gave de Forest many helpful suggestions on the experimental work.

The advent of the triode in the communications field via a technology crossover put in the hands of all science a tool of undreamed of worth. Of course, in its early months—and years—its wide range of uses were unrecognized. In his autobiography de Forest (21) remarks on it as follows:

> The Audion, in a measure, is to the sense of sound what a microscope is to that of sight. ...But when the first steps were taken in the work which eventually resulted in the Audion of today, I no more foresaw the future possibilities than did the ancient who first observed magnification through a drop of water.

Goode Develops a Continuous Reading Titration Apparatus

This new development, the triode, was not immediately used. The first report, some 16 years later, of its use in connection with the hydrogen ion was in the construction of a continuous reading titration apparatus which was published as a paper in the *Journal of the American Chemical Society* in 1922. It is interesting to note that this research was done for the M.S. degree. Goode stated both in the paper (22) and in his thesis (23):

> The investigation described in this paper has shown that the 3-electrode vacuum valve ("audion") presents almost the ideal case of a "voltmeter" which draws no current from the source to be measured, and can therefore be employed as a continuous-reading instrument for determining the concentration of the hydrogen ion.

Goode recognized and reported that this was the ideal instrument for the high resistance case.

Elder and Wright Develop a pH Meter

However, there was another delay of several years before a paper appeared in the literature (1928) applying vacuum tube technology to the glass electrode (24). In the summer of 1928 Walter H. Wright, a young radio-ham undergraduate student at the University of Illinois, had the good fortune to take a course, Advanced Analytical Instruments, under Lucius Elder. His memories of the class are that it was wide-ranging and somewhat similar to what we would now call a "Think Tank." As a result of the interest he generated in this class and his knowledge of radio and wireless telegraphy, he was prompted to propose for a B.S. degree thesis project (25) the construction of a device for pH measurement. (The historic little thesis that resulted is held in the rare books collection of the Archives of the University of Illinois at Champaign-Urbana, Illinois.) The device would employ a vacuum tube potentiometer and a glass electrode. Then, in an amazing piece of research, he overcame both the electronics and electrode construction problems and put together a workable instrument for measuring pH. He cannibalized his one-tube radio receiver and used its audion (UV199) to build the potentiometer. He then prepared a soda-lime glass for electrode production. A portion of his account (26) is as follows:

> The first attempt at pure soda-lime glass preparation was a disaster! The melt was made in a platinum dish and heated in a 20KW carbon granule furnace equipped with a tap-off transformer and protected by a circuit breaker in the power supply. I had been assured that the furnace would not melt Ferro-Silicon (approx. 1590°C) and was therefore safe for platinum (M.P. 1753°C). I reasoned that the furnace power supply maximum could be 20KW so

> I set the tap-off transformer at the highest voltage to keep under this figure. When the breaker tripped I set the tap-off transformer at the next lower tap, reset the breaker and repeated the process. As the furnace temperature rose, the carbon resistance fell, thus drawing more current and tripping the breaker. I followed the temperature rise by observing a range of segar cones placed alongside the platinum dish. After several cycles of [the] above steps I removed the lid to observe the melt. White light came from the furnace, the segar cones had melted, also the platinum dish and its charge! One explanation was that the platinum dish ($300 worth) had been purchased during World War I and therefore might not be pure platinum. I then looked up the thesis (CA 1915) covering the original use of the furnace for preparation of Illium type alloys for S. W. Parr. The original user apparently was afraid to trip the breaker and thus never got the full 20KW available and hence did not reach the temperatures possible. In any event the contents of the furnace bottom were removed and turned over to an inorganic grad student (S. C. Ford) for platinum recovery. The next move was to use a graphite crucible and risk some impurity in the glass being picked up from the binder used in the crucible. The melt from the graphite crucible did have a slight purple color possibly due to manganese or cobalt (?). Finally, a large Nickel crucible was used resulting in a melt that was drawn into small rods with a slight brown color, probably due to a trace of dissolved nickel oxide.

Mr. W. H. Wright, after a long career with E. I. du Pont de Nemours and Co., now (1988) lives in Wilmington, DE in retirement. He still has in his possession the little UV199 vacuum tube and some of the original soda glass that he used to fabricate his electrode.

Thus the glass electrode-vacuum tube pH measuring machine made its way into the literature. The novelty of the Elder and Wright study led to its early publication.

Beckman Builds a Practical pH Measuring Instrument

Then, as we all know, A. O. Beckman(27) a few years later (1935) built and commercialized a sturdy and stable pH measuring device, with a two-stage amplifier, that was to make the name Beckman synonymous with pH meter. This line of instruments with improved electrodes and electronics led to papers by the score and has played a very positive role in the development of all phases of science.

In Conclusion

One cannot pore over the records of this development—which delineate the many clever experiments, interminable hours of work, and seemingly endless frustrations endured by our colleagues of the past while enroute to the highly satisfactory solution of the glass electrode problem—without feelings of humility, joy, pride, and deepened respect for our forbearS .

Literature Cited

1. Berry, A. J. Henry Cavendish, His life and Scientific Work; Hutchinson of London, 1960; p 13.
2. Buff, H. Liebigs Ann. Chem. **1857**, 90, 257-283.
3. Thomson, Sir W., F.R.S. Proc. R. Soc. London **1875**, 23, 463-464.
4. Giese, W. Ann. Phys. (Wied. Ann.) **1880**, 9, 161-208.
5. Professor Helmholtz; XLII. J. Chem. Soc. London **1881**, 39, 277-304.
6. Meyer, G. Wied. Ann. **1890**, 40, 244-263.
7. Cremer, M. Z. für Biologie **1909**, 47, 562-607.
8. Haber, F.;Klemensiewicz, Z. Z. phys. Chem. **1909**, 67, 385-431.
9. Schiller, H. Ann. Phys. **1924**, 74, 105-135.
10. Dole, M. J. Chem. Educ. **1980**, 57, 134.
11. Freundlich, H.; Rona, P. Sitz. Ber. Preuss .Akad. Wiss. **1920**, 20, 397.
12. Horovitz, K. Z. Phys. **1923**, 15, 369.
13. Schiller, H. Ann. Phys. **1924**, 74, 105-135.
14. Hughes, W. S. J. Am. Chem. Soc. **1922**, 44, 2860-2867.
15. Brown. W. E. L. J. Sci. Instrum. **1924**, 2, 12-17.
16 Michaelis, L. Naturwissenschaften. **1926**, 14, 33-42.
17. Kerridge, P. M. T. J. Sci. Instrum. **1924**, 3, 404-409.
18. MacInnes, D. A.; Dole, M. J. Am. Chem. Soc. **1930**, 52, 29-36.
19. Trümpler, G.; Z. Elektrochem. **1924**, 30, 103-109.
20. de Forest, L. Father of Radio: the Autobiography of Lee de Forest: Wilcox & Follett: Chicago, 1950; p 214.
21. *ibid.* "Extracts from a paper— 'The Audion. Its Action and Some Recent Applications'—which I read before the Franklin Institute at Philadelphia, January 15, 1920."; p 477.
22. Goode, K. H. J. Am. Chem. Soc. **1922**, 44, 26.
23. Goode, K. H. M.S. Dissertation: Continuous Reading Electro-titration Apparatus, The University of Chicago, 1924.
24. Elder, L. W.; Wright, W. H. Proc. Nat. Acad. Sci. **1928**, 14, 936-939.
25. Wright, W. H. B.S. Thesis: pH Measurement with the Glass Electrode and the Vacuum Tube Potentiometer, University of Illinois, 1929.
26. Letter from W. Wright to C. E. Moore, November 7, 1985. Archival Document, Loyola University of Chicago, Chicago, IL 60626.
27. Tarbell, D. S.; Tarbell, A. T. J. Chem. Educ., **1980**, 57, 133.

RECEIVED August 12, 1988

Chapter 20

pH Glass Electrode and Its Mechanism

K. L. Cheng

Department of Chemistry, University of Missouri–Kansas City, Kansas City, MO 64110

The past major mechanisms are briefly reviewed. The generally accepted ion exchange theory failed to explain the origin of the electrode potential. The newly proposed double-layer and double-capacitor theory emphasizes the importance of electrode surface, double layers, adsorption of both cations and anions on the zwitterionic surfaces, surface active sites, charge density, Boltzmann distribution, etc. The acid and alkali errors, the suspension effect, the membrane thickness, the charging and discharging, and the pH relationship to the membrane capacitance are discussed. It is a pH electrode in acid solutions but a pOH electrode in basic solutions. A challenging view on the validity of applying the Nernst equation to the non-faradaic pH electrode is presented.

Since 1906 when Cremer (1) first observed the potentiometric relationship of a glass membrane with a pH solution and the first systematic study of the glass electrode for pH measurement was by Haber and Klemenciewicz (2) in 1909, many investigators have attempted to explain the glass membrane potential. Table I lists major investigators and their mechanisms for the past 80 years. The pH glass electrode is one of the most widely used analytical tools, yet it has been perhaps one of the least understood. Many quantitative analysis textbooks have presented its theories, but none have given a clear or definitive explanation of the mechanism relating to the origin of the glass membrane potential. The development and the theories of the pH glass electrode have been discussed in many books (5-8). Previous mechanisms have been briefly reviewed by Dole (3), Durst (4), Buck (44), and Cheng (9). Dole explained that the potential of a thin glass film is attributed to the selective permeability or mobility of the H^+ ion across the glass-aqueous solution interface. The concept of an

0097–6156/89/0390–0286$06.00/0

Table I

Major Mechanisms for pH Glass Electrode

Year	Proposed by	Origin of Potential	Reference
1909	Haber	Phase boundary potential or Donnan potential	2
1925	Gross, Halpern		24
1926	Hughes		25
1923	Horovitz	Ion adsorption or ion exchange between Na^+ and H^+	16
1924	Schiller		26
1937	Haugaard		27
1906	Cremer	Membrane or diffusion potential, glass membrane permeable to H^+	1
1926	Michaelis		28
1928	Quittner		12
1931	Lengyel	Adsorption and capacitor of quartz electrode	43
1941	Haugaard	Disproved H^+ diffusion through membrane by chemical analysis	10
1961	Schwabe and Dahms	Disproved H^+ diffusion by tritium	11,14
1941	Dole	Disproved Lengyel's quartz capacitor No OH^- ions taking part	3,5
1984	Cheng	pH electrode in acid medium, pOH electrode in alkaline medium.	21
1985	Cheng	Double-layer and double-capacitor model	29

actual penetration through the glass membrane by hydrogen ions has been definitely disproved by the work of Haugaard (10) and the work of Schwabe and Dahms with tritium (11). It is surprising that in 1985 some chemists still believed that the glass membrane was permeable to the H^+ ion (15). The adsorption-potential theory postulates an adsorbed layer of hydrogen ions on the glass surface causing a potential drop at the glass-solution interface, corresponding to the difference in chemical potential between the free and adsorbed ions (12). The most generally accepted ion exchange equilibrium theory proposes that the gel layer of the glass membrane acting as an ion exchanger produced a phase boundary potential at which the H^+ ion exchanges with the Na^+ ion (13,16,43). Such an ambiguous proposal has never been supported by experimental results. If such an exchange did occur, there would be no net change in interfacial charges. Furthermore, the sodium ion in the gel layer of membrane would be eventually depleted after a long time use, resulting in the failure of the glass electrode. But the pH glass electrode can be used for a long time. It is also known that a quartz glass membrane containing no sodium could act as a pH glass electrode (43). Addition of sodium oxide to the glass is only to increase the conductivity of the membrane (10,17,18). Prior to the ion exchange theory, the mechanism of a concentration gradient of the proton suggests that the potential arose due to the proton concentration difference on both sides of the thin semipermeable glass membrane. The Donnan equilibrium with mathematical formulas and equations was used to support the theory where a Donnan theory term was proposed. It seems that the failure of previous theories is caused by erroneous assumptions. Many investigators claimed that their theories were based on thermodynamics using the Nernst equation to explain the logarithm relationship between the potentials and the concentrations. Previous investigators devoted too much time interpreting the glass electrode in terms of thermodynamics. Haugaard commented that,

> ". . .most of the previous experiments in this direction have failed because the investigators have tried to deduce their theories from the thermodynamic treatment of systems in equilibrium. But we know that thermodynamics alone can not tell us the mechanism of a process. . . Potentials of glass electrode systems not in equilibrium have been measured and compared. . ." (10).

In the past, it seemed fashionable to explain the mechanism with thermodynamics. As a whole, thermodynamics is always right. However, its usefulness depends on how it is applied to a particular system. In commenting on the historic development before 1947 in the treatment of electrochemical reactions across interfaces, Bockris (19) stated that most electrochemists were still trying to do the impossible, i.e., to treat the highly thermodynamically irreversible electrode reactions by a series of misconceptions and approximations on the basis of reversible thermodynamics. This fundamental error and lack of conceptualization held a dead hand on the mode of achieving electrochemical reactions and on the electrochemical energy conversion for 4 to 5 decades. He called this period in electrochemistry

"The Great Nernst Hiatus." In direct potentiometry using membrane electrodes, a similar situation has occurred in the past 8 decades. Every quantitative and instrumental analysis textbook has treated the glass electrode as a battery obeying the principles of reversible thermodynamics and the Nernst equation. We may follow the similar case and call the period of 8 decades as "The Great Nernst Hiatus of Direct Potentiometry." This paper summarizes a new concept based on capacitances as an attempted explanation.

Battery and Capacitor

A battery is a device containing no insulator and producing an electric current through chemical redox reactions that occur in cells that are placed in series. Each cell contains an anode and a cathode that are immersed in an electrolyte medium. Connecting the anode and the cathode to the external circuit causes an electric current flow until chemical reactions cease. Mostly it is reversible, following the principles of equilibrium thermodynamics, i.e., the Nernst equation. On the other hand, a capacitor is a device for storing electric charges through two conducting plates between which there is a dielectric in which no reversible redox reaction takes place. By connecting two charging plates, no significant current flows. For a capacitor, $C = q/V$, where C is the capacitance, q is the charge, and V is the potential difference.

In the literature the pH electrode is generally represented in a cell as follows (30,44).

Ref. electrode | H^+ unknown | Glass membrane | H^+ (known) | Ref. electrode

$$E_{glass} = \text{constant} + 0.05916 \log \frac{a_{H^+}\ (\text{unknown})}{a_{H^+}\ (\text{known})} \qquad (1)$$

It has been known for a long time that there are no redox reactions involved in the potential development of glass electrode (3,6,10). Based on the definitions of battery and capacitor shown above, it is difficult to accept the statement, "The electrode component of a pH measuring system is comparable to a battery whose voltage changes with pH" (15). We would like to point out that the pH glass electrode is comparable to a capacitor rather than to a battery. It is extremely important to have a correct assumption and concept before deducing any mechanism.

Capacitor

Figure 1 (a) illustrates a capacitor used in a common electric device that is able to hold charges on applying a potential with a battery or power source. In Figure 1 (b) and (c), the membrane electrodes may have their arrangements as shown. This depends on what functional group is present on the surface and whether the cations or anions are adsorbed. The glass electrode or other membrane electrodes may be represented by Figure 1(c) which is the zwitterionic surface structure where both cations and anions may

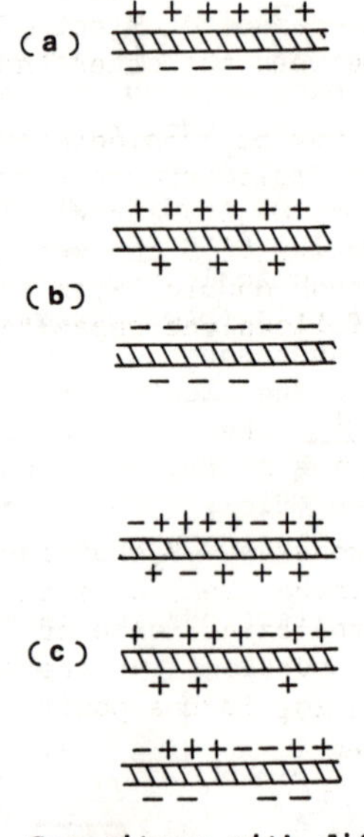

Figure 1. Capacitors with different charges.

be adsorbed simultaneously. For instance, Na^+ and OH^- can be adsorbed simultaneously on the pNa electrode as discussed in the latter part of this paper. Each side of membrane interface potential exists due to the presence of the net charges. When the glass electrode is immersed in an electrolyte solution a double layer is formed that contains surface charges.

Since both layers (inner and outer) of the membrane are in touch with the electrolyte solutions, two double layers are formed on both sides with the same membrane. Helmholtz considered the double layer as a capacitor (20). A membrane with double layers may be considered as double-capacitor electrode. The potential difference (Δ E) is that between the outer and inner interfacial potentials (Figure 2).

An electrode capacitor may be considered as a membrane material made from either a dielectric or a semiconductor which can adsorb cations and anions on its active sites. The membrane potential is believed to derive from the two interfaces that hold charges on the surface through double layer adsorptions. The glass electrode potential follows the capacitance law.

$$E = \frac{q}{C} = \frac{q\,d}{k\,A\,\varepsilon} \tag{2}$$

there k is the dielectric constant of the glass membrane, ε is the permittivity, A is the membrane area, d is the membrane thickness, q is the charge, and C is the capacitance of the membrane. When ε , K, A, d, and C are constant for the same glass membrane, E is then proportional to q (q_+ is the positive charge, and q_- is the negative charge). Then,

$$E = K\,(\Sigma q_+ - \Sigma q_-) \tag{3}$$

where K is a new constant. If only cations (H^+) are adsorbed on the membrane surface, $\Sigma\, q_- = 0$, resulting in the potential increasing with increasing positive charges; if only anions are adsorbed (OH^-), $\Sigma\, q_+ = 0$, resulting in the potential decreasing with increasing negative charges; and if both cations and anions are adsorbed simultaneously, the result is the sum of positive and negative charges (net charges).

From equation (3), besides the charges, the membrane potential is affected by the surface area (A) and the thickness (d), which have been demonstrated (21,22). We have measured the capacitance of the glass membrane as a function of pH as discussed in the last part of this paper. The membrane electrodes behave definitely as capacitors.

Tubular pH Glass Electrode

Tests have been made with a tubular pH glass electrode for quantitative determination of the charge density effect on potential. A cylindrical and pH sensitive glass electrode of 1.0 cm X 1.5 cm was specially prepared and used as a substitute for a pH bulb electrode. The whole section of the tube was sensitive to pH, when sealed, and the tube contained a pH 7 phosphate buffer

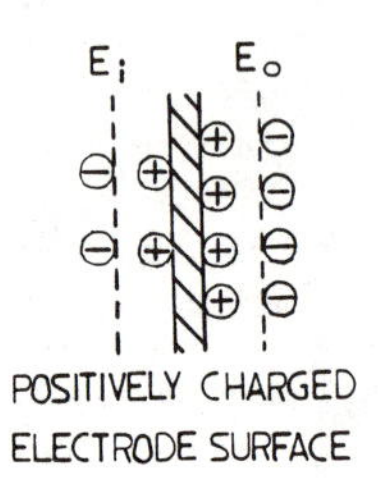

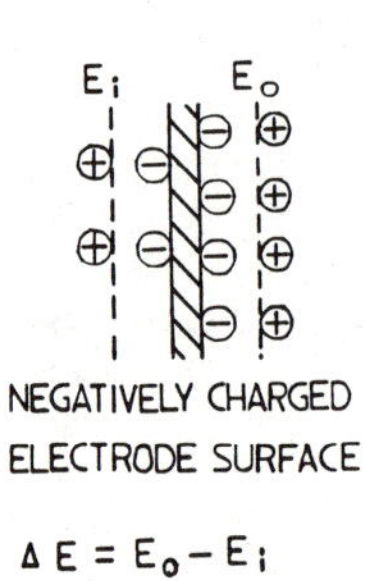

$$\Delta E = E_o - E_i$$

Figure 2. Electrode double layers.

solution and a Ag/AgCl reference electrode. The outer surface of the tubular electrode exposed to the test solution was controlled by the depth of immersion, i.e., the amount of charge on the same whole membrane surface area varied as a function of immersion depth. The results are shown in Figure 3. Different amounts of H^+ or OH^- ions are adsorbed on the surface as a result of varying depth of immersion of the tubular electrode into the same solutions. In acid solutions the potentials increase with increasing depth of immersion. Similarly, in basic solutions the potentials decrease with increasing depth of immersion. At approximately pH 5.5 (isoelectric point of the glass membrane), the potential remains the same regardless of the depth of immersion (the point of zero change, pzc). In acid media, the increased positive potentials are the result of the increased adsorption of H^+ ions on the electrode surface. In basic media, the increased negative potentials are the result of the increased adsorption of OH^- ions on the electrode surface (or possibly neutralizing the surface proton). It emphasizes the fact that the pH glass electrode is a pH electrode in an acid medium, but it is a truly pOH sensor in a basic medium. It should be logically called as a pOH electrode in basic solutions. This has been evidenced by the negative charges of the glass membrane in basic solutions, (Figures 3 and 6). This is one of the fundamental differences between the capacitor theory and the past theories which do not consider the role of OH^- ions in the potential development. The concentration of OH^- has never been included in the cell diagram and the Nernst equation. A correct concept dealing with the actual measurements using a pH glass electrode is very important; otherwise, misleading conclusions may be made (9, 23).

Tubular pNa Electrode

A similar experiment with a tubular pNa electrode at different depth of immersion was carried out in a pH 10.5 buffer (ethylenediamine) with different concentrations of sodium nitrate. For dilute sodium solutions, the potentials decreased to more negative values with increasing depths of immersion, at higher sodium ion concentrations in the same buffer solutions (above 1 M) (Figure 4), the slope changed to positive values. This shows that the pNa glass electrode is also a pH and pOH electrode with a zwitterionic surface which can adsorb both OH^- and Na^+ ions (9). The potential response by the pNa electrode is the result of Equation (3). This explains the reason that the determination of sodium ion with a pNa electrode must be done in a basic buffer solution.

Double Layers

The double layer or triple layer has been known for a long time (19). When an ionic solid, in particular an oxide, is placed in an electrolyte solution, cations or anions are adsorbed on the surface depending on the surface charges or the functional groups at the surface. For a quartz or glass surface, when it is

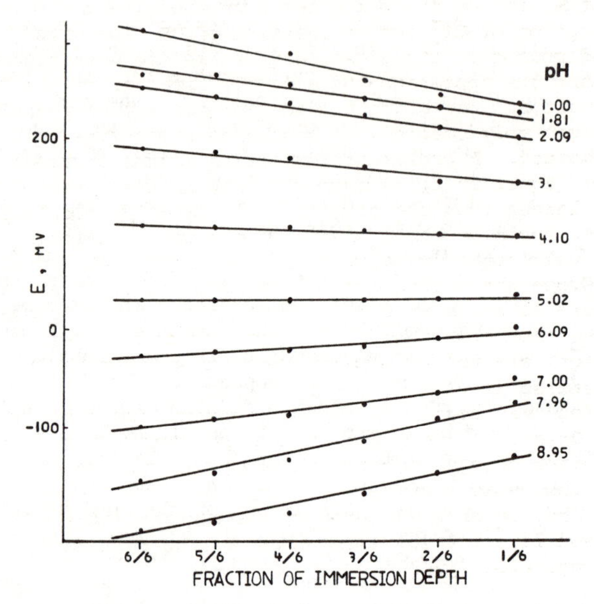

Figure 3. Effect of electrode surface contact in the solution on the potential.

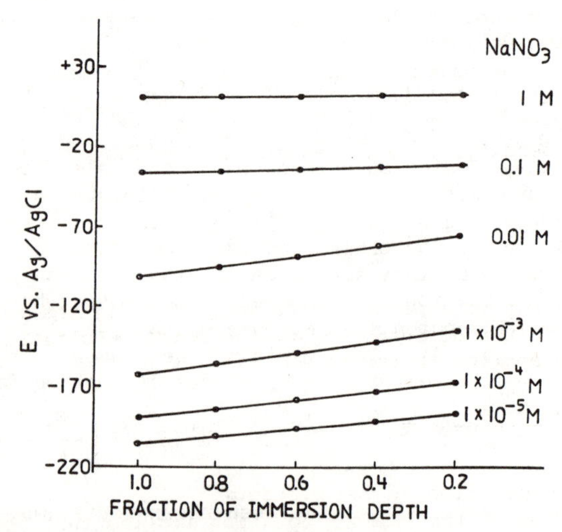

Figure 4. Effect of pNa electrode surface contact in the solution on the potential.

hydrated, the surface contains polymerized silicic-acid which is weakly ionized and can interact with H^+ and OH^- ions. The adsorption of H^+ ion on the glass has been known in the past (10), but the adsorption of OH^- ion or reaction of OH^- ion with undissociated proton on the glass surface has not been reported. We believe that the appearance of a double layer or triple layer yields a capacitor. The glass surface becomes either positively charged or negatively charged, while the solution (the bulk) is oppositely charged. A preliminary estimate of the distance (d) of the capacitor (Equation 2) is approximately in the range of 0.03-0.8 μm, much larger than the commonly calculated double layer thickness. It could be that the distance which we calculated differs from what other investigators calculated with different equations. Since the membrane has both inner solution and outer solution, there is an inner double layer and an outer double layer. The E_o (emf from the outer capacitor) and E_i (emf from the inner capacitor) are formed through the double layers. The measured potential is $\Delta E = E_o - E_i$ (Figure 2).

An interface may acquire a charge by several possible means. According to Ottewill, the charge on a glass surface is controlled by the ionization of chemical groupings on the surface (35). It is called an ionogenic surface.

The distribution of ions close to the surface (in the so-called diffused layer) is commonly given by the Boltzmann distribution equation (35,36),

$$\frac{n_i}{n_{io}} = \exp\left(\frac{-z_i e \psi}{kT}\right) \quad (4)$$

where ψ is the potential difference, n_i is the number of ions of type i per unit volume in the vicinity of the surface, and n_{io} is the concentration of ions far from the surface (bulk concentration). The valence number z_i is either a positive or negative integral, e the fundmental unit of charge, k the Boltzmann constant, and T the absolute temperature.

The charge density, ρ^*, is related to the ion concentration as follows:

$$\rho^* = \sum_i z_i e n_i = \sum_i z_i e n_{io} \exp\left(\frac{-z_i e \psi}{kT}\right)$$

Confining attention to a planar interface, if ρ^* is the volume charge density in the diffused layer, the Poisson-Boltzmann equation may be applied in the form

$$\nabla^2\psi = \frac{\partial^2 \psi}{\partial \chi^2} = -\frac{e}{\varepsilon} \sum_i z_I n_{io} \exp\left(\frac{-z_i e \psi}{kT}\right)$$

where x is the separation of two plates.

The derivation of the Poisson equation implies that the potentials associated with various charges combine in an additive manner. On the other hand, the Boltzmann equation involves an exponential relationship between the potential and the charges. We should realize that these equations are approximations (36).

The Boltzmann equation has been applied to the derivation of relationship betwwen potential and pH (9,45).

Adsorption of Hydrogen and Hydroxyl Ions by Glass

The adsorption of hydrogen ion by Corning glass powder has been studied by Haugaard (10). Since the previous investigators (3) did not believe that the OH^- ion could take part, they never tried to measure the OH^- ion adsorption by the glass. They used the term "absorption" instead of "adsorption". We carried out the experiments for adsorption of both H^+ and OH^- ions (9). The results are shown in Figure 5. At pH 6.2, there was no significant adsorpton of H^+ or OH^- ions. It was also noted that at the same pH and the same pOH, equal number of H^+ and OH^- were adsorbed. At pH 3 or pOH 3, approximately 13 H^+ and OH^- ions per nm^2 were adsorbed. The pH 6.2 is therefore isoelectric point of the glass, and no ions can be adsorbed on the surface (42).

Acid and Alkaline Errors

The acid and alkaline errors of the pH measurements by the glass electrode have baffled chemists ever since the glass electrode was created. After many studies, the conclusion has been that it is too complicated to be understood (30). The Nernst equation cannot explain the nonlinearity at the very low and very high hydrogen ion concentrations. If we accept the following concepts, the acid and alkaline errors are easily understood.

1. The glass surface has a limited number of active sites, and cannot accommodate an unlimited mnumber of H^+ and OH^- ions.
2. Concentrated alkali solutions damage the glass surface.
3. The H^+ and OH^- adsorptions follow the ion adsorption isotherm.
4. At very high concentrations of NaOH, not only the adsorption of OH^- is nonlinear with the concentration, the Na^+ ion may also be adsorbed to offset the negative charges. As mentioned previously, the glass can act as a zwitterionic surface.
5. The electrode is a capacitor.

Chemical Amplification of Potential

A membrane electrode acting as a parallel capacitor should follow the capacitance law as shown in Equation (2). The E_{total} resulting from connecting the capacitors in series with multimembrane electrodes is the sum of each capacitor potential,

$$E_{total} = E_1 + E_2 + E_3 + E_4 + \ldots\ldots$$

for the same solution. The results of 4 pH glass electrodes in series connections are shown in Figure 6. The emf after amplification for both pH 1.0 and 13.0 exceeded one volt. This is a remarkable chemical potential amplification. We further verify that the increase in the positive potential is due to the

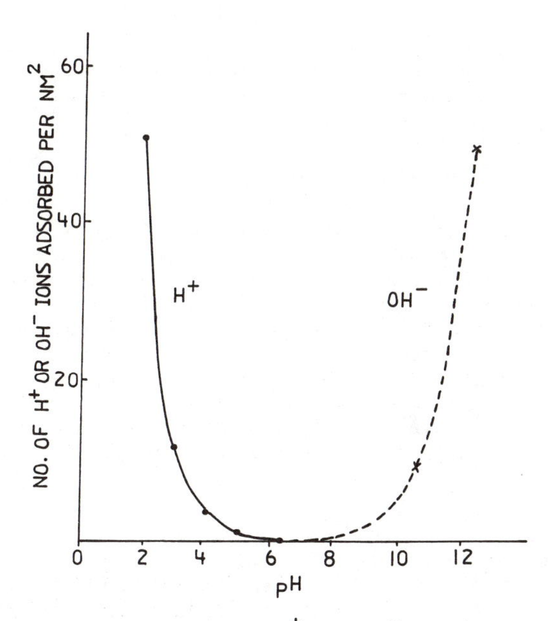

Figure 5. Adsorption of H^+ and OH^- ion on pH glass.

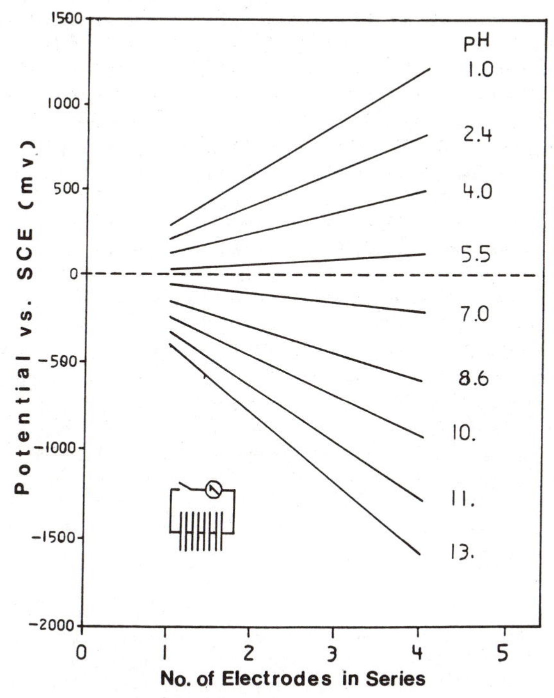

Figure 6. Chemical potential amplification through electrode connection in series.

adsorption of an increased number of H^+ ions, and the increase in the negative potential is due to the adsorption of an increased number of OH^- ions. A multimembrane pH glass electrode can be constructed with higher sensitivity.

Charge and Discharge of pH Glass Electrode

The results in Figure 7 indicate that by grounding the glass electrode, it took less than 3 seconds to discharge completely to zero potential. On the other hand, it took 1-2 min. to charge the glass electrode because of the relatively slow process of chemical charging by the migration of ions to the electrode surface. Qualitatively, one can see a dramatic charging effect by injecting the positive or negative particles on the electrode surface through a zerostat gun (available from Aldrich Chemical Co.) which is used to neutralize the static charges.

Suspension Effect

The suspension effect on the pH measurements by the pH glass electrode has been known for a long time (31, 32). This effect can not be explained by the Nernst equation or ion exchange theory. It is generally attributed to the change in the junction potential of the reference electrode (33,34). We carried out an experiment to separate the reference electrode of Ag/AgCl in a clear NaCl solution from the colloid solution in which the glass electrode was immersed. The two solutions were connected through a piece of copper wire. We found that the colloid particles affected the potential measurements instead of the reference electrode. The suspension effect is caused by the charged colloid particles which have their own double layers and potential from their single capacitors. These tiny colloid particles nearby the glass electrode surface will play an important part in the development of the whole electrode potential. The remedy for eliminating the suspension effect is to measure the pH of the supernatant portion or to coagulate the suspended charged particles with a neutral and polyvalent salt (43).

Difficulties encountered in the pH measurement of emulsions have been reported (37). Glass is capable of taking up other substances than water, for instance, alcohol and acetone (10). It is plausible that some interfering substances in the emulsions may be adsorbed on the glass, reducing the effective surface area, or changing the charge density. The emulsion effect may be the same in nature as the suspension effect caused by the charged particles.

Solid State pH Glass Electrode

Solid state electrodes for copper (38) and for fluoride (39) have been reported. A solid state pH glass electrode has also been reported (40), contrary to the previous belief that a membrane electrode must have an internal solution which contains an ion to be determined in the external solution. Now, we know that this is not so. The above reports further show that even the internal

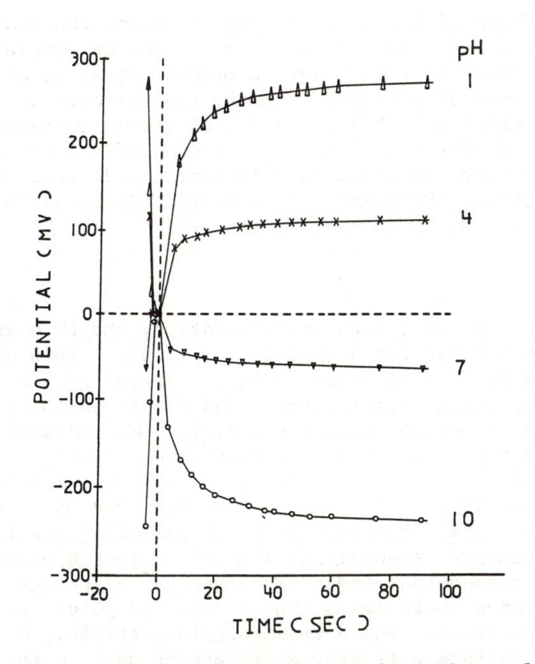

Figure 7. Charge and discharge of pH glass electrode.

solution is not required. This means that the cell diagram and the Equation (1) lose their significance. According to the capacitor theory, the presence of an internal solution is immaterial, so long as there is a charge contact at the inner layer of the membrane for the emf measurement. The coated wire electrodes are another example without an internal solution.

Capacitance of Glass Electrode at Different pH

The capacitance of the flat pH glass membranes was measured at pH 1, 4, and 7, with an internal solution of pH 1, 4, 7, 10 at 140 Mohms. When both the external and internal solutions were pH > 7, a same value of 90 pF/cm^2 was obtained. When both the external and internal solutions were pH < 7, the same value of 60 pF/cm^2 was obtained. If the pH of the internal solution was pH < 7 and the external solution was pH > 7, or vice versa, the capacitance was from 72 to 75 pF/cm^2. Based on the capacitance relationship,

$$\frac{1}{C_s} = \frac{1}{C_1} + \frac{1}{C_2} + \ldots = \frac{1}{90} + \frac{1}{60} = \frac{5}{180}$$

$C_s = 36$, for half of the average capacitance.
The average of individual capacitance will be 36 x 2 = 72 pF/Cm^2. This is in good agreement with the experimental results.

Conclusion

The historical development of the pH glass electrode mechanism has been briefly reviewed. Special attention is directed to the fact that membrane electrodes including the pH glass electrode are not a battery as previously believed, but a capacitor. The correct concept is of utmost importance in the understanding of the origin of electrode potentials. Experimental results are presented to support the notion of ionic adsorptions and double layers in electrode potential development. The pH glass electrode is truly a pH electrode in acid media, but it is a pOH electrode in basic media. An electrode with both the internal and external solutions contains a double-layer and double-capacitor membrane. The chemical amplification of potential has been demonstrated, suggesting the device of multi-membrane electrodes for higher emf signals. The acid and alkaline errors, the suspension effect, and the emulsion effect have been discussed. The capacitance of pH glass membrane is measured as a function of pH. The Nernst equation was misused in the past for the pH glass electrode. Its partial success in the quantitative relationship between the potential and the pH may be coincidental. The development of the capacitor theory for the membrane electrodes should stimulate further studies in the double layers and membrane electrodes.

Acknowledgments

The author is grateful to H. A. Droll, G. D. Christian, and E. W. Hellmuth for helpful discussions and to H. Z. Song, Susie X. R. Yang, and Cynthia Ferrendelli for their potential measurements.

Literature Cited

1. Cremer, M. Z. Biol. 1906, 47, 562.
2. Haber, F.; Klemenciewicz, Z. K. Z. physik. Chem. 1919, 67, 385.
3. Dole, M. J. Am. Chem. Soc. 1931, 53, 4260.
4. Durst, R. A. J. Chem. Educ. 1967, 44, 175.
5. Dole, M. The Glass Electrode; Wiley: New York, 1941.
6. Bates, R. G. Determination of pH. Theory and Practice; 2nd ed. Wiley: New York, 1964.
7. Eisenman, G. Glass Electrodes for Hydrogen and Other Cations. Principles and Practice; Eisenman, G., Ed.; Dekker: New York, 1967.
8. Durst, R. A. In Ion Selective Electrodes; National Bureau of Standards, Special Publ. 314, Washington, DC 1969.
9. Cheng, K. L. Proc. 31st IUPAC Congress, Anal. Chem. Div. 1987, p. 173.
10. Haugaard, G. J. Phys. Chem. 1941, 45 148.
11. Schwabe, K.; Dahms, H. Z. Elektrochem. 1961, 65 518.
12. Quittner, F. Ann Physik. 1928, (4) 35, 745.
13. Eisenman, G. Biophys. J. 1962, 2, 259.
14. Schwabe, K.; Suschke, H. D. Angew. Chem. 1964, 76, 39.
15. Rothstein, F.; Fisher, J. E. Am. Lab. 1985, 17, 124.
16. Horovitz, K. Z. Physik. 1923, 15, 369; Z. Physik. Chem., 1925, 115, 425.
17. MacInnes. D. A., Belcher. D. J. Am. Chem. Soc. 1931, 53, 3315.
18. Warburg, O. Ann. Physik. 1884, 21, 622; Ber., 1884, 17, R193.
19. Bockris, J. O'M.; Reddy, A. R. N. Modern Electrochemistry; Plenum: New York, 1970, Vol., p. 16.
20. Helmholtz, H. L. von. Wied. Ann. 1879, 2, 337.
21. Cheng, K. L.; Chang, H. P. Mikrochim. Acta 1985, I, 95.
22. Cheng, K. L..; Kar Chaudhari, S. Mikrochim. Acta 1981, I., 185.
23. Licht, S. Anal. Chem. 1985, 57, 514.
24. Gross, F.; Halpern, C. Z. Physik. Chem., 1925, 115, 54; 1925, 118, 255.
25. Hughes, N. S. J. Chem. Soc. 1928, 491.
26. Schiller, H. Ann Physik (4), 1924, 74, 105.
27. Haugaard, G. Nature 1937, 140, 66.
28. Michaelis, L. Die Naturwiss. 1936, 14, 33.
29. Cheng, K. L. Seminar at Emory University, March 13, 1984.
30. Skoog, D. A.; West, D. M.; Holler, F. J. Fundamentals of Analytical Chemistry; Saunders: Philadelphia, 1988, p. 367-368.
31. Chernoberezhskii, Yu. M. In Surface and Colloid Science; Matijevic, E. Ed.; Plenum: New York, 1982, Chapter 5.
32. Brezinski, D. P. Talanta 1983, 30, 347.
33. Illingworth, J. A. Biochem. J. 1981, 195, 259.
34. Peck, T. R.; Bock, R. H. Proc. 9th Soil-Plant Analysis Workshop. North Bridgeton, Missouri, 1983.
35. Ottewill, R. H. In State Electrification; Inst. Phys. Conf. Ser. No. 27; Press Roman: 1975, Chapter 1, p. 56.
36. Hiemenz, P. C. Principles of Colloid and Surface Chemistry; 2nd Ed., Dekker: New York, 1986.

37. Berthod, A.; Saliba, C. Analusis 1986, 14 414.
38. Vlasors, Yu. G.; Kocheregin, S. B.; Ermolenko, Yu. E. Zh. Anal. Khim. 1977, 32 1843.
39. Komljenovic, J.; Krka, S.; Radic, N. Anal. Chem. 1986, 58, 2693.
40. Parr, R. A.; Wilson, J. C.; Kelly, R. G. Anal. Proc. 1986, 23, 291.
41. Harris, W. E.; Kratochvil, B. An Introduction to Chemical Analysis; Saunders: Philadelphia, 1981, p. 285.
42. Alois, B.; Stanislava, H. Slevarenstvi, 1976, 24, 361.
43. Lengyel, B. v. S. physik. Chem., Abt. A. 1931, 153, 425.
44. Buck, R. P. In Comprehensive Treatise of Electrochemistry; White, R. E.; Bockris, J. D'M.; Conway, B. E.; Yeager, E. Plenum: New York, 1984. Chapter 3, p. 137-242.
45. Morrison, S. R. The Chemical Physics of Surfaces; Plenum: New York, 1977, p. 267.

RECEIVED August 3, 1988

Chapter 21

Ion-Selective Electrodes

From Glasses to Crystals and Crowns

J. D. R. Thomas

School of Chemistry and Applied Chemistry, Redwood Building, University of Wales College of Cardiff, P.O. Box 13, Cardiff CF1 3XF, Wales, United Kingdom

This paper summarizes the history of development of ion-selective electrodes from the first description of glass pH electrodes at the beginning of the century.

A Long Beginning

The history of ion-selective electrodes (ISEs) began at the beginning of this century with the discovery in 1906, by Cremer (1) of the glass electrode sensitive to hydrogen ions. This was soon followed by the detailed studies in 1909 of Haber and Klemensiewicz (2). By the late 1930s, glass membrane electrodes, as part of electronic pH meters, were seen as standard items of laboratory equipment when, along with the polarograph and photoelectric colorimeters (absorptiometers), they marked the beginning of the era of the use of physico-chemical instrumentation in chemical laboratories. Such instrumentation has evolved by advances in electronics and the phenomenal developments in computerization for experimental control and data processing. This sophistication has called for improvements in sensors and detectors. The history of ISEs shows what has been achieved in the way of an increase in the number of ions which can be selectively determined, improvements in selectivity and sensitivity, an ever-widening range of applications, and better understanding of functional mechanisms in order to promote further development.

Non-glass membranes were introduced in 1935-36 by Tendeloo (3,4) who made measurements with slices of barium sulphate and calcium fluoride (fluorite). This was followed in 1937 by the use of pressed silver chloride polycrystals on platinum discs by Kolthoff and Sanders (5). Twenty years later, Tendeloo and co-workers (6-9) went on to describe membranes of calcium salts of low water solubility in low melting point paraffin.

In the late 1950s, Eisenman and co-workers exploited the

0097–6156/89/0390–0303$06.00/0

deviation from the Nernst equation in alkaline solutions (alkaline error) observed in 1922 by Hughes (10). They showed (11-12) that glass compositions could be formulated to give metal ion function at relatively low pH. To realize this cation selectivity relative to hydrogen ions, it is necessary to introduce a structural element into the glass lattice in a coordination state higher than its oxidation state. For example, when an element R, in oxidation state -3 is introduced into glass in four-fold coordination in place of silicon(IV), the structure is

$$[- \mathrm{R(III)} - \mathrm{O} - \mathrm{Si(IV)} -]^-$$

in which the replacement for Si(IV) has resulted in a site which can participate in cation exchange. When alumina is incorporated in alkali silicate glasses, cation responses are obtained. Most cation-selective glasses are alkali metal aluminosilicates and examples of useful compositions for cation selective glass electrodes are shown in Table I.

TABLE I. Some Compositions of Glasses for Cation Sensitive Glass Electrodes

Cation to be measured	Percentage glass composition			Selectivity (approximate)
	Na_2O	Al_2O_3	SiO_2	
Na^+	11	18	71	$k^{pot}_{Na,K}$ 3.6×10^{-4} at pH 11
K^+	27	5	68	$k^{pot}_{K,Na}$ 0.05
Ag^+	28.8	19	52.2	$k^{pot}_{Ag,H}$ 10^{-5}
Ag^+	11	18	71	$k^{pot}_{Ag,Na}$ 10^{-5}

Note: Adapted from Reference 13.

It is generally accepted that the mechanism of operation of a glass electrode depends on an ion-exchange process (14). However, although hydrogen ions undergo exchange across the solution/hydrated layer interface, these ions do not penetrate the membrane. This was demonstrated by coulometric experiments (15) involving prolonged electrolyses in glass electrode bulbs filled with tritium-labelled sample solutions when no increase in tritium was found on the non-labelled side of the membrane, even after 20 h of electrolysis. Thus, it was concluded that the charge-carrying mechanism across the membrane was based on ion-exchange at the solution/hydrated layer interface, associated with diffusion within the hydrated layer and an interstitial charge carrying mechanism (by sodium ions for sodium silicate glass) through the dry glass layer. In this middle layer, each charge carrier merely needs to move a few atomic diameters before transferring its energy to another carrier.

A Fast Awakening

The real move towards ISEs based on non-glass membranes came in 1961 when Pungor and Hollos-Rokosinyi (16) produced a membrane by incorporating silver iodide into paraffin. This led on to the development of heterogeneous membrane electrodes, especially those based on silicone rubber matrices (17-20).

The research activity on ISEs was greatly enlivened with the development of the homogeneous crystal membrane fluoride electrode in 1966 by Frant and Ross (21) followed a year later by a liquid membrane electrode for calcium by Ross (22). Another significant development at this time was the work of Stefanac and Simon (23-24) on neutral carrier antibiotics which showed valinomycin to be very selective for potassium. After evaluating various membrane support materials, such as filter paper, polyethylene film, Thixcin and nylon mesh, they settled on membranes of low porosity and antibiotic solutions of high viscosities for obtaining high selectivity potassium ISEs from valinomycin (24). Others were active in the field, and prominent among those who contributed to wider use and appreciation of the utility of the selective properties of ISEs were Rechnitz and co-workers (13,25-27).

By the mid-1960s, the author of this paper was assembling a team for ISE researchers at UWIST (Cardiff, Wales), where the early work was based on the prospect of resacetophenone oxime as an ion-sensor for copper(II) ions, this being based on its selectivity as an ion-exchanger (28). Concurrently, recognition was given (29) to earlier membrane potential data (30) whereby membrane potentials (E_M), in a cell consisting of two compartments containing a salt (AY) solution of different activities (a' and a" for A of valence z_A) on each side of the membrane (each containing a reference electrode) could be determined as

$$E_M = \frac{2.303RT}{z_A F} \left(\log \frac{a''_A}{a'_A}\right) \quad (1)$$

However, it was recognized (29-30) that equation (1) is an over-simplification since, in addition to the Donnan phase boundary potential, it should include a term to allow for diffusion phenomena arising from the fact that the membrane is more permeable to counter ion A than to co-ion Y:

$$E_M = \frac{2.30RT}{z_A F} \left[\log \frac{a''_A}{a'_A} - (z_Y - z_A) \int_{'}^{''} t_Y d \log a^{\pm}\right] \quad (2)$$

where t_Y is the transport number of the co-ion in the membrane and z_Y its valence and $a^{\pm}$ is the mean activity of the electrolyte. Essentially, the first term of equation (2) gives the thermodynamic limiting value of the concentration potential, while the second term gives the deviation due to the co-ion flux. For a membrane ideally permselective towards the counterion, the second term vanishes ($t_Y = 0$) and equation (2) reduces to the Nernst type equation (1).

In support of permselectivity, later radiotracer diffusion studies of calcium chloride through selective PVC matrix calcium ion-selective membranes based on organophosphate sensor showed calcium counterion permselectivity with negligible permeation by chloride co-ions (31). The PVC matrix membrane permeation studies also showed selective permeation of calcium ions when compared with other cations (31).

The First PVC Matrix Membrane Electrodes

Considerable interest was focussed on ISEs based on liquid ion-exchangers absorbed on an inert support. However, these early liquid membrane electrodes were affected by short life-time and poor character of the inert support needed to define the interface between the membrane and the analyte solution. These disadvantages were neatly overcome in the author's laboratories by entrapment of the liquid membrane in a PVC matrix membrane (32) by the straightforward procedure of simply dissolving PVC in a solution of liquid ion-exchanger (or neutral carrier sensor system) in tetrahydrofuran (THF) and slowly evaporating the THF at room temperature to give a master membrane sufficient for making several ISEs (32-34). This procedure is far more straightforward than an earlier proposal for a calcium ISE whereby a mixture of thenoyltrifluoracetone calcium ion-sensor plus tributyl phosphate plasticizer and PVC was knife-coated on a woven fabric and heat-cured (35).

The PVC matrix membrane electrodes offer good mechanical resistance to stirring and pressure effects and, although several other materials have been assessed, PVC is still by far the most commonly used support matrix.

Conventional electrodes are relatively large (3-10 mm membrane diameter), and are not suitable for small samples and applications like *in vivo* measurements. This has prompted research into alternative designs and of miniature ISEs. Microelectrodes have tip sizes generally in the 0.25-5 μm range, that is, small enough to penetrate a single cell. A large number of such electrodes have been constructed using different carriers and designs (36-37) but they tend to be of high resistance and noisy due to their small size. Several designs have been developed to reduce this problem and among the newer ones in an all-solid-state construction based on a backfill of electrically conducting epoxy (38). In general, though, microelectrodes are in widespread use.

With regard to simple miniaturization, coated-wire ISEs are an attractive approach (39-40). This involves coating the corresponding conventional polymer membrane onto a platinum or other wire. Numerous coated-wire electrodes selective to different ions have been prepared (41-42), despite the lingering controversy over the performance of these devices in terms of stability, reproducibility and the question as to the mechanism at the interface between the metal wire and polymer film. Cattrall, Drew and Hamilton (43) proposed that, since PVC is permeable to both oxygen and water, an oxygen electrode is set up at the platinum/PVC interface. Hulanicki and Trojanowicz (44) used this model for interpreting improvements in the stability of solid-state calcium ISEs using silver or graphite contacts to the membrane compared with similar membranes contacted by silver chloride coated silver. Schindler and co-workers (45-46), also concluded that a Pt/O_2 system existed in these coated-wire type ISEs.

According to Buck (47-49) there is no basis for expecting the potentials of the membrane/metal boundary of coated wire electrodes to be stable with time. He calls this a "completely blocked system", that is, it is "blocked" for carrying both electrons and ions. Buck considers such a completely blocked interface as a condenser. Nevertheless, coated-wire electrodes are widely used, especially for potentiometric titrations (41,50-51).

Coated-wire electrodes are late entrants into the all-solid-state electrode realm, for crystal membrane electrodes of this mode were already common. The advantages of all-solid-state construction are summarized by Nikolskii and Materova (52). In general, the most important matter for all ISEs with solid internal contact is that of ensuring reversibility and equilibrium stability of the transition from ionic to electronic conductivity. Buck (47-49) as well as Nikolskii and Materova (52), have reviewed the field, covering many electrode types, including Selectrodes, where a graphite rod is covered with an ion-selective membrane of either solid crystal active substances (53-56), or liquid ion-exchanger, or chelate (53-59).

Selectivity Considerations

The name of Nikolskii is synonymous with selectivity of ISEs (60-61), through the celebrated Nikolskii equation

$$E = \text{constant} + \frac{2.303RT}{z_A F} \log \left(a_A + k^{pot}_{A,B}(a_B)^{z_A/z_B}\right) \qquad (3)$$

which gives the parameters of the electrode potential, E, measured with respect to a reference electrode for Nernstian response. Thus, an ISE is taken to respond selectively to an ion, A, of activity a_A and charge z_A in the presence of an interfering ion, B, of activity a_B and charge z_B, where $k^{pot}_{A,B}$ is the selectivity coefficient. The various approaches to determining selectivity coefficients were summarized (29, 62-63), in the early heady days of ISE researches, but more recently some additional proposals have been made, such as, the matched potential method of Gadzekpo and Christian (64).

Essentially, there are two separate-solution methods with the respective solutions producing e.m.fs., E_1 and E_2. The equations used depend, respectively, on equality of the primary and interferent ion activities (equation 4) or of the e.m.fs. (equation 5):

$$\frac{E_2 - E_1}{2.303RT/z_A F} = \log k^{pot}_{A,B} + \left(\frac{z_A}{z_B} - 1\right) \log a_A \qquad (4)$$

$$a_A = k^{pot}_{A,B} (a_B)^{z_a/z_B} \qquad (5)$$

A more common procedure for determining selectivity coefficients is to measure the e.m.f. in solutions containing a fixed amount of interferent, B, with varying activities of the primary ion, A, for which the electrode is designed. The general idealised pattern (65) involves the calculation of $K^{pot}_{A,B}$ by equation (5), when the values of a_A and a_B appertain to the intercept at that part of the calibration curve (of zero slope) corresponding to complete interference by the interferent with that of Nernstian or near-Nernstian slope for a more or less individual response by the primary ion. This procedure has been recommended for assessing ISEs (66-67).

Srinivasan and Rechnitz (62) used two rather complicated equations according to whether high $k^{pot}_{A,B}$ or low $k^{pot}_{A,B}$ values were being determined. The matched-potential method of Gadzekpo and Christian (64) is an alternative mixed solution method. Here, the potential of a reference solution of the primary ion is measured and then again after addition of a further known amount of primary ion. To a second aliquot of the reference solution, a solution of the interfering ion is added to produce the same potential change as the added primary ion. The ratio of the concentrations of the primary ion and interfering ion needed to give the same potential change represents the selectivity coefficient. Among the claimed advantages (64) are the fact that the method makes no assumption about the slope of the response curve for the interfering ion, and that there is freedom of interference from drift.

Ion-Selective Field Effect Transistors (ISFETs)

An alternative approach to miniaturization of ISEs has been through ion-selective field effect transistors (ISFETs). Essentially, these integrate the ISE with the field effect transistor (FET) of the input stage of a pH/millivoltmeter, which takes the high-impedance signal of an ISE and outputs a low-impedance signal which is less sensitive to noise pick-up. In effect, ISFETS eliminate the connecting cable between the ISE and the input FET by coating the ion-selective material directly on to the gate area of the FET as innovated by Janata and co-workers (68). Important contributions have since been made as reviewed by Janata and Huber (69-70). More recently, Sibbald, Covington and Carter (71) have described a four-function ISFET assembly for hydrogen, potassium, sodium and calcium ions, as well as certain other advances at Newcastle (72). Progress also continues at other centres, such as, in a flatpack mounted ISFET device set in a Perspex flow cell which permits flow injection analysis (FIA) and rapid substitution of one ISFET sensor for another (73) and in photopolymerisable butyl methacrylate membranes (73) which offer prospects of mass produced membranes on the FET gate (73).

Bergveld (74) was, of course, the first to report an ISFET as a chemically sensitive semiconductor device (CSSD). A broader grouping is that of chemically sensitive electronic devices CSEDs), sub-classified by Bergveld and van der Schoot (75) into ISFETs as "active electronic components" as one member of a four-class system. Other categories are passive electronic components (resistor, capacitor, etc.), electronic and opto-electronic systems (oscillating crystal sensors, etc.), and systems with a chemical feedback (dynamic oxygen sensor, and coulometric sensors).

Exploitation of Ion-Selective Electrodes

The above is a rather condensed historical survey of the development of glass, solid and liquid membrane types of ISEs. The further developments have been modifications for actual applications of which micro electrodes are but one example. Other modifications include incorporation of ISEs into flowing systems as in the FIA mode pioneered by Ruzicka and co-workers (76-78), the use of complementary membranes, such as, gas-permeable membranes for gas sensing (79), and enzyme membranes for utilizing the additional selectivity available from enzyme reactions (80-81).

Among variations in approaches for electrolyte measurements with ISEs is the Kodak Potentiometric Dry Chemistry System, where batchwise measurements are made on disposable slides. These are manufactured by coating an appropriate polymeric membrane over an internal reference gel layer on a conductive substrate. Slides for potassium, sodium, chloride and carbon dioxide are available. A drop of sample is placed on one half of the slide and a drop of reference solution on the other half. The drops are connected by an electrolyte bridge and the e.m.f. measured via metallic contacts.

Liquid Membrane Type Sensors

Without in any way negating the wide-ranging utility of glass membrane electrodes for pH, pNa and pNH_4 measurements and the good solid-state membrane electrodes for fluoride, sulphide and other ions, the main focus of development work of the last twenty years has been in the search for better liquid membrane systems. This has been stimulated by the convenience of PVC matrices for trapping and supporting liquid membranes (32-34,83).

The original calcium bis didecylphosphate used with dioctyl-phenyl phosphonate (22) as a liquid ion-exchanger for calcium ion-sensing was improved by incorporating a phenyl group between the alkyl and phosphate oxygen of the ion-selective component (84). Attention to purification of the ionophore yielded a conveniently synthesized (85) long-ranging calcium ion sensor (86). The calcium ion-selectivity function is dependent on the synergism of dioctylphenyl phosphonate as plasticizing solvent mediator (22,87), while decan-1-ol when used in place of the phosphonate reduces selectivity to yield a divalent electrode (87-89). Under appropriate conditions, such as low calcium ion levels, the divalent electrode can be used as a magnesium ISE. This has been nicely shown for physiological and intracellular model systems for speciation using chemometrics approaches (90-93). The use of ligands for ion-buffer systems (84) is also nicely demonstrated for extending the calibration range of calcium ISEs down to 10^{-8}M, that is, to well within the intracellular range (87,90-94). Of considerable interest in this respect is a single pH buffered multi-ligand system based on HEDTA, EGTA and NTA for setting up $[Ca^{2+}]$ standards between 10^{-4} and 10^{-8}M by mere spiking with 0.02M calcium chloride (90,91,93).

Liquid membrane anion ion-exchanger type sensors are best illustrated by the nitrate ISE (95-96). This has seen a role in the environmental screening of waters and effluents (97) for nitrate, but such use has been somewhat diminished by the high sensitivity and facility for multi-ion analysis of the newer approach of ion-chromatography.

The ion-exchanger type ISEs can be readily converted to other forms, as illustrated by conversion of a nitrate ion-exchanger to the chlorate form, and assembly into a chlorate ISE (98). Much work continues to be done on anion liquid ion-exchangers for anion ISEs (see Recent Titles in each Volume of Ion-Selective Electrode Reviews). For these tetraalkylammonium and -phosphonium, tetraphenyl-arsonium and -phosphonium, and dye salts are a popular basis (99).

Cation liquid ion-exchanger sensors are much more varied in type (99). Some examples are: dialkyl- and di(alkylphenyl) phosphates for calcium and divalent ions (as discussed above); tetraphenylborates for onium ions and for drug type cations either by directly or by potentiometric titration; lead diethyldithiocarbamate in trichloroethane for lead(II) (100); and mercury(II) PAN chelate in trichloromethane for mercury (101).

Crown Ether and Neutral Carrier Sensors

Considerable work has been directed to the search for better neutral carriers, motivated by the very successful valinomycin sensor for potassium (23-24). Impetus was given to the development of this electrode by the observation of Moore and Pressman (102) that the antibiotic is capable of actively transporting potassium across rat mitochondria membranes. There are numerous synthetic macrocyclic polyethers capable of forming complexes with monovalent cations. Some of the earliest ISE studies were by Rechnitz and Eyal (103) on dicyclohexyl- and dibenzo-18-crown-6 ligands synthesized by Pedersen (104). Soon afterwards, Petranek and Ryba (105-106) assessed various types of crown-6 and larger crown compounds for potassium ion-sensing and promoted dimethyldibenzo-30-crown-10 with dipentyl phthalate in PVC as the best of the series. Generally, these and later studies on crown ether sensors have indicated that the selectivity for potassium does not exceed that of valinomycin.

Researches on crown ether derivatives are now much more widespread. Thus, Pungor and co-workers (107) have studied the synthesis and application of different mono- and bis-crown ether derivatives with urethane and urea linkages. Moody, Owusu and Thomas (108-110) have studied dibenzo-30-crown-10 and its derivatives as ISE sensors for diquat (DQT) and paraquat (PQT), the best being based on dibenzo-18-crown-10 with DQT.2TPB and 2-nitrophenyl phenyl ether in PVC. These studies integrate the molecular receptor researches of Stoddart at Sheffield and structural work of Williams at Imperial College, London, with the ISE work at UWIST (111). On the fundamental side, the researches showed (111-112), that DB30C10 changes its shape dramatically to accommodate the guest molecules, while bis para-phenylene-34 crown-10 (BP34C10) hardly modifies its shape when it engulfs PQT. However, this feature is not carried through to improving the PQT^{2+} electrodes which are best based on PQT.2TPB without crown ether, and activated by a charge transfer interaction (113) between PQT^{2+} and TPB^-. There is no advantage gained in replacing TPB by an alternative anion (109).

Apart from Pungor and co-workers (107), others have prospected in the area of bis crown ethers, e.g., Shono and co-workers have shown good potassium ion-selectivity for some bis (15-crown-5) derivatives (114-115) where one containing a dodecyl group in the link exhibits good lipophilicity and longer lifetime for the resulting ISEs (115).

Passing onto smaller crowns, Shono and co-workers (116) have studied a very useful dodecylmethyl-14-crown-4 which shows good ISE selectivity towards lithium over sodium. Lithium ISEs have to function over a small range around 1 mM $[Li^+]$ in the rather stringent conditions of blood electrolyte levels: only relatively few sensors approach the required selectivity specifications (117). Among other sensors that have received recent attention, and having reasonable prospects of meeting the requirements in respect of lithium selectivity over sodium, and of being functional for use with blood serum, are a barium propoxylate complex which functions by its affinity for lithium (118-119), and some sensors based on non-crown diamide carriers studied by Simon and co-workers (120-122) and by Christian and co-workers (123-124).

The function of neutral carriers, such as the crown ether compounds discussed above, is to bind ions reversibly and to transport them across the aqueous/organic interface. The neutral carrier forms a solvation shell around a cation of appropriate charge and size, thereby solubilizing it in the organic phase. Some influencing factors are important, such as, steric interaction and the effect of the solvent on the metal cations (125). Ideally, these induce a selective permeation of one type of ion into the membrane, although the detailed mechanism for a particular ion appears not to be fully resolved (126). However, despite mechanistic permselectivity controversies, a large number of neutral ligands have been shown to be good ionophores for use in ISEs. An early, tremendously successful, case of the synthetic systems is that of ETH 1001 for calcium (127) which has been widely adopted in commercial calcium ISEs instead of the earlier organophosphate calcium ion-sensors. Simon and co-workers (128) have reviewed neutral carrier ISEs, while some more recent reviews have taken on more specialized aspects, e.g., for lithium (117).

More recent work has pressed towards even more selective sensors for calcium, stimulated by a 2:1 ligand:calcium stoichiometry found by Petranek and Ryba (129) for a calcium crown ether which exhibited high calcium ion selectivity over magnesium. Thus, Shono and co-workers (130) have studied a bicyclic polyether amide derivative with enhanced selectivity for calcium, while Simon and co-workers (131) have reported a non-crown neutral carrier for calcium which forms a 3:1 ligand:calcium complex with very high selectivity for calcium over sodium and potassium, that is, by factors of $10^{7.4}$ and 10^{8}, respectively. This indicates that macrocyclic systems are not a prerequisite for high selectivity.

A further prospecting area for neutral carrier ligands is in the field of anion binding, where Simon and co-workers are making some significant inroads (132-133).

Conclusion

The numerous studies on developing ion-selective electrodes far exceed those which can be covered in this short historical review. The full extent of researches are reflected in a huge volume of references given in more comprehensive reviews, such as, an earlier article by Moody and Thomas (134), the various articles in Ion-Selective Electrodes Reviews (135), and in many published reviews. The recent Database on Electrochemical Sensors (136), with bibliography covering 1983-86 emphasizes the continuing additions to the literature; here, there are more than 2500 individual entries from more than 3500 authors. Not quite all are on ISEs, but four of the five most active electrochemical sensor groups cited are of the ISE fraternity, namely, Pungor (Budapest), Simon (Zürich), Rechnitz (Delaware, USA) and Thomas (Cardiff, Wales, UK); the fifth group is that of Karube (Tokyo, Japan) who are mainly engaged on non-potentiometric biosensors.

Finally, it has not been a prime aim of this review to discuss applications. Detailed information on specific application areas can be obtained from reviews cited in this article and elsewhere. Nevertheless, the widest application area concerns the clinical chemical use of ISEs for blood electrolyte analyses where Working Group on Ion-Selective Electrodes, affiliated to the International Federation of Clinical Chemistry, have devoted their energies to promoting the use of ISEs in this way and to equating that data obtained so that interpretation can be based on the procedures used for flame spectrometric data. The published Proceedings of the Annual Meetings are an invaluable record of deliberations in this area (137).

Literature Cited

1. Cremer, M. Z. Biol., 1906, 47, 562.
2. Haber, R; Klemensiewicz, Z. Z. Physik. Chem., (Leipzig), 1909, 67, 385.
3. Tendeloo, H. J. C. Proc. Acad. Sci., Amsterdam, 1935, 38, 434.
4. Tendeloo, H. J. C. J. Biol. Chem., 1936, 113, 333.
5. Kolthoff, I. M.; Sanders, H. L. J. Am. Chem. Soc., 1937, 59, 416.
6. Tendeloo, H. J. C.; Krips, A. Rec. Trav. Chim., 1957, 76, 703.
7. Idem, ibid., 1957, 76, 946.
8. Idem, ibid., 1958, 77, 406.
9. Tendeloo, H. J. C.; van der Voort, F. H. ibid., 1960, 79, 639.
10. Hughes, W. S. J. Am. Chem. Soc., 1922, 44, 2360.
11. Eisenman, G.; Ruskin, D. C.; Casby, J. H. Science, 1957, 126, 831.
12. Eisenman, G. Biophys. J., 1962, 2, 259.
13. Rechnitz, G. A. Chem. and Eng. News, 1967, 45, 146.
14. Eisenman, G. (Ed) Glass Electrodes for Hydrogen and Other Cations; Edward Arnold: London, 1967.
15. Schwabe, K.; Dahms, H. Monatsber Deut. Akad. Wiss. Berlin, 1959, 1, 279.
16. Pungor, E; Hollos-Rokosinyi, E. Acta Chim. Acad. Sci. Hung. 1961, 27, 63.
17. Pungor, E.; Havas, J.; Toth, K. ibid., 1964, 41, 239.
18. Pungor, E.; Toth, K.; Havas, J.; ibid., 1966, 48, 17.
19. Pungor, E. Anal. Chem., 1967, 39, 28A.
20. Pungor, E.; Toth, K. Analyst, 1970, 95, 625.
21. Frant, M.; Ross, J. W. Science, 1966, 154, 1553.
22. Ross, J. W. Science, 1967, 156, 1378.
23. Stefanac, Z.; Simon, W. Chimia, 1966, 20, 436.
24. Stefanac, Z.; Simon, W. Microchem. J., 1967, 12, 125.
25. Redhnitz, G. A.; Branner, J. Talanta, 1964, 11 617.
26. Rechnitz, G. A.; Zamochnik, S. B. ibid., 1964, 11, 1061.
27. Rechnitz, G. A.; Kresz, M. R.; Zamochnik, S. B. Anal. Chem., 1966, 38, 973.
28. Moody, G. J.; Oke, R. B.; Thomas, J. D. R. Unpublished work.
29. Moody, G. J.; Thomas, J. D. R. Selective Ion-Sensitive Electrodes; Merrow Publishing Co.: Watford, 1971.
30. Helfferich, F. Ion-Exchange; McGraw-Hill: New York, 1952.
31. Craggs, A.; Moody, G. J.; Thomas, J. D. R.; Willcox, A. Talanta, 1976, 23, 799.
32. Moody, G. J.; Oke, R. B.; Thomas, J. D. R. Analyst, 1970, 95, 910.
33. Craggs, A.; Moody, G. J.; Thomas, J. D. R. J. Chem. Educ., 1974, 51, 541.

34. Moody, G. J.; Thomas, J. D. R. In Ion-Selective Electrodes in Analytical Chemistry; Freiser, H., Ed.; Plenum Press: New York, 1978; Vol. 1, p. 339.
35. Shatkay, A. Anal. Chem., 1967, 39, 1378.
36. Walker, J. L.; Brown, H. M. Physiol. Rev., 1977, 57, 729.
37. Brown, H. M.; Owen, J. D. Ion-Selective Electrodes Revs., 1979, 1, 145.
38. Khalil, S. A. H.; Lima, J. L. F. C.; Moody, G. J.; Thomas, J. D. R. Analyst, 1986, 111, 611.
39. Cattrall, R. W.; Freiser, H. Anal. Chem., 1971, 43, 1905.
40. Davies, J. E. W.; Moody, G. J.; Price, W. M.; Thomas, J. D. R. Laboratory Practice, 1973, 21, 20.
41. Cattrall, R. W.; Hamilton, I. C. Ion-Selective Electrode Revs., 1984, 6, 125.
42. Moody, G. J.; Thomas, J. D. R. Laboratory Practice, 1978, 27, 285.
43. Cattrall, R. W.; Drew, D. M.; Hamilton, I. C. Anal. Chim. Acta, 1975, 76, 269.
44. Hulanicki, A.; Trojanowicz, M. ibid., 1976, 87, 411.
45. Schindler, J. G.; Stork, G.; Struth, H. J.; Schmid, W.; Karaschinski-Fresenius, K. D. Z. Anal. Chem., 1979, 295, 248.
46. Heidecke, G.; Kropt, J.; Stork, G.; Schindler, J. G. ibid., 1980, 303, 364.
47. Buck, R. P. In Ion-Selective Electrodes in Analytical Chemistry; Freiser, H., Ed.; Plenum Press: New York, 1973; p. 1.
48. Buck, R. P. In Ion-Selective Electrode Methodology; Covington, A. K., Ed.; CRC Press: Boca Raton, Florida, 1979, Vol. 1, p. 175.
49. Buck, R. P.; Shepard, V. R. Anal. Chem., 1974, 46, 2097.
50. Cosofret, V. V.; Buck, R. P. Ion-Selective Electrode Revs., 1984, 6, 59.
51. Vytras, K. ibid., 1985, 7, 77.
52. Nikolskii, B. P.; Materova, F. A. ibid., 1985, 7, 3.
53. Ruzicka, J.; Tjell, J. Anal. Chim. Acta., 1970, 51, 1.
54. Idem. ibid., 1970, 49, 346.
55. Rujicka, J.; Lamm, C. G. ibid., 1971, 53, 206.
56. Idem. ibid., 1972, 62, 15.
57. Lamm, C. G.; Hansen, E. H.; Ruzicka, J. Anal. Chim. Acta. 1972, 5, 451.
58. Hansen, E. H.; Lamm, C. G.; Ruzicka, J. Anal. Chim. Acta. 1972, 59, 403.
59. Hansen, E. H.; Ruzicka, J. Talanta, 1973, 20, 1105.
60. Nikolskii, B. P.; Schultz, M. M. Zh. Fiz. Khim., 1962, 36, 704.
61. Nikolskii, B. P.; Schultz, M. M.; Belijustin, A. A.; Lev, A. A. In Glass Electrodes for Hydrogen and Other Cations; Eisenman, G., Ed.; Dekker: New York; 1967.
62. Srinivasan, K.; Rechnitz, G. A. Anal. Chem., 1969, 41, 1203.
63. Moody, G. J.; Thomas, J. D. R. Laboratory Practice, 1971, 20, 307.

64. Gadzekpo, V. P. Y.; Christian, G. D. Anal. Chim. Acta., 1984, 164, 279.
65. Ross, J. W. In Ion-Selective Electrodes; Durst, R. A., Ed.; National Bureau of Standards: Washington, D. C., 1969, Special Publications 314, p. 57.
66. Moody, G. J.; Thomas, J. D. R. Talanta, 1972, 19, 623.
67. Guilbault, G. G.; Durst, R. A.; Frant, M. S.; Freiser, H.; Hansen, E. H.; Light, T. S.; Pungor, E.; Rechnitz, G. A.; Rice, N. M.; Rohm, T. J.; Simon, W.; Thomas, J. D. R. Pure and Appl. Chem., 1976, 48, 127.
68. Moss, S. D.; Janata, J.; Johnson, C. C. Anal. Chem., 1975, 47, 2238.
69. Janata, J.; Huber, R. J. Ion-Selective Electrode Revs., 1979, 1, 31.
70. Janata, J.; Huber, R. J. In Ion-Selective Electrodes in Analytical Chemistry; Freiser, H., Ed.; Plenum Press: New York, 1980; Vol. 2, p. 107.
71. Sibbald, A; Covington, A. K.; Carter, R. F. Med. Biol. Eng., 1985, 23, 329.
72. Covington, A. K.; Whalley, P. D. J. Chem. Soc., Faraday Trans. I, 1986, 82, 1209.
73. Moody, G. J.; Slater, J. M.; Thomas, J. D. R. Analyst, 1988, 113, 103.
74. Bergveld, P. IEEE Trans., BME-17, 1970, 70.
75. Bergveld, P.; van der Schoot, B. H. Selective Electrode Revs., 1988, 10, 5.
76. Ruzicka, J.; Hansen, E. H. Anal. Chim. Acta., 1975, 78, 145.
77. Hansen, E. H.; Ghose, A. K.; Ruzicka, J. Analyst, 1977, 102, 705.
78. Hansen, E. H.; Ruzicka, J.; Ghose, A. K. Anal. Chim. Acta., 1978, 100, 151.
79. Ross, J. W.; Riseman, J. H.; Kreuger, J. A. Pure and Appl. Chem., 1973, 45, 473.
80. Moody, G. J.; Thomas, J. D. R. Analyst, 1975, 100, 609.
81. Guilbault, G. G. Ion-Selective Electrode Rev., 1982, 4, 187.
82. Costello, P.; Kubasik, W. P.; Brody, B. B.; Sine, H. E.; Bertsch, J. A.; D'Souza, J. P. Clin. Chem., 1983, 29, 129.
83. Moody, G. J.; Thomas, J. D. R. In Ion-Selective Electrode Methodology; Covington, A. K., Ed.; CRC Press: Boca Raton, Florida, 1979; Volume 1, p. 111.
84. Ruzicka, J. H.; Hansen, F. H.; Tjell, J. C. Anal. Chim. Acta., 1973, 67, 155.
85. Craggs, A.; Keil, L.; Key, B. J.; Moody, G. J.; Thomas, J. D. R. J. Inorg. Nucl. Chem., 1978, 40, 1483.
86. Craggs, A.; Moody, G. J.; Thomas, J. D. R. Analyst, 1979, 104, 412.
87. Craggs, A.; Keil, L.; Moody, G. J.; Thomas, J. D. R. Talanta, 1975, 22, 907.
88. Orion Research Inc., 92-32 Divalent Electrode.
89. Hassan, S. K. A. G.; Moody, G. J.; Thomas, J. D. R. Analyst, 1980, 105, 147.
90. Otto, M.; Thomas, J. D. R. Anal. Proc., 1984, 21, 369.

91. May, P. M.; Murray, K.; Otto, M.; Thomas, J. D. R. Anal. Chem., 1985, 57, 1511.
92. Otto, M.; Thomas, J. D. R. ibid., 1985, 57, 2647.
93. Otto, M.; Thomas, J. D. R. Ion-Selective Electrode Revs., 1986, 8, 55.
94. Craggs, A.; Moody, G. J.; Thomas, J. D. R. Analyst, 1979, 104, 961.
95. Orion Research Inc., Bibliography, April 15, 1970.
96. Davies, J. E. W.; Moody, G. J.; Thomas, J. D. R. Analyst, 1972, 97, 87.
97. Sommerfeld, T. G.; Milne, R. A.; Kozub, G. O. Comm. Soil. Sci. Plant Anal., 1971, 2, 415.
98. Hiiro, K.; Moody, G. J.; Thomas J. D. R. Talanta, 1975, 22, 918.
99. Moody, G. J.; Thomas, J. D. R. Ion-Selective Electrode Revs., 1982, 3, 189.
100. Coetzee, C. J.; Basson, A. J. Anal. Chim. Acta., 1977, 92, 399.
101. Cosofret, V. V.; Zugravescu, P. G.; Baiulescu, G. E. Talanta, 1977, 21, 461.
102. Moore, C.; Pressman, B. C. Biochem. Biophys. Res. Commun., 1964, 15, 562.
103. Rechnitz, G. A.; Eyal, E. Anal. Chem., 1972, 44, 370.
104. Pedersen, C. J. J. Am. Chem. Soc., 1967, 89, 7017.
105. Petranek, J.; Ryba, O. IUPAC International Symposium on ISEs, UWIST, Cardiff, 9-12 April 1973, paper 13.
106. Ryba, O.; Petranek, J. J. Electroanal. Chem., 1973, 44, 425.
107. Lindner, E.; Toth, K.; Orvath, M.; Pungor, E.; Agai, B.; Bitter, I.; Toke, L.; Hell, Z. Fresenius' Z. Anal. Chem., 1985, 322, 157.
108. Moody, G. J.; Owusu, R. K.; Thomas, J. D. R. Analyst, 1987, 112, 121.
109. Idem., ibid., 1987, 112, 1347.
110. Idem., ibid., 1988, 113, 65.
111. Stoddart, J. F.; Thomas, J. D. R. Chemical Sensors Club News, 1987, 5, 6.
112. Colquhoun, H. M.; Stoddart, J. F.; Williams, D. J. New Scientist, 1 May 1936, 44.
113. Moody, G. J.; Owusu, R. K.; Slawin, A. M. Z.; Spencer, N.; Stoddart, J. F.; Thomas, J. D. R.; Williams, D. J. Ang. Chemie Int. Ed. Engl., 1987, 26, 890.
114. Kimura, K.; Mazeda, T.; Tamura, H.; Shono, T. J. Electroanal. Chem., 1979, 95, 91.
115. Kimura, K.; Tamura, H.; Shono, T. J. Chem. Soc., Chem. Commun., 1983, 492.
116. Kitazawa, S.; Kimura, K.; Yano, H.; Shono, T. Analyst, 1985, 110, 295.
117. Gadzekpo, V. P. Y.; Moody, G. J.; Thomas, J. D. R.; Christian, G. D. Ion-Selective Electrode Revs., 1986, 8, 173.
118. Gadzekpo, V. P. Y.; Moody, G. J.; Thomas, J. D. R. Analyst, 1985, 110, 1381.

119. Idem., ibid., 1986, 111, 567.
120. Zhukov, A. F.; Erne, D.; Amman, D.; Güggi, M.; Pretsch, E.; Simon, W. Anal. Chim. Acta, 1981, 131, 117.
121. Metzger, E.; Ammann, D.; Asper, R.; Simon, W. Anal. Chem., 1986, 58, 132.
122. Metzger, E.; Aeschimann, R.; Egli, M.; Suter, G.; Dohner, R.; Ammann, D.; Dabler, M.; Simon, W. Helv. Chim. Acta, 1986, 69, 1821.
123. Gadzekpo, V. P. Y.; Hungerford, J. M.; Kadry, A. M.; Ibrahim, Y. A.; Christian, G. D. Anal. Chem., 1985, 57, 493.
124. Gadzekpo, V. P. Y.; Hungerford, J. M.; Kadry, A. M.; Ibrahim, Y. A.; Xie, R. Y.; Christian, G. D. ibid., 1986, 58, 1948.
125. Simon, W.; Morf, W. E.; Meier, P. Ch. Structure and Bonding, 1973, 16, 113.
126. Morf, W. E.; Simon, W. Helv. Chim. Acta, 1986, 69, 121.
127. Amman, D.; Bissig, R.; Güggi, M.; Pretsch, E.; Simon, W.; Borowitz, I. J.; Weiss, L. Helv. Chim. Acta, 1975, 58, 1535.
128. Amman, D.; Morf, W. E.; Ander, P.; Meier, P. C.; Pretsch, E.; Simon, W. Ion-Selective Electrode Revs., 1983, 5, 3.
129. Petranek, J.; Ryba, O. Anal. Chim. Acta, 1981, 128, 129.
130. Kimura, K.; Kumani, K.; Kitazawa, S.; Shono, T. Anal. Chem., 1984, 56, 2369.
131. Schefer, U.; Ammann, D.; Pretsch, E.; Oesch, U.; Simon, W. ibid., 1936, 58, 2232.
132. Wuthier, U.; Pham, H. V.; Zund, P.; Welti, D. H.; Funck, R. J. J.; Bezegh, A.; Amman, D.; Pretsch, E.; Simon, W. ibid., 1984, 56, 535.
133. Meyerhoff, M. E.; Pretsch, F.; Welti, D. H.; Simon, W. ibid., 1987, 59, 144.
134. Moody, G. J.; Thomas, J. D. R. Selected Annual Reviews in the Analytical Sciences, 1973, 3, 59.
135. Thomas, J. D. R. (Editor). Ion-Selective Electrode Revs., 1979-87; Vol. 1-9; Selective Electrode Revs., 1988; Vol. 10.
136. Okazaki, S.; Nozaki, K.; Hara, H. Database on Electrochemical Sensors; Bibliography I, 1983-1986, SAN-Ei Publishing Co.: Kyoto, Japan.
137. Maas, A. H. J. (Editor). Proceedings International Federation of Clinical Chemistry, 1st-8th European Working Group on Ion-Selective Electrodes, 1979-1987.

RECEIVED October 26, 1988

Chapter 22

Modification of Solid Electrodes in Electroanalytical Chemistry, 1978–1988

Brenda R. Shaw

Department of Chemistry, University of Connecticut, Storrs, CT 06268

In 1978 the intentional formation of films of electroactive polymers on electrode surfaces marked the beginning of a new era in electroanalytical chemistry. The theory created to describe and explain their behavior has led to further advances in the design of surface layers, and in turn, to additional theoretical models. Early work was primarily with polymer films. Later work involved use of solid-state materials and more sophisticated microscopic assemblies. The ever-expanding role of ultramicroelectrodes in studies of modified electrodes is also discussed. Modification of electrodes with monolayers and biological macromolecules is discussed briefly in the context of multilayered films, but their individual histories are not included. The focus of this ten-year, conceptual history is on the development of multilayered electrode coatings and related microstructures; 135 references are included.

Electroanalytical chemistry entered a new era in 1978 with the publication of the first paper on polymer-modified electrodes by Miller and Van De Mark (1). In the decade that followed, a foundation was provided for the creation of devices in which electron transfer will be controlled on a molecular level; and selectivity and sensitivity will rival that of redox enzymes *in vivo*. Electroanalytical chemists have begun to think differently. Measurement of micromoles and microcoulombs using classical microelectrodes has been replaced by thoughts of detecting tens of molecules, ions, and electrons with "organized molecular assemblies" at the surfaces of ultramicroelectrodes.

This transformation occurred because electrons were shown to be shuttled from one site to the next through multilayered polymeric films, and finally to species in solution, some distance from the surface of the electrode. This mechanism is similar to that for a redox enzyme linked to an electron transport chain. Substrate species can also enter the polymeric phase and exchange

0097–6156/89/0390–0318$06.50/0

electrons with species within the film. These new possibilities led to design of new methods and applications, and improved the understanding of electron transfer processes. Theory developed to describe these processes mathematically opened electroanalytical minds and laboratories to prospects of intricate control of interfacial electron transfer on a molecular level. Electroanalytical chemistry had expanded its realm once again.

Concurrent advances in the use and understanding of ultramicroelectrodes (*vide infra*) led to studies of electron transfer in unusual phases: gases (*2*), solids (*3*), and electrolyte-free solutions (*4*). The eventual exploitation of knowledge gained in electrode modification and ultramicroelectrodes will lead to creation of near-molecularly-sized electrodes with chemical properties built in to control electron transfer to a degree that was previously unimaginable.

Electroanalytical advances have always been coupled with advances in physical electrochemistry (*5*). Theoretical constructs were expanded immediately upon introduction of polymer-modified electrodes, and new theoretical descriptions provided a framework for expanding the concept of a modified electrode. The cycle continues with theoretical and experimental efforts supporting each other on a phone-call basis among practitioners in the field.

The factual history of modified electrodes is readily available in reviews (*6-15*) and a myriad of original research papers. The purpose of the present paper is to tell the story of "what has happened in the life and development of [electroanalytical chemists]...," over the ten-year period 1978-1988, and to show that the metamorphosis of thought among electroanalytical chemists will prove to be "something important enough to be recorded" (quotations are from *Webster's* definition of HISTORY, Guralnik, D. B., Ed., *Webster's New World Dictionary*, The World Publishing Co.: New York, 1970; pp. 665-6).

Although some attempt was made to select examples from a number of the research groups who have made the most extensive and significant contributions in the field, this is not a comprehensive review of the subject area, or of its contributors. A conceptual history of the body of knowledge about modified electrodes developed by electroanalytical chemists in the past decade is presented using illustrative examples.

From Dirty Electrodes to Covalent Attachment

Before examining the detailed development of the field of modified electrodes in the past decade, it is useful to consider the general foundations of this work in earlier decades. Before the 1970's clean, inert surfaces were considered ideal; chemical reactivity was undesirable, as indicated by the appellation "ideally polarizable electrode" for the theoretical electrode surface, which did nothing but transfer electrons by an outer-sphere-like mechanism through the pristine and well-behaved double layer.

Adsorption and polymerization ("electrode fouling") were known to occur and known to influence electron transfer, but were generally to be avoided. Strategies for eliminating polarographic maxima proliferated, based first on black magic, and later on an increased understanding of adsorption phenomena. Adsorption and electropolymerization ("fouling") often interfered severely with analysis or mechanistic studies (and they still do). However, the ingenuity of analytical chemists is at its best in the face of such disasters. Each of these processes is now exploited extensively in the study of modified electrodes. Theory developed to describe adsorption of monolayers formed the basis for new theories required to describe electron transfer at intentionally modified electrodes (16,17).

After making use of the intrinsic nature of the electrode surface, the next step was to try to outdo nature by purposely adsorbing, and finally covalently attaching species to the surfaces of electrodes to effect the desired electrode kinetics or analysis. Adsorption of olefins (18) and silanization (19) were early examples of this work. Optically active phenylalanine was also covalently bonded to carbon electrodes and used to reduce 4-acetylpyridine to the corresponding, and optically active, alcohol (20). Matsue, Fujihira, and Osa used cyclodextrins as surface-attached inclusion compounds to obtain stereoselectivity in organic electrosynthesis (21,22). A bis(bipyridyl)LL' ruthenium(II) complex bound via an amide linkage to a pendent alkylamine group on a silanized platinum electrode was reversibly oxidized and reduced during cyclic voltammetry (23). Silanization of electrode surfaces in the presence of water led to covalently attached films that were polymeric (24).

Work with clean surfaces and surfaces coated with submonolayer or monolayer amounts of species adsorbed or covalently attached, either intentionally or unintentionally, is not only indispensable to workers in the area of modified electrodes, but is an ongoing and vital research area in itself. However, the present paper will focus only on multilayer modification of electrode surfaces, and new modification techniques that are based on understanding gained from use of multilayered films.

Polymer-Modified Electrodes

The field of polymer-modified electrodes was established with the publication of five papers in 1978 (1,25-8). These were by Van De Mark and Miller (1,25), Merz and Bard (26), Doblhofer, Noelte, and Ulstrup (27), and Nowak, Schultz, Umana, Abruña, and Murray (28). These workers reported the intentional adsorption of electroactive polymers to the surfaces of solid electrodes. For example, a film of poly(vinylferrocene) on the surface of a solid electrode exhibited chemically reversible electro-oxidation that was consistent with a surface coverage of at least 20 monolayers (26). In this and contemporaneous reports, electron transfer was shown to occur among sites in a rather complex and thick film on the surface of an electrode. New theory had to be developed to explain these electron transfer events. Experiments and theoretical studies were carried out to address questions generated by the performance of the new electrode coatings: 1. What effect does polymer morphology have on electron transfer? How is the morphology of the film affected by choice of solvent? 2. What is the mechanism of charge transfer through the film? What factors control the rate of charge transfer? How is charge compensated within the polymeric film? 3. How can parameters be selected to optimize performance of the system for a given application, such as electroanalysis or electrocatalysis? 4. How will polymer morphology and proximity of electroactive sites affect the energetics of electron transfer? Intense activity aimed at answering these questions led to publication of hundreds of papers, which collectively demonstrated the tremendous prospects for such multilayered films in controlling electron transfer. This work continues steadily today.

Polymers offer the full range of familiar reactions useful to analytical chemists: separation (i.e. partitioning between solution and polymeric phases), derivatization, ion-exchange, acid-base, coordination, precipitation, and redox. The redox reactions may involve simple electron transfer, which is useful in characterizing the films themselves; they may also be used to mediate or catalyze electron transfer reactions. A tremendous and impressive variety of polymers has been examined.

In some polymers, such as poly(vinylferrocene), the monomeric units are intrinsically electroactive. In other polymers, monomeric units may serve as ligands for metal ions. Oyama and Anson (29,30) showed that Ru(III)(EDTA) was bound via coordination to poly(vinylpyridine). Anson and co-workers also showed that electroactive ions could be bound to oppositely-charged polymers, as in the case of ferricyanide and protonated poly(vinylpyridine) (31), or hexammineruthemium (III) by poly(vinylsulfate) (32).

Useful reviews of work done at The University of North Carolina present the evolution from covalent attachment of pyridine and bipyridine ligands to surfaces of electrodes, through physical adsorption, of poly(vinylpyridine) and related polymers, to electropolymerization of metal complexes containing vinylpyridine and vinylbipyridine ligands (33,34).

The Influence of Polymer Morphology on Electron Transfer. Polymer morphology and proximity of electroactive species have a strong influence on charge transfer within polymeric films. Electroactive sites may be segregated into hydrophilic and hydrophobic domains as in Nafion (the trade name for DuPont's poly(perfluorosulfonic acid) cation exchange resin), for example. The local environment provided by the various regions of the polymeric film has a strong influence on the formal potential of redox species found within it (35). Changing the oxidation state of sites within the polymer will also change morphology, which causes dynamic changes in the polymer during oxidation or reduction processes. Charging of a neutral polymer such as poly(vinylferrocene) *via* partial oxidation may lead to structural changes or changes in redox potential for the remaining unoxidized electroactive species (36). These effects can lead to complex electrochemical behavior for films that are relatively simple in their primary structure and composition.

The morphology of polymer films is determined partially by the surrounding solution. Several groups have studied the role of both solvent and supporting electrolyte in controlling charge transport through films (35-40). Charge transport is more facile through films that interact favorably with solvent to become swollen and porous. Nonswelling solvents lead to collapse of the film to form a resistive layer that impedes penetration of electrons, electrolyte ions, or substrates. Though highly swollen films are ideal for electron transfer, they can dissolve when composed of linear polymers.

Crosslinking. Crosslinked polymer films allow swelling without dissolution. Since films of polymers are often prepared by casting linear polymers from solution, crosslinks are usually introduced after film formation. Incorporation of highly charged ions, such as ferricyanide automatically leads to crosslinking of cationic polymers such as protonated poly(vinylpyridine) ($PVPH^+$) since the anion binds to multiple ion-exchange sites.

Other strategies include deposition of a suspension of crosslinked polymer (41), radiofrequency plasma polymerization (42), gamma (43) or ultra-violet (Funt, L.R.; Hoang, P.M., 161st Electrochemical Society Meeting, Abstract 599, May 9-14, 1982.) irradiation, chemical (44) or electrochemical reaction (45), or electro-co-polymerization of bifunctional monomers (46). Slabs of crosslinked polyacrylamide gels containing copolymerized vinylpyridine have been prepared, sliced with a microtome and applied intact to the surface of an electrode (47). Crosslinking is now routinely considered when semi-rigid structure or longevity of a film is important.

Electroanalysis and Electrocatalysis. Polymer-film electrodes can revolutionize electroanalytical techniques because of their ability to provide a matrix for selective pre-concentration of analytes, and to immobilize high concentrations of electrocatalysts. Many reports have appeared to demonstrate such effects. For example, Guadalupe and Abruña reported preparation of copolymers of vinylferrocene with vinylpyridine or vinylbipyridine (48). The ferrocene groups facilitated electron transfer that led to precipitation of the polymer at the surface of the electrode. The bipyridyl groups could be used as coordination sites, while pyridyl sites could be protonated to incorporate specific negatively charged chelating ligands. These films were used to analyze for iron and copper complexes in solution.

In practice, relatively few analytical techniques make use of polymer-modified electrodes because of the mechanical instability of the films, and difficulty in regenerating the films either conveniently or reproducibly. Disposable modified electrodes provide one way around this problem. Also, the considerable knowledge gained so far with films is applicable to the renewable modified carbon paste and composite electrodes described in a later section.

Prospects for efficient electrocatalysis are the strongest driving force behind the development of modified electrodes. Ascorbic acid is often used to test new catalytic systems because of its overpotential, amenability to electrocatalysis, and biological importance. Plasma polymerized poly(vinylferrocene) (49) and pentachloroiridate bound to protonated plasma polymerized poly(vinylpyridine) (50) were each shown to serve as electrocatalysts for oxidation of ascorbic acid. Several other examples of electrocatalysis are described in later sections, but these are only a small sample of work going on in this field.

Theory. Since electrocatalysis is one of the major uses for modified electrodes, theoretical work has centered around optimization of electrocatalytic processes. When electron transfer between catalyst and substrate is slow, it is useful to increase the concentration of catalytic sites at the surface of the electrode. When this is done via adsorption of a catalyst-containing polymer, the thickness and permeability toward substrate and electrolyte of the polymeric film will help determine the efficiency of electrocatalysis (51,52).

The theory of electron transport through polymer films at the surfaces of electrodes has blossomed under the guidance and development of Savéant and his group, and many others (51-60). Savéant's major contribution was to provide a general, mathematical description of charge transfer during electrocatalytic oxidation or reduction of a substrate in solution.

The rate of electron transfer to a substrate in solution may be limited by any one, or combinations, of the following processes. 1. Heterogeneous electron transfer to or from electrode. 2. Electron self-exchange among catalyst sites within the polymer film, or physical diffusion of catalyst. 3. Crossing of the solution/polymer interface by substrate. 4. Permeation of substrate through the polymer film. 5. Cross-exchange reaction in which catalyst transfers electrons to or from the substrate within the polymer film, or at the polymer/solution interface. The first step is generally fast because of the potential selected to achieve electrocatalysis under steady-state conditions. Any of the remaining steps may be rate limiting. Optimization of the electrocatalytic process relies upon determining the rate-limiting process(es), and varying the composition, morphology and thickness of the film to enhance electron transfer, with the aid of the mathematical description.

Buttry and Anson showed that this theory could be applied qualitatively to redesign an inefficient electrocatalytic system (61). Although cobalt porphyrins adsorbed to the surface of an electrode serve as electrocatalysts for reduction of oxygen, the reduction process could not be sustained when the porphyrin was incorporated into a Nafion film. The rate was slow due to poor electron self-exchange within the film. This rate was increased dramatically by adding hexammineruthenium(II) to the film to serve as the electron shuttle, thereby increasing the [mediated] "self" exchange rate.

How do electrons traverse the polymeric layer at the surface of the electrode? Electrons may "hop" (41) or move by "redox conduction" (62-4), or electroactive species that carry the electrons from site to site may deliver them from one region of the film to another by physical diffusion. Daums-Ruff theory was shown to apply to polymer films in 1981 and 1982 (61,65,66). This theory accounts for contributions to electron transfer by both diffusional and exchange mechanisms.

Nafion, a polymer known to form hydrophilic and hydrophobic domains (67), was shown by Buttry and Anson to transport electrons partially by "single-file" diffusion of electroactive species as they competed for ion-exchange sites within the polymeric film (68). The heterogeneity of Nafion domains is important in applications such as the electrocatalytic system described above, and represents one of the earliest moves toward architectural design of microstructures at the surfaces of electrodes (vide infra).

Control over electron transfer may also be obtained by clever selection of electroactive species. Facci and Murray incorporated both hexachloroiridate(III/IV) and hexacyanoferrate(II/III) into copolymeric films containing protonated pyridyl groups (65). The formal potential for the iridium couple is more positive than that for the iron couple. Therefore, Ir(IV) can mediate oxidation of Fe(II), and Fe(II) can mediate reduction of Ir(IV). Facci and Murray observed that the cross-exchange contribution to electron transfer [eg. Ir(IV) + Fe(II) $\rightleftharpoons$ Ir(III) + Fe(III)] could be turned on and off by controlling the redox state of the iridium. Without the cross-exchange reaction, diffusional electron transport was the only mechanism available, and the rate of electron transfer was much slower.

However electrons move, there must be concomitant motion of electrolyte ions, and there is often polymer, catalyst, or substrate motion as well (69). Therefore, effective diffusion coefficients obtained using common transient techniques may not reflect the steady-state phenomena described by the model of Saveant and co-workers.

Attempts to separate the roles of electron exchange from other processes have been successful. Majda and Faulkner used luminescence quenching to measure rates of electron transfer independently of the motion of electrolyte ions (70). More recently, steady-state electrochemical methods have been used. Pickup, Kutner, Leidner, and Murray used sandwich electrodes (71). Chen, He, and Faulkner presented a similar method for obtaining effective diffusion coefficients for polymer-bound species, by experiments designed to examine lateral diffusion from a disk to a concentric ring electrode (44). Their data show that diffusion coefficients are two orders of magnitude larger than those predicted by chronocoulometric experiments. The new steady-state methods reported above are good news since theoretical descriptions all depend upon accurate values for effective diffusion coefficients under steady-state conditions.

Films containing pyridyl groups have been used to test the theoretical constructs. Van Koppenhagen and Majda (47), and

Chen, He, and Faulkner (44) used crosslinked copolymers of vinylpyridine. In the former case, pyridine was protonated to provide a crosslinked, anion-exchanging matrix for incorporation of ferricyanide, which served as an electrocatalyst for oxidation of ascorbic acid; in the latter case, the pyridine groups were used as sites for crosslinking via reaction with 1,2-dibromoethane. The copolymerized styrene sulfonate provided sites for binding of tris(2,2'-bipyridine)osmium (III/II). Both groups varied several parameters to test the Saveant theory (44,47,72). Both showed that with some assumptions, the theory described the systems very well.

Designer Surfaces

Experimental confirmation of the theory of electron transfer with polymer films, and the ramifications of this new knowledge demonstrate that there is a tremendous opportunity to control interfacial electron transfer via surface and thin-film chemistry. The ideal, polarizable electrode, once an icon for electroanalytical chemists, now serves as a hypothetical support for molecular-scale "organized assemblies" that carry out the business of controlled electron transfer. Examples of early prototype assemblies are given below.

Miller and Majda examined lateral electron transport among electroactive species by an approach different from that reported above (73,74). They prepared an insulating support for immobilization of electroactive species that consisted of a film of alumina with cylindrical pores leading to the surface of the electrode. Here electroactive species associated with thin films of $PVPH^+$ or self-assembled bilayers along the inner walls of the cylindrical alumina surface showed electroactivity that resembled thin-layer behavior. The diameter of the cylindrical pores through the alumina was large enough (in the range of 200-1500 angstroms) to allow ready access of electrolyte ions to the modification layer.

Leddy and Vanderborgh showed that deposition of Nafion into the cylindrical pores of Nucleopore polycarbonate membranes enhanced mass transport of electroactive cations to the surface of the electrode by an order of magnitude (75). It appears that morphological changes on the 0.01-0.1 micron scale can lead to chemical effects on the atomic/molecular/ionic level for this and other systems. It will take some time to understand the factors controlling such features of a microstructure. Interfacial chemical and physical interactions control such phenomena, and are some of the most difficult to examine experimentally.

Returning to the concept of lateral electron transport, the Majda group also observed electron transfer at the air-water interface when a line electrode was placed in contact with an electroactive Langmuir-Blodgett film at the surface of a trough (76). Although this system does not constitute a modified electrode, it demonstrates that charge transport can be fed from an electrode across an extended, highly structured interfacial zone, some distance from the electronically conductive electrode.

Current and potential control over such processes are still maintained, *via* the redox states of electroactive species in the two-dimensional phase in electrochemical contact with the electrode, despite the considerable distance that is traversed by the electron. Diffusion control of the charge transfer event was also observed.

This move from the modification of electrodes to the modification of nonconducting surfaces in contact with electrodes is also seen in systems in which electrodes are modified using nonconductive particles. For example, Zak and Kuwana showed that the basic surface of alpha-alumina imbedded in glassy carbon electrodes caused catalysis of the electrooxidation of catechol and other organic species requiring loss of a proton for their oxidation (77). Similar results were obtained by Shaw and Creasy (78) using alumina or layered-double hydroxides in composite electrodes (*vide infra*).

Results described above show that it is possible to control or monitor electron transfer that occurs among electroactive sites on non-electronically conductive surfaces. This knowledge may be exploited in the future design of microstructures, as well as in understanding electron transfer at surfaces, such as biological membranes.

This focus on lateral electron transfer along interfaces came directly from a drive to understand electron transfer within polymeric coatings on electrodes. The nuance is in the geometric dimensionality of electron transfer, and its spacial/direction separation from the electrolyte ion motion required to compensate charge. This separation of the direction of electron transfer from that of ion motion leads to freer mobility of solution species to electroactive sites, while maintaining electrochemical control over redox states. Studies carried out using porous aluminum oxide films and Langmuir-Blodgett films show that lateral charge transport perpendicular to the electrode and along a surface with adsorbed electroactive species is indeed possible, and will likely be shown to have mechanistic features in common with electrocatalysis at electrodes modified with nonconducting solids. These concepts also apply to charge propagation at the surface of modified carbon paste and modified composite electrodes described below.

The Concept of Molecular Design

Sometime in the past decade, probably around 1982-1984, the collective imagination of electroanalytical chemists absorbed the connections among such diverse areas as solid-state chemistry, fabrication of solid-state devices, semiconductors, polymer morphology, surface and interfacial chemistry, membrane chemistry and technology, biochemistry, and catalytic mechanisms. Before the advent of electrodes modified with multilayers of polymers, these areas of endeavor were each distinct within the electroanalytical mind. Once charge transport could be observed in such a simultaneously simple and complex system as a redox polymer film, the relevance of charge transfer in biological systems, and at the surfaces of solids and membranes, became apparent. A 1984 report by Faulkner in Chemical and Engineering News reviews the field of "Chemical Microstructures on Electrodes" (79). The work accomplished to date on electrochemistry at electrodes modified with sophisticated assemblies is just the first evidence that dreams will come true. The next two decades will yield exciting progress as electroanalytical chemists learn to exploit chemical, biological, and geological systems for our own purposes. Any material that can point an electron in a particular direction, or stop one from following a particular undesirable path may be put to work in devices designed for selective and sensitive analysis, and electronic, synthetic, and energy systems.

The remainder of this discussion will focus on the application of new concepts of electron transfer to new electrode modification methods. These include use of conductive polymers, solid inclusion compounds and other nonconducting solids, biological macromolecules, and composite materials. Theoretical descriptions do not exist yet for the majority of these systems. A sidetrip into the area of ultramicroelectrodes is necessary to appreciate the possibilities for exploiting modified electrodes.

Ultramicroelectrodes

The development of modified electrodes is now intimately linked with practical and theoretical progress in the use of ultramicroelectrodes, electrodes with dimensions in the μm range. A good introduction to the subject of ultramicroelectrodes is available (80), and several new papers in this field appear each month.

There are practical advantages of these electrodes because of physical factors associated with the smaller size. The introduction of ultramicroelectrodes allows achievement of steady-state currents in unstirred solution because of enhanced mass transport to a small electrode. This leads to improved signal-to-noise ratios and better precision for analytical applications. The small surface area requires a smaller absolute number of ions for establishment of the double layer at a given potential, and the iR drop due to uncompensated resistance is diminished considerably by reducing the size of the electrode. These latter effects allow use of ultramicroelectrodes in unusual environments with low conductivity as cited above. It is therefore possible to make sensors using modified ultramicroelectrodes for monitoring gases and other low-conductivity media that would not be amenable to analysis using larger electrodes.

In addition, the trend toward miniaturization of sensors for *in vivo* and other applications is enhanced by advances in the use of ultramicroelectrodes. The prospects for combining the physical advantages of ultramicroelectrodes with the chemical advantages of modified electrodes are very exciting for the future of electroanalytical chemistry. Modification of ultramicroelectrodes with polymers has already begun (81), and will be discussed in a later section.

Nonconducting Solids

Since the desirable properties of polymer-modified electrodes are often offset by their morphological instability, and structural uncertainty, solid inclusion compounds are being examined by several groups. These clay, zeolite, Prussian blue, layered double hydroxide, and related films offer structural rigidity and longevity when exposed to solution. In addition, rigorous selection of analytes may be obtained based on crystallographic dimensions of each of these solids, except unpillared clays. Initially, electrochemists hoped for electron transfer among species ionically bound within pores and interlayers of these inclusion compounds. This would allow shape- and size-selective electrocatalysis.

No direct evidence is available in the literature to date to show net charge transfer among electroactive sites within nonconductive inclusion compounds, except in the case of Prussian blue and its analogues, where the crystalline framework itself is electroactive (82). Rudzinski and Bard (83), and King, Nocera, and Pinnavaia (84) showed that electron transfer at montmorillonite clay-modified electrodes occurs primarily among species found at the surfaces of the clay particles, not within the interlayer regions. Although theoretical studies are absent, it can be postulated that the confinement of the interlayer region hinders mobility of electrolyte ions to the extent that electron transfer is too slow to be observed on the timescale of usual electroanaytical experiments. Instead, the role of the clay appears to be to provide an ion-exchanging surface capable of storing high concentrations of electroactive species, such as catalysts or substrates (85,86).

There is still the possibility of finding clay systems that do allow electron transfer from site to site within the interlayers. A preliminary report by Inoue and Yoneyama describes the electropolymerization of aniline that was previously intercalated into montmorillonite (87). The mechanism of the reaction and exact nature of the product will confirm whether or not electron transfer is facile within the clay interlayer in this case.

The surface properties of other solids also have potential applications. For example, electro-oxidation of catechol was catalyzed on a layered double hydroxide, $Zn_x^{2+}Al_y^{3+}(OH^-)_{2x+3y-z}(Cl^-)_z$ which has a basic surface (78). Similar behavior was observed on alumina-modified glassy carbon as described above (77). Voltammetric data suggest that catechol and related hydroxy compounds adsorb to the surface of these basic solids. Charge transfer may occur across the surface of the solid among adsorbed electroactive sites. If heterogeneous electrocatalysis at the surfaces of nonconducting solids turns out to be general, the possibilities in electroanalytical and electrosynthetic applications are endless.

Zeolites offer a somewhat different matrix than layered clays and layered double hydroxides; the pores are three-dimensional and the framework is rigid upon ion-exchange. (However, pillared clays (83) resemble zeolites more closely than do ordinary clays.) In 1978 pressed pellets of zeolite NaA were examined electrochemically at elevated temperature (320°C) where they are solid ionic conductors. No electroactive species was added and no solvent was used; the intrinsic irreversible electroactivity was of interest (88). In 1980 a cesium-selective potentiometric electrode was prepared using the zeolite mordenite (89). Zeolites also have been employed as host materials to provide chemical stability for electroactive species used in solid-state electrochemical cells (90). The first report of a zeolite-modified electrode *per se* used for dynamic techniques appeared in 1983 by Pereira-Ramos, Messina, and Perichon (91). This work was undertaken to examine the prospects of using electrolytic methods for supporting finely divided metal particles on zeolites. A year later electroanalytical and electrochemical aspects of zeolite-modified electrodes were discussed by Murray, Nowak, and Rolison (92). Use of zeolites shows considerable promise for future electrocatalytic applications, and possibly for selected analytical methods. Recent work has focused on mechanistic and electrocatalytic possibilities, as well as analytical aspects (93-6).

Although there is no direct evidence so far of electron transfer deep within zeolite channels, such evidence will probably be forthcoming by careful design of the system. As with clays, a major factor is charge compensation upon electron transfer within the solid state structure. In cases where electroactive species may diffuse freely into and out of the zeolite during electrochemical experiments, it is difficult to demonstrate charge transfer within the cavities of the zeolite. However, Li and Mallouk (95), and Persaud, *et al*. (97) have demonstrated trapping of charge by zeolite-bound species generated via electrochemical and photochemical processes, respectively. These results are very promising, and demonstrate the great prospects for use of zeolites as templates for developing highly sophisticated structures for control of electron transfer.

A serious difficulty in developing theoretical models for electroactivity at films like those described above, is the heterogeneity and random distribution of particle sizes. In addition, the means of holding solid particles to the surface of an electrode can interfere with charge transport in a way that is difficult to describe mathematically. Films of colloidal clay particles can be cast to form smooth, rather robust films, but non-lamellar solids and larger particles are not as easily held at the surface of an electrode. Improvements in this area are discussed below.

Empirical techniques, *eg*. simplex optimization, may be used to optimize the performance of modified electrodes for which no theoretical model is yet available (96). Improvements in microstructural design of electrodes modified with solids will make it possible to obtain mathematical descriptions similar to those that have been so successful in the case of polymer-modified electrodes as described above.

Composite Electrodes

Conductive plastics were prepared at least as early as the 1950's by incorporating electronically conductive materials such as conductive carbon black or metallic particles into a polymeric matrix. There is a precipitous drop in resistivity in these composite materials upon reaching a critical concentration of the electronically conductive particles (98). An unmodified composite surface may be treated mathematically as a partially blocked electrode, or as a random array of ultramicroelectrodes (99-105). These electrodes demonstrate the high signal-to-noise ratio expected for ultramicroelectrodes, yet give a larger absolute signal because of the large number of electrode sites.

Several groups have reported work with polymer films that resemble composite materials in that conductive particles are dispersed within the film. These include TTF-TCNQ (106), carbon (107), and metallic particles (108-12). In 1980 Doblhofer and Durr reported "polymer-metal composite thin films on electrodes" (108). Wrighton's group deposited platinum particles into viologen-containing films for electrocatalytic generation of hydrogen (109). Liu and Anson used hexammineruthenium(II/III) as a mediator to poise the potential of colloidal platinum particles dispersed within a Nafion film (110). Kuwana's group has also been active in dispersing metallic particles within polymeric films (111). For example, poly(vinylpyridine) films prepared by three different methods were exposed to acidic solutions of potassium hexachloroplatinate. The hexachloroplatinate was exchanged into the protonated PVP film where it was reduced electrochemically. Coche and Moutet used precious metal microparticles dispersed in a viologen-containing polymer on high-surface area carbon felt for electrochemical hydrogenation of nitrobenzene and olefinic and acetylenic compounds on the mmol scale (112).

Mixtures of conductive particles with an electrode modifier have also been carried out using carbon paste electrodes. A recent example is the incorporation of PVP, which was protonated and exposed to ferrocyanide (113). The PVP-modified carbon paste electrode behaved similarly to a surface-modified electrode. The advantage of such a system is that the surface could be renewed simply by exposing a fresh layer of modified carbon paste by mechanically removing the surface layer.

A recent advance that combines some of the attributes of surface-modified electrodes, modified carbon paste electrodes, and arrays of ultramicroelectrodes is the use of modified composite electrodes. Composite electrodes can be prepared from monomers of electroanalytical interest, including vinylferrocene (114) and vinylpyridine (Park and Shaw, submitted). For example, vinylpyridine was copolymerized with styrene and divinylbenzene (DVB) in the presence of carbon black to give a chemically modified, shiny, black plastic electrode material. Such electrodes may be polished to renew the surface, and give reproducibilities in the range of 1-2%. Their properties depend upon the loading of vinylpyridine and on the degree of crosslinking, but their behavior is largely similar to that of other types of electrodes modified with PVP.

More recently bundles of carbon fibers have been immobilized in copolymers of vinylferrocene or vinylpyridine with crosslinked polystyrene. Alternatively, the fibers were coated by electro-co-polymerization of vinylferrocene and divinylbenzene before immobilization in crosslinked polystyrene. These electrodes are also polishable and present an array of polymer-modified ultramicro disk electrodes to the solution (Creasy and Shaw, submitted).

Particles may also be incorporated into composite-type electrodes. Zeolites have been incorporated into carbon paste (91,94), and TTF-TCNQ has been incorporated into polymer paste (115). Solid crosslinked-polystyrene-composite electrodes modified with alumina, layered double hydroxides, and zeolites were shown to have useful analytical and electrocatalytic properties (78).

Although the surfaces of such composite electrodes are very complicated, it should be possible to describe these systems mathematically by adding partitioning, diffusional, and migrational terms to existing models used for unmodified composite electrodes discussed above. The major advantages over other systems utilizing solid particles are the improved reproducibility, longevity, and increased control over morphology of the particle-modified surface.

Because modified composite electrodes are rigid and polishable they will be generally useful as "surface" modified electrodes in practical macroscopic sensors and detectors. Wang and co-workers showed the high selectivity and sensitivity of positively-charged PVP-coated electrodes *via* a charge-exclusion mechanism (116). The electrodes were used in detectors for high pressure liquid chromatography and flow injection analysis. The major drawback of coated electrodes is that the films often will not stand up to flowing solvent, due to partial solubility. Strategies to remedy this problem are available by changing the chemistry of the polymer itself. However, use of composite electrodes employing analogous chemical characteristics provides a prospective *generic* solution to the problem, which should allow more rapid development of a variety of sensors and detectors. For example, the crosslinked PVP composite electrode described above has the same electrochemical properties as a coated electrode, but is polishable, and not susceptible to dissolution in flowing streams, even in the presence of organic solvents. This new class of electrodes and their miniaturized descendants (Shaw, Wang, and Creasy, work in progress) will serve well in electrochemical detectors and sensors of the future.

Although films of polymers played a pivotal role in the advancement of concepts in electroanalysis that dominate the field today, the often-fragile, resistant, and nonrenewable films are not likely to revolutionize practical analytical methods. Instead, variations on the theme that make use of new understanding of electron transfer itself will take over. The monotypic mercury electrode of electroanalysis past will be joined by a bazaar of compact sensors and detectors used solo or in combination with separation methods.

Conductive Polymers and Nonmetal Conductors

The examples of composite electrodes and composite films above represent an extension of the electrode itself into the chemically modified interphase with the solution. Another means of extending the dimensions of the electrode is by using films of conductive polymers (117), such as electropolymerized poly(pyrrole) (118), in electrode modification, or by simply replacing the electrode with a conducting polymer or semiconductor. This allows expansion of the intrinsic surface chemistry that is available from electrode materials.

Much of the work done with conductive polymers follows the same trends as that with non-conductive polymers, so will not be described here. There will be important roles for conductive polymers as sophisticated microstructures are designed. One example of early work in this area is the use of poly(pyrrole) as a support for montmorillonite at the surface of electrodes, and in free-standing films (119). Also Kittlesen, White and Wrighton reported in 1984 that electropolymerized poly(pyrrole) and poly(N-methylpyrrole) could be prepared at electrodes with widths of only 1.4 microns (120). These electrodes were part of an ultramicroelectrode array used to demonstrate the possibility of combining surface chemistry with microelectronics technology to prepare microsensors. Since the conductivity of poly(pyrrole) (and many other conductive polymers) depends on redox state, the authors suggest that miniaturized redox sensors may be prepared from systems such as theirs.

Biological Macromolecules

The modification of electrodes with enzymes and other biological macromolecules was well underway before 1978, and a detailed history of this field is beyond the scope of the present paper. A brief discussion of biological systems is given, however, to place them in context with other modification layers. A recent review by Frew and Hill (121) discusses past and future strategies for design of electrochemical biosensors. Topics discussed were enzyme electrodes, electron transfer mediators, conducting salts, electrochemical immunoassay, enzyme labels, and cell-based biosensors. In general, the bioactive molecule or cell is immobilized in proximity to an electrochemical transducer and exposed to the analyte solution for real-time analysis.

The specialization that is built into molecular microstructures in biological systems will be exploited more fully in the coming decade. Present systems make use of enzymes (122) and antibodies, or combinations of these in immunosorbant electrochemical assays (123). Recently, several species of algae were examined by Gardea-Torresdey, Darnall, and Wang for their ability to bind Cu(II) ion from solution (124). The algae were first mixed with carbon paste and used in a preconcentration step. These alga-modified electrodes were selective for copper in the presence of other metal ions and allowed detection of Cu(II) in the micromolar range with a reproducibility of 5% (r.s.d.).

Work with algae described above shows that the most simple to the most complex (antibody/antigen with enzymic amplification) biological systems provide ready-made microstructures and molecular systems for coupling to electrodes for highly sensitive and selective electroanalytical methods.

Coupling of modified-electrode and microelectrode techniques is already leading to modified microelectrode sensors such as the pH microelectrode reported by Oyama *et al.* (125).

Modified Microelectrode Arrays or Electrochemical "Devices"

Murray and his group showed in 1984 that bilayer polymeric films and films sandwiched between two electrodes may be used as diodes or triodes *via* "redox conduction" that is potential dependent (126,127). Since then, Wrighton's group has carried this concept into the microscopic realm. As discussed above, microelectronic technology was used to prepare arrays of ultramicroelectrodes. These were modified using electrochemical techniques such as electropolymerization or electrodeposition (120,128,129).

A device prepared by depositing poly(vinylpyridine) over an array of ultramicroelectrodes was sensitive to pH in a solution containing ferri- and ferro-cyanide (128). Protonation of the PVP led to binding of the redox couple. This in turn yielded redox conduction between source and drain terminals of the device held at a potential difference of 100 mV. When pH was raised to a value of 9, too high for protonation of poly(vinylpyridine), the device "turned off" due to the loss of the redox couple.

This field is in its infancy with the future open to growth in new directions. The geometric dimensions of electronic devices and molecular assemblies are moving closer together. As this trend continues it will be possible to prepare solid-state sensors for detection of small numbers of atoms, molecules, and ions in gases, liquids, and in some solids, as well as in living systems.

Just Imagine

What does the future hold for modified electrodes? Three major areas of electrochemistry will continue to make use of modified electrodes. First, specialized microstructures can be prepared to gain increased understanding of electron transfer in solutions and polymers, along surfaces, and in biological systems. Second, electroanalysis will benefit as devices are created for detection of minute amounts of species in various environments in "the field" and at the outflow of miniaturized devices used for separations.

Third, all of this improved understanding can be put to use in electrosynthesis and energy systems. The control that will become possible over electrochemical reactions should make it economically beneficial to prepare both high-value, and bulk chemicals by new electrochemical methods based on modified electrodes (112,130, Osa, T.; Akiba, U.; Segawa, I.; Bobbitt, J. M. *Chem. Lett.*, accepted.) or using fluidized bed systems in which modified ultramicrospherical electrodes are suspended in solution between feeder electrodes (131-4), or ultramicroelectrodes are supported on zeolites (135) or other solid materials. Energy systems, especially fuel cells, will also benefit from the increased efficiency of electrocatalytic processes.

Acknowledgment

The author wishes to acknowledge Dr. Johna Leddy, Queens College, City University of New York, for her contribution to this paper via useful discussion. Reviewers' comments were also helpful.

Literature Cited

1. Miller, L. L.; Van De Mark, M. R. J. Amer. Chem. Soc. 1978 100, 639-40.
2. Ghoroghchian, J.; Sarfarazi, F.; Dibble, T.; Cassidy, J.; Smith, J. J.; Russell, A.; Dunmore, G.; Fleischmann, M.; Pons, S. Anal. Chem.1986, 58, 2278-82.
3. Reed, R. A.; Geng, L.; Murray, R. W. J. Electroanal. Chem. Interfac. Electrochem. 1986, 208, 185-93.
4. Bond, A. M.; Fleischmann, M.; Robinson, J. J. Electroanal. Chem. Interfac. Electrochem. 1984, 168, 299-312.
5. Laitinen, H. A.; Ewing, G. W., Eds. A History of Analytical Chemistry; American Chemical Society: York, PA 1977; Chapter IV.
6. Snell, K. D.; Keenan, A. G. Chem. Soc. Rev. 1979, 8, 259-82.
7. Murray, R. W. Acc. Chem. Res. 1980, 13, 135-41.
8. Albery, W. J.; Hillman, A. R. Annu. Rep. Prog. Chem. Sec. C, 1982, 78(C), 377-437.
9. Murray, R. W. In Electroanalytical Chemistry; Bard, A. J., Ed.; Marcel Dekker: New York, 1984; Vol. 13, p. 192.
10. Fujihira, M. In Topics in Organic Electrochemistry; Fry, A. J. and Britton, W. E., Eds.; Plenum: New York, 1986; p. 255
11. Linford, R. G., Ed. Electrochemical Science and Technology of Polymers I; Elsevier: New York, 1987.
12. Murray, R. W. Phil. Trans. R. Soc. Lond. A, 1981 302, 253-65.
13. Miller, J. S., Ed. Chemically Modified Surfaces in Catalysis and Electrocatalysis; ACS Symposium Series 192, American Chemical Society: Washington, DC, 1982.
14. Zak, J.; Kuwana, T. J. Electroanal. Chem. Interfac. Electrochem. 1983, 150, 645-64.
15. Redepenning, J. G. Trends in Analytical Chemistry 1987 6, 18-22.
16. Wopschall, R. H.; Shain, I. Anal. Chem. 1967, 39, 1514-27.
17. Laviron, E. In Electroanalytical Chemistry: A series of Advances; Bard, A. J., Ed.; Marcel Dekker: New York, 1982; Vol. 12, p. 53.
18. Lane, R. F. ; Hubbard, A. T. J. Phys. Chem. 1973, 77, 1401-10.
19. Moses, P. R.; Wier, L; Murray, R. W. Anal. Chem. 1975, 47, 1882-6.
20. Watkins, B. F.; Behling, J. R.; Kariv, E.; Miller, L. L. Amer. Chem Soc. 1975, 97, 3549-3550.

21. Matsue, T; Fujihira, M., Osa, T. J. Electrochem. Soc. 1979, 126, 500-1.
22. Matsue, T.; Fujihira, M., Osa, T. Bull. Chem. Soc. Japan 1979, 52, 3692-6.
23. Abruña, H. D.; Meyer, T. J.; Murray, R. W. Inorg. Chem. 1979, 18, 3233-40.
24. Lenhard, J. R.; Murray, R. W. J. Amer. Chem. Soc. 1978, 100, 7870-5.
25. Van de Mark, M. R.; Miller, L. L. J. Amer. Chem. Soc. 1978, 100, 3223-5.
26. Merz, A.; Bard, A. J. J. Amer. Chem. Soc., 1978, 100, 3222-3.
27. Doblhofer, K.; Noelte, D. Ulstrup, J. Ber. Bunsenges. Phys. Chem. 1978, 82, 403-8.
28. Nowak, R.; Schultz, F. A.; Umana, M.; Abruna, H.; Murray, R. W. J. Electroanal. Chem. Interfac. Electrochem. 1978, 94, 219-25.
29. Oyama, N.; Anson, F. C. J. Amer. Chem. Soc. 1979, 101, 739-41.
30. Oyama, N.; Anson, F. C. J. Amer. Chem. Soc. 1979, 101, 3450-6.
31. Oyama, N.; Anson, F. C. J. Electrochem. Soc. 1980, 127, 640-7.
32. Oyama, N.; Shimomura, T.; Shigehara, K.; Anson, F. C. J. Electroanal. Chem. Interfac. Electrochem. 1980, 112, 271-80.
33. Abruña, H. D.; Calvert, J. M.; Denisevich, P.; Ellis, C. D.; Meyer, T. J.; Murphy, W. R., Jr.; Murray, R. W.; Sullivan, B. P.; Walsh, J. L. In Chemically Modified Surfaces in Catalysis and Electrocatalysis; American Chemical Society: Washington, DC, 1982; p. 133.
34. Calvert, J. M; Sullivan, B. P.; Meyer, T. J., ibid, p. 159.
35. Naegeli, R.; Redepenning, J.; Anson, F.C. J. Phys. Chem. 1986 90, 6227-32.
36. Peerce, P. J.; Bard, A. J. J. Electroanal. Chem. Interfac. Electrochem. 1980, 114, 89-115.
37. Daum, P.; Murray, R. W. J. Electroanal. Chem. Interfac. Electrochem. 1979, 103, 289-94.
38. Schroeder, A. H.; Kaufman, F. B. J. Electroanal. Chem. Interfac. Electrochem. 1980, 113, 209-24.
39. Oyama, N.; Anson, F. C., J. Electrochem. Soc. 1980, 127, 640-7.
40. Ewing, A. G.; Feldman, B. J.; Murray, R. W. J. Phys. Chem. 1985, 89, 1263-9.
41. Kaufman, F. B.; Engler, E. M. J. Amer. Chem. Soc. 1979, 101, 547-9.
42. Nowak, R. J.; Schultz, F. A.; Lam, R., Murray, R. W. Anal. Chem. 1980, 52, 315-21.
43. Decastro, E. S.; Smith, D. A.; Mark, J. E.; Heineman, W. R. J. Electroanal. Chem. Interfac. Electrochem. 1982, 138, 197-200.
44. Chen, X.; He, P.; Faulkner, L. R. J. Electroanal. Chem. Interfac. Electrochem. 1987, 222, 223-43.
45. Kaufman, F. B.; Schroeder, A. B.; Patel, V. V.; Nichols, K. H. J. Electroanal. Chem. Interfac. Electrochem. 1982, 132, 151-61.
46. Shaw, B. R.; Haight, G. P. Jr.; Faulkner, L. R. J. Electroanal. Chem. Interfac. Electrochem. 1982, 140, 147-53.
47. Van Koppenhagen, J. E.; Majda, M. J. Electroanal. Chem. Interfac. Electrochem. 1987, 236, 113-38.

48. Guadalupe, A. R.; Abruña, H. D. Anal. Chem. 1985, 57, 142-9.
49. Dautartas, M. F.; Evans, J. F. J. Electroanal. Chem. Interfac. Electrochem. 1980, 109, 301-312.
50. Facci, J.; Murray, R. W. Anal. Chem. 1982, 54, 772-7.
51. Andrieux, C. P.; Savéant, J. M. J. Electroanal. Chem. Interfac. Electrochem. 1978, 93, 163-8.
52. Andrieux, C. P.; Dumas-Bouchiat, J. M.; Savéant, J. M. J. Electroanal. Chem. Interfac. Electrochem. 1980, 114, 159-63.
53. Andrieux, C. P.; Dumas-Bouchiat, J. M.; Savéant, J. M. J. Electroanal. Chem Interfac. Electrochem. 1982, 131, 1-35.
54. Andrieux, C. P.; Savéant, J. M. J. Electroanal. Chem. Interfac. Electrochem. 1982, 134, 163-6.
55. Ikeda, T.; Leidner, C. R.; Murray, R. W. J. Electroanal. Chem. Interfac. Electrochem. 1982, 138, 343-65.
56. Laviron, E. J. Electroanal. Chem. Interfac. Electrochem. 1982, 131, 62-75.
57. Anson, F. C.; Savéant, J. M.; Shigehara, K. J. Phys. Chem. 1983, 87, 214-19.
58. Andrieux, C. P.; Dumas-Bouchiat, J. M.; Savéant, J. M. J. Electroanal. Chem. Interfac. Electrochem. 1984, 169, 9-21.
59. Anson, F. C.; Savéant, J. M.; Shegehara, K. J. Electroanal. Chem. Interfac. Electrochem. 1983, 145, 423-30.
60. Leddy, J.; Bard, A. J.; Maloy, J. T.; Savéant, J. M. J. Electroanal. Chem. Interfac. Electrochem. 1985, 187, 205-27.
61. Buttry, D. A.; Anson, F. C. J. Electroanal. Chem. Interfac. Electrochem. 1981, 130, 333-8.
62. Laviron, E. J. Electroanal. Chem. Interfac. Electrochem. 1980, 112, 1-9.
63. Laviron, E.; Roullier, L.; Degrand, C. J. Electroanal. Chem. Interfac. Electrochem. 1980, 112, 11-23.
64. Pickup, P. G.; Murray, R. W. J. Amer. Chem. Soc. 1983, 105, 4510-4.
65. Facci, J.; Murray, R. W. J. Phys. Chem. 1981, 85, 2870-3.
66. White, H. S.; Leddy, J.; Bard, A. J. J. Amer. Chem. Soc. 1982, 104, 4811-7.
67. Yeager, H. L.; Steck, A. J. Electrochem. Soc. 1981, 128, 1880-4.
68. Buttry, D. A.; Anson, F. C. J. Amer. Chem. Soc. 1983, 105, 685-90.
69. Mortimer, R. J.; Anson, F. C. J. Electroanal. Chem. Interfac. Electrochem. 1982, 138, 325-41.
70. Majda, M.; Faulkner, L. R. J. Electroanal. Chem. Interfac. Electrochem. 1982, 137, 149-56.
71. Pickup, P. G.; Kutner, W.; Leidner, C. R.; Murray, R. W. J. Amer. Chem. Soc. 1984, 106, 1991-8.
72. Jones, E. T. T.; Faulkner, L. R. J. Electroanal. Chem. Interfac. Electrochem. 1987, 222, 201-22.
73. Miller, C. J.; Majda, M. J. Amer. Chem. Soc. 1985, 107, 1419-20.
74. Schmidt, P. P. Selected Reviews (Marcin Majda) In Ultramicroelectrodes; Fleischmann, M.; Pons, S.; Rolison, D. R.; Schmidt, P. P., Eds.; Datatech Systems: Morgentown, NC, 1987; p. 127.

75. Leddy, J.; Vanderborgh, N. E. J. Electroanal. Chem. Interfac. Electrochem. 1987, 235, 299-315
76. Widrig, C. A,; Miller, C. J.; Majda, M. J. Amer. Chem. Soc. 1988, 110, 2009-11.
77. Zak, J.; Kuwana, T. J. Amer. Chem. Soc. 1982, 104, 5514-5.
78. Shaw, B. R.; Creasy, K. E. J. Electroanal. Chem. Interfac. Electrochem. 1988, 243, 209-17.
79. Faulkner, L. R. Chem. Eng. News 1984, 62, 28-45.
80. Fleischmann, M.; Pons, S.; Rolison, D. R.; Schmidt, P. P., Eds. Ultramicroelectrodes; Datatech Systems: Morgentown, NC, 1987.
81. Mallouk, T. E.; Cammarata, V.; Crayston, J. S.; Wrighton, M. S. J. Phys. Chem. 1986, 90, 2150-6.
82. Itaya, K.; Uchida, I.; Neff, V. D. Acc. Chem. Res. 1986, 19, 162-8.
83. Rudzinski, W. E.; Bard, A. J. J. Electroanal. Chem. Interfac. Electrochem. 1986, 199, 323-40.
84. King, R. D.; Nocera, D. G.; Pinnavaia, T. J. J. Electroanal. Chem. Interfac. Electrochem. 1987, 236, 43-53.
85. Ghosh, P. K.; Mau, A. W. H.; Bard, A. J. Electroanal. Chem. Interfac. Electrochem. 1984, 169, 315-7.
86. Oyama, N.; Anson, F. C. J. Electroanal. Chem. Interfac. Electrochem. 1986, 199, 467-70.
87. Inoue, H.; Yoneyama, H. J. Electroanal. Chem. Interfac. Electrochem. 1987, 233, 291-4.
88. Susic, M.; Petranovic, N. Electrochim. Acta 1978, 23, 1271-4.
89. Johansson, G.; Risinger, L.; Falth, L. Anal. Chim. Acta 1980, 119, 25-32.
90. Thackery, M. M.; Coetzer, J. Solid State Ionics 1982, 6, 135-8.
91. Pereira-Ramos, J. P.; Messina, R.; Perichon, J. J. Electroanal. Chem. Interfac. Electrochem. 1983, 146, 157-69.
92. Murray, C. G.; Nowak, R. J.; Rolison, D. R. J. Electroanal. Chem. Interfac. Electrochem. 1984, 164, 205-10.
93. De Vismes, B.; Bedioui, F.; Devynck, J.; Bied-Charreton, C. J. Electroanal. Chem. Interfac. Electrochem. 1985, 187, 197-202.
94. Shaw, B. R.; Creasy, K. E.; Lanczycki, C. J.; Sargeant, J. A.; Tirhado, M. J. Electrochem. Soc. 1988, 135, 869-76.
95. Li, Z.; Mallouk, T. E. J. Phys. Chem. 1987, 91, 643-8.
96. Creasy, K. E.; Shaw, B. R.; Electrochim. Acta 1988, 33, 551-6.
97. Persaud, L.; Bard, A. J.; Campion, A.; Fox, M. A.; Mallouk, T. E. J. Amer. Chem. Soc. 1987, 109, 7309-14.
98. Gul', V. Y. Polym. Sci. USSR 1979, 20, 2427-41.
99. Anderson, J. E.; Tallman, D. E.; Chesney, D. J.; Anderson, J. L. Anal. Chem. 1978, 50, 1051-6.
100. Gueshi, T.; Tokuda, K.; Matsuda, H. J. Electroanal. Chem. Interfac. Electrochem. 1979, 101, 29-38.
101. Weisshaar, D. E.; Tallman, D. E. Anal. Chem. 1983, 55, 1146-51.
102. Amatore, C.; Saveant, J. M.; Tessier, D. J. Electroanal. Chem. Interfac. Electrochem. 1983, 147, 39-51.
103. Rusling, J. F.; Brooks, M. Y. Anal. Chem. 1984, 56, 2147-53.
104. Engstrom, R. C.; Weber, M.; Werth, J. Anal. Chem. 1985, 57, 933-6.
105. Kovach, P. M.; Deakin, M. R.; Wightman, R. M. J. Phys. Chem. 1986, 90, 4612-7.

106. Henning, T. P.; White, H. S.; Bard, A. J. J. Amer. Chem. Soc. 1981, 103, 3937-8.
107. Burgmayer, P.; Murray, R. W. J. Electroanal. Chem. Interfac. Electrochem. 1982, 135, 335-42.
108. Doblhofer, K.; Durr, W. J. Electrochem. Soc. 1980, 127, 1041-4.
109. Dominey, R. N.; Lewis, N. S.; Bruce, J. A.; Bookbinder, D. C.; Wrighton, M. S. J. Amer. Chem. Soc. 1982, 104, 467-82.
110. Liu, H. Y.; Anson, F. C. J. Electroanal. Chem. Interfac. Electrochem. 1983, 158, 181-5.
111. Bartak, D. E.; Kazee, B.; Shimazu, K.; Kuwana, T. Anal. Chem. 1986, 58, 2756-61.
112. Coche, L.; Moutet, J. C. J. Amer. Chem. Soc. 1987, 109, 6887-9.
113. Geno, P. W.; Ravichandran, K.; Baldwin, R. P. J. Electroanal. Chem. Interfac. Electrochem. 1985, 183, 155-66.
114. Shaw, B. R.; Creasy, K. E. Anal. Chem. 1988, 60, 1241-4.
115. McKenna, K.; Brajter-Toth. A. Anal. Chem. 1987, 59, 954-8.
116. Wang, J.; Golden, T.; Tuzhi, P. Anal. Chem. 1987, 59, 740-4.
117. Waltman, R. J.; Bargon, J. Can. J. Chem. 1986, 64, 76-95.
118. Diaz, A. F.; Kanazawa, K. K.; Gardini, G. P. J. Chem. Soc. Chem. Commun. 1979, 635-6.
119. Castro-Acuna, C. M.; Fan, F. R. F.; Bard, A. J. J. Electroanal. Chem. Interfac. Electrochem. 1987, 234, 347-53.
120. Kittlesen, G. P.; White, H. S.; Wrighton, M. S. J. Amer. Chem. Soc. 1984, 106, 7389-96.
121. Frew, J. E.; Hill, H. A. O. Anal. Chem. 1987, 59, 933-44.
122. Smith, V. J. Anal. Chem. 1987, 59, 2256-9.
123. Gyss, C.; Bourdillon, C. Anal. Chem. 1987, 59, 2350-5.
124. Gardea-Torresdey, J.; Darnall, D.; Wang, J. Anal. Chem. 1988, 60, 72-6.
125. Oyama, N.; Hirokawa, T.; Yamaguchi, S.; Ushizawa, N.; Shimomura, T. Anal. Chem. 1987, 59, 258-62.
126. Pickup, P. G.; Murray, R. W. J. Electrochem. Soc. 1984, 131, 833-9.
127. Pickup, P. G.; Leidner, C. R.; Denisevich, P.; Murray, R. W. J. Electroanal. Chem. Interfac. Electrochem. 1984, 164, 39-61.
128. Kittlesen, G. P.; White, H. S.; Wrighton, M. S. J. Amer. Chem. Soc. 1985, 107, 7373-80.
129. Belanger, D.; Wrighton, M. S. Anal. Chem. 1987, 59, 1426-32.
130. Bobbitt, J. M.; Colaruotolo, J. F.; Huang, S. J. J. Electrochem. Soc. 1973, 120, 773.
131. Kastening, B.; Schiel, W.; Henschel, M. J. Electroanal. Chem. Interfac. Electrochem. 1985, 191, 311-28.
132. Kastening, B.; Spinzig, S. J. Electroanal. Chem. Interfac. Electrochem. 1986, 214, 295-302.
133. Letord-Quemere, M. M.; Coeuret, F.; LeGrand, J. J. Appl. Electrochem. 1987, 17, 965-76.
134. Busscher, N.; Kastening, B. Proceedings of the 4th International Carbon Conference 1986, p. 364.
135. Rolison, D. R.; Nowak, R. J.; Pons, S.; Ghoroghchian, J.; Fleischmann, M. Proceedings of the 3rd Molecular Electronics Device Workshop 1987.

RECEIVED August 18, 1988

Chapter 23

With the Drop of Mercury to the Nobel Prize

Petr Zuman

Department of Chemistry, Clarkson University, Potsdam, NY 13676

History of the development of the polarographic method from the birth of Professor Jaroslav Heyrovský in 1890 until the Nobel Prize award in 1959 is described. Background of the Nobel Laureate's family and education is given. Factors contributing to the discovery and development of the method are discussed. Polarography is unique among modern analytical methods, as its development is strongly centered around a single personality - J. Heyrovský - and as the timing of experiments leading to its discovery can be narrowed to a single day, February 10, 1922.

On December 20, 1890 a son, Jaroslav (Figure 1), was born in Prague into the family of Leopold Heyrovský, Professor of Roman Law at Charles University. The juristic tradition had run in his family since the eighteenth century. His wife, born Klára Hanzlová, was a daughter of a professor of administrative law. Young Jaroslav had three sisters (later married to a painter, a lawyer and an officer) and a brother (who became an entomologist). Leopold was an advocate of Czech autonomy, friend of T.G. Masaryk (who later became the first president of Czechoslovak Republic) and author of the textbook "The History and System of Roman Law" which went through five editions.

Jaroslav Heyrovský as a child had great fantasy - together with his brother Leopold he created complex novel fairy-tales, the action of which took place in the streets of the Old Town quarter of Prague. He loved music and became an accomplished piano player. At high school he participated in chamber music, but his great love was Wagner's music. Whenever he had an opportunity, he attended the festivals at Bayreuth. In his youth he enjoyed sports, in particular soccer, biking, skiing, and swimming.

In 1901 Jaroslav entered high school (Akademické Gymnasium) in Prague, taking courses in Latin for eight years and Greek for five years. His interests first encompassed botany and zoology, mineralogy and anthropology, but later turned to physics, mathematics and chemistry. He became interested in the newly developing interdisciplinary field of physical chemistry and made up his mind to study

0097–6156/89/0390–0339$08.75/0

this discipline. In 1909 he passed his final examination (called "maturity examination") at the high school and matriculated in the Faculty of Philosophy of the Czech University in Prague. During his freshman year he took courses in chemistry, physics and mathematics and was most strongly influenced by the lectures of B. Brauner on inorganic chemistry, as well as those of F. Záviška and B. Kučera on physics.

As there was no opportunity to study this discipline in Prague, young Heyrovský became attracted by British educational institutions. He was impressed by the work of Sir William Ramsay and was therefore thankful to his rather strict father (who was at the time Rector of the Czech University in Prague and was rather feared as examiner by the students of law) when he gave young Heyrovský permission to continue his studies in London. In the fall of 1910 he matriculated at University College, part of the University of London, and attended the lectures on general and physical chemistry by Sir William Ramsay and William C. McLewis, on physics by F.T. Trouton and A. Porter, and on mathematics by L.N.G. Filon. In 1913 he received the B.Sc. degree of the University of London. In that year Sir William Ramsay retired. One wonders how Heyrovský's career would have developed, if this had not happened.

Ramsay was succeeded by the eminent physical chemist F.G. Donnan whose main research area was electrochemistry. Young Heyrovský (Figure 2) was appointed demonstrator (teaching assistant) for the school year 1913/14 and started experimental research under Professor Donnan on a project involving determination of the electrode potential of aluminum. The use of the simple Nernst equation for treatment of the potential of this electrode is prevented by a layer of oxides and other passivity effects. This adsorbed layer hinders the exchange reactions, and the measured potential is not a simple function of aluminum ion concentration. Moreover, the evolution of hydrogen causes fluctuations of the measured potential. Heyrovský grasped well several fundamental scientific problems involved and formulated in his laboratory notebook (which he was accustomed to keep carefully into advanced age) the following questions: "What is the mechanism of hydrogen evolution?" "In which way is the potential of an electrode established?"

When experiments with solid aluminum electrodes did not yield satisfactory results, Professor Donnan advised Heyrovský to use a dilute aluminum amalgam and suggested that he let the amalgam flow slowly out of a glass capillary. This approach was based on the assumption that the continuous renewal of the surface would prevent passivation phenomena. This was a device similar to that used earlier by Donnan in his study of membrane equilibria, where he determined the activity of sodium ions by measuring the potential of a dilute sodium amalgam, that was slowly flowing out of a thick-walled capillary tube.

Heyrovský's experiments with aluminum amalgam were only slightly better than those with solid aluminum electrodes, as the evolution of hydrogen affected the measurements. But even when the results were negative for the solution of the particular problem, these studies strongly influenced Heyrovský's future studies. He saw

Figure 1. Jaroslav Heyrovský (1890-1967), picture taken in 1962.

Figure 2. J. Heyrovský, about 1913.

the advantages of liquid metallic electrodes, in particular their periodically renewable surface and he learned how to use capillary electrodes.

Laboratory notebooks bear witness that in London as later in Prague, Heyrovský spent almost all of his time carrying out experimental research. Days upon days were spent in the laboratory. The stay in London had a profound effect both on his way of thinking, on his conduct of research and on his life style. The visible signs were his laboratory notebooks, which until the end of his life were written in English, afternoon teas with his co-workers and a courtly approach to both visitors and subordinates.

The experimental work in London was interrupted in the summer of 1914. Heyrovský was visiting his parents in Prague when World War I broke out and he was unable to return to London. For a short time he was able to carry out some experiments using facilities made available by Professor J.S. Štěrba-Böhm at the Chemical Institute of the Czech University at Prague. In January 1915 he was called up for military service in the Austro-Hungarian Army, but because of his weak physical constitution (tuberculosis) he was posted as dispensing chemist and roentgenologist in a hospital at Tábor in South Bohemia. Here he was unable to carry out experiments, but his notebooks indicate that he was able to carry out evaluation and discussion of the experimental data obtained in London. In 1916 the hospital was transferred to Igls (close to Innsbruck in Austria) where Heyrovský was able to carry out limited experiments in the pharmacy, dealing mainly with dissolution of aluminum in acids and alkali metal hydroxides. He was also able to prepare his thesis, entitled "The Electro-Affinity of Aluminum", which he submitted to the Philosophical Faculty of the Czech University in Prague. He passed his final examinations, and on September 26, 1918 was granted the Ph.D. degree.

Final examinations consisted in a longer oral examination in the major field -chemistry- and a shorter examination in the minor field-physics. It was the examination in physics which strongly affected Heyrovský's scientific career. The examiner was B. Kučera (1874-1921), Professor of Experimental Physics, who gave him a question dealing with the electrocapillarity of mercury. He knew that examiners at this level often ask questions that have some relation to their own work and was prepared for it. Professor Kučera had developed a new experimental technique for the measurement of electrocapillary curves (1), based on weighing of the mercury drops fallen from a glass capillary, connected to a mercury reservoir. Such measurements of electrocapillary curves showing the dependence of the surface tension of mercury on the potential of a mercury electrode were complementary to those using a capillary electrometer, introduced in 1870 by G. Lippmann. As Heyrovský was knowledgeable of Kučera's research, the examination became a discussion, during which Professor Kučera described to the candidate some of his unpublished results, obtained recently. In those experiments, carried out with solutions exposed to atmospheric oxygen, some of the electrocapillary curves obtained with the dropping mercury electrode showed maxima that were absent on electrocapillary curves recorded in the same solution using the capillary electrometer. Professor Kučera expressed the view that such anomalies could be explained only by a physical chemist and proposed to the candidate that he should carry on the research on the surface tension of mercury electrodes, to which a voltage has been applied.

Professor B. Kučera invited Heyrovský to visit him the following day in his institute and showed him how to construct the dropping mercury electrode, using a glass capillary, connected to a mercury reservoir placed so high above the capillary that a drop of mercury would fall from its orifice every few seconds. He also gave Heyrovský reprints of his paper (2) on the anomalous maximum (3) and advised him to collaborate on the tedious weighing of mercury drops with Dr. R. Šimůnek, who was at that time an assistant professor of experimental physics. For the next two years the two young scientists spent their free hours collecting drops of mercury which had fallen from the capillary electrode at varying voltages, weighing them accurately and plotting the weight as a function of the applied voltage.

The electrochemical work proceeded slowly. In January 1919 Heyrovský was appointed Assistant Professor in the Department of Inorganic and Analytical Chemistry. The head of this department was Professor B. Brauner, a former intimate friend of D.I. Mendeleev and R. Abegg, one of the early proponents of the periodical system and proposer of the use of the value 16.00 for the atomic weight of oxygen. Brauner, who was particularly interested in the chemistry of lanthanides, directed Heyrovský's attention to the problems of chemical affinity and valency. The work on aluminic acid, the structure of aluminates and amphotericity, reflecting the influence of the environment in Professor Brauner's laboratories, was submitted by Heyrovský as a "Habilitation Thesis". Based on this research Heyrovský was appointed on August 2, 1920 as the first Docent (Associate Professor) in Physical Chemistry at the Czech University in Prague, henceforth called Charles University. In 1920 Heyrovský published three papers summarizing his studies on aluminum (4) and submitted them for the D.Sc. to the University of London which conferred the degree upon him in 1921.

This new appointment demanded from Heyrovský organization of a laboratory and a series of lectures on physical chemistry. Nevertheless, all of his spare time in the Institute of Physical Chemistry was devoted to the study and interpretation of electrocapillary curves. These experiments were rather time-consuming. Voltage was applied to a dropping mercury electrode and a reference electrode using a simple potentiometer. Fifty drops of mercury fallen from the capillary were collected, dried and weighed. The applied voltage was varied and the whole procedure repeated. To obtain the electrocapillary curve the measured weight was plotted as a function of the applied voltage.

Heyrovský developed a modification, based on measurement of the drop-time which shortened the time spent by drying and weighing. Furthermore, Heyrovský observed the changes in electrocapillary curves in the presence of some metal ions, such as Zn^{2+}, Cd^{2+}, Mn^{2+}, and Ba^{2+} and studied the possibility of measuring their decomposition voltages using such an approach. He reported his results at a meeting of the Czech Mathematical and Physical Society in the Fall of 1920, at which Professor B. Kučera participated, who unfortunately did not live to see the full success of his pupil, because he passed away in 1921.

The laboratory notebooks of J. Heyrovský (5) indicate that during 1921 he came to the conclusion that electrocapillary curves are not suitable for investigation of the processes occurring at the dropping mercury electrode. Sometime towards the end of 1921 the idea to measure the current flowing through the electrolysis cell was developed. The first experiment to measure such current was carried out on January 1, 1922 but was not fully successful because he did not have a sufficiently sensitive galvanometer. He spent the rest of January remeasuring electrocapillary curves, but tried to acquire the needed instrumentation. Because limited funds did not allow purchase of a galvanometer, he contacted his former teacher of physics, Professor F. Záviška, who loaned him a sensitive galvanometer and a potentiometer. On February 10, 1922, J. Heyrovský built a circuit consisting of a potentiometer, an electrolytic cell and the galvanometer. He placed a solution of 1 M NaOH into the cell, immersed a dropping mercury electrode into this solution and added some metallic mercury to form a mercury pool electrode which was used as a reference electrode. The current flowing between the dropping mercury and mercury pool electrodes was indicated by the galvanometer. Already, when a small voltage was applied, Heyrovský observed and recorded that the galvanometer indicated a weak current, the intensity of which oscillated rhythmically with replacement of drops. With stepwise increase in voltage the current increased somewhat and at a more negative voltage increased again. These current increases corresponded to reduction of oxygen, which was understood in detail by Heyrovský only later (6). In the region between -1.9 and -2.0 V, Heyrovský observed on February 10, 1922 substantial increase in current, corresponding to deposition of alkali metal ions (Na^+) forming an amalgam and that became the center of his interest. Thanks to his background in electrochemistry, Heyrovský clearly recognized that he was on the track of an important scientific discovery. Over the weekend of February 12 and 13 he recorded electrocapillary curves for the same solution and compared them with his current-voltage curve and measured the electrode potential of the sodium amalgam from collected drops.

During the following weeks his normally high intensity of work was raised to a feverish pitch. Every page of a 200 page thick notebook was filled with laboratory notes during a period of seven weeks. One week after he recorded the first current-voltage curve he restricted his experimental work to obtaining such curves. Soon he realized that oxygen is reduced at the dropping mercury electrode (D.M.E.) in two steps and also that the current due to oxygen reduction interferes with measurements of currents due to other reduction processes. From the beginning of April Heyrovský removed the dissolved oxygen (present in solutions in contact with air) by bubbling through the solution in the electrolytic cell a stream of hydrogen which he generated in a Kipp apparatus.

Heyrovský was aware that it is of importance not only to make a discovery, but that it is necessary to make ones colleagues aware of the results. He described the results of his first experiments on electrolysis with a dropping mercury cathode in Czech in the October issue of the journal "Chemické listy" (7). He, nevertheless, realized that publication in the Czech language limits the news to a relatively small circle and so he prepared an English version dealing with electrodeposition of alkali and alkaline earth metals, which was published in Philosophical Magazine (8). He followed Newton's motto "A man must resolve either to put out nothing new or to become a slave to defend it." In this year Heyrovský was promoted to extraordinarius (Associate) Professor. He also became the head of the newly established Department of Physical Chemistry at Charles University. The excitement and feverish activity during this period took its toll and Heyrovský had to take a leave on July 12, 1922 from which he returned only on October 27.

When Heyrovský resumed his teaching duties in the fall semester, he was joined by his first five graduate students. Among the topics assigned to these students were problems of hydrogen evolution at the D.M.E., reduction of arsenic (III) ions and of plumbates. He was also joined by the Japanese physical chemist M. Shikata (Figure 3), who came from his studies in Berlin. Heyrovský reported on November 26, 1926 in London at the General Discussion of the Faraday Society dealing with electrode reactions and equilibria about his own research together with that obtained by his research group (9,10). In the first of these contributions (9) Heyrovský recognized that at sufficiently low concentration of the metal ion a limiting current (which over a range of applied voltage remains practically independent of applied voltage) can be observed. The S-shaped increase of the current followed by the limiting current was denoted as "wave". He also concluded that at such concentrations the ions in the vicinity of the electrode are exhausted by reduction and that the current intensity depends mostly on the number of metal ions transported into this space by diffusion. He also pointed out that the position of the wave on the potential axis is characteristic of the reduced species and can be used for qualitative analysis. From occurrence of waves at -0.4 and -0.6 V in solutions prepared from metallic zinc, it was possible to conclude the presence of two impurities, attributed tentatively to indium and gallium. M. Shikata at the same meeting reported (11) the possibility of carrying out the reduction at the D.M.E. in a nonaqueous system - in sodium ethoxide. In the following year the Japanese chemist reported also the first reduction of an organic compound, nitrobenzene (12,13).

The progress of the extension of investigation of the electrolysis with the D.M.E. by J. Heyrovský and his research group to other species and systems was limited by the fact that the point-by-point measurements and plotting of current-voltage curves was tedious and time consuming. Considerable improvement was achieved in 1924, when the cooperation of J. Heyrovský and M. Shikata led to construction of an apparatus (14) which registered current-voltage curves automatically. The new instrument recorded photographically such a curve in several minutes, whereas manual recording took an hour or longer. Development of an automatic method of recording, a short time after the discovery of the new technique, played an important role in the development and dissemination of methods based on electrolysis with a D.M.E. Currently, most physico-chemical and analytical methods are carried out using automated techniques, but in the early twenties an automated instrument represented very progressive and advanced instrumentation (e.g., recording spectrophotometers became generally accessible only two decades later).

Figure 3. Masuzo Shikata

In their joint paper, Heyrovský and Shikata (14) proposed for the instrument the name "polarograph" and for the studies of electrolysis with the dropping mercury electrode coined the term "polarography". Their description of the first polarograph was published (14) in a special volume of Recueil de Travaux Chimique de Pays Bas commemorating the seventieth birthday of B. Brauner together with several other papers on polarography, which became the first collection of papers on this technique. Publication activity reflects Heyrovský's early recognition of the duty of a creative scientist to share his results and ideas with the scientific community. In accordance with his belief a printed inscription of Faraday's words "Work, finish, publish" was found on the walls of laboratories in which he worked and followed him wherever he moved.

It is not known where Heyrovský found this quotation, but in an obituary notice of Faraday, W. Crookes (15) stated: "Certainly, no more golden words were ever uttered than those in which he delivered to a young enquirer the secret of his uniform success -'The secret' said he, 'is comprised in three words - Work, Finish, Publish'."

The papers (16-24) published in the special volume of Recueil reflected understanding of some basic processes occurring at the DME. It was recognized that the reduction of metal ions, which can form amalgams inside the dropping mercury electrode, is facilitated when compared with that of metal ions, the reduced form of which is insoluble in mercury and forms a compact layer or are reduced to a lower oxidation state remaining in solution. Furthermore, the effect of labile complex-formation on polarographic reduction was observed. It was also recognized that the reduction is not restricted to metal cations, but also that uncharged molecules and even some anions can be reduced.

In 1926 Heyrovský was appointed Professor of Physical Chemistry at the Charles University and in the same year he married Marie Kořánová, the daughter of a brewer, whose mother was the daughter of a cousin of J. Heyrovský's father. They married when Heyrovský was thirty six and he indicated occasionally to his graduate students that this is the right age for a polarographer to marry. This particular advice was not heeded by a considerable fraction of his students. His devoted and charming wife Marie supported and encouraged her husband during the remainder of Heyrovský's scientific career and lightened his burden by conscientiously taking care of his correspondence and collecting a bibliography of polarography, among other tasks. She gave him two children, daughter Jitka (Czech for Judith), who became a biochemist and worked in a research institute for food science, and a son Michael (Figure 4) (named after Michael Faraday), who received his Ph.D. in Cambridge in 1966 and became an electrochemist working in the institute bearing his father's name. In 1926 Heyrovský received a Rockefeller Fellowship which enabled him to stay in the laboratory of Professor G. Urbain at the Sorbonne in Paris for six months.

Towards the end of the twenties and the beginning of the thirties about forty co-workers were in Heyrovský's research group. Among them were four - R. Brdička, W. Kemula, G. Semerano and D. Ilkovič, who particularly distinguished themselves in the development of polarography. Of those W. Kemula (Figure 5) started an active center of polarographic research in Poland. The second generation of

Figure 4. Michael and Jaroslav Heyrovský carrying out experiments in oscillopolarography.

Figure 5. W. Kemula (left), well-known Polish physico-chemist W. Swiętoslawski, J. Heyrovský and the rector of the University of Warsaw (Figurowski) at a meeting in Warsaw.

Kemula's co-workers contributed to the flourishing of Polish electrochemistry in the seventies and eighties. G. Semerano had a similar impact on Italian electrochemistry; he founded a polarographic institute in Padua in the thirties that had a profound influence on electrochemistry in the Mediteranean region.

The work of the Prague school led to early recognition of several types of currents that can flow during an electrolysis with the DME. First of these was the migration current (25,26) which results from transport of ions in the electrical field between the electrodes. It was soon recognized that the effect of the migration current can be minimized, when the electrolyzed solution contains high concentration of an electrolyte, the components of which are neither reduced nor oxidized over a wide potential range. Thus "supporting electrolyte" was added to all solutions studied for examination by polarography, voltammetry and related techniques and only recently the use of microelectrodes enables studies in its absence.

When the concentration of an electroactive species at the electrode surface is depleted by electrolysis and when the effect of convection is minimized, the principal mode of transport of the electroactive species from the bulk towards the electrode surface is by diffusion. Based on experimental observations by Kemula (27) (Figure 5), indicating that capillaries with equal flow-rates give approximately equal linear currents, and that the current increases with increasing height of the mercury column, and that the dependence has a shape of a parabola, Ilkovič (Figure 6), (28,29) successfully solved the problem of a current governed by the rate of diffusion towards the electrode surface.

Considering diffusion to a planar electrode with a surface area increasing with time for description of the transport to a growing mercury drop, Ilkovič derived for the mean limiting diffusion current ($\bar{i}_d$) Equation 1, bearing his name:

$$\bar{i}_d = 607\ n\ C\ D^{1/2}\ m^{2/3}\ t_1^{\ 1/6} \tag{1}$$

(where n is the number of transferred electrons, C concentration in millimoles/liter, D the diffusion coefficient in cm^2/s, m the mercury flow rate in mg/s and t_1 the drop-time in seconds).

An analogous equation was derived later by MacGillavry and Rideal (30) who considered spherical diffusion, but used a simplified treatment, and by v. Stackelberg (31)(Figure 7) who combined an expression for the thickness of a diffusion layer with Fick's law. Small differences between accurate experimental data and equation 1 have more recently (32) been interpreted as due to the spherical diffusion, change in mercury flow-rate with time and a small local depletion of the solution at the beginning of the drop formation.

The current needed to charge the electrode and form the electrical double-layer at its surface is called the charging current. For this current, which is observed even in the absence of species undergoing reduction or oxidation, Ilkovič (Figure 6) derived (33) the equation 2:

$$\bar{i}_c = 0.85\ E^*\ C'\ m^{2/3} t^{-1/3} \tag{2}$$

Figure 6. D. Ilkovič

Figure 7. B. Breyer and M. v. Stackelberg, on board a ship on Elbe during the Conference on Polarography, Dresden, 1957.

(where $\bar{i}_c$ is the mean charging current, E* is the potential of the electrode relative to the point of zero charge, C' is the specific capacity per cm^2 and m and t_1 have the same meaning as in equation 1). Presence of this current limits the sensitivity of the simple (d.c.) polarographic method. The first attempt to minimize its effect was carried out by Ilkovič and Semerano (34).

The dissemination of the knowledge of polarography in the thirties was in particular influenced by two publications in book form by Heyrovský and by his two trips abroad.

The first Czech book (35) was revised, extended and translated into Russian, whereas an Italian monograph was published by Semerano (36). The Heyrovský's chapter in the prestigious monograph on physical methods in analytical chemistry edited by Böttger (37) had a strong impact. In 1932, Böttger, a well-known analytical chemist from Leipzig, spent two weeks in Heyrovský's laboratories to get acquanted with the new technique before asking Heyrovský for his contribution. Inclusion of polarography in this volume represented its official acceptance as an acknowledged analytical method, particularly among German chemists.

A visit to the United States in 1933 contributed substantially to the spread of the knowledge of polarography in the English speaking world. Heyrovský was awarded a visiting professorship by the Carnegie Foundation. He crossed the Atlantic on "Europa", and the American continent by rail. He lectured for six months at the University of California at Berkeley, where he gave two seminars per week for the physical chemists, and one per week for the biochemists. He also gave talks at Stanford University, at the California Institute of Technology, and at the Universities of Minnesota and Wisconsin, at Ohio State University and at Princeton and Cornell Universities. He was impressed by the American scientists and found their approach to research similar to his, but different from that common in Central Europe.

"In addition to technical culture, the modern American scientists possess high awareness in natural sciences and in general culture. Their working habits are superior to ours mostly because they are more systematic in their approach and their work is better organized. The chemists from all areas meet at least once a week for seminars and discussions about their individual research projects. All help each other altruistically, ready to assist their colleagues by sharing their knowledge and to collaborate experimentally. In this way, their approach differs sharply from ours, which is highly individualistic. Each of us is used to work as an independent unit, circumvallated by a Chinese Wall against all influences and suggestions from one's surrounding, which we treat rather with mistrust." (38).

The thriving research spirit of the Department of Physical Chemistry at the Charles University in Prague was broken up by the closing of the Charles University by the Nazi German occupants of Czechoslovakia in November 1939. The Czech professors were sent into early retirement; buildings, furnishings and equipment were confiscated by the German University of Prague. Thanks to friendly efforts of Heyrovský's colleague, a quiet but personally audacious German anti-Nazi Professor J. Böhm, his laboratory remained at his disposal during World War II. He could carry on his experiments, even though he was without students and co-workers. During this period Heyrovský was able to finish his large textbook on polarography (39), which was considered of such importance that it was reprinted in Ann Arbor, Michigan in 1944. He also initiated (40) investigation of electrode processes, which take place at the dropping or streaming mercury electrodes during very short periods of time of controlled current electrolysis, using an oscilloscope (Figure 4).

Fortunately some of Heyrovský's closest co-workers found it possible during the German occupation to continue polarographic research in laboratories in hospitals, in the pharmaceutical and in the chemical industry. Perhaps the most important contributions during this period were made by K. Wiesner who at that time worked in the research laboratory of B. Fragner's pharmaceutical company in Prague-Měcholupy and R. Brdička, who was able to work in the laboratories of the Radiological Division of the hospital in Prague-Bulovka, while actively involved in cancer research. They investigated the problems of electrode processes accompanied by a chemical reaction. When the reaction occurs before or parallel to the electron transfer, the rate of such reactions governs electrolytic currents.

The first example of a current governed by a rate of a chemical reaction was observed by Wiesner (41) for a heterogeneous system: the anodic wave of a hydroquinone was increased in the presence of colloidal palladium saturated by hydrogen. This increase was due to a chemical reduction occurring in the vicinity of the electrode surface. In this reaction the electrochemically formed quinone was reduced by atomic hydrogen generated at a palladium catalyst. The majority of the systems investigated subsequently in more detail involved an interaction of an oxidizing agent with an electroreduction product, resulting in an increase of a cathodic wave - e.g. of hemin (42) or iron(III) (43) in the presence of hydrogen peroxide as oxidant or of oxygen in the presence of catalase (44). Such currents are denoted as regeneration or catalytic currents.

Even more frequently encountered are kinetic currents where the chemical reaction occurs before electron uptake. Chemical reactions involved are equilibria which are perturbed by electrolysis, which at a given potential converts one of the participating electroactive species into an electrolytic product. The observed current is governed by the rate by which the electroactive species is regenerated from the other form in equilibrium. In the majority of the cases studied the electroactive species is formed in a bimolecular process. To simplify the treatment, one reactant is kept in excess so that a pseudo-first order kinetics is followed. The measured current is a function of concentration of the component present in excess. The treatment of such systems involves evaluation of the dependence of the current on concentration of the component present in excess. In the initial stage Brdička and Wiesner developed an approximate treatment (42,44), assuming the existence of a thin reaction layer in the vicinity of the electrode, where the conversion of the electroactive species and its regeneration proceed under steady state conditions. The thickness of the reaction layer depends on the value of the rate constant involved (45). This method yields satisfactory results for very fast reactions, but its application has limitations which were recognized and discussed later (46,47). Three types of processes were recognized at this early stage: reduction of organic acids, where the governing reaction was the protonation of the anion (48), reduction of some aldehydes, where the current was governed by the rate of dehydration of the geminal diol form (49) and reduction of aldohexoses, where the reactions involved opening of cyclic forms (50).

Another important factor which can affect electrode processes and result in changes in polarographic current-voltage curves was recognized in the forties, namely the role of adsorption. Characteristic features on current voltage curves in the case when one component of the electroactive oxidation-reduction couple is adsorbed at the surface of the dropping mercury electrode were recognized by Brdička (51-53). Effects of adsorbed electroinactive surfactants were first observed by Wiesner (54) and Heyrovský (55).

Anodic waves were shown (56) to occur on polarographic current-voltage curves not only when oxidation of the electroactive species occurs, but also when mercury is oxidized and resulting mercury ions form slightly soluble or complex salts with components of the studied solutions, (e.g., Cl^- or CN^-).

Finally, attention has been drawn to currents of hydrogen evolution which in the presence of some substances occurs at more positive potentials than in their absence. Such substances catalyze hydrogen evolution and result in high currents which are denoted catalytic hydrogen waves. Such waves are observed in the presence either of platinum group metals (57,58), where the catalysis is attributed to clusters of metals deposited on mercury or of compounds which possess acid-base properties. Catalytic effects of the latter type in solutions of simple buffers have been observed for low molecular weight compounds (59,60), as well as for proteins (61,62). Similar catalytic effects in ammoniacal cobalt (III)-solutions (63) found utilization in Brdicka reaction (64-66), used in cancer diagnosis.

Industrial development during World War II and immediately afterwards resulted in the need for sensitive and rapid methods for analysis of raw meterials, intermediates and products in metallurgy, in heavy chemical, pharmaceutical and food industry, in synthetic rubber manufacture and in particular in development of atomic energy and numerous other areas. Polarography proved in many instances to be better suited to meet such requirements than other analytical methods available during that period and soon was among the five most frequently used analytical methods. The increased use of polarography was facilitated by availability of commercially produced polarographs. The first such instruments were manufactured by D.V. and J. Nejedlý in Prague in 1932; and the number of instruments produced increased yearly. Since 1939 polarographs have been produced by E.H. Sargent & Co. in the U.S.A. (67). With an increasing number of available instruments the number of papers devoted to polarography also increased, surpassing 100 per year in 1937. The adoption of polarography in the English-speaking world was considerably influenced by the interest of the eminent American analytical chemist, I.M. Kolthoff (Figure 8). He began to publish in the area in 1939, and together with J.J. Lingane (Figure 9) produced a widely accepted monograph (68), whose two editions served as an important source of information.

The end of World War II and liberation of Czechoslovakia from German occupation enabled Charles University to reopen in the summer

Figure 8. I.M. Kolthoff (left) and J. Heyrovský in front of the Chemical Institute (Charles University) in late forties.

Figure 9. J.J. Lingane with G. Patriarche and their wives.

of 1945. The pent-up energy of young people who had been deprived for six years of the opportunity to study resulted in a phenomenal upsurge in activity both in teaching and research. Papers which could not have been published for years, because Nazis did not allow publication of Czech scientific journals, appeared suddenly in print in 1947 (see e.g., references (42,44,48-50,53-55 and 60).

The Department of Physical Chemistry again became a center of polarographic research. The difficult task of the new organization, of teaching and re-equipment of the department, was undertaken by two of Heyrovský's prominent co-workers, Professor R. Brdička and M. Kalousek. After the war, the friendly help of Professor J. Böhm was for some time misrepresented by some of Heyrovský's colleagues, yet soon thorough rehabilitation followed and Heyrovský's attitude was fully justified. Delicate health due to the privations suffered during World War II forced Professor Heyrovský to limit his teaching to lectures on polarography. He, nevertheless, took part actively in seminars (Figure 10), in supervision of graduate students, and in travel to lecture abroad, e.g., in 1947 to England, Sweden and Denmark. His main research interest in this period was development of derivative and differential methods of polarography and studies of electrode processes by oscillopolarography.

The driving force behind polarographic research, nevertheless, became R. Brdička. His main field of interest were currents, where the electron transfer was accompanied by a chemical reaction. K. Wiesner soon left the group to eventually become a respected natural product chemist in Fredericton (New Brunswick, Canada). It was with the help in particular of V. Hanuš and J. Koutecký (who joined the group because of his expertise in heat and mass transfer) that rigorous treatment has been developed for all known types of chemical processes accompanying the electrode process proper (69-72). I. Kössler studied dissolution of amalgams (73) and together with J. Koryta (74,75) applied polarography to the studies of the formation of substitution labile complexes. Among other innovative approaches M. Kalousek developed a commutator method (76) which enabled one to follow the products formed under conditions of polarographic electrolysis.

The atmosphere in the department during the period between 1948 and 1950 was the most vigorous in the development of scientific ideas that this author has ever encountered. Weekly seminars were inspiring and scientific discussions continued incessantly throughout the long working days -in the laboratories, in corridors, even in the lavatories. Progress was made by continuous reevaluation of ideas, reinterpretation of experimental results, as well as by carefully planned experiments. In all these activities all members of the group participated as equals, including first year graduate students.

Progress in polarography in this period was demonstrated by the First International Congress of Polarography held in Prague in 1951 and the Second in Cambridge (England) in 1959. Nevertheless, the latter demonstrated that the progress in the fifties also encompassed other electroanalytical techniques, related to polarography. The closest related technique is a.c. polarography, where a dropping mercury electrode is also used, but a sinusoidal voltage of a small amplitude is superimposed on the gradually increasing d.c. voltage. The first instrument based on such a technique was patented early

Figure 10. J. Heyrovský at a seminar at the Polarographic Institute of the Czechoslovak Academy of Sciences, in Prague, Vlašská 9. Front row (from left): R. Brdička, J. Mašek; sedond row: P. Zuman, M. Nováková, B. Matyska, O. Exner, V. Volková; last row: J. Tenygl, A. Kejharová.

(77), but fell into oblivion. Since 1944, the early development in this field was based on the work of B. Breyer (Figure 7) and his group (78).

When the voltage applied to the electrodes is increased linearly with time using a sweep that is much faster than in d.c. polarography, the surface of the electrode is not renewed and the technique is called linear sweep voltammetry. For this technique an equation for the current governed by linear diffusion of the electroactive substance towards the electrode was (for a reversible electrode process) solved independently by Ševčík (79) and Randles (80) and corrected by Matsuda (81). The first instruments in which a single sweep was applied late in the life of the mercury drop were constructed independently by Randles (82) and Airey (83). When the direction of the voltage scan is reversed at a chosen potential, a triangular sweep results and the technique is called cyclic voltammetry. Instruments in which several triangular sweeps were applied were devised by Ševčík (79) and Delahay (Figure 11) (84), but the first instrument in which a single triangular-voltage sweep is applied during a short interval late in the drop-life was constructed by Vogel (85). Cyclic voltammetry later became one of the most widely used electroanalytical techniques, mainly due to the activities of the groups of R. Adams, I. Shain and J.M. Savéant in the sixties. Among techniques using controlled current and measuring voltage or its function, chronopotentiometry was re-introduced in the fifties by Delahay (Figure 11) (86,87). A method using alternating current, denoted oscillopolarography was further developed by Heyrovský. Both the principles and theory of most of these techniques were summarized in the monograph by Delahay (88), which became the text used in education of the new generation of American electrochemists.

But the developments which affected most profoundly future applications of polarography originated from the need of the British atomic energy researchers for ultrasensitive analytical methods, enabling determination of traces of elements in various materials. As mentioned above, the principal factor limiting sensitivity of d.c. polarography is the presence of the capacity (or charging) current. Attempts to use a counter-current, derivative and differential methods or storage and subtraction of charging current using artificial memory proved to be only partly successful as were methods measuring the current only during later stage in the life of a mercury drop (Tast or sampled polarography), when a linear voltage ramp was used. G.C. Barker (Figure 12), working at the Atomic Energy Research Establishment in Harwell (England) developed in the fifties four new techniques, in which the linear voltage ramp was superimposed either by a square-wave alternating voltage (square-wave polarography, (89, 90), (Figure 13), by voltage pulses of constant amplitude applied towards the end of the drop-life [pulse polarography (91)], (Figure 14), or by a radio frequency current modulated by a square wave [radio frequency polarography (92)]. Alternatively, voltage pulses of gradually increasing amplitude were applied to the cell towards the end of the drop-life (normal or integral pulse polarography). In all four techniques, the current is measured only during the last third of the applied pulse or square-wave. During this period the capacity current becomes small when compared to the faradaic

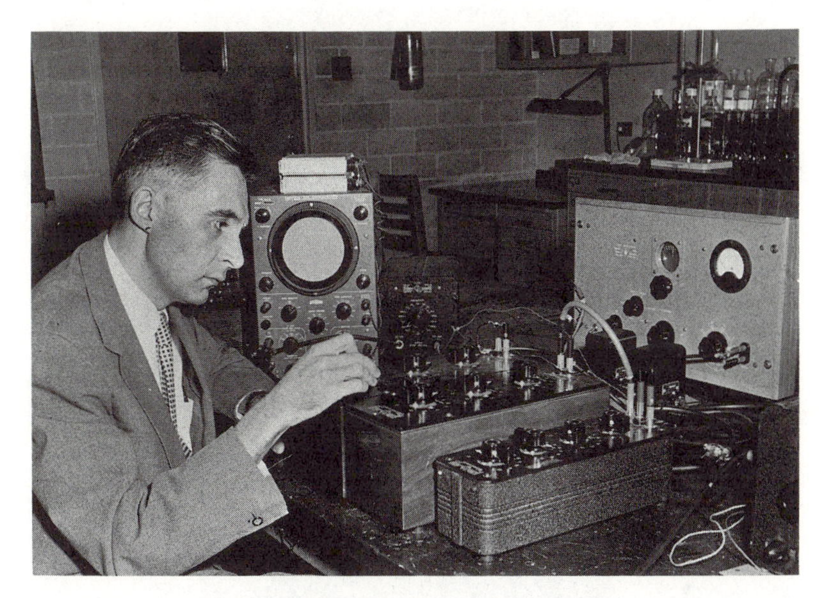

Figure 11. P. Delahay carrying out a double-layer experiment using an a.c. bridge (Louisiana State University, Baton Rouge, early sixties).

Figure 12. G.C. Barker (left) and B. Nygård, followed by H. Berg and A.A. Vlček.

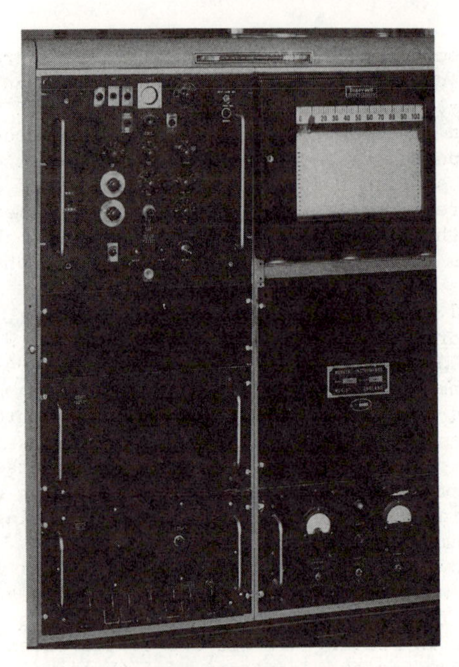

Figure 13. First commercial square-wave polarograph, Harwell 1953.

Figure 14. First pulse polarograph, Harwell, mid-fifties.

(electrolytic) component, and this results in an increase in sensitivity by two or even three orders of magnitude (when compared with d.c. polarography), allowing concentrations up to 10^{-8} M to be reached. Introduction of circuitry allowing the use of such techniques, particularly of pulse polarography, into inexpensive commercial instruments in the sixties resulted in availability of analytical procedures which were more sensitive than spectrophotometric ones and could compete with those using atomic absorption or chromatography. Such applications led to a renaissance of polarography as an analytical method. This and further developments will be discussed in the contribution by J. Osteryoung (93).

Another important development was pioneered by W. Kemula (Figure 5), who combined polarography with chromatography (94-98). In this method (called chromatopolarography) when a dropping mercury electrode was used as a sensor in the effluent from a chromatographic column, the current at a chosen potential was followed as a function of time. It was probably the first of the "hyphenated" methods of analysis. W. Kemula also modified (99) the hanging mercury drop electrode first introduced by H. Gerischer (100) using a pressure applied to a volume of mercury (101).

In Czechoslovakia, the scientific and practical importance of polarography was well recognized and hence in May 1950, the Center for Research and Technological Development founded a Central Polarographic Institute in Prague. It was one of the seven institutes founded as a kernel for the development of the re-organized Czechoslovak Academy of Sciences. Heyrovský became the first Director of this Institute, but he remained an Honorary Professor at the Charles University, where he taught theoretical and practical courses on polarography for another decade.

Professor Heyrovský was accompanied in his move to the Polarographic Institute by twelve of his most recent graduate students. They were soon joined by some of his older pupils. In this way one of his ambitions became fulfilled - to see his co-workers, whom he taught and interested in polarography, able to carry on research in this and related fields of electrochemistry. The group, nevertheless, suffered some losses - of R. Brdicka who first stayed at Charles University, but in 1952 became the Director of the Institute of Physical Chemistry, of J. Koutecký who turned his interest to quantum chemistry and of V. Hanuš and V. Čermák who started a new career in mass spectrometry.

In 1952 the reorganization of the Czechoslovak Academy of Sciences was realized and the Institute became the Polarographic Institute of the Czechoslovak Academy of Sciences. As the number of the research associates in the Institute increased, the space at the original site at 25 Opletalova Street in Prague became insufficient. New laboratories were established in separate buildings in various, rather distant parts of Prague. This affected the coherence of the group. To minimize effects of geographical separation, seminars were held each Thursday morning (8-11 a.m.) at one location (Figure 10). That enabled Professor Heyrovský to follow the progress of the research in the Institute and to keep in touch with the new work carried out in other parts of the world. In this period J. Heyrovský received numerous honors.

The scientific impact of Heyrovský's work was such that he was nominated several times for the Nobel Prize. At several occasions (1938, 1939, 1948) the nomination was unsuccessful because of geopolitical factors. But on September 26, 1959 it was officially announced by the Czechoslovak radio that the Royal Academy of Science in Stockholm awarded Jaroslav Heyrovský the Nobel Prize for Chemistry. The vice-president of the Czechoslovak Academy of Sciences, V. Laufberger (a physiologist), and the Chairman of the Chemistry Division of the Academy, R. Brdička, congratulated Professor Heyrovský (Figure 15). Later in Stockholm, J. Heyrovský received the Prize from the hands of the King of Sweden.

In his Nobel address Heyrovský pointed out not only the achievements of d.c. polarography, but also of oscillopolarography, which during the preceding decade was closest to his heart. Even in his busy schedule he often found short breaks in which he was able to carry out experiments using this technique, whereas the available time was too short to record several polarographic current-voltage curves (102). He was under the impression that linear sweep voltammetry would yield results too similar to d.c. polarography to result in new break-throughs in the understanding of electrode processes. He hoped that oscillopolarography, with controlled ac current, would bring discovery of new phenomena. Further development of analytical methods proved him wrong on this point, mostly because of complex conditions of electrolysis. On the other hand, he was very impressed by the achievements of square-wave polarography, the potential of which he fully appreciated. His enthusiasm was only dampened by the sometimes erratic performance of the expensive commercial square-wave polarograph (Figure 13), which in the fifties carried about 90 vacuum -tubes. The probability that one of them would not function properly was relatively high.

Professor Heyrovský was naturally very happy about the appreciation of his life's work as manifested by the Nobel Prize. Unfortunately, the recognition came late in his life when poor health (circulatory and digestive problems) limited his creativity and energy, even though his reasoning remained logical and piercing. He demonstrated his capability repeatedly by his provoking questions at seminars, leading often to the heart of a complex problem.

The Nobel Prize award was very well deserved. Heyrovský was a discoverer who was scientifically well trained at the time when the opportunity of the discovery occurred. His power of observation and detection of new phenomena was remarkable and he had the ability to sort out fundamental from peripheral information. One of his collaborators compared him to an extraordinary mushroom picker, who is able to find a fine mushroom even on a highway. He soon realized both the theoretical and practical possibilities offered by the discovered technique and with all his energy tried to make these a reality. He found collaborators who were able to deal with the mathematical aspects, but considered the theory useful only when it was shown to be in agreement with experiments and that it could lead to the design of new experiments. He mistrusted abstract theorizing which to him seemed distant from experiments. He always favored the application of polarography in analysis, for solution of practical problems. He spent all his adult life in fostering the growth of polarography in all its aspects. He did this by lectures, partici-

Figure 15. Professors (from left) R. Brdička and V. Laufberger congratulating J. Heyrovský at the occasion of the announcement of the awarding of Nobel Prize.

pation at meetings, visits to foreign centers of research, invitation of guest-workers from all over the world into his laboratories, and by publication of original papers, reviews and monographs (103). He was an outstanding teacher, predominantly at the graduate level. His personality made a deep impression on students in spite (or because ?) of his quiet voice and he could generate great enthusiasm about polarography. He was a conscientious supervisor, who visited all the research laboratories at least once a day and showed keen interest in the work in progress.

Professor Heyrovský always worked very intensely (Figure 16). His working day in the laboratory always was from 8 a.m. to 7 p.m., followed in his younger years by working at home during the evening. In his late years he allowed himself a short nap after lunch. He always spent weekends in the laboratories and he insisted that weekends were the only time when he could be certain not to be disturbed during work.

Not only did he work hard, but he expected his co-workers to follow suit. At the University, discussion of research projects was sometimes arranged on Saturday afternoon; at the Institute he could be seen with his watch in hand, standing a few minutes past eight on the staircase and watching the late-comers arrive! He had the belief that the daytime in the laboratory is for experimental work and the evaluation and reading should be done in the evenings. He hated dust on the instruments ("you have to brush it every morning, like your teeth"), reading of newspapers in the laboratory, and in particular smoking. The smokers in the Institute had to go out of the building, into the garden, to smoke. Even then occasionally they had to bear some sarcastic remarks.

But while Professor Heyrovský expected some sacrifice from his co-workers, he was himself ready to give up most of his early interests for polarography. Once a pianist and a member of a student chamber orchestra, later he was only able to listen to music. He was fond of attending the opera and knew lengthy parts of many of the operas by heart. At one time a well-known reviewer of books, when he became enchanted with polarography he hardly found time to read novels. Only when he wished to brush-up his knowledge of languages, such as before a trip abroad, did he read crime stories. However, during the sixties he again found more time for reading and often returned to the books which impressed him in childhood. In his youth he had, with his father, made several unforgettable tours in the high Alps. Always interested in sports, a soccer and tennis player, skier and swimmer, Heyrovský never missed making the kick-off at the traditional soccer matches between the Institute of Polarography and the Institute of Physical Chemistry (in the early days at the Charles University these matches took place between his students and those of Professor Dolejšek). In the garden of the Institute in Vlašská Street he rarely forgot to feed the squirrels.

Heyrovský's life was possibly unique in the single-minded devotion with which he pursued the subject of polarography and later applied it to numerous problems in pure and applied chemistry. Although this method attracted the attention of a large number of workers in many countries, he remained the center of these developments and continued to exert a profound influence on them, not only because of his scientific eminence but by the force of his personality,

Figure. 16. J. Heyrovský at work.

which made an unforgettable impression on the many scientists who had the good fortune to be in contact or to work with him at some time in their careers. He taught his students and co-workers by example more than most of them realized - mainly that the goal of research is not in pursuing a career, but in seeking the scientific truth.

Literature Cited

1. Kučera, B. Ann. Physik (Drude) 1903, 11, 529, 698.
2. Kučera, B. Bull. Acad. Sci. de Bohemie 1903.
3. Although the observation of the maximum on electrocapillary curves in solutions in which oxygen was dissolved was one of the starting points in the discovery of polarography, details of the origin of this phenomenon are still unclear. Many factors governing the appearance and intensity of this phenomenon, resulting from a streaming of the solution in the vicinity of the dropping electrode, are understood, but attempts to present a rigorous mathematical treatment describing this phenomenon so far failed. These anomalous maxima represent a complex natural phenomenon; attempts to interpret it did not yield per se an extension of our knowledge of scientific principles, but their investigation led to development of polarography which had profound effect on evolution of modern electroanalytical chemistry.
4. Heyrovský, J. Trans. Chem. Soc. Lond. 1920, 97, 11, 27 and 1013.
5. Koryta, J. Chem. listy 1972, 66, 113.
6. Heyrovský, J. Čas. čes. lékarn. 1927, 7, 242.
7. Heyrovský, J. Chem. listy 1922, 16, 256.
8. Heyrovský, J. Phil. Mag. 1923, 45, 303.
9. Heyrovský, J. Trans. Faraday Soc. 1924, 19, 692.
10. Heyrovský, J. Trans. Faraday Soc. 1924, 19, 785.
11. Shikata, M. Trans. Faraday Soc. 1924, 19, 721.
12. Shikata, M. Trans. Faraday Soc. 1925, 21, 42, 53.
13. Shikata, M. J. Agric. Chem. Soc., Japan 1925, 1, 81.
14. Heyrovský, J.; Shikata, M. Rec. Trav. Chim. Pays-Bas 1925, 44, 496.
15. Crookes, W. Chemical News, August 1868.
16. Heyrovský, J. Rec. Trav. Chim. Pays-Bas 1925, 44, 488, 499.
17. Herasymenko, P. Rec. Trav. Chim. Pays-Bas 1925, 44, 503.
18. Bayerle, V. Rec. Trav. Chim. Pays-Bas 1925, 44, 514.
19. Březina, J. Rec. Trav. Chim. Pays-Bas 1925, 44, 520.
20. Emelianova, N.V. Rec. Trav. Chim. Pays-Bas 1925, 44, 528.
21. Sanigar, E.B. Rec. Trav. Chim. Pays-Bas 1925, 44, 549.
22. Smrž, J. Rec. Trav. Chim. Pays-Bas, 1925, 44, 580.
23. Podroužek, V. Rec. Trav. Chim. Pays-Bas 1925, 44, 591.
24. Gosman, B.A. Rec. Trav. Chim. Pays-Bas 1925, 44, 600.
25. Šlendyk, I. Collect-Czechoslov. Chem. Commun. 1931, 3, 385.
26. Heyrovský, J. Ark. Hemiju i Farmaciju 1934, 8, 11.
27. Kemula, W. Trabajos IX Congr. Internat. Quim. Pura y Applicada, Madrid 1934, Vol. 2, p. 297.
28. Ilkovič, D. Collect. Czechoslov. Chem. Commun. 1934, 6, 498.
29. Ilkovič, D. J. Chim. Phys. 1938, 35, 129.
30. MacGillavry, D.; Rideal, E.K. Rec. Trav. Chim. Pays-Bas 1937, 56, 1013.

31. v. Stackelberg, M. Z. Elektrochem. Ber. Bunsenges. 1939, 45, 466.
32. Kůta, J.; Smoleř, I. Progress in Polarography; Zuman, P.; Kolthoff, I.M., Eds.; Interscience: New York, 1962; Vol. 1, p. 43.
33. Ilkovič, D. Collect. Czechoslov. Chem. Commun. 1936, 8, 170.
34. Ilkovič, D.; Semerano, G. Collect. Czechoslov. Chem. Commun. 1932, 4, 176.
35. Heyrovský, J. Application of Polarography in Practical Chemistry (in Czech), Českoslov. Svaz pro výzkum, a zkoušení technicky důležitých látek: Prague, 1933; (in Russian) Onti Chimteoret: Leningrad, 1937.
36. Semerano, G. Polarography, its Theory and Applications (in Italian), Draghi: Padova, 1932.
37. Heyrovský, J. Polarography in Physical Methods in Analytical Chemistry (in German); Bottger, W., Ed.; Akademische Verlagsgesellschaft: Leipzig, 1936; Vol. II, pp. 260-322, 1939; Vol. III, pp. 422-477.
38. Koryta, J.; Berg, H. (Eds.) Polarography, the Work of Jaroslav Heyrovský (in German); Ostwalds Klassiker der Exakten Wissenschaften, No. 266, Akademische Verlagsgesellschaft:Leipzig, 1985, p. 25.
39. Heyrovský, J. Polarography, Theoretical Foundations, Practical Execution and Applications of the Electrolysis with the Dropping Mercury Electrode (in German). Springer: Vienna, 1941; reprint by Alien Custodian Property, Edwards Brothers, Ann Arbor, Michigan, 1944.
40. Heyrovský, J.; Forejt, J. Z. physikal. Chem. 1943, 193, 77.
41. Wiesner, K. Z. Elektrochem. 1943, 49, 164.
42. Brdička, R.; Wiesner, K. Naturwissenschaften 1943, 31, 247; Collect. Czechoslov. Chem. Commun. 1947, 12, 39.
43. Kolthoff, I.M.; Parry, E.P. J. Am. Chem. Soc. 1951, 73, 3718.
44. Brdička, R.; Wiesner, K. Collect. Czechoslov. Chem. Commun. 1947, 12, 138.
45. Wiesner, K. Chem. listy 1947, 41, 6.
46. Brdička, R. Collect. Czechoslov. Chem. Commun. 1954, 19, 41; Suppl. II.
47. Hanuš, V. Chem. Zvesti 1954, 8, 702.
48. Brdička, R. Collect. Czechoslov. Chem. Commun. 1947, 12, 212.
49. Veselý, K.; Brdička, R. Collect. Czechoslov. Chem. Commun. 1947, 1947, 12, 313.
50. Wiesner, K. Collect. Czechoslov. Chem. Commun. 1947, 12, 64.
51. Brdička, R.; Knobloch, E. Z. Elektrochem. 1941, 47, 721.
52. Brdička, R. Z. Elektrochem. 1942, 48, 278.
53. Brdička, R. Collect. Czechoslov. Chem. Commun. 1947, 12, 522.
54. Wiesner, K. Collect. Czechoslov. Chem. Commun. 1947, 12, 594.
55. Heyrovský, J.; Šorm, F.; Forejt, J. Collect. Czechoslov. Chem. Commun. 1947, 12, 11.
56. Revenda, J. Collect. Czechoslov. Chem. Commun. 1934, 6, 453.
57. Šlendyk, I. Collect. Czechoslov. Chem. Commun. 1932, 4, 335.
58. Herasymenko, P.; Šlendyk, I. Collect. Czechoslov. Chem. Commun. 1933, 5, 479.
59. Pech, J. Collect. Czechoslov. Chem. Commun. 1934, 6, 126.
60. Knobloch, E. Collect. Czechoslov. Chem. Commun. 1947, 12, 406.

61. Brdička, R. Collect. Czechoslov. Chem. Commun. 1936, 8, 366.
62. Babička, J.; Heyrovský, J. Collect Czechoslov. Chem. Commun. 1930, 2, 370.
63. Brdička, R. Collect. Czechoslov. Chem. Commun. 1933, 5, 112, 148, 238.
64. Brdička, R. Nature 1937, 139, 1020.
65. Brdička, R. Klin. Wochenschr. 1939, 18, 305.
66. For review see Březina, M.; Zuman, P. Polarography in Medicine, Biochemistry and Pharmacy; Interscience Publi.: New York, 1958.
67. Ewing, G.W. J. Chem. Ed. 1968, 45, 154.
68. Kolthoff, I.M.; Lingane, J.J. Polarography, Polarographic Analysis and Voltammetry, Amperometric Titrations; Interscience Publ.: New York, 1941; 2nd Ed., 1952.
69. Brdička, R.; Koutecký, J. Collect. Czechoslov. Chem. Commun. 1947, 12, 337.
70. Koutecký, J. Collect. Czechoslov. Chem. Commun. 1953, 18, 11, 183, 311, 597.
71. Brdička, R. Z. Elektrochem. 1960, 64, 16.
72. For review see Brdička, R.; Hanuš, V.; Koutecký, J. Progress in Polarography; Zuman, P.; Kolthoff, I.M., Eds.; Interscience: New York, 1962; Vol. 1, p. 145.
73. Kössler, I. Collect. Czechoslov. Chem. Commun. 1950, 15, 723.
74. Koryta, J.; Kössler, I. Collect. Czechoslov. Chem. Commun. 1950, 15, 241.
75. Koryta, J. Collect. Czechoslov. Chem. Commun. 1959, 24, 3057.
76. Kalousek, M. Collect. Czechoslov. Chem. Commun. 1948, 13, 105.
77. MacAleavy, C. Belgian Patent 443, 003 (1941); French Patent 886, 848 (1942).
78. For review see Breyer, B.; Bauer, H.H. Alternating Current Polarography and Tensammetry; Interscience Publ.: New York, 1963.
79. Ševčík, A. Collect. Czechoslov. Chem. Commun. 1948, 13, 349.
80. Randles, J.E.B. Trans. Faraday Soc. 1948, 44, 327.
81. Matsuda, H.; Ayabe, Y. Z. Elektrochem. 1955, 59, 494.
82. Randles, J.E.B. Analyst 1947, 72, 301.
83. Airey, L. Analyst 1947, 72, 304.
84. Delahay, P. J. Phys. Colloid Chem. 1949, 53, 1279.
85. Vogel, J. Proc. 1st Internat. Polarograph. Congr., Přírodověd. Nakl.: Prague, 1952; Vol. III, p. 731.
86. Delahay, P.; Berzins, T. J. Am. Chem. Soc. 1953, 75, 2486.
87. Berzins, T.; Delahay, P. J. Am. Chem. Soc. 1953, 75, 4205.
88. Delahay, P. New Instrumental Methods in Electrochemistry; Interscience Publ.: New York, 1954.
89. Barker, G.C.; Jenkins, I.L. Analyst 1952, 77, 685.
90. Barker, G.C. Anal. Chim. Acta 1958, 18, 118.
91. Barker, G.C.; Gardner, A.W. Z. anal. chem. 1960, 173, 79.
92. Barker, G.C. Progress in Polarography; Zuman, P.; Kolthoff, I.M., Eds.; Interscience Publ.: New York, 1962; Vol. II, p. 411.
93. Osteryoung, J. this volume, p.
94. Kemula, W. Roczn. Chem. 1952, 26, 281, 694.
95. Kemula, W.; Górski, A. Roczn. Chem. 1952, 26, 639.
96. Kemula, W.; Sybilska, D.; Geisler, J. Roczn. Chem. 1952, 29, 643.

97. Kemula, W. Chem. Zvesti 1954, 8, 740.
98. Kemula, W. Progress in Polarography; Zuman, P.; Kolthoff, I.M., Eds.; Interscience: New York, 1962; Vol. 2, p. 397.
99. Kemula, W.; Kublik, Z. Anal. Chim. Acta 1958, 18, 104.
100. Gerischer, H. Z. physikal. Chem. (Leipzig) 1953, 202, 302.
101. For review see Říha, J. Progress in Polarography; Zuman, P.; Kolthoff, I.M., Eds.; Interscience: New York, 1962; Vol. 2, p. 383.
102. Last systematic study using d.c. polarography J. Heyrovský carried out involved electroreduction of carbon disulfide, about which he reported at the Congress of Practical Polarography in 1952 at Bratislova; cf. Anal. Chem. 1952, 24, 915.
103. For a bibliography see Heyrovská, M. Collect. Czechoslov. Chem. Commun. 1960, 25, 2949; Butler, J.A.V.: Zuman, P. Biographical Memoirs of Fellows of the Royal Society 1967, 13, 167.

RECEIVED August 9, 1988

Chapter 24

Past and Future of the Dropping Electrode

Michael Heyrovský, Ladislav Novotný, and Ivan Smoler

The J. Heyrovský Institute of Physical Chemistry and Electrochemistry, Czechoslovak Academy of Sciences, Dolejškova 3, 182 23 Prague, Czechoslovakia

The dropping electrode (DE) appeared towards the end of 19th century in European science in discussions about the electrochemical potential. In 1903 Kučera attempted using it for measuring surface tension of mercury in solutions; his little success was due to the scarce knowledge of electrochemistry in his time. By the spontaneous renewal of its surface the DE paved the way for polarography; its unique properties were utilized also in other physico-chemical measurements. As the main instrument in polarography, the DE went through various modifications and other renewed electrodes have been derived from it. The use of a special spindle capillary led to introduction of a highly sensitive method of electrocapillary measurements with the DE, 75 years after Kučera's attempt.

In general, any polarizable electrode formed by drops of a conducting liquid can be called the dropping electrode. Such a liquid is usually mercury, but various amalgams, carbon paste, aqueous solutions of electrolytes, gallium or bismuth at higher temperatures and other liquids have been used as well. The dropping electrode is characterized by several specific properties which make it a unique instrument for physico-chemical studies of interfaces and for electroanalytical applications. Above all, it has a liquid, i.e., homogeneous and isotropic, surface and it is spontaneously, periodically renewed, its surface area periodically increases and its period of dropping depends on its surface tension. The frequency of renewal of the electrode, and the extent of the change of its surface area, can be varied within wide limits. In the most common case, i.e., when the liquid is mercury, the dropping electrode gains the further advantages of easily attainable high purity and of high overvoltage in the electrolytic evolution of hydrogen.

Since its appearance in scientific laboratories more than a hundred years ago the dropping electrode has undergone a complex

0097–6156/89/0390–0370$06.00/0

development, especially as the main component of polarographic apparatus, and recently significant improvements have been achieved in its design. Some aspects of this development are the subject of the present article.

Early History of the Dropping Electrode

Probably the first documented attempt to combine the phenomenon of drop formation with electrical measurement is due to W.Thomson (Lord Kelvin) (1). In his letter to H.Helmholtz of 2.1.1859 he sketches and describes a "water dropping collector" for measuring atmospheric electricity in which the charge carried by finely dispersed droplets of water is collected in a metallic funnel. The first dropping mercury electrode in a stricter sense was designed and tried in 1871 by a coworker of Thomson, the English inventor C.F.Varley (2). In an experiment intended to find the connection between the surface and the electric energies he used mercury, rapidly dropping into a vessel with diluted sulphuric acid. He observed that, under certain conditions, an electric current flowed between the mercury in the reservoir and that collecting at the bottom of the vessel. Two years later the French physicist G.Lippmann published a short account of his research on the same fundamental problem (3) and, in 1875, he treated the matter in a detailed paper (4). He showed for the first time an electrocapillary curve obtained with his original instrument, the capillary electrometer. He also described an experiment, essentially identical with that of Varley, in which mercury, dripping from a drawn-out stem of a funnel into a beaker with an acidic solution, can produce electric current flowing between the mercury in the funnel and that at the bottom of the beaker. This effect was reexamined and confirmed with mercury dropping into various liquids by the German physicist G.H.Quincke (5). In this way the electrocapillary curve and the dropping electrode emerged in science at the same time as products of the search for understanding the way how one form of energy can be converted into another - a classical problem, typical for the second half of nineteenth century.

H.Helmholtz (6) interpreted Lippmann's electrocapillary curve by assuming that the electrode surface in contact with the electrolyte behaves like a condenser. He also forecast that, between an independent dropping mercury electrode and the electrolytic solution, there ought not to be any permanent potential difference: should there have been one, initially, it would be carried away gradually by dropping. Helmholtz's student A.König (7) found an experimental corroboration of his teacher's statement. This spurred many experimentalists to repeat the measurements, because if Helmholtz was right, the dropping mercury electrode would provide a convenient zero for an electrochemical scale of potentials. W.Ostwald (8), F.Paschen (9,10) and others measured the potential difference between the dropping mercury electrode and Lippmann's electrometer, and observed that it approached zero only when the electrode was dropping rapidly. Such electrodes were prepared, without any attempt at standardization, simply by drawing out in a flame a tube of thermometer glass, and cutting the end so that, under pressure of mercury, many small drops or a coherent jet were formed in the solution.

O.Warburg (11,12) pointed out that mercury dissolves in solutions in the presence of air oxygen. He tried to interpret the electrocapillarity and the potential of the independent dropping electrode by considering the role of mercury ions thus introduced into the system. G.Meyer (13) claimed to have proved Warburg's theory by experiments with dropping mercury and amalgam electrodes. W.Nernst (14), on the other hand, predicted that the potential of the dropping electrode would be determined by the very small equilibrium concentration of mercury ions present in solution according to his formula. This prediction was supported experimentally by Nernst's pupil W.Palmaer (15).

Many other examples can be quoted to demonstrate that the problem of the potential of the dropping mercury electrode occupied the minds of leading scientists in Europe towards the end of the last century and beyond. From the present point of view we understand that no satisfactory simple answer to the problem could be found because, at that time, the electrochemical phenomena determining the behavior of the dropping electrode were not fully known. As a rule, the various suggested theories were strongly disputed and opposed; nevertheless, each of the participating scientists contributed either by experimental result or by theoretical principle, or even by a critical remark, to our present understanding of the matter. Summarizing, this earliest period could be characterized by the heading: dropping electrode as a problem.

Dropping Electrode as a Tool

In 1903 a new use of the dropping mercury electrode was tried, this time based on systematic quantitative measurements. G.Kučera (16) was aware of the drawbacks of measuring the surface tension of mercury by the capillary electrometer - the solution in the narrow capillary in contact with the stationary meniscus often differs in concentration from that in the bulk, and the wetting angle between mercury and the solution changes with potential in an unknown way. In order to avoid them, he decided to take as a measure of surface tension of mercury the weight of its drop. He saw a special advantage of such a method in the circumstance that the measurements would be taken at a completely fresh, always renewed mercury surface. The realization of the idea was simple: in the capillary electrometer he would increase the pressure until mercury would start dropping from the conical capillary into the solution. In this way he introduced - in contrast to the static method of Lippmann - a dynamic method of measuring surface tension and, in analogy with the capillary electrometer, he called his apparatus the "drop electrometer" (das Tropfenelektrometer).

For the dropping electrode he prepared capillaries by drawing out pieces of thermometer tube of about 0.2 mm inner bore, and by cutting them so that the inner diameter of the orifice was about 0.04 mm. From such capillaries he selected only those with a regular orifice. The capillary was then connected by rubber tubing to a reservoir of mercury, of which the height was adjusted each time to give a drop-time of 2 seconds. In order to increase precision he collected as many as 80 drops of mercury from the electrode in each measurement: the gathered mercury was then washed, dried and weighed in a standard way. By his method Kučera obtained curves of the dependence of drop weight on applied voltage which, in some cases, showed the same course as the electrocapillary curves obtained with the capillary electrometer. In other cases, however, various anomalies appeared, in the form of secondary maxima, which he could not explain.

Here again we know today that Kučera was trying to cope with a complex of electrochemical problems of which he could not have been fully aware in his time. Hence he was unable to bring his method, based on sound principles, to its due perfection. In 1918 he suggested to his student, J.Heyrovský, that he take up the surface tension measurements with the dropping mercury electrode as a subject for research.

Dropping Electrode Introduces Polarography

After having gained some experimental experience, Heyrovský simplified Kučera's method by measuring the drop-time of several drops under constant mercury pressure instead of collecting and weighing each time 80 drops of 2-second duration. However, even after 3 years of tedious work, he could not reconcile the results of the dynamic surface tension measurements with those of the static method.

At the beginning of 1922 he tried to measure, in addition to the drop-time, the current passing through the cell when the dropping electrode was polarized to various potentials with respect to the potential of the mercury layer at the bottom of the cell. He realized that, while the electrocapillary curve ideally would give thermodynamic information on the surface energy of mercury, the reproducible current - voltage curves might supply information about the kinetics of various electrode processes. Encouraged by his first results (17,18), he began a systematic study of electrolysis with the dropping mercury electrode. He measured the values of the mean current by means of a damped galvanometer, and plotted these as a function of the voltage applied to the electrodes. In the first stage, electrocapillary and current - voltage curves were measured in parallel but, later, the former were abandoned, as the measurements were more time-consuming and less reproducible, and appeared less informative than the latter. The situation was, indeed, disadvantageous for electrocapillarity, as the dependence of drop-time on surface tension is best measured when the drop-time is long, whereas the then measured changes in short drop-times were not much greater than the errors of reproducibility of drop-times. On the other hand, the measured currents were practically insensitive to the drop-time, as the diffusion current is proportional to drop-time to the power 1/6. The dropping electrode had thus lost the function of surface tension indicator and began serving as a mere surface renewer.

In 1924, in order to speed up the tiresome plotting of the current - voltage curves, Heyrovský and Shikata (19) designed an automatic apparatus which continuously increased the voltage applied to the electrodes, and simultaneously provided a photographic recording of the current as it changed with changing voltage. The apparatus was introduced for recording the course of electrochemical polarization of the dropping mercury electrode - the ideally polarizable electrode, and hence it was called the "polarograph". The name of the instrument gave name to the method: by 1930 the subject of studies with the dropping mercury electrode became known as polarography. Studies of polarization of solid electrodes had already been carried out by several authors in the 19th century, and hence Heyrovský, for fear he might be accused of patronizing other people's work, preferred to limit this name first to the dropping, later also to other renewed mercury electrodes.

One of the first problems studied by means of the polarograph was that of Kučera's anomalous electrocapillary curves. Heyrovský and Šimůnek (20) demonstrated that air oxygen present in solutions is electrolytically reduced at a positively charged mercury surface, and that this reduction from dilute solutions is accompanied, in the same potential region, by a maximum on the polarographic curves and by an anomaly on Kučera's electrocapillary curves. Nowadays it is known that the uptake of the second electron by the oxygen molecule occurs in a surface reaction, which is the primary cause of the maximum as well as of the anomaly. In this way one of the important factors pertaining to electrocapillary measurements has been found.

Other Application of the Dropping Electrode

The story of the development of polarography is sufficiently known and documented in text-books and review articles (for a recent review see (21)). While the polarographic research consisted in its "classical" stage mainly in interpretations of current - voltage curves obtained with the dropping mercury electrode, the unique qualities of this electrode also spurred scientific development in other directions. In the following some examples of such researches are given.

D.C.Grahame used the easily renewed, clean surface of the dropping electrode for precise measurements of the electric capacitance of the mercury-solutions interface (22). Capacitance measurements were utilized also for studies of oriented interfacial layers and their properties (23,24). H.Berg and coworkers (25) introduced photopolarography, outlining ways of studying homogeneous photochemical reactions in solutions by means of the dropping mercury electrode. With the same electrode, G.C.Barker (26) discovered and investigated the phenomenon of photoemission of electrons into solutions. A.Calusaru and J.Kůta (27) observed on its surface the formation of crystals of electrodeposited metals. It also inspired J.Koryta, P.Vanýsek and M.Březina (28) to construct a dropping electrolyte electrode in order to study the interface between two immiscible electrolyte solutions. These were the applications making use of the spontaneously renewed surface of the dropping electrode. I.Oref and coworkers, on the other hand, used the electrode in a study of the gas - mercury interface for measuring the surface tension changes (29,30).

Development of the Dropping Electrode in Polarography

The first polarographic capillaries were of the type used by Kučera. They were conical, about 0.03 mm wide at their orifice, giving drop-times of 2-4 seconds in solutions of electrolytes under a mercury pressure of 50 cm. The reproducibility of such dropping electrodes was sufficient to yield reproducible current - voltage curves, on the basis of which the elementary polarographic theory could be developed. It was a fortunate circumstance, that the Kučera's type of capillaries with the relatively short drop-times were taken over for polarography. Had Kučera used capillaries with long drop-times which would have been more convenient for surface tension measurements, the currents measured would have been affected by the extension of the diffusion layer from the electrode into the solution bulk and by various adsorption and secondary chemical processes, and the results would have been much more difficult to interpret than with the short drop-times.

Polarographic capillaries were first characterized only by the drop-times t_1 they provided under given conditions. W.Kemula (31) drew attention to the importance of the rate of flow of mercury through the capillary, $\underline{m}$, for polarographic currents: after the Ilkovič equation was derived (32), $\underline{m}$ became the second important parameter for characterizing polarographic capillaries. When the quantitative expressions for the main types of d.c. polarographic currents became known, it was established that the direct proportionality of $\underline{m}$ and the indirect proportionality of t_1 with the height of the mercury level above the capillary tip, $\underline{h}$, of the spontaneously dropping electrode allowed a simple distinction of d.c.polarographic currents to be made according to their dependence on $\underline{h}$. After the cause of polarographic maxima of the second kind was explained (33), the upper limit of $\underline{m}$ for polarographic purposes was set at 2 mg/s. From the experimental data of Kučera it follows that some of his results were obtained with $\underline{m}$ higher than 4 mg/s which means that the surface of the dropping electrode was streaming under the high rate of flow of mercury into the drop. This was another phenomenon of which Kučera was unaware and which, together with the reduction of oxygen, contributed to the anomalous results of his electrocapillary measurements.

After 1938 the original, drawn-out conical capillaries were gradually replaced by thick-walled capillaries of uniform cylindrical bore, of about 0.05 mm, which were readily available, less fragile and could be replaced easily (34).

In 1943 Heyrovský and Forejt (35) introduced the mercury jet electrode, originally in order to provide stationary polarization curves in oscillographic polarography. From a conical capillary dipped in the solution and inclined upwards with the orifice submerged a few millimeters under the surface the mercury begins to jet above the surface through a thin layer of solution after the pressure of mercury has been risen sufficiently high. The jet electrode represents the extreme case of a renewed electrode where the fast renewal allows only fast processes to be followed. It has the unique advantage of an approximately constant surface area: its disadvantages are a high consumption of mercury and a high charging current due to the continuous rapid change of its surface.

The more sophisticated techniques derived from d.c. polarography, which started appearing in the nineteen-fifties and were directed towards higher sensitivity, selectivity and rapidity of polarographic measurements, set new requirements for the dropping electrode. While the spontaneously dropping electrodes were further used predominantly in d.c. polarography, for some advanced techniques a mechanical detachment of the drop (36) was found necessary, and various modifications of the renewable hanging mercury drop electrode (37) were introduced. In these cases, the function of the dropping electrode has been totally reduced to mere reproduction.

The refinement of polarographic and basic electrochemical theories necessitated verification by precise experimental measurements: the instantaneous current passing during the life-time of individual drops had to be measured (38), which raised the question of the degree of reproducibility of the dropping electrode. Various authors tried various ways of changing the polarographic capillary in order to improve its performance, either by reshaping its end (see, e.g., (39-44)) or by coating its inner walls by a hydrophobic agent (45-47). However, none of the suggested methods would lead to a substantial, lasting improvement in the electrode reproducibility for general use.

New Type of the Dropping Electrode

Our experience has taught us, first, that the performance of the cylindrical capillaries was notably inferior to that of the old type drawn-out conical ones (48). A critical assessment of possible causes of disturbance of the regular laminar flow of mercury through the glass capillary, and of the growth and detachment of the drop, together with practical knowledge gained from several years of experimental research on electrocapillarity with the capillary electrometer, led us finally to the introduction of the so-called "spindle capillary" for the dropping mercury electrode (49-52). This original type of capillary provides the best utilization of the advantages of the dropping mercury electrode so far.

The spindle capillary is made from a cylindrical capillary, of which the lower end is blown out and drawn into the shape of an unsymmetrical spindle, with its longer end pointing downwards. The end of the spindle is terminated by a steep cone leading to the orifice of the capillary. This shape minimizes the penetration of the solution to the interior of the capillary, which is the main cause of irregular dropping. After detachment of the drop from the spindle capillary there is a mercury meniscus left at the orifice, which can serve as a stationary yet renewable mercury electrode. It can be reproduced with an accuracy of 2%. While the inner bore of the stem of the capillary determines the rate of flow $\underline{m}$, the size of the drop, or the drop-time, depends on the diameter of the orifice of the capillary. Hence several capillaries can be made from one long glass tube, each giving drops of a different size.

The drops growing from a spindle capillary are more stable than those from cylindrical or conical capillaries, and their reproducibility is very high. These factors led to the development of the "controlled convection" method of electrocapillary measurements (53-55). With a drop-time of, say, 100 s, the reproducibility under constant conditions is better than 0.02 s, i.e. 0.02 %. The stability of the drop allows the solution to be stirred over 90 % of the drop life without affecting its reproducibility, providing the solution is entirely free of surface-active impurities. If this is not the case, the stirring accelerates the transport of low concentrations of surfactants to the electrode until adsorption equilibrium is established. The resulting decrease in drop-time then indicates adsorption which, without stirring, might not have been detected. By increasing the drop-time, we can further increase the sensitivity of this method, which thus allows adsorption studies to be carried out in concentration regions inaccessible by any other method. By changing the time of stirring with different drop-times, full attainment of adsorption equilibrium can be ascertained, and a distinction can be made between reversible and irreversible adsorption. When measured with a dropping electrode with stirring, strongly adsorptive substances show surface saturation in solutions where the capillary electrometer only begins to indicate adsorption (56). In this way, after 75 years of development of the dropping electrode, Kučera's conviction about the fundamental superiority of the renewed over the stable surface in surface tension measurements could be fully confirmed. (In this conviction Kučera was not alone - personalities like Ostwald (8) were of similar opinion.) With a certain exaggeration it can be said that Heyrovský introduced polarography as an emergency escape in order to avoid answering Kučera's question whether his belief in dynamic principle was justified. The development of polarography did, ultimately, answer the question in the affirmative, as Heyrovský himself would have wished to do.

When the flow of mercury into the capillary is interrupted, the drop, stopped in its growth, remains hanging from the capillary, and hence the hanging, stationary mercury drop electrode is produced. This can be done automatically in various ways (50,57). With the spindle capillary again the drop, supported by the meniscus, is very stable and highly reproducible. If a spindle capillary of appropriate parameters is chosen, and the flow of mercury is stopped at short intervals, then by means of a simple arrangement a reproducible mercury microelectrode of desired and, if necessary, gradually increased size is formed (58). The small size makes the drop invisible to the naked eye. The consumption of mercury for the functioning of the microelectrode is thus drastically reduced.

Future Development

The spindle capillary provides a multipurpose renewed mercury electrode, which utilizes all the features of the dropping electrode. It should consist of a stand containing the necessary electronic controls, with an appropriate, safely closed reservoir of mercury, an exchangeable cell, and an electrode holder into which special capillaries can be easily fitted according to the purpose of the measurement. In addition to other accessories, a wide selection of capillaries of various parameters should be readily available, the less usual ones obtainable on order.

As demonstrated by the example of the spindle capillary, the development of the dropping electrode is far from exhausted. By its origin it is predestined for research of interfacial electrochemical processes, where it can render invaluable services in analyzing or effectuating complex reactions at molecular or macroscopic levels. As the ideal tool for measuring adsorption it has wide perspectives in systematic fundamental studies as well as in analytical applications, especially after on-line automation is completed. In analytical chemistry it has become a highly versatile sensor, with possibilities for microscale use, and in combination with specialized solid electrodes.

Like the dropping electrode itself, its vitality in science appears to be ideally reproducible.

Literature Cited

1. Thompson, S. P. The Life of William Thomson; MacMillan: London, 1910; p 399.
2. Varley, C. F. Phil. Trans.1871, 101, 129.
3. Lippmann, G. Pogg. Ann. Phys. 1873, 149, 547.
4. Lippmann, G. Ann. chim. phys. 1875, 5, 494.
5. Quincke, G. H. Pogg. Ann. Phys. 1874, 153, 161.
6. Helmholtz, H. Wiss. Abh. Phys. Techn. Reichsanst. 1897, 1, 925.
7. König, A. Wied. Ann. Phys. 1882, 16, 1.
8. Ostwald, W. Z. physik. Chem. 1887, 1, 583.
9. Paschen, F. Wied. Ann. Phys. 1890, 40, 36.
10. Paschen, F. Wied. Ann. Phys. 1890, 41, 42.
11. Warburg, O. Wied. Ann. Phys. 1889, 38, 321.
12. Warburg, O. Wied. Ann. Phys. 1890, 41, 1.
13. Meyer, G. Wied. Ann. Phys. 1899, 67, 433.
14. Nernst, W. Über Berührungselektrizität; Beiblätter zu Wied. Ann. Phys. 1896, p 4.
15. Palmaer, W. Z. physik. Chem. 1901, 36, 664.
16. Kučera, G. Drude Ann. Phys. 1903, 11, 529.
17. Heyrovský, J. Chem. listy 1922, 16, 256.
18 Heyrovský, J. Phil. Mag. 1923, 45, 303.
19. Heyrovský, J.; Shikata, M. Rec. trav. chim. 1925, 44, 496.
20. Heyrovský, J.; Šimůnek, R. Phil. Mag. 1929, 7, 951.
21. Vlček, A. A.; Volke, J.; Pospíšil, L.; Kalvoda, R. Polarography; In Phys. Methods of Chemistry; Vol. 2, Electrochemical Methods; Rossiter, B. W.; Hamilton, J. F., Eds.; J. Wiley: New York, 1986; p 797.
22. Grahame, D. C. J. Am. Chem. Soc. 1946, 68, 301.
23. Vetterl, V. J. Electroanal. Chem. 1968, 19, 169.
24. Pospíšil, L.; Müller, E.; Emons, H.; Dörfler, H. D. J. Electroanal. Chem. 1984, 170, 319.
25. Berg, H.; Stutter, E.; Schweiss, H.; Weller, K. J. Electroanal. Chem. 1967, 15, 415.
26. Barker, G. C.; Gardner, A. W.; Sammon, D. C.; J. Electrochem. Soc. 1966, 113, 1182.
27. Calusaru, A.; Kůta, J. Nature 1966, 211, 1080.
28. Koryta, J.; Vanýsek, P.; Březina, M. J. Electroanal. Chem. 1977, 75, 211.
29. Oref, I.; Nemirovsky, Y. J. appl. Phys. 1975, 46, 2057.
30. Labin, R.; Oref, I. J. appl. Phys. 1977, 48, 406.
31. Kemula, W. Proc. 9th IUPAC Congr. Madrid 1934, 1935, Vol. 2, p 297.

32. Ilkovič. D. Coll. Czech. Chem. Commun. 1934, 6, 498.
33. Kryukova, T. A. Zh. fiz. khim. 1947, 21, 365.
34. Siebert, H.; Langer, T. Chem. Fabrik 1938, 11, 141.
35. Heyrovský, J.; Forejt, J. Z. physik. Chem. 1943, 193, 77.
36. Tereshchenko, P. N. Zavod. lab. 1948, 14, 1319.
37. Říha, J. In Progress in Polarography; Zuman, P.; Kolthoff, I.M., Eds.; Interscience: New York, 1962; Vol. 2, p 383.
38. Kůta, J.; Smoler, I. In Progress in Polarography; Zuman, P.; Kolthoff, I. M., Eds.; Interscience: New York, 1962; Vol. 1, p 43.
39. Barker, G. C. Anal. chim. Acta 1958, 18, 118.
40. Cooke, W. D.; Kelley, M. T.; Fischer, D. J. Anal. Chem. 1961, 33, 1209.
41. DeLevie, R. J. Electroanal. Chem. 1965, 9, 117.
42. Koryta, J.; Němec, L.; Pivonka, J.; Pospíšil, L. J. Electroanal. Chem. 1969, 20, 327.
43. Smoler, I. Coll. Czech. Chem. Commun. 1966, 31, 703.
44. Flemming, J.; Berg, H. J. Electroanal. Chem. 1964, 8, 291.
45. Stackelberg, M.; Toome, V. Leybold Polarogr. Ber. 1953, 1. Heft 4, p 55.
46. Lawrence, J.; Mohilner, D. M. J. Electrochem. Soc. 1971, 118, 1596.
47. Christie, J. H.; Jackson, L. L.; Osteryoung, R. A. Anal. Chem. 1976, 48, 242.
48. Smoler, I. J. Electroanal. Chem. 1974, 51, 452.
49. Novotný, L.; Smoler, I.; Kůta, J. Czech. Patent PV 5 193, 1976, AO 185 982, Prague.
50. Novotný, L. Proc. J. Heyrovský Mem. Congr. Prague, 1980, Vol. II, p 129; Czech. Patent PV 9 612, 1979; PV 9 611, 1979, Prague.
51. Novotný, L. Ph.D. Thesis, Czechosl. Acad. Scis, Prague, 1980.
52. Novotný, L.; Heyrovský, M. Tr. Anal. Chem. 1987, 6, 176.
53. Novotný, L.; Smoler,I. J. Electroanal. Chem. 1983, 146, 183.
54. Novotný, L.; Smoler, I.; Kůta, J. Coll. Czech. Chem. Commun. 1983, 48, 963.
55. Novotný, L.; Smoler, I. Coll. Czech. Chem. Commun. 1985, 50, 2525.
56. Heyrovský, M.; Novotný, L. Coll. Czech Chem. Commun. 1987, 52, 54.
57. Peterson, W. M. Internat. Lab. 1980, 10, 51.
58. Novotný, L. Czech. Patent 220 439, 1980, Prague.

RECEIVED August 9, 1988

Chapter 25

Development of Pulse Polarography and Voltammetry

Janet Osteryoung and Carolyn Wechter

Department of Chemistry, State University of New York at Buffalo, Buffalo, NY 14214

Stepwise changes in potential at a dropping mercury electrode were first employed by Kemula in 1930 while doing postdoctoral work at Charles University. Subsequently Ishibashi and Fujinaga in Japan pursued a line of development based also on mechanical switching of potential, while in England Barker and coworkers built the first electronic instruments. The advent of solid-state electronics made possible broad commercial development of instruments which in turn extended pulse techniques to other electrodes and stimulated applications. Computer- and microprocessor-controlled instruments have expanded the use of pulse techniques and encouraged development of specialized waveforms.

The term pulse polarography was initially used by Geoffrey Barker to describe the techniques of normal and differential pulse polarography. In his implementation these techniques made use of a constant potential or a slow linear ramp (1-2 mV/s) to which was added one square pulse during the life of each drop of a dropping mercury electrode (dme). The pulse and the drop were synchronized so that a new drop was mechanically initiated at the end of each pulse. Current was sampled over a small fraction of the pulse duration both before the pulse was applied and near the end of the pulse. In the case of pulses of successively increasing amplitude added to a constant potential the resulting difference current-potential curve is "normal", i.e. it has the general shape of the dc polarogram, and in the case of pulses of constant amplitude added to a slowly-increasing ramp the resulting curve is peak shaped.

In the instrumental and computational setting of today's digital computers a suite of techniques including "pulse polarography" and based on arbitrary stepwise changes in potential are conveniently and usefully treated as one from experimental and theoretical points of view. These include normal and differential pulse, staircase, square wave, and differential normal pulse

0097–6156/89/0390–0380$06.00/0

voltammetry. In each case the technique may be particularized to the dme and is then called polarography rather than voltammetry. Waveforms for these techniques are shown in Figure 1.

Unified experimental and computational approaches to these techniques based on the concept of arbitrary stepwise changes in potential have been discussed in the literature (1,2). In the digital environment the most simple conceptual view is that each stepwise change in potential yields a transient response which in general depends on previous history and the nature of the system. This contrasts with previous views arising out of techniques based on analog electronics (or even electromechanical devices) which synthesize a waveform by adding various components of potential. This focuses attention on various "components" of the current, e.g., the "dc current" and the "pulse current". The tools we use shape our concepts so profoundly that it is useful for understanding pulse techniques to trace their linked instrumental and theoretical development.

The reader should realize that the following is merely an account. In particular, constraints of time have prevented closer inquiry into aspects of these developments with which we are not personally familiar.

Early Developments

The first report of a pulse polarographic experiment is that of Kemula in 1930 (3) for the purpose of investigating the cathodic reaction of mercuric cyanide at the dme. This work was carried out at Charles University in Prague where Kemula had gone for post-doctoral work in the laboratory of Heyrovsky. Kemula had two clearly stated purposes. The first was to shorten the time scale of the polarographic measurement and thereby to investigate the kinetics of faster processes. The second was to reestablish the initial conditions frequently during the life of each drop and thus to examine closely the initial response to the change in potential. These two purposes continue to motivate technique development in voltammetry. Intermittent polarization was accomplished by means of a star-shaped connector with sliding contact. As the connector is rotated, the contact is made only during the times when the slider is passing over an arm of the star. The minimum contact period was 10 ms. The ratio of connect to disconnect time was variable from 5:1 to 1:20 by moving the contact of the slider along the radius of the circle inscribing the star. Coupled with a recording polarograph employing a galvanometer of period 10 ms, this produced a sophisticated instrument the power of which far exceeded the prevailing ability to model reaction mechanisms.

In the context of the time Kemula's work was truly remarkable, especially for a young person at the beginning of a career. We find no evidence through citation that anyone recognized its significance. It is not possible at this distance to know why this important work apparently did not influence contemporary workers in the field. We can speculate, however, that the experimental results were too complicated to interpret in any but qualitative terms, and that, given the poor base in theory, it was very difficult to design conditions which would yield fairly unambiguous results. Thus here

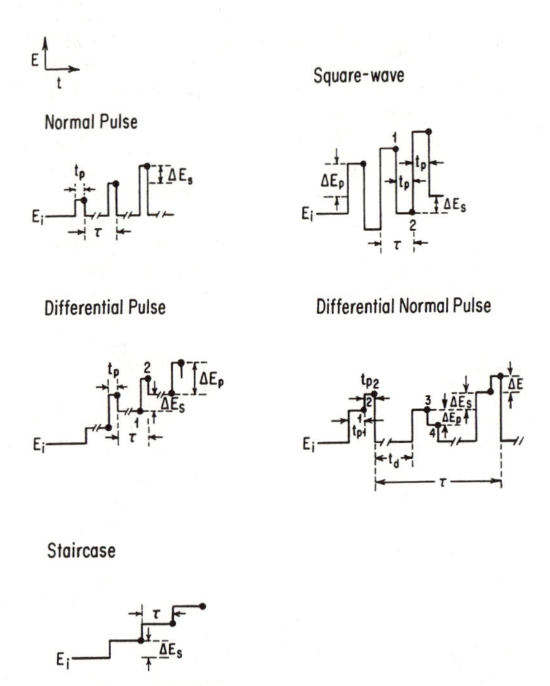

Figure 1. Potential-time waveforms for pulse voltammetry techniques. In each case the time at which current is sampled is indicated by a filled circle. The initial potential is E_i, potential increment in each cycle is ΔE_s, pulse amplitude ΔE_p, period τ, and pulse width t_p. (Reproduced with permission from Ref. 1. Copyright 1988 CRC Press, Inc.)

we have an isolated technical development, stimulated by a scientific problem but unsupported by the scientific context.

In the 1940's Kalousek described his well-publicized commutator (4). This work was also carried out in the Physico-Chemical Institute of Charles University. The Kalousek polarograph employed a rotating switch designed to apply alternatively a fixed potential and a second fixed potential combined with a potential changing slowly and linearly with time. The purpose was to investigate directly, by detecting products, the reversibility of electrochemical reactions on the conventional polarographic time scale. It seems unlikely that Kalousek could have been unaware of Kemula's previous work (3), yet for whatever reason he does not mention it in his paper (4).

The next important development was the commutator-based square-wave polarograph of Ishibashi and Fujinaga (5-7). The aim of these authors was to obtain a differential signal employing only one electrode by means of a square-wave modulation of the usual slow linear potential change. They refer to the work of Heyrovsky on differential polarography employing two electrodes and also to Kalousek (4). Ishibashi, according to his publications dating back to the early 20's, was primarily an analytical chemist, and the motivation for the square-wave polarograph appears to be mainly analytical. A second important feature of this development is that it coincided with wide-spread interest in theories for describing the voltammetric response for such devices. Kambara most notably treated the theory of square-wave polarography for reversible systems (8-10) and showed the essential unity of square-wave and ac techniques.

The First Electronic Instrument

During the same period on the other side of the world Geoffrey Barker and coworkers were building an electronic instrument for square-wave voltammetry which employed for the first time synchronous sampling of the current (11,12). Barker states clearly (11) that the objective of the design was to eliminate from the signal the capacitative current so troubling in kinetic studies employing ac polarography. However Barker and his coworkers realized immediately that square-wave voltammetry yielded such improved signal-to-noise ratio in comparison with dc polarography that it could be used for analysis at the 10^{-7} M level.

These developments were all based on the dropping mercury electrode and in each case the central feature is the instrument which is the embodiment of the technique. The developments of Barker are particularly significant because his was the first electronic instrument and because it was soon commercialized by Mervyn Instruments as the Mervyn-Harwell Square Wave Polarograph. A photograph of this instrument is shown in Figure 2. This pattern was to prove increasingly important because the electronic implementation of pulse voltammetric techniques required expertise and time outside the reach of the average scientist. Thus the use of these techniques would depend on the availability at an acceptable price of reliable commercial instruments.

Figure 2. The first electronic instrument: the Mervyn Instruments Square-Wave Polarograph, circa 1958.

Development of Pulse Polarography

Barker was also unusual among contemporary scientists in that he published less in the open literature. To follow his thinking it is necessary to read his government reports, which may be difficult to obtain. The following considers some aspects of this work taken from these reports which are central to the development of pulse techniques (11,13-22). The rather odd spacing of these reports is explained by their classification by the British government. The manuscript dates of the reports issued in 1957 are in 1953 or 1954.

Because Barker in this period confined his work to studies at the dme, he focused on and attempted to solve problems peculiar to that electrode. One of those problems is "capillary noise" or "capillary response". These terms refer to the current required to charge the interface between mercury and the film of solution which penetrates some distance into the capillary. Charge must be supplied when the potential is changed, but also the interface itself is unstable, so that the charging current varies from drop to drop. In order to understand the reasoning which led Barker to normal and differential pulse polarography, it is useful to describe in more detail the operation of his square-wave instrument.

The Barker square-wave instrument operated at a fixed frequency of 225 Hz with a symmetrical square wave superimposed on a slow linear ramp with a rate of potential change of *ca* 2 mV/drop. Pulse amplitudes were 4, 12, or 35 mV peak-to-peak. Each cycle was initiated by means of a high-frequency measurement of impedance to detect the fall of a drop. After a delay of 1.75 s, the current was sampled after each change of potential at 19/24 of the half-period (i.e., 1.76 ms after the potential change). The resulting differential signal was sampled using a capacitor for 20 ms. The continuous dc electrolysis made this technique, like ac polarography, insensitive to irreversible reactions. The measurement of current only 2 ms after change in potential emphasized the problem of capillary response. For analytical purposes the background current level could be reduced and stability increased by going to much lower frequencies (*ca*. 10 Hz). The improvement in the background more than compensated for the accompanying 4- or 5-fold decrease in signal for reversible systems.

However the complex scheme for sampling and reading out the current signal depended on the square-wave frequency of 225 Hz. Thus the capillary response problem could not be solved by simply operating the square-wave instrument at a lower frequency. Faced with this difficulty, Barker decided to limit the experiment to one pulse per drop. Having made that decision, it became clear that one could employ either the normal pulse or differential pulse waveform as described above. This created an interesting technical problem, for now the instrument required two precisely timed intervals of different magnitude. Barker solved this by using a mechanical timer based on a slowly rotating cam to time the 2 s interval between the fall of a drop and the initiation of the pulse and measurement sequence, and electronic timing to establish the 20 ms intervals over which current was sampled and the 40 ms interval of the pulse. A consequence of this scheme, perhaps unanticipated by Barker, is that the precise timing of the pulse and measurement sequence required for an accurate experiment can be done simply, reliably, and cheaply with modern electronics. An ancillary benefit is that for reasonable pulse amplitudes ($\geq$35 mV) the relative sensitivity for irreversible reactions in the differential mode is not diminished to the extent observed for square wave polarography.

The "pulse polarograph" designed and built by Barker at Harwell served as the prototype for an instrument marketed by Southern Analytical, Ltd. in the 1960's. This vacuum-tube based instrument reportedly sold for about $25,000 in 1969 (23). In 1961 Milner and coworkers at Harwell demonstrated beautifully the power of the differential pulse polarographic technique as implemented by the Southern Analytical instrument for the determination of uranium in seawater. After separation the determination was done at the level of 10 nM with precision of several per cent. A few years later in the United States, Gilbert reported on using this instrument for determination of nickel and vanadium in petroleum stocks (25).

Meanwhile, also in the United States, Barker's work had captured the attention of E.P. Parry and R.A. Osteryoung. Parry had studied with Kolthoff and thus was well-versed in polarography, whereas Osteryoung had worked with Laitinen on molten salt electrochemistry. The conjunction of these backgrounds with a practical problem led to two interesting developments. Parry needed to determine the composition of some films and proposed to do it by (dc) polarography. Osteryoung had read some of Barker's work and suggested that the added sensitivity of pulse polarography would better handle these small samples. Osteryoung worked in collaboration with G. Lauer and H. Schlein to develop an instrument based on solid-state amplifiers (which had only recently become available) which would be capable of carrying out pulse polarographic experiments (26-28). This instrument was patented by Schlein (28). Use of the instrument for evaluation of analytical pulse polarography was later described by Parry and R. Osteryoung in *Analytical Chemistry* (29).

The second interesting aspect is that Osteryoung had been and continued to be interested in electrochemistry at solid electrodes. This considerably broadened the scope of thinking, from pulse polarography to pulse voltammetry, and stimulated the initial work on application of pulse techniques to solid electrodes. A preliminary report on the electronic pulse instrument includes the first example of pulse voltammetry at solid electrodes (30).

This work stimulated commercial interest on the part of Beckman Instruments, which built a prototype instrument under agreement with North American Aviation, then the employer of R. Osteryoung, Parry, Lauer, and Schlein. The project was abandoned without commercial production (23). Melabs (31) and National Instrument Laboratories (32) carried similar projects to the stage of production and sales, but neither was successful in achieving the sort of acceptance in the scientific and technical community required for developing a profitable market.

It is important in evaluating this situation to remember the context. In the '50's fundamental work on pulse polarography decreased polarographic detection limits from about 10^{-5} M (dc) to about 10^{-7} M (dp). The advent of solid-state circuitry and the beginning of an applications literature for pulse polarography set the conditions for rapid scientific and technical advances with these new techniques. However this coincided with the discoveries (and rediscoveries) in the areas of separations and optical spectroscopy in the '50's and their aggressive and successful commercialization in the '60's. In this environment the much improved performance of pulse polarography with respect to the established dc polarographic mode opened up interesting possibilities for analytical applications, but did not confer unique advantages of detection limit, selectivity, and so on.

With the developing sophistication regarding markets during the '60's, instrument manufacturers came more clearly to see that instruments were best viewed as solutions to problems, i.e., that the thing actually being sold was a method or procedure based on the technique as embodied in the instrument. Virtues of simplicity and ease of operation were generally (and correctly) viewed as not merely desirable but rather essential. Too often in the chemist's kitchen pulse polarography seemed the equivalent of the magical brown sauce requiring three days of fussing about with buckets of fresh starting materials; atomic absorption spectroscopy, by contrast, was the bouillon cube. Thus, whereas the scientist-specialist might be fascinated by the analytical and electrochemical possibilities suggested by these new electrochemical techniques, this did not provide sufficient impetus for their widespread use in the scientific community.

These points are elaborated in a most interesting way by Flato in A History of Analytical Chemistry (33). He describes there the development by Princeton Applied Research of the first successful pulse polarographic instrument in the United States. The description is brief and apparently candid, showing the tight interconnection of design, function, marketing, and price. The denouement was the model 174, introduced in 1971 at a price of $1925 including drop-time controller. Now, 17 years later, the 174 has the status of a classic instrument, functional, robust, reliable, and inexpensive. Its introduction marked the beginning of pulse polarography as a practical technique for analytical and electrochemical work.

Elaboration of Pulse Techniques

In the early 1960's there was a great flowering of theory for pulse voltammetry and for other voltammetric techniques. For the study of mechanism or for measurement of diffusion coefficients the simplicity of normal pulse voltammetry was especially attractive. Under conveniently accessible experimental conditions, normal pulse can be approximated accurately by the theory for single potential step chronoamperometry. Thus when commercial instruments became available for implementing normal pulse it become possible for the average experimenter to carry out kinetic studies and to analyze the results using text-book material. Differential pulse proved theoretically far more complicated and thus not useful for kinetic studies. However, the decrease in background signal due to the differential current measurement resulted in much lower signal-to-noise ratios, hence lower detection limits, than possible for normal pulse. Thus differential pulse became a favorite for trace analysis, especially for fairly reactive organic substances. These points are elaborated on in a recent review of pulse voltammetry (1). In that context we focus here on the key element which controls the evolution of technique, i.e , the development of commercial instruments. Early instruments such as Barker's were large, slow, expensive, and based upon vacuum-tube technology (1960's). Advances in electronics led to the production of instruments designed around operational amplifiers (early 1970's), integrated circuits (mid 1970's), those with dedicated, built-in microprocessors (late 1970's), and finally entire computer systems (1980's). At each level, instruments for pulse voltammetry became smaller, faster, and had more capabilities. Most instruments were sold with a variety of accessories, including recorders and electrode stands.

Instruments in the U.S. National Instrument Laboratories, Inc. in the late 1960's produced their solid-state NIL Electrolab (9) which was capable of performing experiments with constant or controlled potential or current. Although the instrument panel contained many dials and knobs, for each function chosen lights adjacent to each control would light up, to show the operator which controls must be set for a given technique. Pulse polarography was among the techniques available.

Princeton Applied Research introduced their first systems in 1969, and EG&G PARC are still one of the major producers of pulse polarographic instrumentation. Both operational amplifiers and digital logic were incorporated into the Model 170 Electrochemistry System and the Model 171 Polarographic Analyzer. Three elements were combined in these machines: a function generator to apply the potential, a measuring circuit, and a data processing system to provide the signal. Each of these functions was controlled from a different section of the front panel of the instrument. Both instruments contained a built-in X-Y recorder, and employed a drop "dislodger" for synchronization of pulse and drop times (32).

The Model 171 (the simpler version) sold for $7500 in 1970, including drop stand. It was capable of performing dc, phase sensitive ac, pulse, derivative and differential pulse polarography, as well as other electrochemical techniques. For the normal pulse waveform, pulses of fixed duration (56 ms) were applied just at the end of the drop life. Currents were measured during the last 16 ms, allowing 40 ms for charging currents to decay. Scan rates ranged from 0.1 mV/s to 0.5 V/s, drop times from 0.5 s to 5 s. For NP polarography, the pulse amplitude increases linearly with time, and this same waveform was applied for derivative pulse. However, for the derivative pulse experiment, each current (measured at the end of a pulse) is subtracted from that for the succeeding pulse, and this difference used as the output. In the case of differential pulse polarography, a pulse of fixed height (50 mV) was superimposed upon a linearly increasing dc ramp. Currents were measured just before pulse application and at the end of each pulse, and their difference was the output current.

The PAR Model 170 Electrochemistry System had all the capabilities of the 171, but also provided more complex waveforms and increased current capabilities required for other techniques.

Programming of the excitation waveform and measurement of the system allowed the user to design a specific experiment. In 1970, the system cost $11,500.

Not only were more techniques accessible with the 170, but there was a wider choice of parameters than those available on the 171. The upper limit on scan rate increased to 500 V/s; pulse durations would be varied from 1 ms to 500 s. For pulse techniques, pulse modulations were possible from 45-65 ms with a 5-20 ms current integration at the end of each pulse, all timing being synchronized with the mechanical drop timer. Full scale current outputs ranged from 1 nA to 0.5 mA. A "tilt" light was used to inform the operator that the sequence of operations could not be performed due to an invalid set of conditions chosen.

The key difference between the complex 170 and the more simple 171 is the choice of values for experimental parameters. Cost and instrumental complexity of these analog instruments were increased by providing variable pulse amplitude, and even more so by variable pulse width. Because the pulse width is the parameter which establishes the experimental time scale, it typically must be varied in any mechanistic study, and it is highly advantageous to have a selection of pulse widths available when optimizing analytical methods. Thus in instrument design there is a clear tradeoff between power of the technique and cost of its implementation. These instruments led to the much more simple PAR 174 Polarographic Analyzer, which through integrated circuit technology provided both classical and pulse polarographic capabilities with fixed pulse width at a price ($1925 in 1971) equivalent to that of simple dc polarographs. This basic instrument is still available today.

In 1975, the PAR Model 374 Polarographic Analyzer was introduced, at a price of $9500. This instrument employed a 16-bit microprocessor to control acquisition and analysis of data. During data collection, scale expansion and autoranging were automatic, and invalid data points could be rejected by comparison with pre-set criteria. In the PLAYBACK mode, the entire curve was reread and plotted on the recorder, with all peaks on scale. If desired, baseline and background corrections were measured, and concentrations calculated from pre-set standards. This was the first pulse voltammetric instrument designed around a microprocessor. Like the Southern Analytical pulse instrument based on vacuum tube technology, it came a bit too soon, before the technology (or the customer) was quite ready.

The PAR 384 Polarographic Analyzer introduced in 1980 is a "second-generation" microprocessor-based instrument. It was designed for routine analytical use, and incorporates a 40-character alphanumeric display, a 69 button touch panel, and a built-in floppy disk used to store and recall up to 9 analytical curves and 9 methods. This instrument's console is in complete control of every step of an experiment: setup, data acquisition, data storage, processing, quantitation and output. Concentrations, calculated from standards, are output in the range 0.001 ppb to 999 ppm. Tangent-fit baselines and background subtraction enhance data analysis. Polarographic techniques that can be performed include dp, np, sampled dc, dp and dc stripping, Osteryoung square-wave, and linear scan voltammetry. This appears to be the first instrument based on digital electronics which actually employs a staircase as a surrogate for the conventional ramp for "linear scan" voltammetry. It was the first successful microprocessor-controlled instrument. In addition to features of data processing, the microprocessor permitted for the first time flexible choice of the parameters of an experiment without unacceptable cost.

The PAR 384B was the first instrument to offer "Osteryoung square-wave" voltammetry (OSWV). "Osteryoung square-wave" (Fig. 1; ref (2)) has advantages of speed, sensitivity, and signal-to-noise ratio with respect to the other pulse techniques, but had been impractical both experimentally and theoretically before the advent of microprocessor-based instruments. When the PAR 384 was introduced, OSWV was not sufficiently well developed to have received commercial attention. However, because of the microprocessor-based design, it was relatively straightforward to incorporate OSWV in the daughter instrument, the 384B.

In 1985, PAR introduced its most versatile instrument: the Model 273 Potentiostat/Galvanostat. This instrument may be operated as a stand alone instrument, interfaced to a computer through a serial GPIB or RS232 port, or connected to a dedicated controller. Almost all forms of voltammetry, amperometry, and coulometry may be performed on this machine, which allows complex waveform programming. Signal averaging, smoothing, and filtering may be performed by the instrument in all three modes of operation. The EG&G PAR Electrochemical Set is composed of over 80 software statements for electrochemistry experiments. Two 14-bit digital-to-analog converters, a 12-bit analog to digital converter, and the built-in microprocessor and memory fulfill most hardware requirements. The Model 1000 System Processor is a dedicated controller for the 273. It contains two disk drives, a touch screen, and a high-speed bus. An optional keyboard enables the user to design specific experiments using the BASIC programming language. In addition a software package is available which is menu-driven and which permits the user to employ a wide variety of techniques through external control of the 273 by a personal computer.

This system represents a compromise among competing goals of flexibility and simplicity, between general purpose and specialized application. The inboard microprocessors create a powerful instrument which will accept instruction, but which also has a mind of its own. The star-shaped switch of Kemula could be understood by a child; even the designers of the model 273 cannot know fully how it operates. Thus critical scientific evaluation of results becomes the only acceptable foundation for using these instruments.

The second important feature of this type of instrument is that it operates exclusively through the principles of pulse voltammetry. That is, all experiments are carried out by specifying a sequence of stepwise changes in potential (or current) via a digital-to-analog converter and specifying a related sequence of windows over which current (or potential) is sampled. This pattern is repeated in theoretical developments. For example, the approximate theory for a linear scan voltammogram is the exact result for the corresponding staircase voltammogram (34). Thus, pulse voltammetry, once an extension of (dc) polarography, is now the foundation of voltammetry.

Princeton Applied Research entered the electrochemical equipment market through the connection between ac polarography and its own line of lock-in amplifiers. E.H. Sargent & Co., on the other hand, first marketed and manufactured dc polarographic equipment in the U.S. They obtained by assignment from Polarograph Laboratories of America the trademark "Polarograph" under which the present Sargent-Welch Scientific Co. introduced a series of voltammetric instruments with plug-in modules in about 1980. Heyrovsky coined the name "Polarograph", and Sargent's trademark has over the years forced other manufacturers into more cumbersome names (e.g., the "polarographic analyzer" of PAR).

A pulse module was standard equipment on the Sargent-Welch Model 4001 Polarograph and optional for other models. Pulse amplitudes could be set from 0 to 99 mV in 1 mV steps, had a duration of 66.8 ms, and sampling time of 2 ms. Mechanical drop times from 0.5 s to 4 s could be chosen. An X-Y recorder was built into all Polarographs. In 1983, the 4001 module sold for $895, the 4001 complete Polarograph for $3865.

A microprocessor controlled instrument, Sargent Welch Model 7001, also performs pulse techniques. A 20-character alphanumeric display and a 30-touch button keypad allow the user to specify the experiment. After recording a voltammogram, smoothing, averaging, and background subtraction are options. The price in 1980 was $8800. Pulse widths from 1 to 1000 ms may be chosen, with sampling from 100 μs after pulse initiation to 500 μs before the end of each pulse.

In 1983, Bioanalytical Systems introduced the BAS 100 Electrochemical Analyzer, which has been updated and improved to the BAS 100A. These are freestanding, interactive, microprocessor-based instruments that can, however, be interfaced with microcomputers. They are intended for general-purpose use and make available over 25 electrochemical techniques. The cell stand is used to control deaeration, drop time, and a magnetic stirrer. Pulse techniques that can be performed include: dp, np, Osteryoung and Barker square-wave, and sampled polarography (or voltammetry). The maximum current is 100 mA, with the current range from 0.1 μA to 10 mA full scale. For np, sample intervals are 1, 17, 83, or 166 ms; pulse widths may vary from 3 to 1000 ms. The frequency ranges of OSWV depend upon the resolution desired. For only a single data sampling point per pulse, frequencies up to 2000 Hz are allowed.

The BAS 100 was developed in the same way as the Southern Analytical instrument, in that the original version was based directly on a prototype which was conceived of and constructed in a research laboratory (35). In many ways the BAS 100A, although quite a different instrument, has the qualities in its range of techniques of the old Tacussel VOCTAN. The comparison shown in Figure 3 gives the correct subjective impression of the difference between these two instruments and in general the difference between the analog instruments of the '70's and the digital instruments of the '80's.

With the infusion of personal computers (PCs) into the workplace, several new companies have introduced systems controlled by PCs. For example, software is available from Delta Instruments to control experiments using an IBM-compatible microcomputer with a Cypress System potentiostat. A "mouse" is used to facilitate data

Figure 3. 1988 vs. 1970: Bioanalytical Systems 100A and the Tacussel VOCTAN.

acquisition; available functions are represented on the computer terminal. An ultrahigh sensitivity system is suitable for work with microelectrodes (10 pA/V output); the standard unit has a limit of 0.5 mA/V. Cyclic voltammetry, swv, dp, chronoamperometry, and chronocoulometry are among the techniques implemented on this system. For swv, the minimum pulse duration is 1 ms, minimum step and pulse heights are 1 mV. Semi-integration of cyclic voltammograms is possible, along with data smoothing and background subtraction. At this writing the basic unit sells for $9,999, the high sensitivity unit for $10,999, including an IBM-AT, software, and the potentiostat.

Instruments in Other Countries. A similar pattern of development is displayed in the pulse voltammetric equipment manufactured in other countries.

A French manufacturer, Tacussel, produced several electroanalytical systems. Rack systems, based on analog electronics and composed of several different units, were able to perform a wide assortment of electrochemical techniques. The UAP 4 pulse polarography unit was employed in the PRG 34 Electroanalytical System, which was referred to as VOCTAN (volt, current and time could be controlled or measured). The UAP 4 unit sold for $8970 in 1973, the total system, including cell stand, for $21,280. Pulse amplitudes between 1 mV and 1000 mV could be applied from 1 ms to 10 s. The current was sampled for an adjustable portion (0% to 100%) of the pulse duration. Full scale currents ranged from 1.25 nA to 5 mA. Drop times as long as 100 s could be employed. A self-contained pulse polarographic instrument, the Tacussel PRG 5, sold for $3990 in 1973. This instrument could be used for dc, np and dp polarography.

When it made its debut in the mid 70's, the Metrohm Recording Polarographic Analyzer E506 had more modes of operation than any other voltammetric instrument then available. These included dc polarography, rapid scan dc, first and second harmonic phase sensitive ac polarography, np, dp, Kalousek polarography, stripping voltammetry, and cyclic voltammetry. The lower cost Metrohm 626 Polarecord is suitable for routine work and teaching purposes.

The Metrohm 646 VA Processor is another microprocessor based instrument. Used in tandem with the 647 VA Electrode Stand and 675 VA Sample Changer, this system is capable of performing automated data acquisition, including the use of the standard additions method. Data analysis features include smoothing and differentiation, and a peak shape analysis routine that performs independently of the base current. Pulse polarographic techniques that can be performed include dp, which can be optimized for reversible and irreversible systems; staircase, with current measurement during the final 20 ms of each current step; and Barker square wave, which employs a waveform composed of five square wave oscillations superimposed upon a staircase, with currents measured for 2 ms at the end of each half cycle of the second, third and fourth oscillations. The 1988 price of this instrument is $14,000.

Eastern Europe has tended to develop its own instruments in each country, for example the Unitra PPO4 pulse polarograph of Poland. These are not generally available, and the instruments used in Western Europe and the U.S. are not widely used in Eastern Europe. The same situation obtains to a lesser extent in Japan. Reliance on different instruments tends to separate the respective research communities and to reinforce tendencies to isolation.

Conclusions

The initial work of Kemula is anomalous in that it had no apparent effect on theory or implementation of voltammetric techniques. The mechanical square wave polarograph of Ishibashi and Fujinaga coincided with a flowering of theoretical development but proved to be a technological dead end. Barker's seminal work is characterized by a unique combination of theoretical competence and technological artistry. Elaboration of the techniques he devised depended critically on the development of solid-state electronics which occurred after the full flowering of his efforts.

In a continuation of the same pattern, the development of pulse techniques in a digital instrumental setting presents new problems and new opportunities. As described in the introduction, the inherently digital character of pulse voltammetry is suited both experimentally and theoretically to the digital instrumental environment, which can provide the setting for new technical and conceptual breakthroughs. On the more practical side, the instrument design (including software) now incorporates more and more insidious traps for the unwary experimenter. The constant pressure to make complex techniques appear simple in implementation tends to introduce behavioral features that corrupt the relation between experimental measurement and the prediction of theory.

The clear lesson of the past is that commercial instruments define techniques and drive research. Furthermore, electroanalytical applications, broadly defined, do not in any way influence dominant trends in instrument design. On the contrary, scientists interested in this area must be prepared to capitalize on new developments.

Literature Cited

1. Osteryoung, J. G.; Schreiner, M. M. Crit. Rev. Anal. Chem., 1988, 19, Supplement 1, S1-S27.
2. Osteryoung, J. G.; O'Dea, J. J. In Electroanal. Chem.; Bard, A. J., Ed.; Vol. 14; Marcel Dekker: New York NY, 1986; pp 255-263.
3. Kemula, W. Collect. Czech. Chem. Commun. 1930, 2, 502-519.
4. Kalousek, M. Collect. Czech. Chem. Commun. 1948, 13, 105-115.
5. Ishibashi, M.; Fujinaga, T. Bull. Chem. Soc. Japan 1950, 23, 261.
6. Ishibashi, M.; Fujinaga, T. Bull. Chem. Soc. Japan 1952, 25, 68-71.
7. Ishibashi, M.; Fujinaga, T. Bull. Chem. Soc. Japan 1952, 25, 238-242.
8. Kambara, T. Bull. Chem. soc. Japan 1954, 27, 523-526.
9. Kambara, T. Bull. Chem. Soc. Japan 1954, 27, 527-529
10. Kambara, T. Bull. Chem. soc. Japan 1954, 27, 529-534.

11. Barker, G. C.; Jenkins, I. L. AERE C/R 924; British Government Report, 1952.
12. Barker, G. C.; Jenkins, I. L. Analyst 1952, 77, 685-696.
13. Barker, G. C. CRC-440; Canadian Government Report, 1950.
14. Barker, G. C. AERE C/R 1553; British Government Report, HMSO: London, 1954.
15. Barker, G. C.; Cockbaine, D. R. AERE C/R 1404; British Government Report, HMSO: London, 1957.
16. Barker, G. C. AERE C/R 1563; British Government Report, HMSO: London, 1957.
17. Barker, G. C.; Gardner, A. W. AERE C/R 1606; British Government Report, HMSO: London, 1957.
18. Barker, G. C. AERE C/R 1567; British Government Report, HMSO: London, 1957.
19. Barker, G. C. AERE C/R 1565; British Government Report, HMSO: London, 1957.
20. Barker, G. C.; Faircloth, R. L.; Gardner, A. W. AERE C/R 1786; British Government Report, HMSO: London, 1958.
21. Barker, G. C.; Gardner, A. W. AERE C/R 2297; British Government Report, HMSO: London, 1958.
22. Barker, G. C.; Nuernberg, H. W.; Bolzan, J. A. Juel-137-CA; Zentralbibliothek der Kernforschungsanlage Juelich: Juelich FRG, 1963.
23. Osteryoung, R. A. Personal communication, 1988.
24. Milner, G.W.C.; Wilson, J.D.; Barnett, G.A.; Smales, A.A. J.Electroanal. Chem. 1961, 2, 25-38.
25. Gilbert, D.D. Anal. Chem. 1965, 37, 1102-1103.
26. Osteryoung, R. A.; Lauer; G.; Schlein, H. American Chemical Society National Meeting, Los Angeles, March 1963.
27. Lauer, G.; Schlein, H.; Osteryoung, R. A. Anal. Chem. 1963, 35, 1789.
28. Schein, H. U.S. Patent 3420764 (Appl. 16 Mar 1964); C.A. 1969, 70, 84076k.
29. Parry, E. P.; Osteryoung, R. A. Anal. Chem. 1965, 37, 1634-1637.
30. Parry, E.P.; Osteryoung, R.A. Anal. Chem. 1964, 36, 1366-1367.
31. Burge, D. E. J. Chem. Ed., 1970, 47, A81-A94.
32. Ewing, G. W. J. Chem. Ed., 1969, 46, A717-A725.
33. Flato, J. In A History of Analytical Chemistry, Laitinen, H. A.; Ewing, G. W., Eds.; Division of Analytical Chemistry of the American Chemical Society, Washington, 1977, pp 275-279.
34. Seralathan, M.; Osteryoung, R.A.; Osteryoung, J. J. Electroanal. Chem. 1986, 214, 141-156.
35. He, P.; Avery, J. P.; Faulkner, L. R. Anal. Chem. 1982, 54, 1313A-1326A.

RECEIVED August 12, 1988

Chapter 26

Fortuitous Experiments

Discoveries of Underpotential and of Quantitative Anodic Stripping Voltammetry

L. B. Rogers

Department of Chemistry, University of Georgia, Athens, GA 30605

Both phenomena were encountered in studies aimed toward using platinum, gold, or graphite electrodes, rather than mercury, to isolate carrier-free radiotracers. Preliminary "polarographic" scans of millimolar silver ion using platinum electrodes followed by reversal of the scan to clean the deposit from the electrode produced large sharp peaks. A quick calculation indicated the promise of that approach for trace analyses. Then, electrodeposition of carrier-free radio-silver found fractional monolayers depositing much sooner on platinum than predicted by conventional use of the Nernst Equation. Later, studies at M.I.T. suggested that structural parameters of the substrate could be related to the size of the underpotential - or its absence.

The electrochemical experiments described below were performed during a 30-month period when the writer was a group leader in long-range research in analytical chemistry at Clinton Laboratories, now known as Oak Ridge National Laboratory. One major goal of the research group was to explore the use of electrodeposition as a means of isolating quantitatively carrier-free radionuclides. Another goal was to explore ways for determining trace amounts of elements; for example, non-radioactive nuclides of silver were also present in carrier-free preparations partly because they, too, were produced in the bombardment with neutrons and be-

0097–6156/89/0390–0396$06.00/0

cause they probably were also present as an impurity in the target prior to the irradiation. A third goal developed later from our early success in continuously recording "polarographic" (voltammetric) data using solid electrodes. That small step in automatic recording, using a Sargent-Heyrovsky polarograph, made possible the rapid survey of many half-cell reactions, especially those run in a series in the same solution.

Background for the developments

Voltammetry. Although Heyrovsky had developed polarography using a dropping mercury electrode in the early 1920s, it was in 1941 that Kolthoff and Lingane (1) forcefully brought the technique to the attention of American chemists. For elements soluble in mercury, a constant halfwave potential, independent of the initial concentration of the reducible ion, made identification relatively easy. That potential can be related to the standard halfcell potential by taking into account a shift in potential as a result of amalgam formation.

Shortly thereafter, Laitinen and Kolthoff (2) showed that the halfwave potential for a metal deposited upon a solid electrode changed with the initial concentration in accord with predictions based upon the Nernst Equation. Voltammograms were obtained manually by changing the potential in small increments and waiting 2-3 minutes after each voltage setting for a steady current to be reached. The deposit was dissolved, after removing the electrode from the solution, before another "polarogram" was run.

Control of electrode potential. Historically, one adjusted the pH and allowed hydrogen evolution to limit the potential while the electrolysis ran at constant current. Hickling (3) was the first to employ instrumental control of the potential. Most recently, Lingane (4) had reported a 3-electrode system for electromechanically controlling the electrode potential in both directions. Automatic control opened up the possibility of performing selective reductions over a wide range of potentials with only a minimum concern for pH.

Radiation and radioactivity. Immediately following World War II, there was a great deal of interest in the isolation of fission products and of carrier-free nuclides produced by neutron bombardments followed by decays into other radioactive species. Although many chemists were using liquid-liquid extraction, ion exchange and, sometimes, copreciptation with a "foreign" carrier, no one at the laboratory was exploring electrodeposition. However, the writer did find a report (then classified as secret) by his predecessor, D. N.

Hume, which stated that the electrochemical behavior of a dropping mercury electrode did not appear to be noticeably affected by a high neutron flux or the presence of a modest level of radioactivity. Presumably the same would be true for other electrode materials.

Experimental goals

Basic idea. Substitution of a platinum electrode for a dropping mercury electrode appealed to the writer as a means of greatly reducing the problem of contamination. Further reflection suggested that, when a carrier-free radiotracer was deposited, the chances were heavily in favor of its forming only a fractional monolayer of deposit. Hence, the conditions would be very similar to those that resulted in a constant halfwave potential for mercury-soluble elements. The possibility of being able to deposit an element, regardless of its initial concentration in solution, at a preselected potential for a given electrode area and volume of solution, was very attractive.

Overall plans. The initial strategy included several steps. First, it was desirable to confirm Laitinen's work using a platinum electrode for 10^{-3} to 10^{-5} M (close to the lowest measureable concentration) reducible ion. Second, a suitable element had to be selected, preferably one with a rapidly reversible one-electron transfer, a relatively long halflife so as to minimize the need for frequent preparations of tracer, and a suitably high specific activity. A thorough search by A. F. Stehney led to selection of Ag-111 prepared by neutron bombardment of palladium. Although nearly 500 times more non-radioactive Ag-110 was produced at the same time, calculations showed that, barring the presence of significant amounts of silver impurity in the palladium (or other sources of contamination), concentrations of 10^{-9} M or less should be sufficiently active for 2-3 weeks. Shortly thereafter, J. C. Griess (5) started work on an isolation procedure based upon dissolution of the palladium target followed by isolation of the radio-silver by electrodeposition from a cyanide solution.

Extensions to the original plan. To speed the gathering of current-voltage "polarographic" data using a platinum electrode, we first explored the use of a Sargent Heyrovsky polarograph which changed the potential continuously and recorded the current on a drum holding photographic film (6). Although the reduction curves for silver sometimes had maxima, the halfwave potentials usually gave data that agreed within 10 mv and also agreed closely with the values predicted from the Nernst Equation.

Second, we modified the procedure for cleaning the deposit from the electrode between runs. The usual procedure of removing the platinum electrode from solution, dissolving the deposit in nitric acid, and reconditioning the electrode surface before running another voltammogram was frustrating because of the time it consumed. Out of desperation, the writer decided to run the potential scan in the opposite direction to clean off the deposit. A large, sharp peak was obtained over a relatively short range of potential. A quick calculation showed that the coulombs checked well with those involved in the deposition step. Furthermore, because of the sharpness of the peak, quantitation was easier, especially for concentrations less than 10^{-5} M. Starting two years later at M.I.T., Lord and O'Neill (7) employing faster scan rates, demonstrated quantitative anodic stripping voltammetry down to 10^{-11} g of silver.

Third, when concerns arose that the silver impurity in the platinum electrodes might distort the deposition data at the trace level, the writer substituted a pencil lead and, later, a graphite spectroscopic electrode in the voltammetric runs. A small (0.1 V or less) overvoltage was found in the first run with a new electrode, but later runs on the same electrode showed no overvoltage. Again, voltammetric behavior of graphite electrodes was later examined in more detail by Lord (8) at M. I. T. while demonstrating their utility for organic oxidations. When the writer presented preliminary data on Lord's organic oxidations in a seminar at Pennsylvania State University for Elving, he immediately saw its broad utility and later applied graphite electrode voltammograms to antioxidants (9).

Fourth, when carrier-free silver in a nitrate or perchlorate medium failed to follow the writer's predictions (instead depositing several tenths of a volt more readily than expected(10,11)), other experimental variables had to be examined. The Oak Ridge studies were continued at M. I. T. on silver, by Byrne (12-14), and on copper, by DeGeiso (15). The latter paper summarized the evidence for using interatomic distances in the substrate and in the deposit as a basis for making rough estimates of the magnitude of the underpotential - or its probable absence.

Related studies. Two or three doors down the hallway from our laboratory was one in which G. E. Boyd and Q. V. Larsen were working on the chemistry of technetium, especially in the form of pertechnetate. Because permanganate might be a useful model as well as being of interest in its own right, voltammograms were run on solutions of alkaline permanganate. Three one-electron steps were found. Controlled-potential electrolysis on the second plateau of current produced a sky-blue solution of Mn(V) with some black dioxide suspended in the stirred solution (16). Later, studies with pertechnetate showed a black precipitate, probably the dioxide, on the electrode. Using the voltammetric data, controlled potential electrolysis under those conditions was shown to produce technetium in a form having much less molybdenum than the original preparation (17). Hence, the voltammetric studies with platinum electrodes paid other dividends.

Fortuitous experiments

Perhaps the reader is wondering which of the above experiments were judged to be fortuitous - and why. The fact is that none of them would have been performed if the initial search of the literature had been successful!

Before deciding to start the electrochemical research, the writer made a very thorough search of Chemical Abstracts and all treatises on electrochemistry, including chapters in books on microchemistry. No previous studies of very dilute solutions were found under titles such as electrochemistry, Nernst Equation, or electrodeposition. So our work began.

About 2-3 months after we had found the phenomenon of underpotential for silver on platinum and gold, Chemical Abstracts had an abstract of an article by Haissinsky (18) dealing with electrochemistry in very dilute solutions. The article detailed the classical studies of Hevesy, Joliot-Curie and others in which the chemistry of protoactinium, RaD (Pb), and RaE (Bi) and the corresponding daughters of thorium had been characterized in a variety of ways, including electrodeposition. Aside from the embarrassment of having missed those references, the important result was that in none of those cases had underpotential been reported! (By hindsight, one could detect marginal evidence for partial underpotential deposition only in some of the more recent data which dealt with extremely low concentrations of RaD and RaE). However, if the writer had found those references earlier, the experiments leading to the discovery of underpotential and of quantitative anodic stripping voltammetry would not have been attempted because the usual Nernst Equation was followed. So if the experiments themselves were not fortuitous, the act of missing the references was.

Acknowledgments

The writer thanks the Department of Energy, Division of Basic Sciences, for support through Contract No.DE-AS09-76ER00854 during the early stages of this manuscript. The writer also thanks his collaborators at Oak Ridge and at M. I. T. for their generous and absolutely essential contributions of ideas, criticisms, and hard work.

Literature cited

1. Kolthoff, I. M.; Lingane, J. J. Polarography; Interscience: New York, 1941.
2. Laitinen, H. A.; Kolthoff, I. M. J. Phys. Chem. 1941, 45, 1061.
3. Hickling, A. Trans. Faraday Soc. 1942, 38, 27.
4. Lingane, J. J. Ind. Eng. Chem.,Anal. Ed. 1945, 17, 332.
5. Griess, J. C.; Rogers, L. B. J. Electrochem. Soc. 1949, 95, 129.
6. Rogers, L. B.; Miller, H. H.; Goodrich, R. B., Stehney, A. F. Anal. Chem. 1949, 21, 777.
7. Lord, S. S.; O'Neill, R. C.; Rogers, L. B. Anal. Chem. 1952, 24, 209.
8. Lord, S. S.; Rogers, L. B. Anal. Chem. 1954, 26, 284.
9. Gaylor, V. F.; Conrad, A. L.; Elving, P.J. Anal. Chem. 1953, 25, 1078.
10. Rogers, L. B.; Krause, D. P.; Griess, J. C.; Ehrlinger, D. B. J. Electrochem. Soc. 1949, 95, 33.
11. Rogers, L. B.; Stehney, A. F. J. Electrochem. Soc. 1949, 95, 25.
12. Griess, J. C.; Byrne, J. T.; Rogers, L. B J. Electrochem. Soc. 1951, 98, 447.
13. Byrne, J. T.,; Rogers, L. b.; Griess, J. C. J. Electrochem. Soc. 1951, 98, 451.
14. Byrne, J. T.; Rogers, L. B. J. Electrochem. Soc. 1951, 98, 457.
15. DeGeiso, R. C.; Rogers, L. B. J. Electrochem. Soc. 1959, 106, 433.
16. Miller, H. H.; Rogers, L. B. Science 1949, 109, 61.
17. Rogers, L. B. J. Am. Chem. Soc. 1949, 71, 1507.
18. Haissinsky, M. J. Chim. Phys. 1946, 43, 21.

RECEIVED August 9, 1988

Chapter 27

Coulometric Titrimetry

Galen W. Ewing

New Mexico Highlands University, Las Vegas, NM 87701

Titration with electrically generated reagents is unique in that it depends for its standardization on precisely measurable electrical, rather than chemical, quantities. Its utility has been expanded drastically over the years since its first description in 1938. Several instruments are now available commercially. The method, however, has not been exploited as far as it deserves as a general analytical tool. Its history, principles, and areas of usefulness are explored.

Coulometry, in the context of electroanalytical chemistry, can be taken to include the theory and practice of any analytical procedures that depend on Faraday's laws of electrochemical equivalence. Specifically, the analytical chemist is interested in the quantitative determination of chemical species by measurement of the quantity of electrical charge (in coulombs) necessary just to convert a known amount of that species from one oxidation state to another. This approach to chemical analysis is attractive because it permits dispensing with traditional primary standards and standardized reagent solutions, both of which are notorious for the amount of time and space required for preparation and storage. Instead, precision can be established through purely electrical standards. Perhaps this can be considered buck-passing, involving physicists rather than chemists in the ultimate establishment of precision, but it must be realized that the standards utilized in electrical work—the standard volt and ohm particularly—are already available with greater precision than is normally required in chemical analysis.

A coulometric analysis can be performed utilizing a direct electrochemical reaction, or it can be accomplished by an indirect process using an electrochemically prepared intermediate. Either of these can be operated with controlled current or at controlled potential. All four variations are useful, and will be detailed below.

0097–6156/89/0390–0402$06.00/0

Historical

Michael Faraday formulated the fundamental laws bearing his name in 1834. The Faraday laws can conveniently be combined into one equation:

$$G = (It/nF)(MW) \tag{1}$$

where G is the mass in grams of a chemical substance oxidized or reduced at an electrode, I is the current in amperes, t is the elapsed time in seconds, n is the number of electrons involved in the reaction (this can also be considered to be the number of gram-equivalents per mole), and F is the Faraday constant, the quantity of electric charge in coulombs associated with one gram-equivalent weight of reactant or product; MW designates the molecular weight in grams per mole (1). These relations have been confirmed many times over since Faraday's initial work, under many different sets of conditions, including high and low temperature, high pressure, various solvents, and transitions across phase boundaries.

Many attempts have been made to determine accurately the value of the constant F. There are two alternative approaches to this problem, the first constituting a direct application of Equation 1, the second involving the combinations of other physical and numerical constants. The latter approach is exemplified by the work of Hipple et al. at the National Bureau of Standards in 1949 (2). Their work was based on a miniature cyclotron, called an "omegatron," which is best known as the forerunner of modern ion-cyclotron resonance (ICR) mass spectrometers.

Early coulometry based on chemical reactions were beset with difficulties, in that the results were not sufficiently reproducible. N. Ernest Dorsey, of the Bureau of Standards, is quoted (3) as saying: "In 1903 it was generally believed that it was not possible to make absolute electrical measurements to a higher accuracy than one in one thousand; [but] by 1910, such measurements had been made with an accuracy of a few parts in 100,000... In 1903, the results obtained with the silver coulometer were distressingly variable; by 1910, the major cause of the variations had been discovered, and several types of coulometers yielding high reproducibility had been designed." This work was especially important in that the silver coulometer was the reference standard for defining the International Ampere.

Bower and Davis, also at the National Bureau of Standards, published in 1980 a meticulous study of all previous determinations of the Faraday constant (4). They found the best value to be 96,486.33$_{24}$ with a standard deviation of 2.5 ppm. This figure, with 7 significant figures, is far better than needed for most purposes in analytical chemistry. The high precision of this quantity, together with highly accurate instruments for measuring current and elapsed time, provide the basis for chemical coulometry.

Some Analytical Applications

Various workers over the years have attempted to utilize the Faraday relations for analysis. Early results were unsatisfactory, largely due to the unrecognized occurrence of side reactions. Clearly, successful applications must involve only reactions that are 100-percent efficient, that is, all the electrons that pass through the electrode *must* contribute only to the desired chemical process. This is not a trivial problem. The great majority of electrochemical reactions cannot be made to conform to this stringent requirement. However, enough reactions *do* conform to make this a highly useful analytical tool.

One of the earliest applications to give acceptable analytical results appeared in a 1917 contribution by G. G. Grower, published in the ASTM Proceedings (5). Grower described a direct method for the determination of the thickness of a layer of tin plated on copper wire, through measurement of the quantity of electricity necessary to remove it electrolytically. A similar application was reported in 1938 by Cambell and Thomas (6), and in 1939 by Zakhar'evskii (7).

Such direct methods, though powerful, are of limited applicability. The more flexible method, coulometric titrimetry, began with the 1938 work of two Hungarian chemists, Laszlo Szebellédy and Zoltan Somogyi (8,9). Szebellédy was a Professor in the Analytical and Pharmaceutical Chemistry Institute of the Royal Hungarian Pázmány-Péter-University, Budapest (now the University of Budapest), and Somogyi was one of his students. They were interested in improving methods of standardizing various reference solutions. Their cell contained a large platinum gauze anode surrounding a spiral platinum cathode. The current was supplied by two or three lead storage cells of large capacity with two regulating resistors (rheostats) of 10- and 150-ohms (Figure 1). Also in the circuit was a 0-50 milliammeter with an optional shunt to permit measurements to 1 ampere. The end point of the reaction was identified in the classical way, by observing the color change of a suitable indicator.

The publications of these chemists are so significant that it is worthwhile to reconstruct an example of their thinking, as described by Swift (10): "These workers were interested in the determination of thiocyanate. They either knew or learned that the anodic oxidation of thiocyanate to cyanide and sulfate with 100% current efficiency is difficult if not impossible of attainment. However, it was known that thiocyanate could be quantitatively, stoichiometrically, and rapidly oxidized by bromine in acid solutions with 100% current efficiency. Therefore they added a relatively high concentration of a soluble bromide to an acid solution containing the thiocyanate, anodically produced bromine, and allowed this bromine to diffuse into the solution and to oxidize the thiocyanate. By working with relatively large samples of thiocyanate, and by measuring the quantity of electricity involved by means of a chemical coulometer, they demonstrated that an accuracy within 1 ppt could be attained."

It is undoubtedly true that the work of Szebellédy and Somogyi was founded on prior work. But as Lingane remarks: "...it seems... that [they] deserve credit for being the first to appreciate clearly the catholicity of the method and its capabilities as a widely applicable improvement on classical titrimetric technique. Henry Ford did not invent the wheel, but everyone agrees that he set it rolling as it never rolled before" (11).

Following the initial invention of the technique, little progress occurred in coulometric titration for some ten years. The method of Szebellédy and Somogyi was cumbersome, and other analysts were loath to even try it. The earliest succeeding reference that I have been able to locate is to a U.S. Government defense document dated 1945 (12), published in the open literature in 1948 (13). This described a continuous analyzer for mustard gas in air by electrolytic generation of bromine. One may quibble about whether this is strictly "coulometric," since it is current rather than charge that is measured, but a modification is also described that permits batchwise determinations of mercaptans, that are clearly "coulometric titrations," in the modern meaning of that term. Antedating the open publication of this work is a paper from Edgewood Arsenal (14) describing the titration of acids by electrogenerated hydroxide ion, to a potentiometric end point utilizing a quinhydrone electrode (this technique also measured current rather than charge).

Also of historical interest is an undocumented remark by Abresch and Claasen (15), in which they state that "Coulometric analytical methods have been developed and used in the chemical laboratories of the August-Thyssen works in Duisberg-Hamborn [Germany] since 1948."

We will now describe some of the apparatus used over the years in coulometric titration, and then mention some typical applications.

Coulometers

In order to make a quantitative measure of the amount of electricity required for a reaction to take place, a reliable coulometer is essential. Much of the early work in the field followed Szebellédy and Somogyi in using a chemical coulometer. This amounts to a circular application of the Faraday relation, so that an exact value of F is not required. On the other hand, it merely replaces an inconvenient or difficult chemical measurement with another one, hopefully less inconvenient.

Take, for example, the dissolution of tin from a plated wire. This determination could be made non-electrochemically, the final result being obtained by a volumetric titration. On the other hand, the dissolution can be carried out electrochemically in one cell, with a second cell, connected in series, in which silver is dissolved from a silver anode. The amount of tin oxidized in the first cell becomes directly related to the amount of silver dissolved in the second, and this can be determined volumetrically. The result is that instead of titrating the tin, one titrates the silver. If silver can be quantified with greater accuracy and convenience than tin, then this becomes a useful procedure, but not otherwise.

Figure 2 shows the silver coulometer used by Richards (16), first described in 1902. The anode, *C*, is a massive piece of silver, preferably coated with electrolytic silver, suspended by its connecting wire, *A*, and surrounded by a porous cup, *D*, which is also suspended from above, *B*. The cathode consists of a platinum bowl or crucible, *E*. The porous cup serves to catch "anode mud" from dropping onto the cathode.

A simple coulometer using two plates of copper dipping into a suitable electrolyte has been used by many generations of physical chemistry students. The two electrodes are weighed both before and after the reaction being studied. (The student considered himself

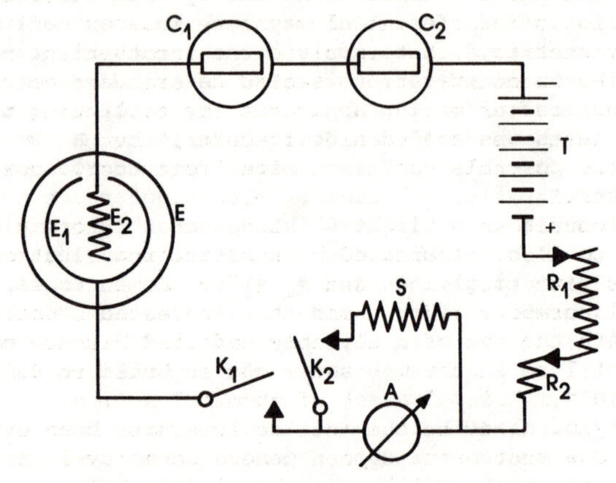

Figure 1. The coulometric titration apparatus of Szebellédy and Somogyi. E_1 and E_2 are the anode and cathode, respectively; C_1 and C_2 are the two coulometers. (Redrawn with permission from Ref. 8. Copyright 1938 Springer-Verlag.)

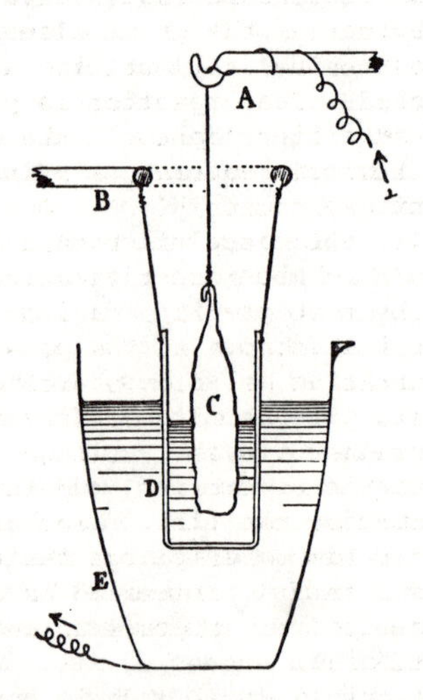

Figure 2. The silver coulometer of T. W. Richards. (Reproduced with permission from Ref. 16. Copyright 1902 American Academy of Arts and Sciences.)

lucky if the loss of weight of one electrode equalled the gain of the other, and if they did not, he had to explain the discrepancy in his report.)

Another chemical coulometer depends on the production of a gas. Classically, this was the combined H_2 and O_2 from electrolysis of water containing an indifferent electrolyte such as sodium sulfate. This is fairly sensitive, but involves the inconvenient measurement of a gas volume, which must be corrected to standard conditions of temperature and pressure. The apparatus for collecting and measuring the mixed gases was called a "voltameter," but T. W. Richards (16), noting the possible confusion with "voltmeter," coined the name "coulometer."

It was found that a slight negative error inherent in the O_2-H_2 coulometer could be eliminated by substituting electrolysis of hydrazinium sulfate to give H_2 and N_2 (17). J. E. Fagel, Jr., in the author's laboratory (18), found that increased sensitivity could be obtained with the use of a slightly modified Warburg monometer; a displacement of 1 cm in the manometer corresponded to 0.05 coulomb, or about 6×10^{-4} milliequivalent of chemical action.

A variety of electromechanical devices have been utilized as coulometers. One successful type depended on a low-inertia DC motor operating a turns counter (19). Another is the ball-and-disk integrator, which has been widely used in connection with a potentiometric strip-chart recorder for chromatographic and other measurements.

The development of electronic instrumentation has effectively displaced mechanical devices as the prime element of coulometers. Those in use today almost universally utilize a capacitor charged by the current being measured. The capacitor is placed in the feedback loop of an operational amplifier (op amp), the output of which registers the time-integral of the current, a principle previously used in analog computation circuitry (20). One of the first to use this method for chemical coulometry was Booman in 1957 (21). The difficulty with this approach is that a capacitor of reasonable size (Booman used a 30-µF non-electrolytic capacitor, a giant of its kind) cannot accomodate sufficient charge at the potentials suitable for chemical use. This limitation has been overcome in several ways. One way is to use a small gas-filled tube (thyratron) to discharge the capacitor when it reaches a preset voltage; continuing current then recharges the capacitor repeatedly, and the discharge events are recorded by an events counter (22). Another possibility is to use a moving-coil meter-relay to discharge the capacitor (23). In present practice a second op amp, connected as a comparator, discharges the integrating capacitor through an analog switch when the potential reaches a set point.

Stock (24) has described an electronic coulometer designed around a voltage-to-frequency converter, an integrated circuit that produces a series of pulses at a frequency proportional to a control voltage. The voltage to be converted originates in the potential drop across a standard resistor through which the generating current flows. The resulting pulse train is made to energize a digital display reading directly in microequivalents. The instrument can be used for either constant-current or controlled-potential coulometry. Stock (25) has also described a simple yet effective potentiostat for controlled-potential coulometry.

Constant Current Coulometry

The coulometers described above can be used to integrate current against time, regardless of whether or not the current is varying. A simplification, without loss of precision, is possible if the current can be maintained at a constant value. The quantity of electric charge passed is given, in coulombs, merely by multiplying the magnitude of the current in amperes by the elapsed time in seconds. This approach is commonly used in indirect determinations ("coulometric titrations").

The accuracy with which elapsed time can be measured is rarely a limitation in modern instruments. The frequency of the AC power lines is of sufficient accuracy in most locations, except for very short times (less than a few cycles). If higher precision is needed, and for battery-operated instruments, a timer based on a crystal-controlled oscillator is capable of precision to better than 1 part per million. Sufficient accuracy for many purposes can conveniently be obtained from the time-base of a strip-chart recorder on which the detector response is plotted. Older instruments depended on a hand-held stopwatch or other mechanical or electromechanical device.

The current passing through an electrolytic cell can be held constant by drawing it from a high-voltage source through a large resistor. This could be done with a battery of 90 volts or more [Cooke et al. used a 250-volt battery (26)], or in the days when vacuum tubes were in their heyday, by a rectified and filtered supply of, say, 300 volts. The required resistor is determined by a simple application of Ohm's law. The precision of such a source depends on the degree of regulation of the voltage supply and the use of a resistor with a low temperature coefficient. It also depends on the relative magnitude of the resistor and the resistance of the solution. For a 100-volt source, if 10 mA of current is desired, the series resistor must be 10,000 ohms, whereas the resistance of a typical solution and cell for titrimetry would be less than 100 ohms. Hence change in the resistance of the solution by as much as 10 percent (a greater change than usually accompanies a titration reaction) would alter the total circuit resistance, and hence the current, by less than 0.1 percent.

Various electronic circuits can be devised for maintaining constant current without the need of a high voltage. In the days of vacuum tubes, many circuits were published, varying from simple to complex. Perhaps the simplest utilized the characteristic curve of a pentode vacuum tube, which shows a wide plateau region over which the current is nearly constant. The value of the current at this plateau is determined by the potential applied to the control grid. This phenomenon was used successfully as a current source for coulometry in a circuit involving only two tubes, a 6J7 pentode and an 885 thyratron (27). At the other extreme, a circuit for a combined current generator and coulometer was published in 1954 using a total of 18 tubes, including a regulated voltage supply (22); the authors comment that prior coulometers were not sensitive enough: 1 coulomb corresponds to only 1.118 mg of silver, or 0.1739 mL of mixed gas, whereas with their apparatus it was possible to measure "quantities of electricity as small as 0.01 coulomb within 0.1%."

Sometimes difficulty is encountered with respect to the requirement that the solution must contain the reagent precursor admixed with the material under examination. In many cases this means that not only the precursor will be oxidized or reduced, but other components may be as well. For example, this requirement precludes the titration of acids by electrolytically generated hydroxide ion in the presence of reducible substances, because these substances, as well as the water that forms the hydroxide ions, will be reduced at the cathode. To circumvent these limitations, DeFord and his students developed a method for generating the reagent externally, allowing it to flow into the stirred solution without appreciable amounts of the precursor (28,29). Figure 3, taken from their first paper, illustrates the principle. Production of the reagent (for example, OH^- ion) occurs at a cathode in the central part of an inverted T-tube, whence it flows into the analytical solution. The anodic products flow through the other arm of the T-tube to waste.

End-Point Detection

As in any titration, a suitable method of detection of the equivalence point is necessary. In general, it can be said that any of the techniques that are useful in classical volumetric titrimetry are also applicable to coulometric titrations. As noted above, Szebellédy and Somogyi used traditional color-change indicators to determine their end points.

Modern coulometric titrators commonly use electrochemical detection, either potentiometric or biamperometric. These are readily implemented with electronic circuitry that is compatible with the devices used to produce the controlled current. Biamperometry was first applied to end-point detection by Swift and co-workers (30), potentiometry by Epstein et al. (14), both in 1947.

The ever-present possibility of deleterious interaction between generation and detection circuits led Wise et al., in 1953, (31) to introduce a photometric end-point detection scheme. Their vacuum-tube apparatus included a constant-current source as well as an automatic circuit to terminate the current at the command of the detecting photocell. Other photometric coulometric titrators have been reported, using more modern electronics; see, for example, Ref. 32.

In common with other forms of titration, coulometric titrimetry must address the problem of how to avoid or correct for overshooting of the end point. In a manual volumetric titration, the operator adds the reagent quickly until the end point is approached, then decreases the incremental addition until, in the immediate vicinity of the end point addition is dropwise, thus avoiding overshoot. It was realized early on that the same approach can be applied to coulometric titrations, using a manual switch to turn the current off and on again as needed. Obviously this requires either a coulometer, or provision for interrupting the timer along with the current flow. Other end-point anticipation techniques have been used. With potentiometric detection, the logarithmic nature of the Nernst equation makes it feasible to utilize a switch closure at the first maximum of a second-derivative curve, so that the lag inherent in the stirring of the solution is compensated by the time interval between the maximum and the true end point.

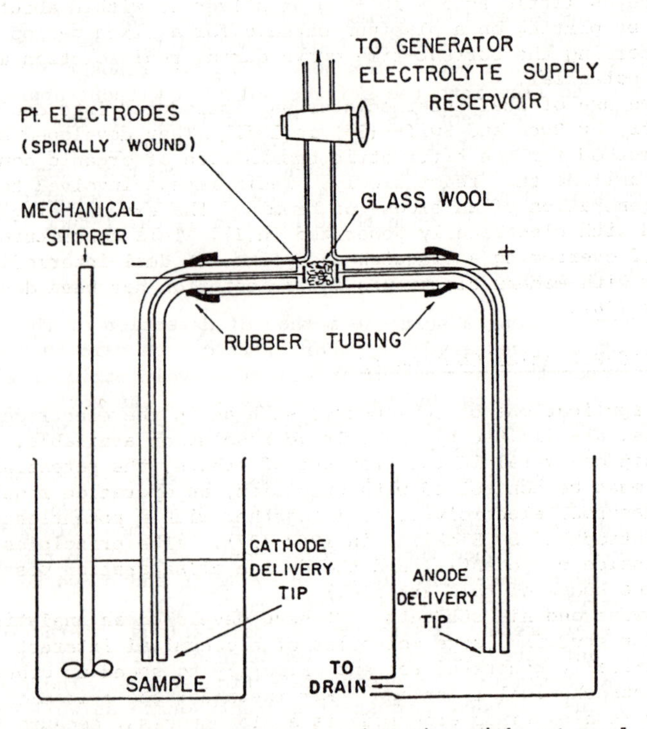

Figure 3. Cell for coulometric titration with external generation of reagent. (Reproduced from Ref. 28. Copyright 1951 American Chemical Society.)

Indirect Titrations

A number of indirect titrimetric applications have been reported. Elema in 1947 determined 10-100 μg of copper by first depositing it from solution, then measuring the quantity of electricity needed to redissolve it (33). A similar principle was used by Lord et al. in 1952 to determine trace amounts of silver (34). They were able to determine as little as 5×10^{-10} g of silver to within about 1×10^{-10} g by plating on a platinum cathode for a fixed period of time, then recording the current-time curve during redissolution with scanned potential.

The use of dual intermediates was introduced into coulometric titrimetry by Buck and Swift (35) in 1952. They developed an improved method for the titrimetric bromination of organic compounds such as aniline that react slowly. Their method involved the electrical generation of an excess of bromine, the excess being back-titrated with electrically generated Cu(I). This constitutes another method of overcoming a sluggish end point. A dual-intermediate titrator with modern electronic instrumentation has been described recently (36).

Controlled-Potential Coulometry

Primary applications of coulometry, such as in the electrodeposition of metals, are limited by the voltage resolution available. In order to deposit one metal in the presence of others, the potential of the cathode must be controlled with precision, an operation requiring a three-electrode electrolytic cell together with a potentiostat, a method introduced by Hickling in 1942 (37). (The principles involved in the choice of potential and the degree of separation possible lie beyond the scope of this article.)

Kuwana and his students (38) have developed an analytical method for studying large molecules of biochemical interest, using a coulometrically generated reagent primarily to speed up otherwise slow electrochemical reactions. For instance, the oxidation of cytochrome-c at a platinum electrode is a slow process, because the oxidizable iron atom is protected by the large protein component. However it can be oxidized readily in a homogeneous solution by ferricyanide, which in turn can be easily generated by electrochemical oxidation of ferrocyanide. This reagent is designated as a "mediator." The process is carried out under controlled-potential conditions, the end point being identified photometrically with transparent electrodes. A listing of many such mediators with their physical properties is presented in Ref. 39.

Since the electrical resistance of the solution in coulometric titration is not under the control of the operator, clearly it is not possible to maintain constant current while also controlling the potential. Because of the logarithmic nature of the governing equations, the electrolysis current, under constant driving potential, will fall off exponentially with time. It was pointed out by MacNevin and Baker (40) that a plot of the logarithm of current against elapsed time could be used to determine the result of a coulometric analysis (Figure 4). As soon as the experiment has run long enough to establish the slope, the number of coulombs for the

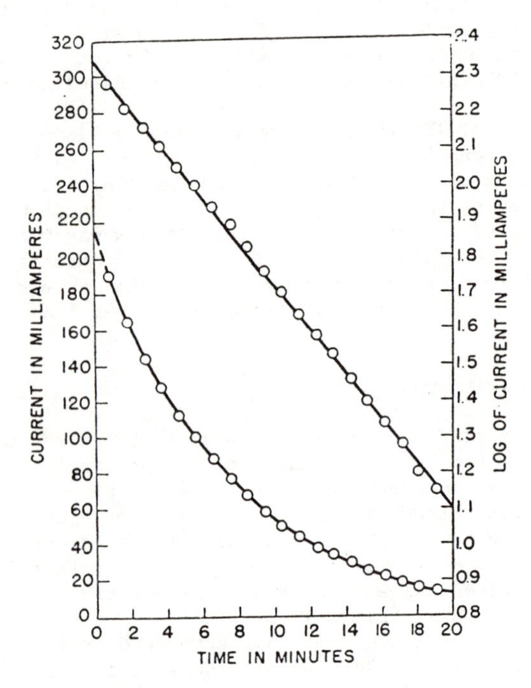

Figure 4. Data for the constant-potential electrolytic oxidation of iron(II), showing plots of current and of log-current against time. (Reproduced from Ref. 40. Copyright 1952 American Chemical Society.)

complete analysis can be calculated. This saves considerable time over an exhaustive electrolysis, with no reduction in accuracy. No separate coulometer is required in the use of this technique.

Data Collection

Stephens, et al., in 1970, first reported computerized data collection for controlled-potential coulometry (41). They used both analog and digital real-time computer techniques to predict the final value of the charge and the rate constant for the reaction. Jaycox and Curran (42), considering constant-current methods, showed that electronic control can permit working with much smaller currents and elapsed times than previously considered possible. Their apparatus did not use a computer as such, but involved extensive analog and digital circuitry. They could work with currents as small as 1 μA and times down to 1 μs. They reported, for example, the determination of 26 ng of manganese with a precision and accuracy of 2 percent.

Applications

Most classes of titration reactions can be implemented using constant-current coulometric techniques. Both H^+ and OH^- ions can easily be generated, thus permitting acid-base titrations. Several strong oxidants, such as Ce(IV) and Ag(II) can be formed [from Ce(III) and Ag(I), respectively, as precursors]. Reductants such as Sn(II) and Cu(I) are available, as are a number of complexogens, including EDTA [generated, for example, by reducing the mercury from $Hg(EDTA)^{2-}$]. Silver cation, Ag^+, can be formed by electrolytic oxidation of a silver anode, providing a convenient method for the titration of halides. The halogens, Cl_2, Br_2, and I_2, are easily generated, leading to a large number of varied applications. Comprehensive tabulations of reported coulometrically-generated reagents have been published (15,43). Other references can be gleaned from the biennial reviews in *Analytical Chemistry*, in even-numbered years since 1952.

Coulometric titrimetry offers many advantages over other types, and it is strange that it is so little used. No instrumental standardization is required beyond ascertaining that current and time circuits are operating within specifications. No method calibration with primary standard samples is needed, as the electron itself acts as primary standard. No standardized titrants need be prepared and stored. Furthermore, the apparatus is relatively inexpensive, considering the high sensitivity and precision possible (in the range of 0.1 percent error) even for small samples. Volumes as small as about 10 μL can be handled. The method is readily suited to automation, without the need for any moving parts other than a stirring mechanism. Another advantage is that potentially interfering substances in the supporting electrolyte can be removed easily by pre-titration, the addition of titrant until an end point is reached, prior to inserting the sample.

Of course the method is subject to limitations, just as is any analytical tool. It is not the method of choice for large samples or concentrations, as either the current must be large (which calls for large area electrodes) or the time will be excessively long.

Undoubtedly the most widely used coulometric titration at the present time constitutes an implementation of the Karl Fischer procedure for the determination of small amounts of water (44,45). This reaction has been incorporated into several commercially available moisture titrators. The reagent is electrically generated iodine in ethylene glycol solution. It can be shown that 10.71 coulombs of generating current corresponds to 1 mg of water, which makes it possible to measure water down to a few micrograms per milliliter of sample.

In the late 1950s there was a considerable movement favoring the international establishment of the coulomb as the primary standard for all analytical procedures [see, for example, Tutundžič (46)]. The general consensus seemed to be that, though this proposal had merit, it should be considered to be just one more valid standardization method, without necessarily displacing the familiar chemical primary standards.

Continuous Titrators

There have been numerous descriptions of instruments purporting to be continuous coulometric titrators intended for monitoring flowing streams. These operate by generating electrolytically just enough reagent to maintain a detector system at an equilibrium value. The current required then becomes a valid measure of the concentration of analyte in the stream.

It is the considered opinion of the present author that such instruments do not fall under the definition of *coulometric* titrators. Clearly a coulometric method must, by definition, involve the measurement of the quantity of electric charge, $Q = \int i\, dt$, in coulombs. In contrast these instruments measure a current, i, in amperes. However, they are commonly called "coulometric," hence deserve to be mentioned here. Apparently the first instrument of this type, as mentioned earlier, was for the determination of mustard gas in air (12,13). The authors (they did not use the term "coulometric") utilized a platinum indicator electrode against a saturated calomel to identify the equilibrium point and control the generating current.

There have since been many similar instruments, largely dedicated to measurement of mercaptans and other sufur compounds by reaction with generated bromine. An early example was the Titrilog, manufactured by Consolidated Electrodynamics Corporation in the 1950s. Others, produced by European or Japanese firms, are intended for the determination of SO_2 or CO_2 from combusion analyzers (47). Stock (48) has summarized the development and present status of continuous "coulometric" titrators.

Literature Cited

1. Diehl, H.: Anal. Chem., 1979, 51, 318A.
2. Hipple, J. A.; Sommer, H.; Thomas, H. A.: Phys. Rev., 1949, 76, 1877; NBS Annual Report 1950, pp. 6, 34.
3. Cochrane, R. C.: Measures for Progress: A History of the National Bureau of Standards, Nat. Bur. Stds., Washington, 1966, p. 104.
4. Bower, V. E.; Davis, R. S.: J. Res. Nat. Bur. Stds, 1980, 85, 175.
5. Grower, G. G.: Proc. Amer. Soc. Testing Mater., 1917, 17, 129.
6. Campbell, W. E.; Thomas, U. B., Nature, 1938, 142, 253.

7. Zakhar'evskii, M. S.: Referat. Zhur., 1939, 2, 84.
8. Szebellédy, L.; Somogyi, Z., Fresenius Z. anal. Chem., 1938, 112, 313, 323, 332, 385, 391, 395, 400.
9. Szabadváry, F.: History of Analytical Chemistry, Pergamon: New York, 1966; pp 316-317.
10. Swift, E. H.: Anal. Chem., 1956, 28, 1804.
11. Lingane, J. J.: Electroanalytical Chemistry, 2nd ed., Interscience: New York, 1958; p. 484, ff.
12. Briglio, A., Jr.; Brockman, J. A., Jr.; Schlinger, W.; Shaffer, P. A., Jr.: U. S. Department of Commerce Office of Publication Board, OSRD Report 1945, 6047, PB 5925.
13. Shaffer, P. A., Jr.; Briglio, A., Jr.; Brockman, J. A., Jr.: Anal. Chem., 1948, 20, 1008.
14. Epstein, J.; Sober, H. A.; Silver, S. D.: Anal. Chem., 1947, 19, 675.
15. Abresch, K.; Claasen, I.: Coulometric Analysis, Franklin: Englewood, N.J., 1966.
16. Richards, T. W.; Collins; Heimrod: Proc. Amer. Acad. Arts and Sci., 1902, 37, 415.
17. Page, J. A.; Lingane, J. J.: Anal. Chim. Acta, 1957, 16, 175.
18. Fagel, J. E., Jr., unpublished work; see Ewing, G. W.: Instrumental Methods of Chemical Analysis (1st ed.), McGraw-Hill: New York, 1954; p. 99, footnote.
19. Parsons, J. S.; Seaman, W.; Amick, R. M.: Anal. Chem., 1955, 27, 1754.
20. Korn, G. A.; Korn, T. M.: Electronic Analog Computers, McGraw-Hill: New York, 1952.
21. Booman, G. L.: Anal. Chem., 1957, 29, 213.
22. Kramer, K. W.; Fischer, R. B.: Anal. Chem., 1954, 26, 415.
23. Kelley, M. T.; Jones, H. C.; Fisher, D. J.: Anal. Chem., 1959, 31, 488.
24. Stock, J. T.: Microchem. J., 1984, 30, 92.
25. Stock, J. T.: Microchem. J., 1984, 29, 361.
26. Cooke, W. D.; Reilley, C. N.; Furman, N. H.: Anal. Chem., 1951, 23, 1662.
27. Ehlers, V. B.; Sease, J. W.: Anal. Chem., 1954, 26, 513.
28. DeFord, D. D.; Pitts, J. N.; Johns, C. J.: Anal. Chem., 1951, 23, 938.
29. DeFord, D. D., Johns, C. J.; Pitts, J. N.: Anal. Chem., 1951, 23, 941.
30. Sease, J. W.; Niemann, C.;Swift, E. H.: Anal. Chem., 1947, 19, 197.
31. Wise, E. N.; Gilles, P. W.; Reynolds, C. A., Jr.; Anal. Chem., 1953, 25, 1344.
32. Lindberg, A. O.: Anal. Chim. Acta, 1978, 96. 319.
33. Elema, B.: Antonie van Leeuwenhoek, J. Microbiol. Serol., Jubilee Vol. Albert J. Kluyver, 1947, 12, 243.
34. Lord, S. S., Jr.; O'Neill, R. C.; Rogers, L. B.: Anal. Chem., 1952, 24, 209.
35. Buck, R. P.; Swift, E. H.: Anal. Chem., 1952, 24, 499.
36. Stock, J. T.: Anal. Chim. Acta, 1981, 124, 85.
37. Hickling, A.: Trans. Faraday Soc., 1942, 38, 27.
38. Hawkridge, F. M.; Kuwana, T.: Anal. Chem., 1973, 45, 1021.
39. Fultz, M. L.; Durst, R. A.: Anal. Chim. Acta, 1982, 140, 1.
40. MacNevin, W. M.; Baker, B. B.: Anal. Chem., 1952, 24, 986.

41. Stephens, F. B.; Jakob, F.; Rigdon, L. P.; Harrar, J. E.: Anal. Chem., 1970, 42, 764.
42. Jaycox, L. B.; Curran, D. J.: Anal. Chem., 1976, 48, 1061.
43. Farrington, P. S.: in Handbook of Analytical Chemistry, Meites, L., ed., McGraw-Hill: New York, 1963; p. 5-187 et seq.
44. Meyer, A. S., Jr.; Boyd, C. M.: Anal. Chem., 1959, 31, 215.
45. Kelley, M. T.; Stelzner, R. W.; Laing, W. R.; Fisher, D. J.: Anal. Chem., 1959, 31, 220.
46. Tutundžič, P. S.: Anal. Chim. Acta, 1958, 18, 60.
47. Nebesar, B., in Topics in Chemical Instrumentation II, Ewing, G. W., ed., American Chemical Society: Washington, 1977; p. 294.
48. Stock, J. T.: Trends Anal. Chem., 1982, 1, 117.

RECEIVED August 3, 1988

Chapter 28

History of Electroanalytical Chemistry in Molten Salts

H. A. Laitinen

Department of Chemistry, University of Florida, Gainesville, FL 32611

The earliest electroanalytical measurements in molten salts were based on potentiometry using the oxygen electrode to study high temperature acidity and basicity. This electrode is interesting in being applicable only at temperatures above 500°C. At temperatures in the 300-500° range, especially in chloride melts, a large number of metal-metal ion electrodes are useful, including the transition metals and noble metals.

Beginning in the 1950's, the scope of electroanalytical measurements has been greatly broadened to include steady state and cyclic voltammetry, coulometry, chronopotentiometry, classical and pulse polarography, and AC impedence measurements. Special problems associated with such measurements include melt purification, metal-glass seals, and porosity effects. Stationary, dropping, and rotating disk electrodes have all been used successfully.

While molten salt electrochemistry dates back to the time of Sir Humphry Davy and Michael Faraday, electroanalytical measurements in such systems are of much more recent origin. In such measurements we are primarily interested in solutes in dilute solutions rather than in properties such as activity coefficients or conductivities of major components. Molten salt solvents are almost always mixtures rather than pure compounds and therefore there is literally an infinite array of possible solvents.

Because of the experimental difficulties associated with molten salts, it is to be expected that the application of electroanalytical techniques lagged behind the corresponding application to aqueous and even non-aqueous solutions. Potentiometry was the earliest electroanalytical technique applied to molten salts, no doubt because it presented no special problems outside of thermoelectric effects. Interestingly, the scope of potentiometry in high tempera-

0097–6156/89/0390–0417$06.00/0

ture systems has turned out to be broader than at ambient temperatures because of the wider variety of reversible electrode reactions at high temperatures.

For the various electroanalytical techniques beyond potentiometry, molten salt solvents present several types of special problems depending on the nature of the solvent. First and foremost is the problem of purity. Salts used as solvents are at a much higher concentration than those used as supporting electrolytes in aqueous or non-aqueous solutions and special methods often had to be developed to decrease moisture and trace metal contamination to acceptable levels. In addition, possible reactions such as pyrohydrolytic decomposition or reactions of the melt constituents with containing vessels or components such as insulating materials or salt bridges need to be considered. In general, the simpler the method from the viewpoint of technique the earlier it was applied, but another factor, namely the importance of the solvent system with regard to applications, determined the amount of effort expended. Methods requiring insulation of electrodes and provision for reproducible mass transport to electrodes of known area tended to be developed later than those without such requirements. Thus chronopotentiometry could be applied using a simple flag type electrode suspended from a wire and this method has proven to be relatively much more important in molten salts than in aqueous solutions, especially in exploratory research. Cyclic voltammetry has proven to be the "workhorse" in more detailed mechanistic studies once the solvent purification has been worked out. In contrast, rotating disk electrodes, thin layer electrochemistry chemically modified electrodes, and spectroelectrochemistry have received relatively few applications because of special problems of technique.

Each solvent system presents its own problems of purification, insulation, and materials of construction, so it is convenient to consider various types of molten salt solvents in turn. Within each type we shall consider the various electroanalytical techniques with emphasis on the historical development rather than on exhaustive coverage.

Oxides, Hydroxides, Carbonates, Silicates, Borates, etc.

The earliest electroanalytical measurements were concerned with studies of high temperature acidity in relatively alkaline systems. The oxygen electrode has been used since 1912 (1), and provides an interesting example of an electrode more useful at temperatures of the order of 1000° C than at room temperature. Pioneering work was done by Treadwell (2) in 1916 and Lux (3) in 1939. Flood and Forlund (4) discussed the dependence of potential on oxide ion concentrations. Flood, Forlund and Motsfeld (5) in 1952 and Antipin (6) in 1955 showed that the potential of a platinum electrode surrounded by oxygen and immersed in an alkaline melt gave reproducible potentials corresponding to oxide activities in the melt.

At lower temperatures, the oxygen electrode has proven to be of more limited applicability. In 1948 Rose et al. (7) used oxygen electrodes at 400-700° in molten NaOH, and in 1958 Hill et al. (8) measured formation potentials of oxides by using an oxygen electrode against a metal coated with its oxide in a eutectic mixture of lithium and potassium sulfates containing dissolved CaO at temperatures of 550-750°C. Attempts to measure reversible oxygen-oxide potentials in LiCl-KCl containing lithium oxide were unsuccessful at 400-450° (9).

In 1963, an important development was the zirconia membrane electrode showing ionic conductivity due to oxide ion (10). This electrode, initially composed of calcium oxide and zirconia, later has appeared in other forms, notably zirconia-yttria and zirconia-thoria. It has proven effective for oxide ion activity measurements over an extremely wide range at temperature of 1000° or higher, but it is of limited use at temperatures below 500° because of excessive resistance.

Relatively few electroanalytical measurements beyond oxide ion activity have been made in alkaline melts, although in 1967 Bartlett and Johnson (11) used a Ag(I)/Ag reference electrode to measure the potentials of a few electrodes in lithium carbonate-sodium carbonate melts and for steady state voltammetry.

Alkali Metal Chloride Melts

Alkali metal chlorides have received a great deal of attention because of their importance in applications such as metallurgy and high temperature batteries. Electrochemically they are interesting because they have moderate melting points, they are excellent solvents for many metal chlorides through the formation of chloro-complexes, they are reasonably stable towards oxidants and reductants, and can be easily purified.

LiCl-KCl eutectic has been widely used because of its moderate melting point, 352°C. Electroanalytical applications were of limited value until a reliable purification method, involving vacuum and fusion in an HCl atmosphere was worked out (12). If the pyrohydrolysis of moist LiCl is prevented the melts can be used in Pyrex or silica vessels. Osteryoung (13) described a Pt(II)/Pt reference electrode that can be easily generated coulometrically and used at temperatures up to 500°C. Liu used this reference electrode to establish an electromotive force series of 26 entries at 450°C (14), and later entries were added by Pankey (15) and Plambeck (16). It was found that a wide variety of metals, including transition metals and noble metals exhibit reversible potentials against their lowest valence ions, and that a number of redox couples in solution can also be measured.. Halogen/halide ion electrodes, and two types of hydrogen electrode, namely hydrogen/HCl and hydrogen/hydride ion were successfully used (17). These measurements are of analytical significance because at low concentrations (below 0.01 mole fraction) the activity coefficients remain constant, so that the Nernst equation can be used to measure ion concentrations. Laitinen, Tischer, and Roe (18) described a novel application of potentiometry to measure the total concentration of metal ions more noble than cadmium by using molten cadmium to displace these metals and measuring the potential of the cadmium electrode. In a similar way, they estimated the metals more noble than zinc by measuring the potential of the Zn(II)/Zn electrode. Magnesium was used to purify the melt of metals more noble than magnesium, but of course at the cost of introducing an equivalent concentration of Mg(II) (of the order of millimolar) into the melt. The ultimate in displacement purification is to use a porous tungsten electrode containing molten lithium to displace less active metals by internal electrolysis (17).

Flengas and Ingraham, beginning in 1956, measured the potentials of 14 metals against their ions using a silver/silver chloride reference electrode in a 1:1 mole ratio KCl-NaCl melt at temperatures of 700-900°C, and reported activity coefficient variations at higher concentrations (19). Delimarskii and Markov (20) and Laity (21) have reviewed other potential measurements in melts.

As compared with potentiometry, voltammetry presents special experimental problems, because of the necessity for insulation of a microelectrode of a reproducible surface area. The most common metal seals are glasses, which become increasingly conductive at higher temperatures. Lead glass compositions have especially good properties for sealing to metals, but must be avoided due to the cathodic deposition of lead (22). Voltammetry also places stringent demands on solvent purity.

In 1954, Black and DeVries (23) used platinum microelectrodes in LiCl-KCl melts to record polarograms (voltammograms) of several metals. From the low decomposition potentials they reported, it is evident that the melts were contaminated with hydroxyl ion and water. Laitinen and coworkers (22) took care to purify the solvent and used simple microelectrodes of platinum sealed glass to record steady state voltammetric curves for a large number of metal ions using a slow polarization rate. During the early 1950's, cyclic voltammetry and linear sweep voltammetry were still in their infancies. While the limiting currents were not highly reproducible for exact work, the results proved to be quite useful in establishing the stable oxidation states of the elements and in estimating the redox potentials of many metal ion systems.

Maricle and Hume (24) were able to extend the temperature range of such measurements to 740°C in molten NaCl-KCl by using a tungsten electrode sealed in Vycor. In this way the limitation of excessive conductivity of the glass seal was avoided. They found the limiting currents to be insensitive to scan rate because of the rapid establishment of a diffusion-convection steady state and reported a precision of 5 to 8%. While tungsten is less noble than platinum and therefore not applicable for anodic processes, it is advantageous in the cathodic region, especially for the deposition of several liquid metals that alloy with Pt but not with W. For example, lithium shows an abnormally low reduction potential at Pt because of alloying. Similar difficulties are observed with several other low melting metals, such as Cd, Pb and Zn and Al (18).

Some disagreements have arisen as to the shapes of rising portions of voltammetric waves for metal deposition of solid electroes. For example, Panchenko (25) reported S-shaped symmetrical curves usually designated as Heyrovsky-Ilkovic curves for the deposition of Ag, Pb, and Cd on Pt from molten KCl, whereas Laitinen et al. (22) and Maricle and Hume (24) reported the unsymmetrical curves expected for the deposition of a product at constant activity. It appears that the apparent symmetry, which is a non-steady state phenomenon, can be caused by impurities in the melt and also by alloying of the plated metal with the electrode material. For example, Gaur and Behl (26) and Delimarskii and Kuz'movich (27) found both types of behavior depending on the metal being deposited.

Cyclic voltammetry was introduced relatively late because its extreme sensitivity places severe demands upon melt purity, especially at the high scan rates required to avoid convection effects at stationary electrodes. The 1959 work of Johnson (28) was limited to relatively low scan rates, while recent work in our laboratory (29) and elsewhere has been successful up to scan rates of at least 10 volts/second.

Chronopotentiometry is an important molten salt technique because it can be used with electrodes of relatively large areas, such as simple flag electrodes without an insulating seal. By using current-reversal chronopotentiometry, preliminary diagnostic work to determine whether the electrode reaction product is soluble or insoluble, and whether the electrode reaction is reversible or irreversible has proven to be convenient, especially for complex reactions such as the reduction of chromate (30). The important restriction of a short transition time to avoid convective disturbance of the diffusion layer was established in 1957 by Laitinen and Ferguson (31) who evaluated diffusion coefficients of several metal ions in LiCl-KCl at 450°C. In 1963, Stromatt (32) used the same technique in NaCl-KCl melts at 716°C to study the reduction of uranium (VI).

Coulometry represents a simple method of quantitative additions to molten salt systems once the electrochemical processes have been established. For example, Laitinen and Liu (14) studied the Nernst equation behavior of a number of metals against their ions by generating the ions coulometrically to form a series of solutions of increasing concentration.

Coulometric titrations can be conveniently carried out by electrolytic generation of a reagent and following the course of the titration potentiometrically or amperometrically. In 1958, Laitinen and Bhatia (33) generated iron(III) as an oxidant for Cr(II) or V(II) in LiCl-KCl and followed the titration curves potentiometrically, amperometrically or biamperometrically. Of the three methods, the biamperometric method proved to be most sensitive. It happens that iron (III) chloride is relatively volatile and that the coulometric method therefore has distinct advantages over the use of a standard solution.

The rotating disk electrode (RDE) was used in LiCl-KCl melts by Delimarskii et al. (34) as early as 1960. The RDE, although more complex mechanically than stationary electrodes, is an "absolute" method which offers an approach to evaluating diffusion coefficients.

In 1955, Laitinen and Osteryoung (35) reported on impedance measurements on platinum electrodes in dilute solutions of metal ions in LiCl-KCl. Platinum(II) showed nearly the behavior expected for a reversible electrode process, while for Co(II) and Ni(II) some anomalous behavior qualitatively attributed to adsorption and underpotential deposition was observed. In 1957, Laitinen and Gaur (36) refined the measurements and used a correction for excessive admittance due to adsorption to estimate exchange currents. In 1960, Laitinen, Tischer, and Roe (18) evaluated exchange currents for several metal ions using a voltage step method and a double pulse galvanostatic method. Again, some anomalies attributable to ion adsorption were observed.

Zajicek and Hubbard (37) appear to have been the only investigators who have used thin layer electrochemistry techniques in molten salts. They successfully studied the Cr(III)/Cr(II) and Pt(II)/Pt couples in LiCl-KCl melts at 450°C and Ag(I)/Ag in nitrate melts. Evidently because of the experimental difficulties of this technique it has not been pursued by others.

Chloroaluminate Melts

Quite a variety of electroanalytical techniques have been used in chloroaluminate melts, dating back to 1942, when Yntema et al. (38) determined deposition potentials for a number of metal ions. Polarography with the dropping mercury electrode was described by Saito et al. in 1962 (39), and numerous papers on steady state voltammetry using platinum indicator electrodes have appeared since the work of Delimarskii et al. (40) in 1948. More advanced techniques have included linear sweep and cyclic voltammetry chronopotentiometry, and chronoamperometry.

Although mixtures of aluminum chloride and alkali metal chlorides are attractive from the viewpoint of moderate melting points, early work in these solvents was plagued by difficulties due to lack of effective purification procedures and a lack of understanding of the acid-base properties of the solvents. Beginning in the late 1960's, removal of the last traces of water and of nobler metals by means of metallic aluminum yielded melts with low residual currents (41). Acid-base equilibria involving chloride ion donor-acceptor reactions were studied in 1968 by Tremillon and Letisse (42) by means of a potentiometric titration. Later electrochemical studies by Torsi and Mamantov (43) and by Osteryoung et al. (44) elucidated the acid-base equilibria involved in chloroaluminate melts containing an excess or deficiency of aluminum chloride over a wide range of temperatures.

In 1974, Gilbert, Brotherton, and Mamantov (45) used chronopotentiometry in chloroaluminate melts to demonstrate the formation of a poorly conducting layer of Al_2Cl_6 at the surface of an aluminum anode and to demonstrate the reversible behavior of the aluminum electrode.

Mamantov et al. (46) studied the oxidation and reduction of sulfur, selenium, and iodine beginning in 1975, and later suggested sulfur(IV) as an oxidant in molten salt batteries. In 1980, Mamantov, Norvell, and Klatt carried out what appear to be the first spectroelectrochemical experiments in chloroaluminate melts using optically transparent electrodes to observe absorption spectra of species formed at the electrode (47).

Nitrate Melts

Nitrate melts are experimentally convenient because of their low melting points and they were historically among the earliest for electroanalytical measurements beyond potentiometry, but they have proved to be of lesser importance than chloride or chloroaluminate melts.

The dropping mercury electrode was used in $LiNO_3$-KNO_3 and $LiNO_3$-NH_4NO_3-NH_4Cl melts at 160°C by Nachtrieb and Steinberg (48) as early as 1948. Its use, of course, is limited to relatively low-melting systems. Conventional polarograms were observed for several metal ions. In 1962, application of conventional polarography was made by Christie and Osteryoung (49) in $LiNO_3$-KNO_3 eutectic at 160°C to study formation constants of chlorocomplexes.

Stationary electrode voltammetry was used in nitrate melts as early as 1948 by Lyalikov and Karmazin (50), using a platinum microelectrode in the form of a "dipping" electrode with bubbles of an inert gas to produce a periodically fluctuating current. A similar electrode was used by Flengas (51) and by Bockris, et al. (52) in 1956. In 1960, Hills, Inman and Oxley (53) described an improved version of the dipping electrode, which of course is limited by relatively ill-defined mass transport conditions. Later work has involved linear sweep voltammetry as described by Hills and Johnson (54) in 1961 or steady state voltammetry with stationary electrodes by Swofford and Laitinen (55) in 1963.

Jordan and coworkers, beginning in 1959 (56) performed several thermochemical titrations in nitrate melts including precipitation of silver halides and silver chromate, measuring the temperature change in an adiabatic system. Zambonin and Jordan (57) used voltammetry to study oxygen and peroxide species in 1967.

The earliest report of the use of AC impedance measurements for measurement of electrode kinetics in melts appears to be that of Randles and White (58), who in 1955 measured charge transfer rate constants for nickel ions in nitrate melts and determined the effects of added moisture.

Interest in nitrate melts has diminished in recent years due to the emergence of other low melting systems.

Fluoride Melts

Although molten fluorides have long been important industrially, electroanalytical measurements have been relatively slow to emerge because of the experimental difficulties of purification and handling these melts.

Grjotheim (59) in 1957 reported measurements of the electrode potentials of metal-metal ion couples of NaF-KF at 850°C using Ni(II)/Ni as a reference electrode. This reference electrode has been used in most subsequent work. Beginning in 1963, Manning and Mamantov (60) made several steady state and linear sweep voltammetric studies in LiF-NaF-KF eutectic (mp 454°C). In 1965, Senderoff, Mellors, and Reinhart (61) used the same solvent to study the chronopotentiometry of tantalum (V) at 650-850°C. In 1970 Jenkins, Mamantov and Manning, (62) measured several redox couples against Ni(II)/Ni as a reference electrode in LiF-NaF-KF and LiF-BeF_2-ZrF_4 at 500°C. The latter solvent, with added U(IV)-U(III), was of importance in the molten salt nuclear reactor and was later the subject of a number of electroanalytical studies, using chronopotentiometry and chronoamperametry as well as cyclic voltammetry and square wave voltammetry (63). Deserving special mention is the use of a lanthanum fluoride single crystal of the type used for ion selective electrodes as a separator for the Ni(II)/Ni reference electrode for the virtual elimination of liquid junction potentials in fluoride melts, described by Bronstein and Manning in 1972 (64) and used by Clayton, Mamantov, and Manning in several electro-analytical studies in fluoride and fluoroborate melts (65).

The use of a rotating disk electrode in LiF-NaF-KF was attempted by Senderoff and Mellors in a study of the electrodeposition of refractory metals (66). The problem of insulation was approached by use of a pressure fitted boron nitride mounting for a tungsten electrode, but the RDE was abandoned in favor of chronopotentiometric measurements with stationary electrodes, which proved to be more practicable because of experimental simplicity. Recently, Tellenbach and Landolt (67) successfully used a similar RDE assembly in molten cryolite at 1020K, plating the tungsten electrode with gold or titanium boride.

Ambient Temperature Molten Salts

The first electroanalytical studies in molten salts at room temperature appear to have been made by Osteryoung et al. in 1975 (68), although the solvent derived from the early work of Hurley and Wier in 1951 (69), who studied mixtures of various N-substituted pyridinium halides with inorganic halides. In particular, they observed the electrodeposition of aluminum from a mixture of aluminum chloride and ethyl pyridinium bromide, which form a eutectic melting at -40°C. Gale and Osteryoung (70) in 1979 used potentiometry with an aluminum electrode to make acid-base studies of a room temperature melt of butylpyridinium choride-aluminum chloride mixtures. More recent work in this melt by Osteryoung et al. (71) has included rotating disk and ring disk electrodes and cyclic voltammetry with stationary electrodes to study the Fe(III)/Fe(II) system and by Mamantov et al. (72) for cyclic voltammetry and for Raman spectroscopic studies of sulfur species. This work has extended the field to applications in ambient temperature molten salt batteries. A related solvent, containing also antimony chloride, has been used by Mamantov et al. for cyclic voltammetric and spectrophotometric studies of the redox chemistry of hydrocarbons such as anthracene and perylene (73). Another ambient temperature melt system consisting of organic cation tetrachloroborates has recently been described by Mamantov et al. (74).

Miscellaneous Melts

In 1963, Caton and Freund (75) made preliminary voltammetric studies of a number of redox systems in molten alkali metal metaphosphates, but the work was limited by the lack of a reference electrode. In 1971 Wolfe and Caton (76) devised a silver reference electrode which was applied to potentiometric and chronopotentiometric studies in equimolar mixtures of sodium and potassium metaphosphates at 700°C of several redox couples. Diffusion coefficients and standard potentials were evaluated.

A limited amount of electroanalytical work has been done in sulfate melts. The oxygen electrode has been mentioned above for alkaline melts. In 1961, Liu (77,78) applied potentiometry, chronopotentiometry and coulometric titrations in lithium sulfate-potassium sulfate melts at 625°C using a silver reference electrode. An electromotive force series of limited scope and voltammetric curves were reported in 1963 by Johnson and Laitinen (79) in a ternary alkali metal sulfate system at 550°C. A complication is the reduction of the sulfate ion at negative potentials.

Melts containing magnesium chloride and alkali metal chlorides have received study going back to 1950 (80). An electromotive force series has been described by Gaur and Behl (26).

Conclusion

Electroanalytical techniques have made significant contributions over the years to our knowledge of the behavior of molten salt systems at temperatures ranging from room temperature to over 1000°C. A wide variety of melts have been investigated both from the viewpoint of fundamental studies of sensors and electrode processes and from the viewpoint of practical applications to battery systems, electroplating, and preparative electrochemistry.

References

1. Baur, E.; Ehrenberg, H., Z. Elektrochem., 1912, 18, 1002-1011.
2. Treadwell, W.D. Z. Elektrochem., 1916, 22, 414-421.
3. Lux, H. Z. Elektrochem., 1939, 45, 303-309; 1948, 52, 220-224; 1949, 53, 41-43.
4. Flood H.; Forland, T. Acta Chem. Scand., 1947, 1, 592-604; Discussions Faraday Soc., 1947, 1, 302-307.
5. Flood, H.; Forland, T.; Motzfeldt, K. Acta Chem. Scand., 1952, 6, 257-269.
6. Antipin, L.N. Zh. Fiz. Khim., 1955, 29, 1668-1677.
7. Rose, B.A.; Davis, G.J.; Ellingham, H.J.T. Discussions Faraday Soc., 1948, 4, 154-162.
8. Hill, D.G; Porter, B.; Gillespie, Jr., A.S. J. Electrochem. Soc., 1958, 105, 408-412.
9. Laitinen, H.A.; Bhatia, B.B. J. Electrochem. Soc., 1960, 107, 705-710.
10. Bauerell, J.; Ruka, R. Paper presented at Pittsburgh meeting of the Electrochemical Society, 1963.
11. Bartlett, H.E.; Johnson, K.E.; J. Electrochem. Soc., 1967, 114, 64-67.
12. Laitinen, H.A.; Ferguson, W.S.; Osteryoung, R.A. J. Electrochem. Soc., 1957, 104, 516-520.
13. Osteryoung, R.A. Ph.D. Thesis, University of Illinois, Illinois, 1954.
14. Laitinen, H.A.; Liu, C.H. J. Am. Chem. Soc., 1958, 80, 1015-1020.
15. Laitinen, H.A.; Plankey, J.W. J. Am. Chem. Soc., 1959, 81, 1053-1058.

16. Laitinen, H.A.; Plambeck, J.A. J. Am. Chem. Soc., 1965, 87, 1202-1206.
17. Plambeck, J.A.; Elder, J.P.; Laitinen, H.A. J. Electrochem. Soc., 1966, 113, 931-937.
18. Laitinen, H.A.; Tischer, R.P.; Roe, D.K. J. Electrochem. Soc., 1960, 107, 546-555.
19. Flengas, S.N.; Ingraham, T.R. J. Electrochem. Soc., 1959, 106, 714-721.
20. Delimarskii, Y.K.; Markov, B.F. Electrochemistry of Fused Salts, translated by A. Peiperl, Sigma Press, Washington, D.C., 1961.
21. Laity, R. in Reference Electrodes, Janz, G.; Ives, B., Eds.; Academic: New York, 1961, p. 524-606.
22. Laitinen, H.A.; Liu, C.H.; Ferguson, W.S. Anal. Chem., 1958, 30, 1266-1270.
23. Black, E.D.; DeVries, T. Anal. Chem., 1955, 27, 906-909.
24. Maricle, D.L.; Hume, D.N., Anal. Chem., 1961, 33, 1188-1192.
25. Panchenko, I.D. Ukr. Khim. Zh., 1955, 21, 468-471; 1956, 22, 153-155.
26. Gaur, H.C.; Behl, W.K. J. Electroanal. Chem., 1963, 5, 261-269.
27. Delimarskii, Y.K.; Kuz'movich, V. Zh. Neorgan. Khim, 1959, 4, 2732-2738.
28. Johnson, K.B.; Ph.D. Thesis, University of London, 1959.
29. Waggoner, J.R.; unpublished experiments, 1982.
30. Propp, J.H.; Laitinen, H.A. Anal. Chem., 1969, 41, 644-649.
31. Laitinen, H.A.; Ferguson, W.S. Anal. Chem., 1957, 29, 4-9.
32. Stromatt, R.W. J. Electrochem. Soc., 1963, 110, 1277-1282.
33. Laitinen, H.A.; Bhatia, B.B. Anal. Chem., 1958, 30, 1995-1997.
34. Delimarskii, Y.K.; Panchenko, I.D.; Shilina, G.Y. Dopovidi Akad. Nauk. Ukr. R.S.R., 1961, 2, 205-208.
35. Laitinen, H.A.; Osteryoung, R.A. J. Electrochem. Soc., 1955, 102, 598-605.
36. Laitinen, H.A.; Gaur, H.C. J. Electrochem. Soc., 1957, 104, 730-737.
37. Zajicek, L.P.; Hubbard, A.T. J. Electrochem. Soc., 1969, 116, 80C-82C.
38. Verdieck, R.G.; Yntema, L.F., J. Phys. Chem., 1942, 46, 344-352.
39. Saito, M.; Suzuki, S.; Goto, H. Nippon Kagaku Zasshi, 1962, 83, 883-886.
40. Delimarskii, Y.K.; Skobets, E.M.; Berenblyum, L.S. Zh. Fiz. Khim., 1948, 22, 1108-1115.
41. Anders, U.; Plambeck, J.A. Can. J. Chem., 1969, 47, 3055-3060.
42. Tremillon, B.; Letisse, G. J. Electroanal. Chem., 1968, 17, 371-386.
43. Torsi, G.; Mamantov, G. Inorg. Chem., 1971, 10, 1900- ; 1972, 11, 1439.
44. Boxall, L.G.; Jones, H.L.; Osteryoung, R.A. J. Electrochem. Soc., 1973, 120, 223-231.
45. Gilbert, B.; Brotherton, D.L.; Mamantov, G. J. Electrochem. Soc., 1974, 121, 773-777.
46. Marassi, R.; Mamantov, G.; Chambers, J.Q. Inorg. Nucl. Chem. Lett., 1975, 11, 245-252.

47. Mamantov, G.; Norvell, V.E.; Klatt, L. J. Electrochem. Soc., 1980, 127, 1768-1772.
48. Nachtrieb, N.H.; Steinberg, M. J. Am. Chem. Soc., 1948, 70, 2613-2614; Steinberg, M.; Nachtrieb, N.H. J. Am Chem. Soc., 1950, 72, 3558-3565.
49. Christie, J.H.; Osteryoung, R.A., J. Am. Chem. Soc., 1960, 82, 1841-1844.
50. Lyalikov, Y.S.; Karmazin, V. Zavodsk. Lab., 1948, 14, 138-143; 144-148.
51. Flengas, S.N. J. Chem. Soc., 1956, 534-538.
52. Bockris, J.O'M.; Hills, G.J.; Menzies, I.A.; Young, L. Nature, 1956, 178, 654.
53. Hills, G.J.; Inman, D.; Oxley In Advances in Polarography, Langmeier, G; Ed., Vol. III, p. 982, Pergamon, N.Y.
54. Hills, G.J.; Johnson, K.E. J. Electrochem. Soc., 1961, 108, 1013-1018.
55. Swofford, H.S.; Laitinen, H.A. J. Electrochem. Soc., 1963, 110, 814-820.
56. Jordan, J.; Meier, J.; Billingham, Jr., E.J.; Pendergrast, J. Anal. Chem., 1959, 31, 1439-1440; 1960, 32, 651-655.
57. Zambonin, P.G.; Jordan, J. J. Am. Chem. Soc., 1967, 89, 6365-6366; 1969, 91, 2225-2228.
58. Randles, J.E.B.; White, W. Z. Elektrochem., 1955, 59, 666-671.
59. Grjotheim, K.; Z. Physik. Chem. (N.F.), 1957, 11, 150-164.
60. Manning, D.L. J. Electroanal. Chem., 1963, 6, 227-233; Manning, D.L.; Mamantov, G. ibid, p. 328-329.
61. Senderoff, S.; Mellors, G.W., Reinhart, W.J. J. Electrochem. Soc., 1965, 112, 840-845; Mellors, G.W.; Senderoff, S. ibid, 1966, 113, 60-65, 66-71; 1967, 114, 556-560, 586-587.
62. Jenkins, H.W.; Mamantov, G.; Manning, D.L. J. Electrochem. Soc., 1970, 117, 183-185.
63. Manning, D.L. J. Electroanal. Chem., 1963, 6, 227-233; Manning, D.L.; Mamantov, G. ibid, p. 328-329.
64. Bronstein, H.R.; Manning, D.L. J. Electrochem. Soc., 1972, 119, 125-127.
65. Clayton, F.R.; Mamantov, G.; Manning, D.L. J. Electrochem. Soc., 1973, 120, 1199-1201.
66. Senderoff, S.; Mellors, G.W., unpublished experiments.
67. Tellenbach, J.M.; Landolt, D. Electrochim. Acta, 1988, 33, 221-225.
68. Chum, H.L.; Koch, V.R.; Miller, L.L.; Osteryoung, R.A. J. Am. Chem. Soc., 1975, 97, 3624-3625.
69. Hurley, F.H.; Wier, Jr., T.P. J. Electrochem. Soc., 1951, 98, 203-206.
70. Gale, R.J.; Gilbert, B.; Osteryoung, R.A. Inorg. Chem., 1979, 18, 2723-2725.
71. Nanjundiah, C.; Shimizu, K.; Osteryoung, R.A. J. Electrochem. Soc., 1982, 129, 2474-2480.
72. Marassi, R.; Laher, M.; Trimble, D.S.; Mamantov, G. J. Electrochem. Soc., 1985, 132, 1539-1543.
73. Chapman, D.M.; Smith, G.P.; Serlie, M.; Petrovic, C.; Mamantov, G.; J. Electrochem. Soc., 1984, 131, 1609-1614.
74. Williams, S.D.; Schoebrechts, J.P.; Selkirk, J.C.; Mamantov, G. J. Am Chem. Soc., 1987, 109, 2218-2219.

75. Caton, Jr., R.D.; Freund, H. Anal. Chem., 1963, 35, 2103-2108.
76. Wolfe, C.R.; Caton, Jr., R.D. Anal. Chem., 1971, 43, 660-663.
77. Liu, C.H. Anal. Chem., 1961, 33, 1477-1479.
78. Liu, C.H. J. Phys. Chem., 1962, 66, 164-166.
79. Johnson, K.E.; Laitinen, H.A. J. Electrochem. Soc., 1963, 110, 314-318.
80. Rempel, S.I., Dokl. Acad. Nauk SSSR, 1950, 74, 331-333.

RECEIVED August 9, 1988

Chapter 29

Electrodeless Conductivity

Truman S. Light[1]

The Foxboro Company, Corporate Research Center (N01-2A), Foxboro, MA 02035

The history of electrodeless conductivity is discussed from its practical beginning in 1947 to the present. This technique for measuring the concentration of electrolytes in solution utilizes a probe consisting of two toroids in close proximity which are immersed in the solution. The toroids may also be mounted externally on insulated pipes carrying the solution. One toroid generates an alternating electric field in the audio frequency range and the other acts as a receiver to pick up the current from the ions moving in a conducting loop of solution. Fouling coatings, suspensions, precipitates or oil have little or no effect. Applications are reviewed for continuous measurements include the mining, pulp and paper, and heavy chemical industries.

The measurement of electrical conductivity of ions in solution has long provided useful quantitative chemical composition information. As a tool for physical chemical studies, it is used for determining the degree of dissociation of weak electrolytes and their ionization constants. Similarly, it is used for the study of precipitation and complex formation reactions and for determination of solubility product and formation constants. Analytical chemistry applications include direct quantitative analysis of strong and weak acids, bases and salt solutions, aqueous and nonaqueous conductometric titrations, and a variety of continuous monitoring determinations such as oceanographic salinity, aluminum and pulp industry processing liquors, pickling, plating, anodizing and degreasing baths and chromatographic detectors (1-3).

The classical technique measures the electrical resistance (or its reciprocal, the conductance) between two inert conducting electrodes contacting the solution. Low frequency alternating current, 10 to 50,000 hertz, is usually employed to minimize

[1]Current address: 4 Webster Road, Lexington, MA 02035

0097-6156/89/0390-0429$06.00/0

electrolytic reactions and polarization at the electrode/solution interface. Physical chemical studies of conductivity and its theory and practice are discussed in many standard reference works (1-9). The chief difficulty with this measurement is associated with electrodes contacting the solution. In addition to the more subtle errors of electrode/solution capacitance and polarization, coatings are an obvious and serious problem. Insulating or diffusion hindering layers may be formed from oils, bodily fluids, metal precipitates and waste streams. Coating is especially prevalent in alkaline solutions where heavy metal anions such as phosphates, carbonates, hydroxides and sulfates, are precipitated, and in suspensions such as latex and the liquors of the pulp and paper industry.

There are two frequency domains in which conductance measurement have been made without electrodes in contact with the solution. The first is a high-frequency method in the megahertz region. The electrodes take the form of a pair of metal sheets or bands on the outside of the sample cell, which is made of an insulating material such as glass. Alternatively, the glass encased sample may be placed inside an induction coil which is part of the circuit. The glass plays the part of a dielectric in a capacitor. The impedance of a capacitor is so low at high frequency that the alternating current passes freely into the sample and the impedance of the solution becomes a complex function of the resistance of the solution, its dielectric constant and the capacitance of the circuit. High-frequency conductometry, also called oscillometry, has been treated extensively by Blake (10), Sherrick et al (11) and Pungor (12) and appears to be seldom used now. The most recent paper known to the author was concerned with the design of a high-frequency oscillometer and appeared in 1981 (13). Oscillometry will not be discussed further in this paper.

A second method of measuring the conductance without the use of contacting electrodes has become popular, especially in the chemical process industries. Usually referred to simply as "electrodeless conductivity", it has also been called "inductive" or "magnetic" conductivity. This method is the subject of this paper and is described below.

Although instruments for electrodeless conductivity measurement have been commercially available since the 1950's for process industry applications, literature review of this subject is lacking. This paper will review and discuss the history and applications of electrodeless conductivity.

Instrumentation

The electrodeless conductivity measuring system utilizes a probe consisting of two encapsulated toroids in close proximity to each other, as shown in Figure 1. One toroid generates an electric field in the solution, while the other acts as a receiver to pick up the small alternating current induced in the electrolytic solution as illustrated in Figure 2. The equivalent electrical circuit may be compared to a transformer with the toroids forming the primary and secondary windings and the core replaced by a coupling loop which is the conducting solution.

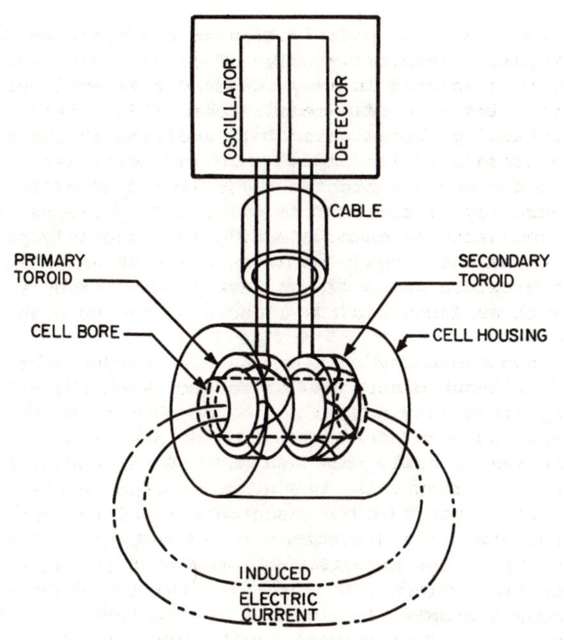

Figure 1. Principle of the electrodeless conductivity cell and instrument.

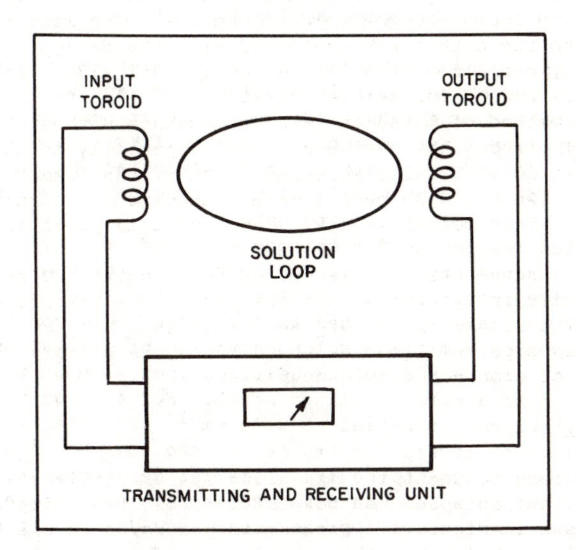

Figure 2. Simple representation of an electrodeless conductivity measuring circuit.

Several configurations for the sensing cells are possible. Cells designed for immersion and available in several sizes are shown in Figure 3. The toroids are covered with a chemically resistant fluorocarbon or other high-temperature resistant non-conducting thermoplastic material. Any precipitates or coatings adhering to this probe have little or no effect on the measured conductance as long as they do not displace a significant fraction of the solution. The generating toroid is energized from a stable audio-frequency source, typically at 20 kHz. The pick-up toroid is connected to a receiver which measures the current through this secondary winding. The current is then amplified and displayed on a meter or an analog strip-chart recorder. It may also be used to control a valve or an alarm. The output is a direct function of the conductance of the solution in the loop, in a manner completely analogous to the traditional measurement with contacting electrodes.

Range and Temperature Effects. The useful range of commercially available instruments extends from 0 to 100 µS/cm to 0 to 2 S/cm, with relative accuracy of a few tenths of one percent of full-scale. (The siemens, S, is the SI unit for conductance and is identical with the "mho" or the reciprocal ohm.) Conductance is temperature dependent and a temperature sensor is incorporated in the toroid probe with a compensation circuit which corrects to the standard reference temperature of 25°C. Many salts have conductivity temperature coefficients of the order of 2 percent per degree Celsius. This temperature coefficient is non-linear and in some cases may vary from 2 to 7 percent per degree Celsius over a 100 degree range (14). Using a microprocessor-based electrodeless conductivity instrument, compensation for this non-linearity may be provided (15).

Limitations. The electrodeless conductivity technique using low-frequency inductive cells is available for analysis and control in the chemical process industries and in other continuous monitoring applications. Although its stability, freedom from maintenance, and accuracy, are superior to contacting conductivity techniques, lack of bench models of this type has hindered its laboratory use and application to date.

One of the reasons that electrodeless conductivity is not favored as a laboratory tool is due to the size of the probes and the sample size requirement. The smallest electrodeless probe is about 3.6 cm in diameter and has an equivalent cell constant of 2.5 cm^{-1}. It requires a minimum solution volume of several hundred milliliters to ensure a complete solution loop without wall effects which distort the apparent cell constant. For a large probe of 8.9 cm diameter, the cell constant is 0.45 cm^{-1} and an effective solution volume of several liters may be needed. For the lower conductance ranges, which require a smaller cell constant, the diameter of the probe and measuring container must increase. Accurate measurement below approximately 10 µS/cm is not practical.

History

Patent Literature. In 1947, Matthew Relis at the Massachusetts Institute of Technology wrote a thesis titled "An Electrodeless Method for Measuring the Low-Frequency Conductivity of Electrolytes" (16). He had begun his work at the Naval Ordnance Laboratory,

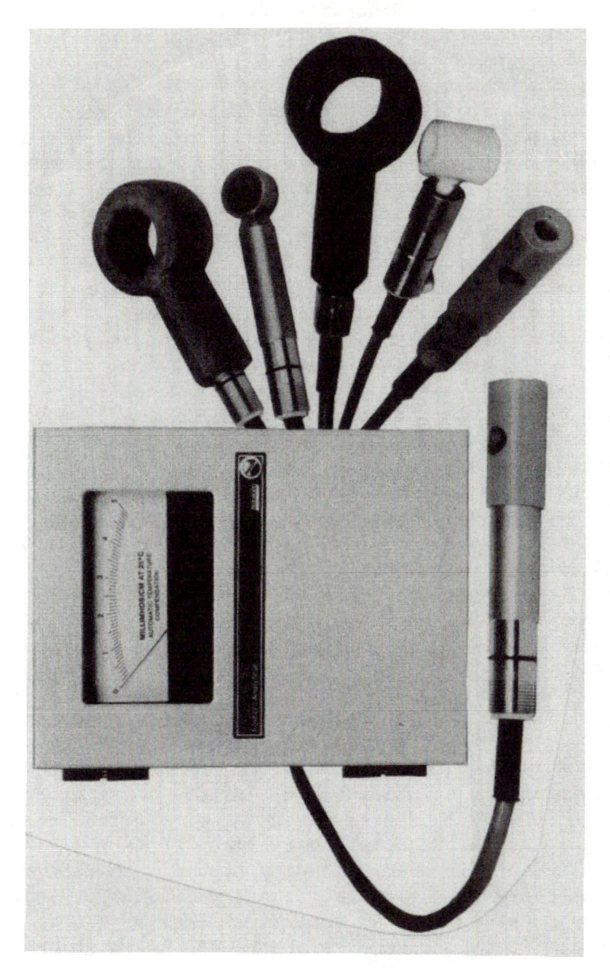

Figure 3. Various types of immersion cells for electrodeless conductivity with an associated instrument (Courtesy of The Foxboro Co.)

Washington, D.C. and subsequently continued it at the Woods Hole Oceanographic Institution in Massachusetts where an instrument based on his work was used as an in situ salinometer. He described an electrodeless conductivity instrument which was based on the immersion of coupled generating and pickup toroids in an electrolyte solution and used audio frequency alternating current. Figure 4, illustrating this principle, is taken from the patent titled "Method and Apparatus for Measuring the Electrical Conductivity of a Solution" that was subsequently granted to Relis in 1951 (17). The patent was licensed to Industrial Instruments, Inc. which was acquired by Beckman Instruments in 1965.

Relis described the principle and an apparatus that measured conductivity with an accuracy of better than 1% over the conductance range of 10 micromhos to 10 mhos (10 microsiemens to 10 siemens). Within experimental error, the electrodeless method yields data identical to that obtained via the conventional contacting method of measuring conductivity but without the interference caused by minor deposits on or polarization of the contacting electrodes. Relis reported that the only earlier description of the principle of this method was given in an account of experiments conducted in 1920, "Demonstration of Induction Currents Produced in Electrolytes without Electrodes", at the Federal Polytechnical School of Switzerland by Piccard and Frivold (18). However, a patent was issued to Ruben in 1927 titled "Electrochemical Sensing Device" based on similar principles (19). Stock (20), in discussing "Two Centuries of Quantitative Electrolytic Conductivity, 1776-1879-1984", has noted that electrodeless conductivity measurements were made much earlier by inducing current in solution from adjacent magnets. These measurements were reported in 1860 by Beetz (21) and in 1880 by Guthrie and Boys (22). The latter workers suspended a solution in a vessel from a torsion wire in the field of a strong magnet and measured the force required to overcome the rotation of the vessel.

Following the Relis patent, several workers and instrument companies patented improvements on electrodeless conductivity equipment (23-32). In 1955, Fielden received a patent which was assigned to the Robertshaw-Fulton Controls Company (23). He described a method for mounting the toroids on a nonconducting pipe external to the electrolyte solution to be tested. This method, illustrated in Figure 5 by the Sperry patent (24), required that the solution be in a loop so that complete electrical contact is maintained between the primary and secondary toroids. Although seemingly an attractive way to measure process solutions, this method has not been extensively used because it created the equivalent of very large cell constants due to the large solution loop path and because there was the possibility of loss of some of the generated current in the pipelines and ground paths of the process streams. This latter problem was addressed in 1968 by patents issued to Sperry (24), Kidder (25) and Rosenthal (26) and assigned to Beckman Instruments. Gross (27-28) in patents assigned to Balsbaugh Laboratories and The Foxboro Company eliminated anomalies in the permeance of the cores of the toroids. Then the only coupling is effected by the fluid loop and results in greater sensitivity and improved linearity. A design for a high temperature and pressure electrodeless conductivity probe is described and

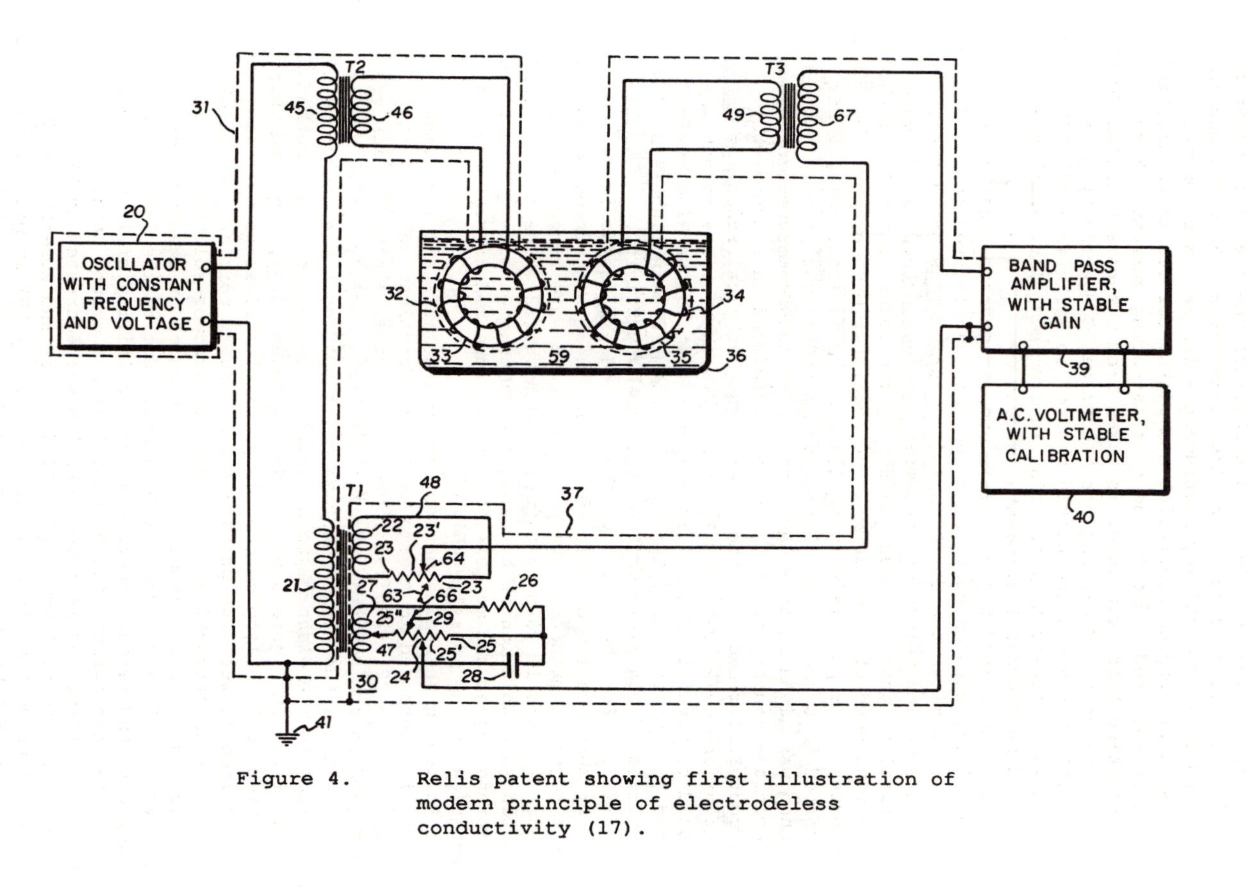

Figure 4. Relis patent showing first illustration of modern principle of electrodeless conductivity (17).

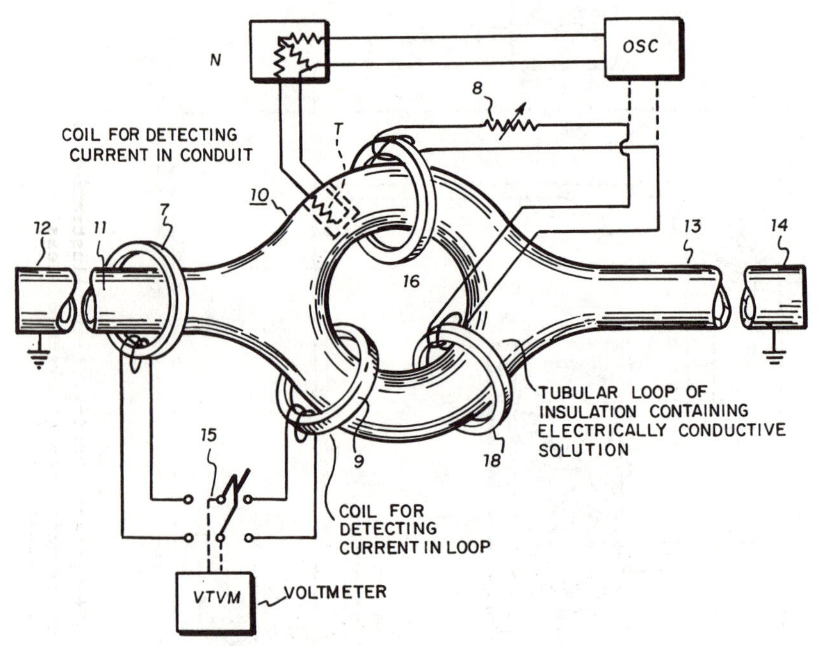

Figure 5. Sperry patent illustrating electrodeless conductivity cells externally mounted to the solution being tested (24).

patented by Koski (29-30). The employment of a pair of electrodeless conductivity probes for differential measurement as part of a continuous process feed forward control technique for washing a slurry stream is described in a patent by Rosenberger (31) assigned to Nalco Chemical Co. A German patent issued to Diebel (32) also discusses an electrodeless conductivity apparatus for determination of electrolyte concentration.

Instrumentation Literature. Starting in 1956, papers describing various aspects of electrodeless conductivity instrumentation have appeared. Gupta and Hills (33) described a transformer bridge circuit which was also discussed by Calvert et al (34) and Griffiths (35). Other papers modifying and discussing the instrumentation aspects have appeared (36-48). Johnson and Hart have described a simplified analyzer (41). Another analyzer capable of measuring dilute solutions of 10 µS/cm with a precision of ±2% has been described (42). The subject of electrodeless conductivity was reviewed in 1969 and 1971 (49-50), and discussed briefly in a few books (1-3, 51). Modern instrumentation using a microprocessor-based electrochemical monitor which provides temperature compensation, curve characterization and calibration, flexible ranging, output expansion and damping and self-diagnostic capability has been described by Queeney and Downey (15).

Application Literature. An excellent dissertation by Martin , "Electrodeless Conductance Measurements Using Toroidal Inductors", (50) reviewed the theory and applications of electrodeless conductivity to 1971. This thesis covers variables affecting the inductive response mechanism, useful response of the electrodeless conductivity system to the micromolar level, and extension to situations where small conductance changes are determined in the presence of large amounts of foreign electrolytes. Conductometric titrations of a redox reaction with small conductivity change in the presence of a high conductivity acid background were demonstrated. Conductometric titrations in the low microsiemens/cm region were also shown.

The earliest application of electrodeless conductivity measurements appears to have been for measurement of salinity at various ocean depths (16, 17, 45, 52). Other early uses have also included determination of the equivalent conductances of salts at high concentrations (33, 34, 46, 53) and the monitoring of nitric acid concentration in radioactive waste (41). Attempts to extend response to solutions of low conductance seem to have met with doubtful response (47-48). As commercial availability of electrodeless conductivity equipment increased, this technique became accepted for the hostile environments of the chemical process industries. Corrosive, elevated temperature, particle and oil or grease laden solutions with suspended solids and fibers, did not interfere with the measurement as they might with contacting conductivity.

Applications in the mining and metallurgical extraction industries were reported starting in 1960 (37,54,55) and in control of lime slurry and alkaline processes in the pulp and paper industry at about the same time (44,56-62). Sulfuric acid and oleum have had special attention (63, 64) and a measurement system for oleum described by Shaw and Light is shown in Figure 6 (64).

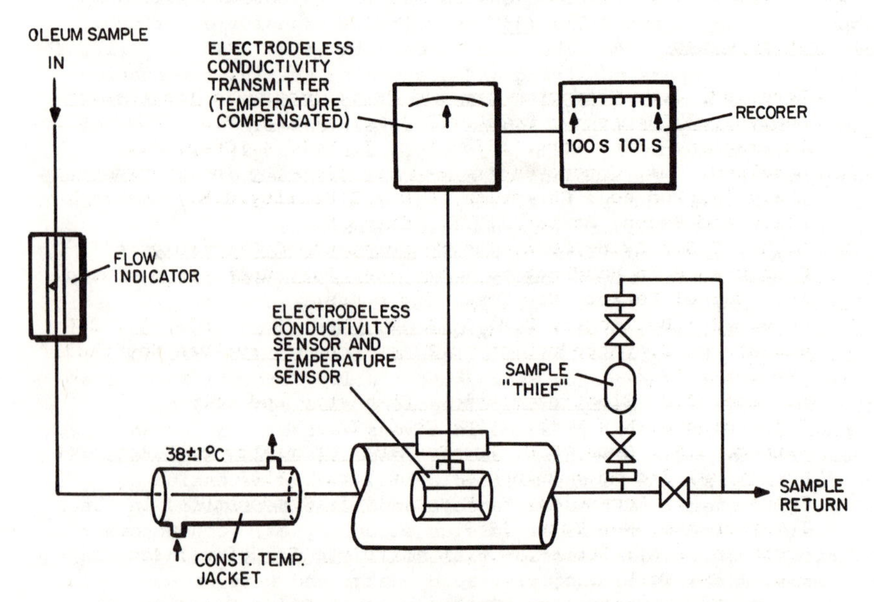

Figure 6. Continuous analyzer, using electrodeless conductivity, for oleum over the range 100 to 102 ± 0.02 equivalent percent H_2SO_4 (Reproduced with permission from Reference 64. Copyright 1982 ISA Transactions.)

Conclusion

Electrodeless conductivity is a technique for measuring the conductance of a solution using the electrical inductance principle at low frequencies. This method does not use contacting electrodes thereby eliminating maintenance and other errors due to surface effects created by coatings and fouling. The measurement enjoys widespread acceptance in the chemical process industries. Its history and a review of the literature have been presented here.

Literature Cited

1. Loveland, J.W. Conductometry and Oscillometry in Treatise on Analytical Chemistry; Kolthoff, I.M.; Elving, P.J.; Eds.; Wiley Interscience: New York, 1963; Part I, Vol. 4, Chap. 51.
2. Loveland, J.W. Conductometry and Oscillometry in Instrumental Analysis, 2nd ed.; Christian, G.D.; O'Reilley,J.E.; Eds.; Allyn and Bacon: Boston, 1986; Chap. 5.
3. Light, T.S.; Ewing, G.W. The Measurement of Electrolytic Conductance in Handbook of Analytical Instrumentation; Ewing, Ed.; Marcel Dekker: New York, (in press).
4. Jones, G.; Bollinger, D.M. J. Amer. Chem. Soc. 1935, 57, 280.
5. Glasstone, S. Introduction to Electrochemistry; Van Nostrand: New York, 1942.
6. Lingane, J.J. Electroanalytical Chemistry 2nd ed.; Interscience: New York, 1958; Chap. IX.
7. Harned. H.S.; Owen, B.B. The Physical Chemistry of Electrolytic Solutions, 3rd ed.; Reinhold: New York, 1958.
8. Fuoss, R.M.; Accascina, F. Electrolytic Conductance; Interscience: New York, 1959.
9. Robinson, R.A.; Stokes, R.H. Electrolyte Solutions; Academic Press: New York, 1959.
10. Blake, G.G.; Conductimetric Analysis at Radio-Frequency; Chemical Publishing Co.: New York, 1952.
11. Sherrick, P.H.; Dawe, G.A.; Karr, R.; Ewen, E.F. Manual of Chemical Oscillometry; E.H. Sargent: Chicago, 1954.
12. Pungor, E. Oscillometry and Conductometry; Pergamon Press: Oxford, 1965.
13. Sher, A.; Yarnitzky, C. Anal. Chem. 1981, 53, 356-358.
14. Light, T.S.; Licht, S.L. Anal. Chem. 1987, 59, 2327-2330.
15. Queeney, K.M.; Downey, J.E. Advances in Instrumentation, 1986, Vol. 41 Part 1, pp 339-352.
16. Relis, M. M.S. Thesis, Mass. Institute of Technology, Cambridge, Mass., 1947.
17. Relis, M. U.S. Patent 2 542 057, 1951.
18. Piccard, A.; Frivold A. (Swiss) Archives des Sciences Physiques et Naturelles, 1920, Ser. 5, Vol. 2, pp 264-265, May-June.
19. Ruben, S. U.S. Patent 1 610 971 1926.
20. Stock, J.T. Anal. Chem., 1984, 56, 561A-570A.
21. Kohlrausch, F.; Holborn, L., Das Leitvermogen der Elektrolyte; Teubner: Leipzig, 1898, p 5.
22. Guthrie, F.; Boys, C.V. Phil. Mag., 1880, 10, 328.
23. Fielden, J.E., U.S. Patent 2 709 785, 1955.
24. Sperry, E.A. U.S. Patent 3 396 331, 1968.

25. Kidder, R.J. U.S. Patent 3 404 335, 1968.
26. Rosenthal, R. U.S. Patent 3 404 336, 1968.
27. Gross, T.A.O., U.S. Patent 3 806 798, 1974.
28. Gross, T.A.O., U.S. Patent 4 220 920, 1980.
29. Koski, O.H. U.S. Patent 3 867 688, 1975.
30. Koski, O.H.; Danielson, M.J., Rev. Sci. Instrum. 1979, 50, 1433-36.
31. Rosenberger, R.R., U.S. Patent 4 096 028, 1978.
32. Diebel, H., German Patent 1 598 075, 1974.
33. Gupta, S.R.; Hills, G.J. J. Sci. Instrum. 1956, 33, 313-314.
34. Calvert, R.; Cornelius, J.A.; Griffiths, V.S.; Stock, D.I. J. Phys. Chem., 1958, 62, 47-53.
35. Griffiths, V.S.; (a) Anal. Chim. Acta 1958, 18, 174; (b) Talanta 1959, 2, 230.
36. Salamon, M. Chem. Techn. (Berlin) 10. Jg. Heft, 1958, 207-210.
37. Eicholz, G.G.; Bettens, A.H. Can. Min. Metall. Bull. 1960, 53, 901-907.
38. Rosenthal, R.; Kidder, R.J. Industrial and Engineering Chem. June, 1961, 53, 55A.
39. Fatt, I. Rev. Sci. Instrum. 1962, 33, 493-494.
40. Williams, R.A.; Gold, E.M.; Naiditch, S. Rev. Sci. Instrum. 1965, 36, 1121-1124.
41. Johnson, C.M.; Hart, G.E. Anal. Instrum. 1967, 4, 23-30.
42. Hackl, A.E.; Deisinger, H. Allgemeine und Praktische Chemie 1968, 19, 229.
43. Gross, T.A.O.; Sawyer, P.B. Measurements & Data Nov.-Dec., 1975, 102.
44. (a) Timm, A.R.; Liebenber, E.M.; Ormrod, G.T.W.; Lombaard, S.L. An Electrodeless Conductivity Meter of Improved Sensitivity and Reliability, report no. 2003, Nat. Inst. Metallurgy, Johannesburg, Feb. 28, 1979; (b) Ormrod, G.T.W., Electrodeless Conductivity Meters in the Measurement and Control of the Amount of Lime in Alkaline Slurries, NIM-SAIMC Symposium on Metallurgical Process Instrumentation, Nat. Inst. Metallurgy, Johannesburg, 1978
45. Brown, N.L.; Hamon, B.V. Deep-Sea Research 1961, 8, 65.
46. Lavagnino, B.; Alby, B. Ann. Chim. 1959, 49, 1272.
47. Lopatnikov, V.I. Soviet Physics (Eng. trans.) 1961, 6, 505.
48. De Rossi, M. Sci. Tec. 1962, 6, 31; Chem. Abstr. 1964, 61, 3754.
49. Pazsitka, L.; Z. Anal. Chem. 1969, 245, 103.
50. Martin, R.A. Ph.D. Thesis, Univ. of Pittsburgh, 1971.
51. Smith, D.E.; Zimmerli, F.H. Electrochemical Methods of Process Analysis; Instrument Society of America: Research Triangle Park, NC, 1972, pp 138-141.
52. Park, K. Anal. Chem. 1963, 35, 1549.
53. Davis, R.L.; Le Master, E.W. Texas Journal of Science, 1972, XXIII, No. 4, 497-501.
54. Kelly, F.J.; Stevens, C.S. Canadian Mining Journal, Jan. 1964, pp 42-43.
55. G.D. Fulford, Use of Conductivity Techniques to Follow Al_2O_3 Extraction at Short Digestion Times, in Light Metals; Bohner, H.O., ed.; The Metallurgical Society of AIME: 1985; pp 265-278.
56. Gow W.A.: McCreedy H.H.; Kelly F.J. Canadian Mining and Metallurgical Bulletin, July, 1965.

57. Moon, A.G.; Vaughn, R.L. Soc. Mining Engineers of AIME, Transactions, 1978, 264, 1727-1730.
58. Farrar, D.; Khandelwal, B. Effective Alkali Measurement Improves Continuous Digester Performance, presented at PAPPI (Pulp and Paper Institute) Conference, Vancouver, Spring, 1983.
59. The Foxboro Co., Foxboro, Mass. 02035, Product Application Data, (a) PAD B2000-001 Electrodeless Conductivity Systems for Clean-In-Place (CIP) Caustic Dilution; b) PAD B2000-007 Electrodeless Conductivity Systems for Interface Detection In Clean-In-Place (CIP) Systems; c) PAD B2000-008 Electrodeless Conductivity Systems for Acid Concentration Control in CIP Systems; d) PAD G2600-006 Reclamation of Black Liquor Spill in the Pulp and Paper Industry; e) AID G2600-007 Digester Alkali Concentration Control; f) PAD G2600-014 Continuous and Batch Kraft Digesters; g) PAD J2200-009 Caustic Concentration Control in Caustic Saturator; h) PAD K2200-011 Controlling the Neutralization of Caustic in Textile Mercerization; i) PAD Q9900-014 Oleum Strength Analyzer System; j) PAD B2030-001 Caustic Control for Vegetable Peeler.
60. Musow, W. On-Line Causticity Sensor and Programmable Monitor Applied to Slaker Lime Addition Control, Canadian Pulp and Paper Assoc., Montreal, 1986.
61. Musow, W.; Bolland, A. Toroidal Conductivity Sensor Technology Applied to Cyanidation of Flotation Tailings Circuits, Instrument Soc. of America 12th Annual Mining and Metallurgy Industries Symposium, Vancouver, 1984.
62. Musow, W.;Montgomery, J. Recausticizing Control Utilizing Toroidal Magnetic Sensor Technology and Controller with Artificial Intelligence Atlanta, Ga., Pulping Conference/TAPPI, Sept., 1985.
63. del Valle, J.L. Measurement of the Concentration of H_2SO_4 Using an Electrodeless Conductivity Method, Technisches Messen Atm. 1977, Vol. 7/8, 263-265.
64. Shaw, R.; Light, T.S. ISA Transactions 1982, 21, 63-70.

RECEIVED August 9, 1988

Chapter 30

Spectroelectrochemistry Using Transparent Electrodes

An Anecdotal History of the Early Years

William R. Heineman and William B. Jensen

Department of Chemistry, University of Cincinnati, Cincinnati, OH 45221-0172

Spectroelectrochemical methods involve the in situ coupling of a spectroscopic technique with an electrochemical technique and may be subdivided into those methods which focus on the spectroscopic characterization of the electrode surface and those which focus on the spectroscopic characterization of the electrogenerated solution species. This paper will provide an anecdotal account of the early history of techniques in the second category based largely on the senior author's personal recollections and on interviews with several of the key participants in the development of optically transparent electrodes.

Strictly speaking, the term spectroelectrochemistry subsumes any technique which involves the in situ coupling of a spectroscopic technique with an electrochemical technique. However, from both a practical and historical standpoint, spectroelectrochemical methods may be further subdivided into those which focus on the spectroscopic characterization of the electrode surface and those which focus on the spectroscopic characterization of the electrogenerated solution species.

Techniques in the first category are much older than those in the second and can be traced back to the pioneering work of L. Tronstad and his students in the late 1920's and early 1930's (1-4). Most of the work in this area was done in Europe and Great Britain and involved measurement of the changes in the polarization of light reflected from the surfaces of metallic electrodes -- a procedure later given the name of ellipsometry (5,6). In contrast, techniques in the second category are of much more recent origin, were largely developed by research groups in the United States, and are essentially coupled to the discovery and development of various kinds of optically transparent electrodes (OTE) during the 1960's.

Though important advances were also made in the 1960's in the area of electrode surface characterization (7), this account will focus almost exclusively on the methods in the second category. In

0097-6156/89/0390-0442$06.00/0

addition, rather than attempt what would, in all probability, turn out to be a premature historical assessment of the importance of these relatively recent events, we will instead provide a more limited anecdotal account of their early development based largely on the senior author's personal recollections and on interviews with several of the key participants.

Transmission Spectroelectrochemistry

The origins of modern transmission spectroelectrochemistry appear to date from a conversation held in the late 1950's between Ralph Adams, who was at the time a young assistant professor at the University of Kansas and just beginning his classic researches on the mechanisms of the oxidation of organic compounds at solid electrodes (8), and Ted Kuwana, who was Adam's first graduate student. As Kuwana later recalled it, Adams, while observing the production of an intense yellow color in the solution near the platinum anode during the coulometric oxidation of o-tolidine, had wishfully mused that ". . . it would be nice to have a 'see-through' electrode to spectrally identify the colored species being formed . . . (9)". This statement, to the best of the authors' knowledge, was the first to introduce the concept of an optically transparent electrode as a means of making *in situ* spectroscopic measurements of electrode processes. It also addressed the important "raison d'etre' for spectroelectrochemical techniques in general -- namely -- the fact that it is often quite difficult to deduce the detailed mechanism of a complicated electrode process unambiguously from the results of electrochemical measurements alone. Consequently, another physical probe acting simultaneously with an electrochemical perturbation is needed to give additional information about the electrogenerated species being formed.

Adams' prophetic statement did not go unheeded by Kuwana (Figure 1). While still a graduate student, he attempted his first spectroelectrochemical experiment by passing a light beam parallel to the surface of a platinum electrode. Inability to focus sufficient beam intensity at the electrode surface stymied the experiment. [It was not until *ca*. 20 years later that this experiment was performed successfully with a laser (10).] Several years later and now as an assistant professor at the University of California at Riverside, Kuwana was visiting the laboratory of David Kearnes, a physical chemist who was studying photoconduction in organic crystals. Kuwana saw two wires emerging from one of the crystals and wondered how electrical contact was being made with the crystal. Closer inspection showed the crystal sandwiched between two pieces of a glass that was coated on one side with tin oxide, a semiconductor. The contact wires were glued to the tin oxide sides of each piece of glass. Since the tin oxide-coated glass was optically transparent, the organic crystal could be irradiated with a light beam through the glass and the resulting conductivity measured by means of the conducting coating of tin oxide. Kuwana returned to his office with samples of the "conducting glass".

These samples became the first OTEs and were used for the first optical spectroelectrochemical characterizations of electrogenerated solution species reported in the literature. These were

Figure 1. Ted Kuwana.

performed by Kuwana, graduate student Keith Darlington, and undergraduate student Don Leedy. From the many organic redox systems with which Kuwana was familiar from his Ph.D. research, he picked o-tolidine, a colorless compound that undergoes a 2-electron oxidation to form an intensely yellow colored species -- the very reaction that had sparked Adams' earlier comment:

$$H_3N^+\text{-}C_6H_3(CH_3)\text{-}C_6H_3(CH_3)\text{-}NH_3^+ \rightleftharpoons {}^+H_2N{=}C_6H_3(CH_3){=}C_6H_3(CH_3){=}NH_2^+ + 2H^+ + 2e^-$$

The formation of the yellow species was monitored at 437 nm during its electrogeneration by a constant anodic current (i.e., chronopotentiometry), and the results were reported in 1964 (11). The optical beam was passed through the OTE, at an angle perpendicular to the plane of the electrode, and the adjacent solution as shown in Figure 2A. This is now referred to as transmission spectroelectrochemistry. A drawing of the cell, which was reported later (12), is shown in Figure 3. The tin oxide glass working electrode and a piece of ordinary glass were clamped against the cylindrical cell body. After Kuwana moved to Case Western Reserve University, the electrochemical and optical characteristics of this OTE were studied in more detail by visiting professors Jerzy Strojek from Poland (12) and Tetsuo Osa from Japan (13) and the fundamental equation for chronoabsorptometry was derived

$$A = \frac{2}{\pi^{\frac{1}{2}}} \varepsilon C D t^{\frac{1}{2}}$$

where A = absorbance, ε = molar absorptivity of the electrogenerated species being monitored optically, and C and D are the concentration and diffusion coefficients, respectively, for the species being electrolyzed. The first use of transmission spectroelectrochemistry to study an electrode mechanism was the study of o-tolidine oxidation at a pH where a semiquinone radical cation is formed (14). The first spectroelectrochemical working curves were calculated by Strojek and Kuwana in collaboration with Stephen Feldberg at Brookhaven, who developed the method of digital simulation of electrode mechanisms (14,15). This work was carried on by postdoc Henry Blount and graduate student Nicholas Winograd (16).

Internal Reflection Spectroelectrochemistry

During the 1960s the North American Aviation Science Center in Thousand Oaks, California, was a hotbed of electrochemical research. Robert Osteryoung, Keith Oldham and Joseph Christie were involved in the development of chronocoulometry and pulse techniques. Wilford Hansen (Figure 4), a physicist working in Osteryoung's group, had initiated research in the area of internal reflection spectroscopy (17,18), a new spectroscopic technique reported by Fahrenfort (19). Hansen was apparently the first to use multiple internal reflection spectroscopy in a "light pipe" to

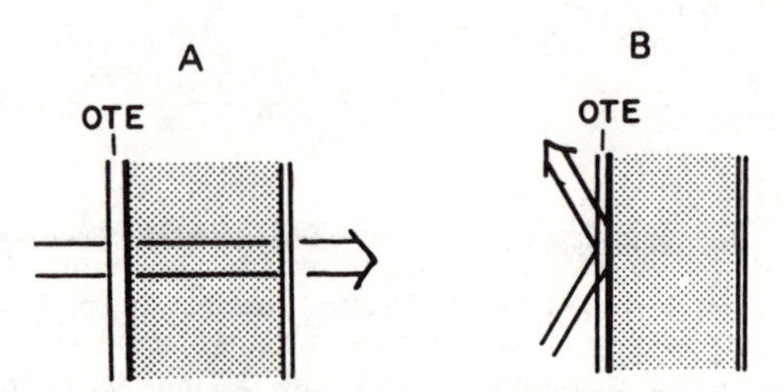

Figure 2. Various optical techniques used in spectroelectrochemistry. A. Transmission. B. Internal reflection.

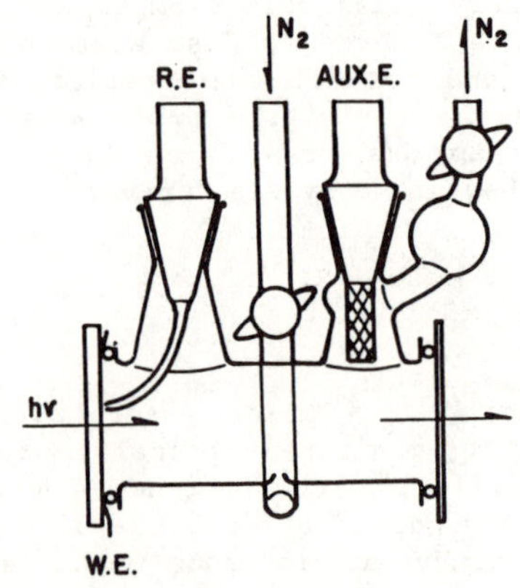

Figure 3. First cell for transmission spectroelectrochemistry. (Reproduced with permission from Ref. 12. Copyright 1968 Elsevier).

Figure 4. Wilford Hansen.

obtain visible spectra of liquids (17). Being in an electrochemistry group, he wanted to apply this new technique to electrochemistry. The optical beam would be directed into the back side of an OTE at an angle greater than the critical angle so that the beam would be totally reflected (Figure 2B). According to Hansen (20), "A key in the whole thing was my discovery that an enhanced evanescent wave could be arranged on the far side of a thin conducting film (metal or otherwise) placed on a prism or multiple reflection plate and operated in ATR mode with the conductor as an electrode. This was totally unexpected but was confirmed by my numerical calculations using Maxwell's equations for stratified media. We then proceeded to do spectroelectrochemistry more seriously. Skeptics kept trying to say that the effect was due to 'cracks in the surface where the solution seeped in,' etc. The work with Kuwana's o-tolidine silenced the critics." Hansen designed a spectroelectrochemistry cell for use with a tin oxide or gold film OTE, which he had discovered independent of Kuwana. However, Hansen needed a good chemical system with which to demonstrate the technique. Hansen and Osteryoung met with Kuwana at the Gordon Research Conference on Electrochemistry in Santa Barbara. Kuwana happened to have in his pocket a sample of o-tolidine, which he gave to Hansen (21). Experiments performed on o-tolidine at the Science Center were subsequently published as the first examples of multiple internal reflection spectroelectrochemistry (22,23). The cell is shown in Figure 5. Hansen continued to study phenomena associated with IRS (24-27) as did Kuwana (28,29).

The first infrared spectroelectrochemistry was reported in 1966 by Harry Mark and Stan Pons (30). As a postdoc at Cal Tech with Fred Anson, Mark (Figure 6) had visited Kuwana at Riverside -- a visit that sparked his interest in spectroelectrochemistry. He also visited the electrochemistry group at the Science Center where he discussed internal reflection spectroelectrochemistry with Hansen. Upon his arrival at the University of Michigan as an assistant professor, Mark decided to try to develop infrared internal reflection spectroelectrochemistry because of the structural information in an infrared spectrum. He was able to persuade Paul Wilks (Wilks Scientific Corp.) to lend him a double beam internal reflectance attachment. A Ge internal reflectance plate-electrode was made by Recticon Corp. from n-type semiconducting Ge, which was already known to function as a working electrode (31,32). Stan Pons, a first year graduate student, performed the experimental work in which the spectra of the reduction products of 8-quinolinol and tetramethylbenzidine free radical were chracterized.

During this time they found out about an organic liquid complex of platinum (Engelhard Industries, Inc.) that could be painted on a glass substrate and reduced at sufficiently high temperatures to an optically transparent film of platinum. Pons, together with graduate student James Mattson and undergraduate Leon Winstrom, showed that these metal films on glass could function as an OTE for internal reflection spectroscopy (33). The chronopotentiometric oxidation of o-tolidine to the yellow product, which by now had become the standard for testing new spectroelectrochemical techniques, was monitored. Absorbance changes much

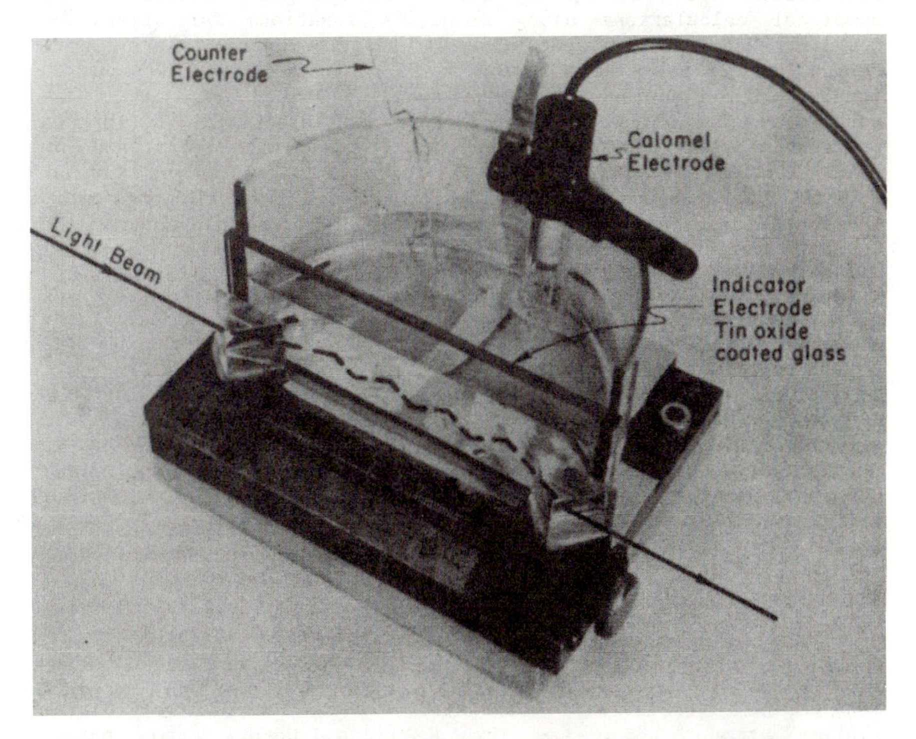

Figure 5. Cell for internal reflection spectroelectrochemistry. Reproduced from Ref. 24. Copyright 1967 American Chemical Society.)

Figure 6. Harry B. Mark, Jr. Mark's attire is not part of his technique for infrared spectroelectrochemistry, but rather reflects his interest at the time in driving race cars. (USAC Publicity Photo circa 1965. Only photo available from that era.)

greater than those at a tin oxide OTE were observed. This apparent anomaly resulted in an argument with the reviewers, who were skeptical of the results (34). Subsequently, a student of Mark's, Arnold Prostak, spent a year with Hansen at the Science Center. Their work explained some of the anomalies of internal reflectance spectra (25,27).

An Aside

During this period, Fred Anson (California Institute of Technology) had founded the San Clemente Surfing and Discussion Society as a means of stimulating informal discussion among electrochemists. The society's name proved to be an impediment to some scientists receiving permission and/or funds to attend the discussions. Consequently, Mark, who was the Society's secretary and editor of its newsletter, held a contest among the membership for a new name. The winner was to receive a platinum-coated bottle of Cutty Sark. Ralph Adams was declared the winner for the name Western Electroanalytical Theoretical Society (i.e., WETS, which is the forerunner of the Society for Electroanalytical Chemistry, SEAC). Mark tried coating numerous empty Scotch bottles by the liquid Pt process. During these experiments, the Assistant Department Head at Michigan was giving some visiting dignitaries a tour of the Department. Mark was standing at a lab bench with about 20 bottles lined up in front of him facing the door, when the door was opened by the Assistant Head with the words "And here we have our laboratory for electrochemistry." He then saw Mark and his bottles and closed the door immediately without a word to anyone. Adams was subsequently presented the trophy at Robert Osteryoung's house in San Clemente at the next meeting of the Society. During this meeting the idea of a Gordon Research Conference on Electrochemistry was approved. Incidentally, the trophy presented to Adams was silver plated. Mark decided that the liquid platinum gave a surface that was inappropriately rough to be a fitting trophy. He claims that the idea of using the liquid platinum for an OTE came before the experiments with the liquor bottles (34).

Optically Transparent Thin Layer Electrodes

In 1967 the first OTE based on light passing through holes in an electrode and the first thin-layer spectroelectrochemical experiment were reported (35). The discovery of this OTE was fortuitous. John Ashley, a graduate student with Charles Reilley, had seen an electroformed mesh (i.e., minigrid) advertised by Buckbee-Mears, Inc. He mentioned this to Larry Anderson (36), a postdoc with Charles Reilley, who sent for a free sample. The minigrid was produced in various dimensions ranging from 5 to 2000 wires/inch from different metals -- Au, Cu, Ni, Ag. Anderson showed the sample around the "Wigwam", which was the Murray-Reilley laboratory for electrochemistry. Royce Murray (Figure 7) took interest in the material in relationship to his studies on the inhibition of electrode processes by adsorbed films (37). One model system for this behavior involved islands of adsorbed material that completely inhibited the electrode reaction and regions in between with no adsorption and at which the electrode process was unaffected. The minigrid could serve as an electrode

Figure 7. Royce Murray.

for this model -- the holes simulating regions of adsorption inhibition and the wires simulating bare electrode. A mathematical model based on this geometry could be constucted and the electrochemical behavior of the minigrid electrode characterized and compared with that for systems in which adsorption inhibition occurred. Consequently, Murray ordered pieces of 100 and 1000 wire/inch gold minigrid. In the meantime, the classic article by Hansen, Kuwana and Osteryoung (23) appeared and stimulated interest in spectroelectrochemistry. Having noticed that the minigrid was transparent to light, Murray realized that this material would function as an OTE. This idea was mentioned to a postdoc, George O'Dom, who had done undergraduate research with Kuwana at Riverside, and a graduate student, William Heineman, both of whom expressed interest in the project. Heineman was given the task of designing and constructing a suitable cell. At this time the Reilley group was heavily involved in the development of thin layer electrochemistry. This probably sparked the idea of thin-layer spectroelectrochemistry. The first optically transparent thin layer electrode (OTTLE) was a piece of minigrid sandwiched between two glass microscope slides that were separated by spacers consisting of layers of Tygon paint along the periphery (Figure 8) (38). The first cell didn't work because it was so thin that air bubbles couldn't be dislodged from around the minigrid. Extra layers of Tygon paint were used to build up the spacer thickness for subsequent cells and the minigrid was moved to the bottom of the cell to minimize iR drop. Kuwana's now classic o-tolidine system was used to characterize the cell.

During the next year, George O'Dom left the group and Heineman worked with another postdoc, Nick Burnett, to construct an OTTLE from an infrared cell with salt plates for the optical windows (39). Multiple minigrids were stacked between spacers to increase the optical pathlength of the cell while retaining thin-layer diffusional distances. This cell was used to demonstrate infrared thin-layer spectroelectrochemistry with ninhydrin. This was chosen for its three carboxyl groups, the reduction of which would be easily observable in the infrared. Another thin-layer cell was constructed with quartz plates in order to illustrate applicability in the UV (40).

The summer after the gold minigrid thin layer work was initiated in Murray's group, two students of Reilley's, Attila Yildiz and Peter Kissinger, learned about a vapor-deposition facility in the Physics Department. Preliminary experiments showed that voltammetric measurements could be made on Pt films vapor deposited on glass. Subsequently, the first thin layer cell with an OTE vapor-deposited on a quartz slide was demonstrated (41). Spectra of oxidized rubrene were recorded and fluorescence spectra recorded of regenerated rubrene. Kissinger subequently designed thin-layer cells whereby specular reflectance measurements could be made in the UV-visible and infrared regions (42).

Electron Spin Resonance Spectroelectrochemistry

During the development of OTE-based spectroelectrochemistry, a totally different type of spectroelectrochemistry, which was also useful for the structural identification of electrogenerated solu-

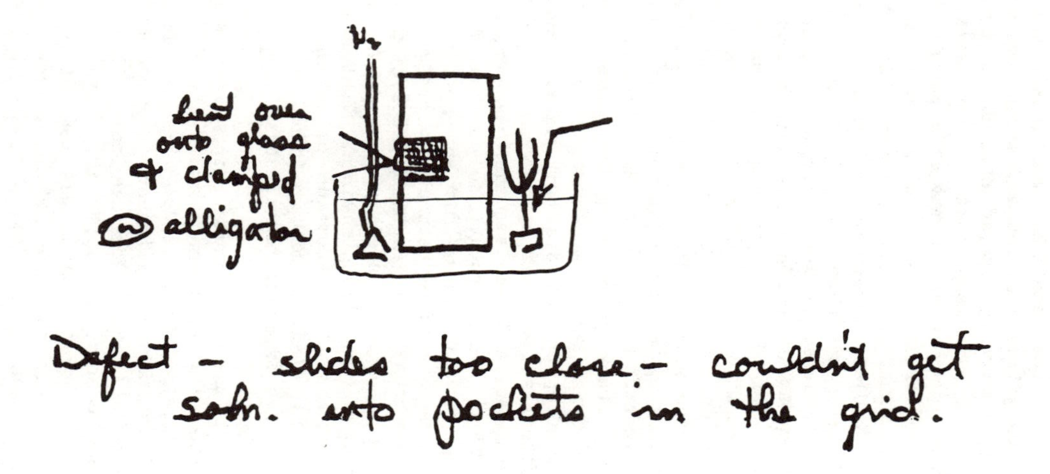

Figure 8. First optically transparent thin layer electrode (38).

tion species, was in the making based on a coupling of electrochemistry with electron spin resonance spectroscopy. Although this technique does not involve the use of an OTE and thus, strictly speaking, lies outside the limits of our historical survey, it has, nevertheless, proven effective in the study of the electrogeneration of transient radical species in solution and consequently deserves some mention. The first electrochemical experiment in an ESR spectrometer was reported in 1957 by A.A. Galkin, I.L. Shamfarov and A.V. Stephanishiva in Russia (43). One year later D.E.G. Austen, P.H. Given, D.J.E. Ingram, and M.E. Peover reported the formation of aromatic radical ions by electrochemical reduction in non-aqueous solvents (44). The samples were frozen prior to introduction into the spectrometer. A.H. Maki and D.H. Geske (Harvard University) are generally credited (45) with introducing the real potentialities of ESR spectroelectrochemistry (46). They carried out electrochemical generation at a mercury pool electrode in the microwave cavity and produced a number of radical ions in non-aqueous media. The technique was subsequently extended to aqueous solutions by R.N. Adams, L.H. Piette and P. Ludwig at the University of Kansas (47).

Conclusions

As noted at the beginning of our survey, spectroelectrochemistry is of fairly recent origin and is still a developing area of research. Consequently, we have had to severely restrict our subject matter in terms of both breadth of coverage and time period. For surveys of current developments, the reader is referred to recent review articles (48,49).

Literature Cited

1. Tronstad, L. Z. Phys Chem. (Leipzig) 1929, A142, 241.
2. Tronstad, L. In Optische Untersuchungen zur Frage der Passivitat des Eisens und Stahls; Kgl. Norske Videnskab Selsk. Skr., 1931, 1.
3. Tronstad, L. Trans. Faraday Soc. 1933, 29, 502.
4. Winterbottom, A.B. In Optical Studies of Metal Surfaces, Kgl. Norske Videnskab. Selskabs Skrifter, 1, F. Bruns Bokhandel, Trondheim, 1955.
5. Reddy, A.K.N.; Genshaw, M.; Bockris, J.O'M. J. Electroanal. Chem. 1964, 8, 406-407.
6. Delahay, P. and Tobias, C.W. In Advances in Electrochemistry and Electrochemical Engineering; Muller, R.H., Ed.; Wiley: New York, 1973; Vol. 9, p. 168.
7. Optical Studies of Adsorbed Layers at Interfaces, Symposia of the Faraday Society, No. 4, The Faraday Society, London, 1970.
8. Adams, R.N. Electrochemistry at Solid Electrodes; Marcel Dekker: New York, 1969.
9. Interview with T. Kuwana, April 1988.
10. Pruiksma, R.; McCreery, R.L. Anal. Chem. 1979, 51, 2253-2257.
11. Kuwana, T.; Darlington, R.K.; Leedy, D.W. Anal. Chem. 1964, 36, 2023-2025.
12. Strojek, J.W.; Kuwana, T. J. Electroanal. Chem. 1968, 16, 471-483.

13. Osa, T. and Kuwana, T. J. Electroanal. Chem. 1969, 22, 389-406.
14. Kuwana, T.; Strojek, J.W. Discussions Faraday Soc., 1968, 45, 134-144.
15. Strojek, J.W.; Kuwana, T.; Feldberg, S.W. J. Am. Chem. Soc. 1968, 90, 1353-1355.
16. Winograd, N.; Blount, H.N.; Kuwana, T. J. Phys. Chem. 1969, 73, 3456-3462.
17. Hansen, W.N. Anal. Chem. 1963, 35, 765-766.
18. Hansen, W.N. Spectrochim. Acta 1965, 21, 815-833.
19. Fahrenfort, J. Spectrochim. Acta 1961, 17, 698-709.
20. Hansen, W.N. Private communication, May 10, 1988.
21. Interview with W.N. Hansen, April 1988.
22. Hansen, W.N.; Osteryoung, R.A.; Kuwana, T. J. Am. Chem. Soc. 1966, 88, 1062-1063.
23. Hansen, W.N.; Kuwana, T.; Osteryoung, R.A. Anal. Chem. 1966, 38, 1810-1821.
24. Horton, J.A.; Hansen, W.N. Anal. Chem. 1967, 39, 1097-1100.
25. Hansen, W.N.; Prostak, A. Phys. Rev., 1968, 174, 500-503.
26. Hansen, W.N. J. Optical Soc. Am. 1968, 58, 380-390.
27. Prostak, A.; Mark, Jr., H.B.; Hansen, W.N. J. Phys. Chem. 1968, 72, 2576-2582.
28. Srinivasan, V.S.; Kuwana, T. J. Phys. Chem. 1968, 72, 1144-1148.
29. Winograd, N.; Kuwana, T. J. Electroanal. Chem. 1969, 23, 333-342.
30. Mark, Jr., H.B.; Pons, B.S. Anal. Chem. 1966, 38, 119-121.
31. Gerischer, H. In Advances in Electrochemistry and Electrochemical Engineering; Delahay, P., Ed.; Interscience: New York, 1961; Vol. I, pp. 139-232.
32. Holmes, P.S. Electrochemistry of Semi-Conductors; Academic Press: New York, 1962.
33. Pons, B.S.; Mattson, J.S.; Winstrom, L.O.; Mark, Jr., H.B. Anal. Chem. 1967, 39, 685-688.
34. Interview with H.B. Mark, Jr. May 10, 1988.
35. Murray, R.W.; Heineman, W.R.; O'Dom, G.W. Anal. Chem. 1967, 39, 1666-1668.
36. Interview with L.B. Anderson, April 1988.
37. Interview with R.W. Murray, April 1988.
38. Heineman, W.R. Research Notebook 2, University of North Carolina, Feb. 9, 1967, p. 38.
39. Heineman, W.R.; Burnett, J.N.; Murray R.W. Anal. Chem., 1968, 40, 1974-1978.
40. Heineman, W.R.; Burnett, J.N.; Murray, R.W. Anal. Chem. 1968, 40, 1970-1973.
41. Yildiz, A.; Kissinger, P.T.; Reilley, C.N. Anal. Chem. 1968, 40, 1018-1024.
42. Kissinger, P.T.; Reilley, C.N. Anal. Chem. 1970, 42, 12-15.
43. Galkin, A.A.; Shamfarov, I.L.; Stefanishina, A.V. J. Exptl Theoret. Phys. (U.S.S.R.) 1957, 32, 1581.
44. Austen, D.E.G.; Given, P.H.; Ingram, D.J.E.; Peover, M.E. Nature 1958, 182, 1784.
45. Adams, R.N. J. Electroanal. Chem. 1964, 8, 151-162.
46. Maki, A.H.; Geske, D.H. J. Chemical Phys. 1960, 33, 825-832.

47. Piette, L.H.; Ludwig, P.; Adams, R.N. J. Am. Chem. Soc. 1960, 83, 2671; 1962, 84, 4212.
48. McCreery, R.L. In Physical Methods in Chemistry; Rossiter, B., Ed.; Wiley: New York, 1987; Vol 2.
49. Heineman, W.R.; Hawkridge, F.M.; Blount, H.N. In Electroanalytical Chemistry; Bard, A.J., Ed.; Dekker: New York, 1984; Vol. 13, pp. 1-113.

RECEIVED September 12, 1988

Chapter 31

Borrowing from Industry

Edgar Fahs Smith's Rotating Anode and Double-Cup Mercury Cathode

Lisa Mae Robinson

Department of Natural Science, Michigan State University, East Lansing, MI 48824

From 1890 to 1910, Edgar Fahs Smith and his students at the University of Pennsylvania brought electrolysis into standard analytical practice by developing two new techniques designed to make electrolytic separation easier and faster. They pioneered the development of several electrochemical analytical techniques, the best known of which are the rotating anode and the double-cup mercury cathode. In 1903, Smith and his doctoral student, Franz Frederick Exner developed a rotating anode expressly for industrial analytical practice that greatly reduced the time needed for analysis. The next year, Smith and another doctoral student, Joel Henry Hildebrand, created a double-cup mercury cathode by explicitly borrowing the basic operating principle of Dow Chemical Company's Castner-Kellner electrolytic cell. These inventions illustrate the interrelationship between industrial and academic science in the early twentieth century that produced a flow of ideas and methods in both directions.

The research of Edgar Fahs Smith and his students at the University of Pennsylvania represents an attempt to bring electrolysis into standard analytical practice and to revolutionize mineralogy. From 1890 until 1910, Smith and his students developed two new electrolytic analytical techniques, the rotating anode and the double-cup mercury cathode that gave a simpler and faster electrolysis with a more complete separation. Smith and his students published a total of twenty-six papers exploring the uses of these techniques, which Smith further promoted through his widely-used textbook, Electrochemical Analysis. (1-2)

Edgar Fahs Smith (1854-1928) had been trained at the University of Göttingen under Friedrich Wöhler in both mineralogy and

0097–6156/89/0390–0458$06.00/0

analytical chemistry. Smith believed that electrolysis would prove to be particularly useful in the analysis of minerals, claiming in 1909 that

> the day is coming when we shall analyze a great many minerals in the electrolytic way. If you ask me how we are going to do it, I do not know, but we are going ahead and shall try to do it. What the method will be no one knows, but something will come. (3)

Smith was confident that he and his students could provide the methods necessary to take mineral chemistry in a new direction.

However, mineral chemists turned to other techniques and derived their inspiration mainly from geology, not chemistry. Smith's electroanalytical techniques did find an appreciative audience among industrial chemists concerned with the electrolytic production of heavy chemicals, especially chlorine. Ironically, this was the one group of chemists that Smith had no interest in reaching, for he despised their interest in profit from scientific research. (4-5)

An examination of Smith's electrochemical research reveals some interesting facets of the relation between academic research and industry. First, it shows that even very academically-oriented research can be useful to the improvement of industrial processes. Second, we see that academic research is not totally isolated from industrial developments and industry can provide valuable models for academic research. Finally, Smith's work reveals that research has unintended and perhaps unwanted influences on academic science. However much Smith wanted to revolutionize mineralogy or analytical chemistry, chemists would use Smith's techniques in ways he never intended or imagined.

Electrolysis began to acquire importance in analytical chemistry in the last quarter of the 19th century. The first American study was done in 1864 by Oliver Wolcott Gibbs, who investigated the electrolytic determination of copper and nickel and described methods for electrolytically analyzing silver, bismuth, lead, and manganese. Although others later claimed prior use of electricity for analytical purposes, most chemists (including Edgar Fahs Smith) saw Gibbs as the father of American electrochemistry. (6)

The Rotating Anode

Smith's first contribution to the development of electroanalytical techniques was the rotating anode. Smith became interested in the idea of anode rotation in 1901 while attempting the separation of molybdenum from tungsten. He could not obtain a satisfactory separation because a tungsten oxide that contaminated the molybdenum always formed during electrolysis . Smith tried rotating the anode to agitate the solution, thereby preventing local reactions around the anode from interfering with the precipitation of the molybdenum. He devised his own agitating apparatus, but did not publish this research at the time. (7-9)

During the next two years, Smith and Franz Frederick Exner (1868-1950), one of his doctoral students, refined the rotating anode for use in the separation of molybdenum from tungsten. They found that its rotation at high speed allowed the application of a more intense current and a higher voltage. The more intense current caused a more rapid precipitation of metals, greatly reducing the time needed for electrolytic precipitation. They tried this new technique on the precipitation of copper, silver and mercury with great success.

The reduction of analysis time was the most important, although unintended, result of Smith and Exner's research. Electrolytic precipitation using the old methods usually took from three to twelve hours, and Smith regularly applied electric current to solutions for the entire night. The rotating anode with its more powerful current accomplished complete electrolytic separations in ten to twenty minutes, a significant reduction indeed.

Exner was very keen on seeing electrolytic methods replace traditional volumetric and gravimetric methods of analysis in industry and believed that the rotating anode would be a breakthrough in industrial analytical practice. In his doctoral dissertation, he stressed the importance of analytical speed for industry, speed that could be best obtained through electroanalysis.

> The time factor is a most important one to the technical man, and has no doubt made many slow to exchange their old and tried methods for the new, on the ground that the advantages of the change were not sufficiently pronounced. This investigation, it is believed, will be sufficient to convince the hitherto most skeptical, concerning the advantages of electrolytic analysis, as far as the time factor is concerned. (10)

Exner's dissertation, which appeared in part in the Journal of the American Chemical Society in 1903, was the first publication on the rotating anode. Using a strong current with an anode rotating at high speed, Exner investigated the effect on the precipitation of various metals. He successfully precipitated copper, nickel, zinc, bismuth, mercury, cobalt, cadmium, iron, lead, molybdenum, tin, antimony, gold, and silver. His attempted precipitations of arsenic and manganese were unsuccessful. (11)

Exner's anode (Figure 1), made by the Electrochemische Werkstatte of Darmstadt, Germany, consisted of a flat platinum spiral two inches in diameter. It could rotate from 300 to 1700 revolutions per minute, although Exner found that 600 to 700 revolutions per minute worked for most precipitations. His critical design problem was maintaining constant contact between the anode and solution. As the flat spiral anode whirled the electrolyte against the side of the dish, the solution formed a vortex. The bottom of this funnel often dipped below the anode, losing contact with it, and thereby increasing the solution's resistance. Exner solved this problem by bending the anode into a

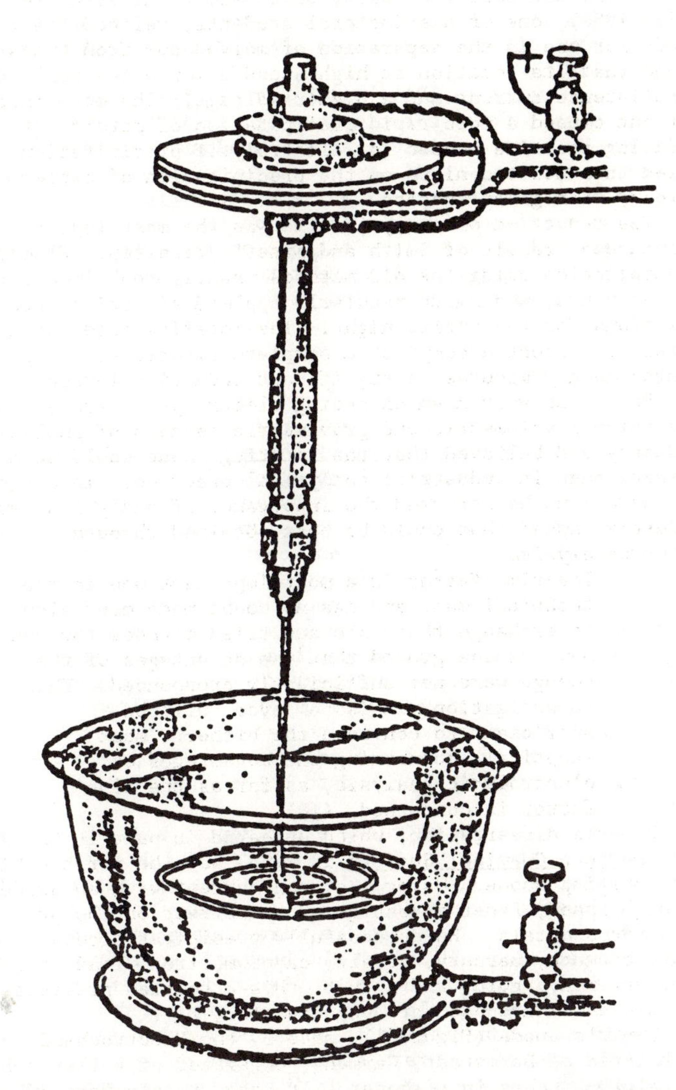

Figure 1. The rotating anode used by Exner and Smith. (Reproduced with permission from the Edgar Fahs Smith Memorial Collection, Van Pelt Library, University of Pennsylvania.)

bowl shape, so that as the funnel formed, the anode remained in contact with it. He also found that this anode shape stirred the solution more effectively and completely, thereby further aiding electrolysis.

The next year, in 1904, Smith himself published a paper on possible uses for the rotating anode. In it he presented the work of George H. West, an undergraduate student, and Lily Gavit Kollock, a doctoral candidate. West successfully used the rotating anode to precipitate nickel, while Kollock produced pure cobalt. Smith also mentioned Exner's thesis on the precipitation of various metals, as well as the dissertations of two other graduate students. He concluded that the rotating anode would be most effectively used in conjunction with the mercury cathode and referred readers to his recent paper on this device. (12-16)

The Mercury Cathode

Smith first published his experiments with the mercury cathode in September 1903 in the Journal of the American Chemical Society. He reported various attempts at the electrolytic separation of various metallic sulfates, nitrates, and halides. Smith used the best available mercury cathode that consisted of a beaker of mercury into which extended a hollow, carbon-tipped glass tube, also full of mercury. A copper wire inserted into the beaker's mercury through the glass tube and connected to the battery's negative electrode made the mercury into a cathode. A thin platinum wire inserted through the wall of the beaker served as the anode. Smith found this arrangement unsatisfactory for two reasons. First, the mercury in the beaker was difficult to dry and required repeated washing with ether and alcohol to remove all traces of water. Second, some of the metal electrolyte usually precipitated on the platinum anode, which then had to be weighed along with the mercury. (17-18)

During the course of these experiments, William H. Howard, an undergraduate student in Smith's laboratory, devised a new mercury cathode (Figure 2) that solved the problem of weighing the mercury. Howard's cathode was a small beaker of mercury penetrated at the bottom by a thin platinum wire that connected the mercury with a copper disk underneath the beaker. Another wire connected the disk to a battery's negative electrode, making the mercury into a cathode. Since the platinum wire was imbedded into the beaker, any precipitant adhering to it would be weighed along with the mercury. However, accurate atomic weight determination still required repeated washing and drying of the amalgam. (19)

Three years later Joel Henry Hildebrand (1881-1983), one of Smith's doctoral students, developed an improved configuration for the mercury cathode during the course of his dissertation research. Hildebrand's thesis attempted the simultaneous electrolytic determination of both components of various electrolytes, especially sodium chloride. He pointed out that little work had been done previously on the electrolysis of salts or anions. Smith's preliminary study in 1903 of simultaneous

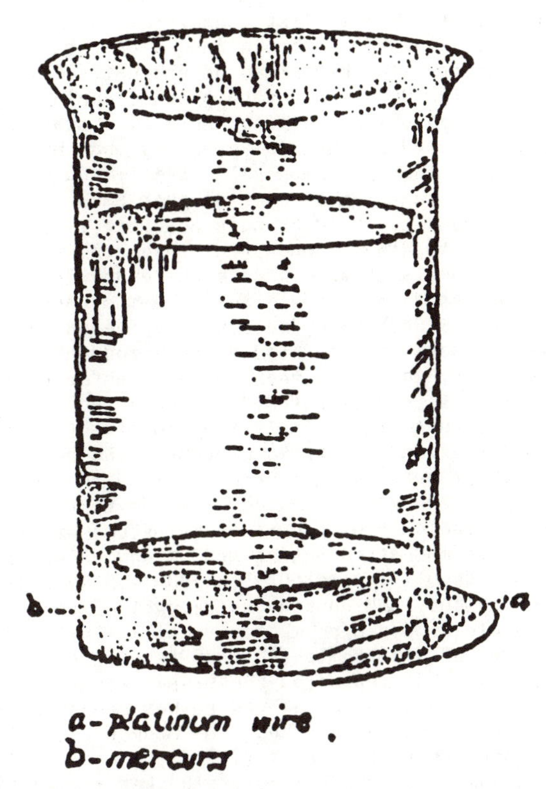

Figure 2. The mercury cathode used by Smith in 1903. (Reproduced with permission from the Edgar Fahs Smith Memorial Collection, Van Pelt Library, University of Pennsylvania.)

electrolysis used a stationary silver anode and a mercury cathode like the one developed by William Howard. In 1905, James Renwick Withrow, also Smith's student, completed his doctoral thesis on the simultaneous precipitation of gold halides and used a similar mercury cathode with a rotating anode. Hildebrand proposed to use an improved mercury cathode, which he called the double-cup, with a rotating anode. (20-22)

The electrolysis of salt (sodium chloride) was also the object of continued interest by the electrochemical industry. The Dow Chemical Company, for example, had a long-standing interest in the electrolysis of salt, and they began an extensive, in-house research campaign in 1908 in order to produce sodium hydroxide (caustic soda) and chlorine. Hildebrand's dissertation was not overtly directed towards industrial problems, but his successful electrolytic separation of common salt legitimated such an industrial research campaign by showing that such an electrolytic separation was possible. Hildebrand's academic research provided Dow with new ideas for their industrial research. (23)

Hildebrand's thesis is also a case where developments in industry provided ideas and models for academic research. Hildebrand based his double-cup mercury cathode on the Castner-Kellner process for the industrial production of caustic soda. This process, which has been called the "most elegant electrolytic process ever invented," was patented in 1894 by Hamilton Castner, an American analytical chemist. By 1902, Castner had increased its efficiency to 90 percent. The heart of the process was Castner's "rocking" mercury cell. The cell had two chambers through which mercury moved back and forth, continuously removed metallic sodium from the decomposing chamber, thus preventing the recombination of the sodium and chlorine. (24)

Hildebrand's double-cup mercury cathode employed two of Castner's innovations, the continuous removal of the product and the use of two electrolytic chambers or cells. Hildebrand's cathode consisted of a bottomless beaker resting on a thin, Y-shaped glass rod that supported the beaker slightly above the bottom of a larger glass dish. Three rubber stoppers fitted between the dish and the beaker kept the beaker securely in position. A thin layer of mercury filled the bottom of the dish, while the arrangement of the bottomless beaker formed two sections for the mercury, which were connected sufficiently to permit passage of current and ions, yet were separate enough to prevent recombination of ions. The solution to be electrolyzed was put into the inner compartment and distilled water covered the mercury in the outer compartment. As the cations slowly moved into the outer compartment, they dissolved in the mercury of the inner compartment. The anions remained behind in the inner compartment and therefore did not recombine with the cations. A platinum wire dipped into the outer compartment connected the mercury to the battery. After completing the electrolysis, the entire contents of the dish (and beaker) were poured off for separation, washing, drying, and weighing. (25)

Hildebrand combined his new mercury cathode with the rotating anode developed by Smith and Exner the previous year. This anode spun freely in the electrolytic solution in the inner compartment. The next year (1907), Thomas Potter McCutcheon, Jr. used both the rotating anode and the double-cup mercury cathode (Figure 3) in his doctoral thesis on electroanalysis. McCutcheon slightly modified the rotating anode by using a wire mesh at the bottom of the spinning rod, rather than a wire coil. He also modified the double-cup cathode by resting the inner beaker on a triangular glass rod for greater stability. (26)

Hildebrand's and McCutcheon's dissertations marked the apex of development for the rotating anode and the double-cup mercury cathode. Smith and his students continued to used these techniques for three more years until Smith became Provost of the University of Pennsylvania in 1910. The responsibilities of this office caused Smith to cut back the time he spent on his scientific research and limit the number of graduate students he trained. Consequently, with the advent of Smith's new administrative duties, the rotating anode and double-cup mercury cathode ceased to be objects of active research.

Conclusion

While the rotating anode and the double-cup mercury cathode were useful analytical techniques and enjoyed a brief popularity with analytical chemists, they did not find an enduring audience. Analytical chemists retained their reliance on chemical means of separation and essentially forgot their brief fascination with electrolysis. Neither did Smith succeed in revolutionizing mineralogy, which again relied more on gravimetric and volumetric methods of analysis. While those physical chemists interested in describing the behavior of ions in solutions used electrolytic techniques, Smith himself deplored their use of mathematical models. Indeed, Smith's most promising student, Joel Henry Hildebrand, left Smith and the University of Pennsylvania amid much bitterness in order to pursue the study of solution theory. The group that most appreciated Smith's work were industrial chemists concerned with the electrolytic production of heavy chemicals. Ironically, this was the one group of chemists that Smith had absolutely no interest in reaching, for he despised their concern with profits. (27-29)

The development of the rotating anode and the double-cup mercury cathode reveals the complex interaction between academic science and chemical industry. At first glance, we see that the research of Smith and his students furthered knowledge about electrolytic separation, knowledge that was greatly needed by the growing electrochemicals industry. But a closer examination of this research reveals three ways in which these academic chemists borrowed from their industrial cousins. The search for speed in electrolytic separations was fueled by the importance of time for industrial analyses. Smith and Exner developed the rotating anode, in part, because they wished to see electrolytic techniques adopted

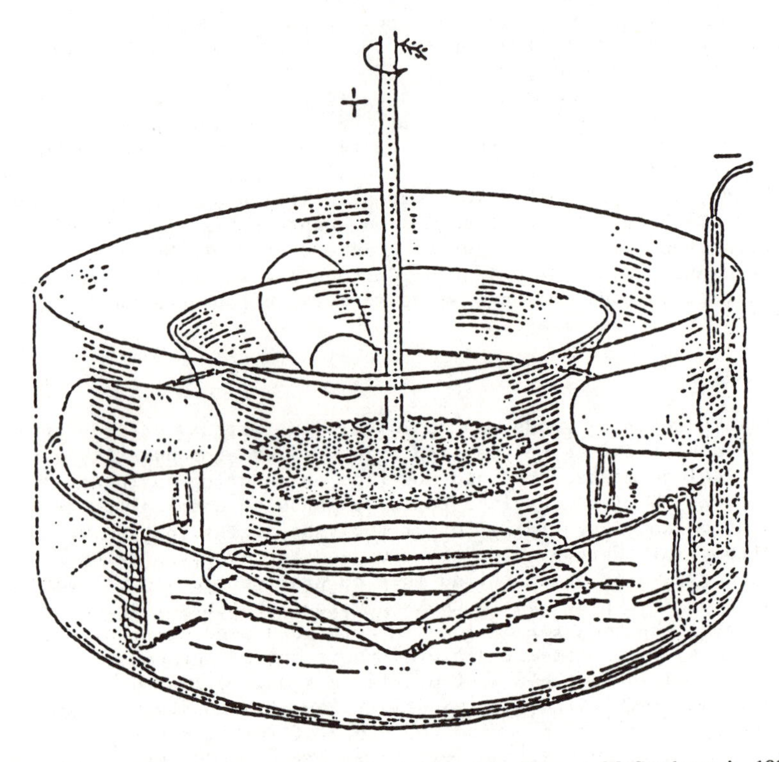

Figure 3. The double-cup mercury cathode developed by Thomas McCutcheon in 1907. (Reproduced with permission from the Edgar Fahs Smith Memorial Collection, Van Pelt Library, University of Pennsylvania.)

by industrial chemists. Another borrowed concern was the choice of problem. Smith and Hildebrand chose to investigate the electrolysis of salt, again in part, because it was a pressing problem for electrochemical producers. Finally, Smith and Hildebrand borrowed directly from industry when they used the Castner-Kellner cell as a model with which to redesign their mercury cathode. The development of the rotating anode and the double-cup mercury cathode reminds us that industrial science is not merely a parasite of academic science, consuming useful ideas to feed profits. Rather, it also provides possible directions for academic research and valuable methods to facilitate that research.

Acknowledgments

I would like to thank P. Thomas Carroll, John Servos, Anthony Stranges, Jeffrey Sturchio, and Arnold Thackray for their comments on this paper.

Literature Cited

1. Smith, E. F. Electrochemical Analysis [6 editions]; Philadelphia, 1890-1918.
2. Robinson, L. M. "The Electrochemical School of Edgar Fahs Smith" Ph.D. Thesis, University of Pennsylvania, 1986.
3. Smith, E. F. Trans. Am. Electrochem. Soc. 1909, 16, 65-77; quote on p 76.
4. Smith,.E. F. "Address at Franklin and Marshall College on the occasion of the Dedication of a Science Hall" (manuscript, 1907) in the Edgar Fahs Smith Memorial Collection, University of Pennsylvania (hereafter referred to as EFSC).
5. Easton, W.; Ross, S. Letters in The John Harrison Letter [March 1906] in EFSC.
6. Ihde, A. The Development of Modern Chemistry; Dover: New York, 1984; pp 292-293.
7. Smith, E. F. Electrochemical Analysis; Philadelphia, 1918; pp 40-41.
8. Von Klobukow, J. Prakt. Chem. 1886, 33, 473.
9. Von Klobukow, J. Prakt. Chem. 1873, 40, 121.
10. Exner, F. F. "The Rapid Precipitation of Metals in the Electrolytic Way" Ph.D. Thesis, University of Pennsylvania, 1903; p 1.
11. Exner, F. F. J. Am. Chem. Soc. 1903, 25, 896-907.
12. Smith. E. F. J. Am. Chem. Soc. 1904, 26, 1595-1615.
13. Ashbrook, D. S. "Electrolytic Separations Possible with a Rotating Anode" Ph.D. Thesis, University of Pennsylvania, 1904.
14. Ingham, L. H. "The Use of a Rotating Anode in the Electrolytic Estimation of Zinc and of Nitric Acid" Ph.D. Thesis, University of Pennsylvania, 1904.
15. Kollock, L. G. "Electrolytic Determinations and Separations" Ph.D. Thesis, University of Pennsylvania, 1899.
16. Smith, E. F. Electrochemical Analysis; Philadelphia, 1918; pp 40-63.

17. Smith, E. F. J. Am. Chem. Soc. 1903, 25, 883-96.
18. Drown and McKenna, J. Anal. Appl. Chem. 1891, 5, 627.
19. Smith, Electrochemical Analysis; 1918; pp 63-71.
20. Hildebrand, J. H. "The Determination of Anions in the Electrolytic Way" Ph.D. Thesis, University of Pennsylvania, 1906.
21. Smith, E. F. J. Am. Chem. Soc. 1903, 25, 890.
22. Withrow, J. R. "The Electrolytic Precipitation of Gold with a Rotating Anode and the Rapid Analysis of Halide" Ph.D. Thesis, University of Pennsylvania, 1905.
23. Campbell, M; Hatton, H. Herbert H. Dow: Pioneer in Creative Chemistry; Appleton-Century-Crofts: New York, 1951; pp 89-101.
24. Trescott, M. M. The Rise of American Electrochemical Industry, 1870-1910; Greenwood Press: Westport, CT, 1981; p 448.
25. Hildebrand, J. H. "The Determination of Anions in the Electrolytic Way" Ph.D. Thesis, University of Pennsylvania, 1906.
26. McCutcheon, Jr., T. P. "New Results in Electro-Analysis" Ph.D. Thesis, University of Pennsylvania, 1907.
27. Hildebrand, J. H. Perspectives in Biology and Medicine Autumn 1972, 16, 93-95.
28. Browne, C. A. "Reminiscences of Edgar Fahs Smith" manuscript, 1926, p 53 in EFSC.
29. Smith, E. F. "Address at Franklin and Marshall College on the occasion of the Dedication of a Science Hall" manuscript, 1907 in EFSC.

RECEIVED August 3, 1988

Chapter 32

Henry J. S. Sand (1873–1944), A Well-Remembered Tutor

John T. Stock

Department of Chemistry, University of Connecticut, Storrs, CT 06268

The name "Sand" usually conjures up only a particular electrochemical equation. Older workers can recall the flow of papers from what is now the City of London Polytechnic, where Sand and his students made notable advances in the practice of electrogravimetric analysis. Actually Sand's Ph.D., obtained under Bamberger, was in organic chemistry. Nevertheless he soon began electrochemical work, but by no means restricted himself to this field. Studies on the dissociation pressures of alkali metal bicarbonates and on fermentation by yeast cells are examples of his diverse interests. He held several patents on glass working and the like. His joint authorship of "Dyestuffs and Coal Tar Products" showed that he had not forgotten his doctoral area. John Stock was in Sand's lecture group in 1936 and had one of Sand's few but very able graduate students as laboratory supervisor.

Nowadays, most chemists know the name "Sand" only because it appears attached to a particular equation in texts that describe chronopotentiometry. This technique can be useful in the diagnosis of electrode reactions (1). A current step i is impressed across an electrochemical cell containing unstirred solution, the potential of the working electrode is measured with respect to time and the transition time τ is noted. According to the Sand equation, the product $i\tau^{1/2}$ should be constant for an uncomplicated linear diffusion-controlled electrode reaction at a planar electrode.

Sand did the work that led to this equation before my time. However, I was around when Sand and his students were developing apparatus and techniques that greatly influenced electrogravimetric analysis. Like other gravimetric processes, electrogravimetry is now overshadowed by more rapid methods. This deposition method, capable of giving high precision and accuracy,

0097–6156/89/0390–0469$06.00/0

has at least one satisfying advantage; if a metal is being determined, it is usually obtained in the elementary state. The technique is faster than "precipitate" gravimetry; routine electrogravimetric depositions can sometimes be run in a few minutes.

Henry Julius Salomon Sand was born on December 7, 1873, in Dundee, Scotland, and was educated in that city. He then went to Germany, where Walther Hempel, Professor at the Technical University of Dresden, was one of his teachers. Hempel is best remembered for his work on gas analysis. Sand then moved to Switzerland, to become a student at the Zurich Polytechnicum. There he obtained his Ph.D degree under the direction of Bamberger. Sand's first publication, jointly with Bamberger and Busdorf, appeared in June, 1898 (2). It was based on his thesis work and was submitted from the Analytical Laboratory of the Polytechnicum! However, it concerns typical "react and isolate" organic chemistry. I never thought to ask Sand about this, but I imagine that his "analytical chemistry" was largely concerned with the tedious business of numerous combustion analyses of his products!

Returning to Britain, he spent a short period with William Ramsay at University College, London. Having by then discovered the noble gases of the atmosphere, Ramsay was famous. Here Sand began the work that eventually led to the "Sand equation." What a switch from classical-style organic chemistry! He became Bowen Research Scholar at Mason University College (later, Birmingham University) in 1899, working under Percy Frankland. Here he continued the work begun in London, publishing a full account in 1901 (3). This contains both theory and an extensive description of the apparatus and experiments that Sand developed to test his ideas on acidified copper sulfate solution. He noted that stirr'ng can arrest hydrogen evolution that occurs when the solution is not stirred. Sand referred to this observation when he began his work on electrogravimetric analysis. This places him at the threshold of the present-day interest in hydrodynamic voltammetry.

In 1901, Sand became Lecturer and Demonstrator at University College, Nottingham, and remained there for about 13 years. Here he continued his electrochemical investigations (4,5), when a definite trend towards analytical chemistry occurred. He studied the trace determination of arsenic (6), the estimation of the acidity of tanning liquors (7,8), including the description of an improved hydrogen electrode, and the absorption of oxygen by effluents (9). Accounts of his work on the rapid electrolytic deposition and separation of metals began to appear in 1907 (10-14). This involved the development of improved electrode systems and other equipment (15, 16). He received the D.Sc. degree from the University of Birmingham in 1905.

Not all of Sand's efforts at Nottingham were devoted to analytical chemistry. With R.M. Caven, he studied the dissociation pressures of various alkali metal bicarbonates (17,18). Diffusion in liquids obviously interested him (19,20). Some other publications from the Nottingham period concern the radiative evaporation of liquids (21) and the sealing of wires into glass (22). He was granted several patents, including one for sealing metal to glass (23) and another for making bubble-free silica glass (24).

Career at the Sir John Cass Technical Institute

In 1914, Sand moved back to London as Senior Lecturer in Chemistry at the Sir John Cass Technical Institute (now the City of London Polytechnic). This photograph (Figure 1) may have been made during the early years of his tenure. He became Head of the Department of Inorganic and Physical Chemistry in 1921 (there was no separate Department of Analytical Chemistry) and retained this position until he retired in 1938.

In the 1920's, Sand wrote a number of papers concerning the phenomenon of overpotential (25-29). His other interests are reflected in publications concerning a steam drying oven (30), the cadmium vapor lamp (31), the anomaly of strong electrolytes (32), an arrangement for alternating current electrolysis (33), and the evaporation of liquids (34). Sand had not forgotten his "organicker" training; he was joint author of the text "Dyestuffs and Coal Tar Products," a fourth edition of which appeared in 1926 (35).

Sand retained his interest in electrolytic deposition (36). However, real emphasis on this field did not return until the end of the 1920's. Then he opened the series of papers on electrogravimetry that came from his Department with accounts of new apparatus (37) and separations by internal electrolysis (38). Sand certainly attracted some very able research students. He and his research group gave great impetus to the practice of electrogravimetric analysis. Interestingly enough, Sand's name appears only on the papers that concern the separation of lead and antimony (with E.M. Collin) (39), then graded potential separations (40), antimony alloys (41), and microelectrolytic techniques (42) (all with A.J. Lindsey). Dr. Lindsey tells me that Sand was most anxious that the credit should go to those who needed it much more than he. Accordingly, other papers by his students do not bear his name, except for an acknowledgment of his interest. Collin wrote three papers on the determination of bismuth (43-45) and one on spelter analysis (46). Lindsey, who became a faculty member of the Institute and was one of my instructors, continued the electrogravimetric program, including the microchemical aspects. His improved microelectrode system became a standard item (47-53). Other active members of Sand's research group were A.J. Fife (54-60), S. Torrance (61-66), and F.C. Kny-Jones (51,67-69).

Sand's retirement in 1938 might have reduced this outflow of results. The collapse was due to the outbreak of World War II in 1939. Lindsey joined the Royal Air Force; doubtless the other members of the group were soon involved in the war effort.

I became an undergraduate in Sand's lecture group in 1936. Holding a full-time job while studying in the evening, I was a bit older than most of the other students. This helped me to understand Sand's obviously failing powers due, I suspect, to ill health. Nearing the end of my list of assigned analytical experiments, I asked "HJS" if I might go on to some of the

Figure 1. Henry J. S. Sand (Reproduced by permission of the Journal of Chemical Education 1977, 54, 637.)

electrogravimetric techniques that were being developed by his graduate students. I remember his penetrating look and his remark, "we'll see!" He was a kindly man, but I think that he had had his fill of undergraduates in general! I never did get to "try out" but, several decades later, did actually publish something on electrogravimetry (70).

In his retirement, Sand wrote "Electrochemistry and Electrochemical Analysis." This appeared in three slender volumes during the War years (71). Volume II, which carries the subtitle "Gravimetric Electrolytic Analysis and Electrolytic Marsh Tests," surveys apparatus and experimental techniques and gives concise instructions for the performance of numerous analyses. It acknowledges Sand's earlier contribution to Lunge and Keane's encyclopedic work (72).

Even in his later years, Sand retained his interest in fundamentals----in his case, heats of reaction (73). Fittingly, his last paper (74), published in the year of his death, reflects a major theme (microelectrolytic methods of chemical analysis) of the research group that he founded. Obviously, Sand was a talented chemist whose activities ranged very widely.

Advances in Electrogravimetry

Electrogravimetry has quite a long history. The technique was successfully performed in the 1860's by both Wolcott Gibbs and by Luckow. By 1903, both Gooch and Exner had shown that analysis could be speeded up by the use of vigorous stirring. Deposition on the large area afforded by a gauze type of platinum electrode had been practiced by Winkler in 1899. Controlled-potential deposition, proposed by Kiliani in 1893, had been demonstrated by Freudenberg some ten years later. Sand did not claim to have discovered any of these phenomena. His great contribution was to tie them all into a theoretically sound and completely practical procedure. He realized that the key was the design of an electrode system that provided maximum stirring efficiency, low electrolytic resistance, permitted the use of a reference electrode for closely monitoring the potential of the working electrode, and allowed easy washing and drying before the reweighing. His answer was to use an inner electrode that could be rotated in a glass sleeve within the outer, or working, electrode. The interelectrode distance of only a few millimeters enabled the electrolytic resistance to be kept small. A glass plate was fixed diametrically within the rotating electrode to increase the stirring efficiency. Because the working electrode surrounds the rotating electrode and thus confines the interelectrode current, the potential of the working electrode can be closely measured by placing the fine tip of a reference electrode near the working electrode. Freudenberg did not use a third, or reference, electrode. More than one half of Sand's long paper (10) is devoted to experimental details for determinative separations and the results obtained.

Sand's original electrodes were constructed from about 2 ounces of platinum. Much later (37), he redesigned the system. The outer electrode, which usually carries the analytical deposit, was reduced in weight to about 10 grams, and was easier to wash, dry, and weigh. This electrode slips over the legs of a glass tripod that is part of the guiding system for the rotating electrode. This electrode system and the stand that carried the drive, meters, and other auxiliary equipment became commercially available.

Sand coined the term "internal electrolysis" to describe a method introduced by Ullgren in 1868. The principle is that of a short-circuited cell; no external power supply is needed. In his apparatus (38), designed originally for the determination of bismuth and copper in lead bullion (44), hollow lead anodes are used. These are immersed in acidified lead nitrate solution contained in parchment thimbles. The lid of the cell supports these anodes, four glass rods that guide the platinum gauze cylindrical cathode, and also the glass tube that serves as a bearing for the centrifugal stirrer.

As far as I can ascertain, the originals of the apparatus developed by Sand and by Lindsey have vanished. In the case of the electrodes, this is not surprising; platinum has a high value in the scrap-metal market. Apparently, we have yet another instance of "a scientist's instruments are his successors' junk!"

Acknowledgment

Part of this work was carried out under the Research Fellowship Program of the Science Museum, London.

Literature Cited

1. Heineman,, W.R.; Kissinger, P.T. Laboratory Techniques in Electroanalytical Chemistry; Dekker: New York, 1984, p. 132.
2. Bamberger, E.; Büsdorf,H.; Sand, H. Ber. 1898, 31, 1513-22.
3. Sand H.J.S. Phil. Mag., 6th series, 1901, 1, 45-79.
4. Sand, H. Z. Elektrochem. 1904, 10, 452-54.
5. Sand H.J.S. Phil. Mag., 6th series, 1905, 9, 20-41.
6. Sand, H.J.S.; Hackford, J.E. Trans. Chem. Soc. 1904, 85, 1018-28.
7. Sand, H.J.S.; Law, D.J. J. Soc. Chem. Ind. 1911, 30. 3-5.
8. Wood, J.T.;Sand, H.J.S.; Law, D.J. J. Soc. Chem. Ind. 1911, 30, 872-77.
9. Sand, H.J.S.; Trotman, S.R. J. Soc. Chem. Ind. 1912, 31, 1166-67.
10. Sand, H.J.S. Trans. Chem. Soc. 1907, 91, 373-410.
11. Sand, H.J.S. Trans. Chem. Soc. 1908, 93, 1572-92.
12. Sand, H. Z. Elektrochem. 1907, 13, 326-27.
13. Sand, H.J.S. Analyst 1908, 33, 395-6.
14. Sand. H.J.S. Chem. News 1909, 100, 269-70.
15. Sand, H.J.S. Trans. Faraday Soc. 1909, 5, 159-64.
16. Sand, H.J.S. Trans. Faraday Soc. 1911, 6, 205-11.
17. Caven, R.M.; Sand, H.J.S. Trans. Chem. Soc. 1911, 99, 1359-69.
18. Caven, R.M.; Sand, H.J.S. Trans. Chem. Soc. 1914, 105, 2752-61.
19. Sand, H.J.S. Proc. Roy. Soc. 1905, 74, 356-69.
20. Slater, A.; Sand, H.J.S. Trans. Chem. Soc. 1910, 97, 922-27.
21. Sand, H.J.S. J. Soc. Chem. Ind. 1907, 26, 1225-26.

22. Sand, H.J.S. Proc. Phys. Soc. London 1914, 26, 127-30.
23. Sand, H.J.S.; Reynolds, E. Brit. Pat. 23 854, 1913.
24. Sand, H.J.S. Brit. Pat. 15 639, 1913.
25. Sand, H.J.S.; Weeks, E.J. Trans. Chem. Soc. 1923, 2896-2901.
26. Sand, H.J.S.; Weeks, E.J. J. Chem. Soc. 1924, 125, 160-68.
27. Sand, H.J.S.; Grant, J.; Lloyd, W.V. J. Chem. Soc. 1927, 378-96.
28. Sand, H.J.S. Rec. Trav. Chim. 1927, 46, 342-49.
29. Sand, H.J.S. Trans. Faraday Soc. 1930, 26, 19-26.
30. Sand, H.J.S. J. Chem. Soc. 1929, 214.
31. Sand, H.J.S. Phil. Mag. 1920, 39, 678-79.
32. Sand, H.J.S., Phil. Mag. 1923, 45, 129-44.
33. Sand, H.J.S.; Lloyd, W.V. J. Chem. Soc. 1926, 2971-73.
34. Sand, H.J.S.; Brown & Son. Brit. Pat. 314 923, 1928.
35. Becall, T.; Challenger, G.; Martin, G.; Sand, H.J.S. Dyestuffs and Coal Tar Products, 4th ed.; Technical Press, Ltd.: London, 1926.
36. Sand, H.J.S. Brit. Assn. Adv. Sci., 4th Rept. 1921, 346-56.
37. Sand, H.J.S. Analyst (London) 1929, 54,, 275-82.
38. Sand, H.J.S. Analyst (London) 1930, 55, 309-12.
39. Collin, E.M.; Sand, H.J.S. Analyst (London) 1931, 56, 90-93.
40. Lindsey, A.J.; Sand, H.J.S. Analyst (London) 1934, 59, 328-35.
41. Lindsey, A.J.; Sand, H.J.S. Analyst (London) 1934, 59, 335-38.
42. Lindsey, A.J.; Sand, H.J.S. Analyst (London) 1935, 60, 739-43.
43. Collin, E.A. Analyst (London) 1929, 54, 654-55.
44. Collin, E.A. Analyst (London) 1930, 55, 312-18.
45. Collin, E.A. Analyst (London) 1930, 55, 680-82.
46. Collin, E.M. Analyst (London) 1930, 55, 495-501.
47. Lindsey, A.J. Analyst (London) 1935, 60, 598-99.
48. Lindsey, A.J. Analyst (London) 1935, 60, 744-46.
49. Lindsey, A.J. Analyst (London) 1938, 63, 159-62.
50. Lindsey, A.J. Analyst (London) 1938, 63, 425-26.
51. Kny-Jones, F.C.; Lindsey, A.J.; Penney, A.C. Analyst (London) 1940, 65, 498-501.
52. Lindsey, A.J. Analyst (London) 1948, 73, 67-73.
53. Lindsey, A.J. Analyst (London) 1948, 73, 99.
54. Fife, J G. Analyst (London) 1936, 61, 681-84.
55. Fife, J.G.; Torrance, S. Analyst (London) 1937, 62, 29-31.
56. Fife, J.G. Analyst (London) 1937, 62, 723-27.
57. Fife, J.G. Analyst (London) 1938, 63, 650-51.
58. Fife, J.G. Analyst (London) 1940, 65, 562-63.
59. Fife, J.G. Analyst (London) 1941, 66, 192-93.
60. Fife, J.G. Metallurgia 1948, 30 (177), 167-69.
61. Torrance, S. Analyst (London) 1936, 61, 688-89.
62. Torrance, S. Analyst (London) 1937, 62, 719-22.
63. Torrance, S. Analyst (London) 1938, 63, 104-07.
64. Torrance, S. Analyst (London) 1938, 63, 488-92.
65. Torrance, S. Analyst (London) 1939, 64, 109-11.
66. Torrance, S. Analyst (London) 1939, 64, 263-64.
67. Kny-Jones, F.C. Analyst (London) 1939, 64, 172.
68. Kny-Jones, F.C. Analyst (London) 1939, 64, 575-77.
69. Kny-Jones, F.C. Analyst (London) 1941, 66, 101-04.
70. Olm, D.D.; Stock, J.T. Mikrochim. Acta 1977, 2, 575-82

71. Sand, H.J.S. Electrochemistry and Electrochemical Analysis; Blackie: London, Vol. I, 1939; II, 1940: III, 1941.
72. Lunge, G.; Keane, C.A., Eds.; Technical Methods of Chemical Analysis; Gurney & Jackson: London, 1914.
73. Sand, H.J.S. Nature 1937, 140, 809-10.
74. Sand, H.J.S. Metallurgia 1944, 30 (176), 107-09.

RECEIVED August 3, 1988

INDUSTRIAL ELECTROCHEMISTRY

Chapter 33

Industrial Electrochemistry

James J. Leddy

Central Research, The Dow Chemical Company, 1776 Building, Midland, MI 48674

The history of industrial electrochemistry parallels the development of the electrical industry itself. Both contain fascinating blends of good science mingled with trial and error methods. These also paralleled the development of physical chemistry and engineering science, the latter being especially influenced by the accelerating demands for the invention of new processes and devices to take advantage of the availability of large amounts of relatively inexpensive electrical energy. Similar incentives also drove the development of electrical storage systems, as well as the growing need for remote energy sources. Most fascinating of all are the entrepreneurs who in their fledgling years gave us the foundations of great commercial enterprises whose roots remain in industrial electrochemistry.

Fueled by accelerating growth in the world's consumption of textiles, paper, fertilizers, soap and glass, the parallel needs for the chemicals used in these enterprises also mushroomed into exciting opportunities for the chemical industry in the second half of the nineteenth century (1-5). In particular, the British development in papermaking in which esparto grass began to supplant rags and straw as sources of cellulose required considerable increases in the production of bleaching powder. British imports of esparto rose from 16 tons in 1861 to 200,000 in 1887, requiring increases in the production of bleaching powder which rose from 13,000 tons in 1852 to 90,000 in 1874 (6). This was a good fit for the Leblanc Alkali producers, who employed various modifications of the Deacon conversion of waste hydrochloric acid into chlorine, and then into bleaching powder. This conversion was also spurred by early environmental problems caused by the Leblanc process plants venting their waste HCl to the atmosphere. Government intervention was required in the form

0097–6156/89/0390–0478$09.00/0

of the British Alkali Act of 1863 to force the introduction of scrubbers (7). Stream pollution then resulted from the direct sewering of these scrubber effluents, necessitating passage of the Alkali Act of 1874 (8). These incentives firmly entrenched the alkali industry in the manufacture of bleach. But the rediscovery and perfection of the ammonia-soda process by the Solvay brothers into an alternative source of soda proved troublesome to the Leblanc interests. Comparatively simpler and less consumptive of raw materials and energy than the Leblanc process, the Solvay approach produced a higher quality product at a lower cost. To compete with this increasing aggravation the Leblanc producers in Britain banded together in 1890 (42 plants in England, 4 in Scotland, 1 in Ireland, and 1 in Wales) to form the United Alkali Co., Ltd. (9). The price of bleaching powder was fixed by them and kept sufficiently high to subsidize the production of soda ash and lye (causticized soda). The world's first open conflict based on imbalanced supplies of chlorine and alkalies was in full swing. Leblanc was in trouble even more than it realized. Lurking in the wings was the ultimate usurper of both Leblanc and Solvay: electrolysis. But even prior to the beginnings of that impact the Leblanc interests should have considered ominous the visit of Ludwig Mond, a renowned Leblanc practitioner, to the Solvay brothers in 1872. The result was the introduction of the ammonia soda process to Britain in 1874 with the formation of Brunner, Mond and Co. By 1890, the Leblanc process was entering the final stages of its death struggle with the Solvay process, and the electrolytic process was about to help Solvay finish off Leblanc before turning its attention to an ultimately similar fate for ammonia soda.

Industrial incentives for the development of electrolytic processes were lacking until the late 1800s brought the commercial availability of electricity, even though as early as 1807 Davy had demonstrated the first electrolytic cell with his isolation of potassium and sodium metals from their molten hydroxides. Watt patented an electrochemical preparation of chlorine and caustic in 1851, but Matthes and Weber are credited with the first electrolysis of brine in a diaphragm cell which was patented in 1886. This cell formed the basis of an 1890 venture in Germany by the Chemische Fabrik Griesheim, who purchased the patent and used the cells to manufacture KOH, eventually expanding operations to Bitterfeld (10). The chlorine in these plants was converted to bleaching powder, and the Germans began impacting the British position in this arena. In the meantime the North American market for bleaching powder had risen by 1895 to a then-respectable two million dollars per year. The United Alkali syndicate was charging $0.035 per pound for bleaching powder delivered to the U.S., which is equivalent to nearly $0.10 per pound of chlorine, a very handsome price in 1895 (11). The result was predictable: North American talents turned their attention to the opportunity.

There is a fascination in studying the early developments of electrochemical processes when it is remembered that before 1846 there was no real understanding of the relationship between voltage, current and resistance. There was as yet no Ohm's Law, even though his basic work had been done some twenty years earlier (12). Volta's battery led to Oersted's discovery of the magnetic effect of current and his construction of the galvanometer, with which Michael Faraday discovered electromagnetic induction. This first recorded conversion of mechanical into electrical energy was reported in 1831 (13). Joseph Henry in the U.S. and the Russian, Heinrich Lenz, also made this discovery about the same time (14). Not only had these three uncovered the basis for future designs and operations of transformers and generators, but two years later in 1833 Faraday published the all-important electrochemical laws which bear his name (15). But the new age, the second industrial revolution, dominated by electric-powered technology, christened neotechnic by Lewis Mumford (16), experienced a long gestation period while awaiting the inventions and entrepreneurship of people like Siemens in Germany; Holmes, Wilde and Ferranti in Britain; Edison, Brush, Sprague, Westinghouse and Tesla in the U.S.; and Gramme in France.

In the meantime electricity was available for limited, small scale commercial and research uses only in the forms of the various hand-cranked generators and the galvanic cells patterned after Volta's original voltaic pile built in 1800, consisting of pairs of copper and zinc discs separated between every other disc by layers of pasteboard soaked in brine (17). A commercial variant of the voltaic pile was the 1836 Daniell cell, which was supplanted in 1867 by the Leclanche dry cell, a much superior device. Plante devised an excellent secondary cell in 1859, but the lack of generators for recharging limited its use. J. C. Fuller improved the Daniell cell for telegraphic use in 1853. This was superseded in 1875 by his own mercury dichromate cell which gained wide acceptance. With the advent of practical, reliable generators in 1870 the lead-acid secondary systems largely replaced the primary cell in telegraphy (18).

The other practical use of electrochemistry in this early period was in electroplating, principally by the contact method used to plate copper or to apply thin coatings of gold or silver on less valuable substrate metals to limit corrosion and for decorative purposes. There was also some limited activity in the plating of Ni and Zn. Electroplating and electrorefining grew rapidly in the variety and importance of their applications only with the advent of the dynamo in 1870 and the eventual availability of relatively inexpensive central station power, paralleling the situation in the electrolytic process industry. Chemistry waited in the wings for the debut of the electric power industry.

The Electrical Industry

By themselves none of these chemical uses of electricity would have forced the huge investments which were made by the electrical industry in developing prime movers, generators, transformers and distribution systems. This impetus stemmed almost exclusively from the commercial advantages obtainable from electric lighting and in motive systems both for driving stationary machines and for traction. The ten-fold growth in the latter from 2200 miles of electric street-railway trackage in the U.S. alone in 1890 to 22,000 miles only twelve years later is illustrative. As well, at the turn of the century there were 3620 electric-light stations in the U.S., representing an investment of about a half of a billion dollars, plus the annual manufacture of eighteen million incandescent lamps and nearly 400,000 arc lamps (19). A scant twenty-five years earlier essentially none existed anywhere.

The conversion of mechanical into electrical energy at reasonable cost on a scale sufficient to permit the construction of central power stations required solutions to many difficult design and engineering problems. Not the least of these were the replacement of permanent magnets with electromagnets and the use of self-excitation based on the residual magnetism in the field poles of a generator. These were achieved by Werner von Siemens in Germany and by Henry Wilde in Britain in the mid 1860s. Another key development was the use of an armature coil of soft-iron wire insulated with bitumen during winding to reduce eddy currents. This 1870 invention by Z. T. Gramme, a Belgian working in France, gave us the first dynamo of practical dimensions capable of sustained operation while producing a continuous current (20). Its success was immediately applied to arc-lighting in France where more progress was made in illumination than anywhere else at that time, thanks to Gramme. The Siemens' interests answered immediately with the drum armature which is more cost-effective than the ring armature. Also, the embedding of the armature conductors in slots within the core, so recognizable to us today, was invented in 1880 by Wenstrom in Sweden (21). In the U.S. in 1879 Edison patented his version of a d.c. generator, claimed to be the most efficient then available, converting 80% of the applied horsepower into useable electricity compared to the Siemens' machine at 55% (22). But then Edison was not noted for his modesty. The World's first electrical exposition was held in Paris in 1881, followed by a second one in London in '82. Edison displayed a 27 ton generator which won top honors. That same year his group put into operation the world's first two d.c. central power stations to serve private consumers, one in London and the other in New York. The latter was run by Charles Clarke, who in 1881 built the first direct-coupled generator. This elimination of the belt drive between the generator and the prime mover was another important advance in the struggle to increase efficiency.

Edison was approached in 1881 by George Westinghouse, who had invented the railway air brake some twelve years earlier, to enter into a partnership to build high speed, direct-coupled turbo-generators, with an eye toward using a.c. electricity for transmission, following the lead of Ferranti and Wilde in Britain and Gramme in France (23-24). Edison declined impolitely. Westinghouse decided to go it alone. In '86 William Stanley had demonstrated transforming 500 volt a.c. up to 3000 volts, transmitting it a mile, and then transforming it back down to 500 volts. He offered his patents to Edison, who turned him down as well, believing a.c. power would never amount to much. Westinghouse then bought Stanley's patents, and in '88 hired the gifted Tesla who had left Edison's employ when the latter refused him a raise from $18.00 to $25.00 per week. Tesla's genius in understanding and working with alternating current paid off handsomely for Westinghouse, who championed the use of a.c. A bitter struggle followed between Edison and Westinghouse, who emerged victorious when his company was awarded the contract for a.c. generators and auxiliaries for the then world's most ambitious hydroelectric project at Niagara Falls, New York. Edison did not distinguish himself in that arena.

Prime Movers

Rapid growth in demand for electrical energy had fueled the design and construction of ever-larger prime movers, both hydraulic and steam driven. Internal combustion engines did not begin to impact the production of electricity until after the turn of the century. At the Chicago Exposition in 1893, for example, the largest internal combustion engine delivered a token 35 horsepower compared to one capable of delivering 1000 hp displayed only seven years later at the Paris Exposition of 1900 (25). By comparison, the famous Corliss steam engine displayed in 1876 at the Philadelphia Centennial was rated at 1400 hp with steam supplied at 20 psig. Fifty years later, in 1926, a Corliss typically delivered 20,000 hp using steam at 600 psig. But the real excitement in the development of prime movers for electrical generation lay in the work on turbine drives with the prime mover mounted directly on the generating shaft. Steam turbines to drive ships had been developed in England by 1890, and along with the earlier-perfected hydraulic turbines there was soon to be a very substantial impact on industrial electrochemistry.

The first hydroelectric plant was a lighting station energized in 1882 on the Fox River at Appleton, Wisconsin, using two Edison d.c. generators powered by a 42-inch water wheel operating under 10 feet of head generating 25 kw. The first a.c. hydro plant began operation in 1890 on the Willamette River in Oregon, with the power transmitted at 4000 volts thirteen miles to Portland. Similar installations were in place by then in Dahlhausen, Germany and in Tivoli, Italy, where a second a.c. hydro plant was built in 1892 to supply Rome from sixteen miles away (26). But none to date could match the plans of the syndicate which had been granted the authority in 1886 to harness sufficient water from the Upper Niagara River to develop 200,000 hp with a 500 ft. wide by 12 ft. deep head race 1.5 miles long, and a drop of 160 ft. to a 6700 ft. long tail race 21 ft. high by 19 ft. wide (27). By 1895 the Niagara Falls Power Co. had the capability of delivering 100,000 hp as well as having its first contracts in hand. These are most important to us in that they were not contracts for lighting or traction, but for the electrolysis of alumina-in-cryolite by the Pittsburgh Reduction Co., later to become Alcoa, and for the electrothermal manufacture of SiC by the Carborundum Co. (28). Charles Martin Hall and

Edward G. Acheson were on the scene. Chemistry and electricity were now joining forces on a respectable scale. Only a year later did the city of Buffalo sign up for power.

The manufacture of neither electrolytic aluminum nor electrothermic silicon carbide required any other resources from Niagara Falls than the inexpensive electricity. But just as the first U.S. Solvay plant had been located at Syracuse because of a favorable position in raw materials, especially salt, there would soon be a number of companies locating at Niagara Falls not only for its inexpensive power, but for the huge subterranean salt deposits in the area. The ingredients essential for the commercialization of electrochemistry on a grand scale were finally coming together. The lowest cost prime mover drove an efficient and controllable electrical system to energize chemical reactors (be they electrolytic cells or electrothermic devices) to convert inexpensive raw materials into higher-valued products which were in demand.

Products and Processes

Electrorefined copper is the "chicken and egg" of industrial electrochemistry. From mid-century on, the copper industry was driven by the ever-increasing needs of the electrical and communications industries. It was not long before the principal use for this metal lay at the feet of the combination of its superb conductivity and workability. Commercially available copper in those days, the so-called fire-refined copper, generally had less than 70% of the conductivity of pure copper because of the impurities it contained. For example, the transatlantic cable which was laid in 1856 was fabricated of copper whose conductivity was half that of the then-known value for pure copper. In Wales in 1865, J. B. Elkington invented a process for refining copper electrolytically (29). Impure anode slabs of copper are interleaved with thin plates of cathode copper suspended in a bath of copper sulfate. The bath is periodically purged to remove impurities, and the sludge which forms around the anodes is further processed to recover gold, silver, tin, antimony, and other metals depending upon the original ore source. Recovery of valuable constituents from the anode sludge is a step of great importance to the economic viability of the overall process (30). Elkington did a thorough job. Not much in his process has required changing to this day. His first refinery was constructed near Swansea in 1869, and it included his own design of a d.c. generator employing 1400 pounds of permanent magnets. He was a real pioneer. The first installations in the U.S. were at Newark, New Jersey in 1883, and Laurel Hill, New York in 1892. Eventually all major copper producers installed electrorefining capability. By 1914, electrolytic copper had become the official basis for price quotations (31).

Electrolytic aluminum was in commercial production only two years after the independent discoveries of the same process were made in 1886 by Charles Martin Hall in Oberlin, Ohio in the U.S., and by Paul Heroult in France. They had found that alumina is soluble in molten cryolite, a mineral found in Greenland, largely composed of sodium fluoroaluminate. Electrolysis of this melt produces aluminum metal at the carbon cathode and carbon dioxide at carbon anodes at 950°C in a semi-continuous cell operation, identical in both inventions and essentially the same as practiced today. The inventors were lucky that mother nature ordained the density of molten cryolite to be less than that of aluminum, which sinks to the bottom of the cell. By strange coincidence these two 22 year-old young men, who knew nothing of each other until after the facts of their inventions brought them into conflict with one another's interests in the U.S. Patent Office, both arrived at the same invention by different approaches, each commercialized his discovery into the beginnings of great industries on both sides of the Atlantic, and both died in the same year at age fifty in 1914 (32). Hall was delayed in getting his patent because of interference from Heroult's claims in the U.S. Patent Office. Finally, on July 9, 1886, Hall was granted the basic U.S. patent (33) on the basis of satisfactory proof of his reduction to practice on Feb. 23, 1886, compared to Heroult's documentation of reduction to practice as of April 23 of that same year. Each of these young inventors took a different approach in commercializing his process. Heroult sold licenses in 1887 to a Swiss company which eventually became Aluminium Industrie A.G., then to the French Societie Ellectrometallurgique (Pechiney), and finally to interests which established the British Aluminium Co. Heroult also distinguished himself in the development of electrothermics, especially in electric-smelting and in steel making.

In 1886-87 Hall contracted to develop his process at the Cowles Electric Smelting and Aluminum Co. in Lockport, N.Y. The Cowles brothers had pioneered electric furnace smelting and were producing aluminum alloys, especially aluminum bronzes, but not primary aluminum. Cowles shortsightedly did little to encourage Hall. But the association was not without value to Hall, however, since it was there that he met Romaine Cole, a young man with the entrepreneur's gift. Cole knew of a prominent young metallurgist in Pittsburgh, A. E. Hunt, who, though in his early thirties had already distinguished himself in the iron and steel manufacturing circles. When presented with samples and details of Hall's process, Hunt secured sufficient venture capital to build a pilot plant and founded the Pittsburgh Reduction Co. in 1888 with himself as president, granting Hall and Cole a 47% interest, three-fourths of which was Hall's. When Hall died in 1914 he left an estate of thirty million dollars. In the meantime, Cowles Aluminum Co. blatantly began electrolyzing alumina in cryolite at Lockport in 1890. A bitter lawsuit followed (34).

But why cryolite? Why would these two young men, one a metallurgist in Europe, the other a chemist in North America, choose a mineral from Greenland with which to experiment? The answer may be related to an 1850 connection between the Pennsylvania Salt Manufacturing Co. in the U.S., and the Kryolith Co. of Denmark, to a process invented by Julius Thomsen of that company (35), and to the earlier discovery of cryolite on Danish soil in Greenland. The Kryolith Co. used Thomsen's process to manufacture both alkali and alum from lime and cryolite. The Pennsalt Co. operated its first plant in Natrona, Pennsylvania, near Pittsburgh. They were one of the few companies in the Americas to manufacture substantial amounts of alkali at the time of the Civil War in the U.S.. In 1865 Pennsalt began to import cryolite from Greenland and to convert it into alkali, alumina and fluorspar. Pennsalt saw an opportunity to market the alumina as aluminum sulfate to the growing paper industry. That business grew so well that they began to import bauxite from France to augment their supply of alumina in order to be able to produce sufficient quantities of alum. It is not unlikely that an investigator dedicated to finding a practical route to aluminum would not have known about Pennsalt if he were doing his work in Ohio; or that one would not have known about the Kryolith Co. in Denmark if he were doing his work in France. In any event it is a fact that the first Bayer process plant for purifying bauxite was installed at Natrona by Pennsalt, who supplied the Pittsburgh Reduction Co. with their raw materials, alumina and cryolite (36). In the latter reference Haynes comments that there was an unsubstantiated rumor that Hall originally offered his process to Pennsalt, who turned it down.

The world's first potline of only a few 4000 ampere Heroult aluminum cells was energized at Neuhausen, Switzerland in 1887 using hydropower generated at the famous falls of the Rhine (37-38). The Hall-Heroult process then required about 12 kwh per pound of aluminum. Most of this energy is required to keep the bath molten, so that the process is actually both electrothermal and electrolytic. This high sensitivity to energy costs reflected itself in the early dash of the Hall-Heroult practitioners for hydroelectric power plants. The Pittsburgh Reduction Co. expanded from Pittsburgh to New Kensington, Pennsylvania in 1890, but ceased electrolytic operations there in 1896 when the new facility was operational at Niagara Falls because the fuel-based power cost at New Kensington could not compete with the low cost hydropower. Half of the energy required to produce aluminum at Niagara Falls in 1896 was purchased as d.c. produced by five 600 kw rotary converters. The other half was purchased as motive power as the forerunner of Alcoa positioned itself to enter the business of making their own electricity on a major scale. By 1899 another operation was begun at Shawinigan Falls, Quebec, where energy was purchased in the form of water in the forebay. This was expanded in 1907, the year the Pittsburgh Reduction Co. changed its name to The Aluminum Company of America.

By 1909 Alcoa had decided to take advantage of the water in the mountain streams of North Carolina and eastern Tennessee. The first pot line was energized in 1914 at Alcoa, Tennessee, and soon became the world's largest aluminum plant. Alcoa's harnessing of water power paralleled the U.S. government's building of Wilson Dam and the eventual establishment of the Tennessee Valley Authority, which became a major supplier of electricity to many electrothermic process plants. The original incentive for the U.S. government was to establish nitrogen fixation on a scale sufficient to cease dependence on imported nitrates for the production of explosives and fertilizers. In fact, Alcoa completed a plant in May, 1918 to fix nitrogen by the carbothermal conversion of alumina to aluminum nitride in an electric furnace. The AlN was to be hydrolyzed to ammonia, and the alumina recycled to process acting as a carrier to fix the nitrogen from air. But the plant was never operated because the armistice was only a few months away in November of 1918 (39).

From 1919 through World War II several dams and power plants were added in that system by Alcoa for a total power capacity of over 500 megawatts. Alcoa spent one dollar on hydroelectric plants for every dollar spent on building process plants. In 1939 they were the first customer of Bonneville Power in the state of Washington, and by WW(II) were producing 100 million pounds of aluminum annually at the Vancouver plant alone. By 1948 Alcoa was producing 3.5% of the total hydroelectric power in the U.S. (40).

But times change, and the hydroelectric sources are not unlimited. Alcoa elected in the 1950s to build a nominal 100 million pound per year plant at Point Comfort on the Texas Gulf Coast. The plant was energized with 120 one megawatt internal combustion engines fueled with natural gas from off-shore gas wells about 12 miles from the plant in the Gulf of Mexico (41). After OPEC such operations have not been without hard times. In fact, in 1986, there were several plant closings, including portions of the plants at Point Comfort and at Alcoa, Tennessee. These rationalizations also reflect the general movement of many basic industries to areas of the world where costs are lower (42).

As it grew to enormous world proportions from a few producers making two million pounds per year in 1893 to many producers making over fifteen billion pounds per year in 1970 (43), the aluminum industry contributed much to the evolution and development of the chemical engineering profession, particularly as competition forced attention to economy of scale. Potlines grew to 150 cells in single 150,000 ampere circuits, while requisite d.c. cell energy dropped from about twelve kwh/lb to less than eight. Such advances are especially important when the need arises for metal of purity greater than that of primary aluminum at 99.5%. This is achieved by electrorefining the primary aluminum alloyed with copper in so-called three-layer cells such as those invented by Hoopes in 1925 in the U.S., and by Gadeau in France in 1934 (44). A molten anode of Cu-Al alloy rests on the cell bottom separated by a layer of cell bath from a top layer of pure aluminum which forms at the carbon cathodes. These cells also use cryolite-based baths, and can produce purity of 99.99%, but not without doubling the overall energy consumption per pound of aluminum.

From its earliest beginnings this industry succeeded in growing by innovatively satisfying market needs, leaving no potential opportunity unexamined. The gauntlet was run from the prosaic, as in getting housewives to purchase aluminum cookware, to the innovative, as in inventing steel-reinforced aluminum cable to compete with copper in electric transmission lines (45). As it grew, the aluminum industry impacted not only the metals and electrical industries, but sister industries as well, especially petrochemical and chlor-alkali concerns. The preferred source of carbon is petroleum coke for making either pre-baked (Pechiney) anodes or the continuously-fed paste anodes (Soderberg) which are baked in-situ in the cells (46). Hall-Heroult cells consume about 0.5 lb of carbon per lb of aluminum. Petroleum coke is also the source of carbon in the manufacture of graphite. Yet, petroleum coke is only a by-product of oil-refining operations, which relegates both carbon and graphite electrode manufacture to by-product status, heavily dependent upon the petroleum industry. These by-products impacted world chlor-alkali operations until about 1980 and continue to impact the steel, aluminum and magnesium producers.

But the impact which the aluminum industry makes on the ever-present problem of chlor-alkali balance in the world is substantial. The electrolysis of brine produces almost equal weights of chlorine and sodium hydroxide as co-products. Typically, when one is in demand the other is not, creating an aggravating problem of imbalance. Even though the Bayer process to purify alumina for aluminum cell feed recycles as much caustic as possible within its own confines, nevertheless the size of that industry's global consumption of caustic soda for process make-up alone is enough to get equal attention in both the board rooms and the production plants of the world's chlor-alkali producers. Not many consumers of caustic soda order it by the shipload as do the Bayer processors. Historically, as the aluminum business goes, so goes caustic, though there are certainly other big players.

But the subject of aluminum manufacture is poorly addressed if it does not include highlights from the all-too-short life of the great industrial chemist and inventor, Hamilton Young Castner, whose fame is not generally connected with aluminum (47-48). Born in Brooklyn, New York, in 1858, he attended Columbia University for three years, leaving without a degree in 1878 to start a chemical consulting and analytical business, which gave him the opportunity he sought to experiment and develop his own ideas. Principal among these was his pursuit of a low cost method for making sodium metal which he intended to use in the long-known St. Claire-DeVille process for making aluminum metal by the reduction of aluminum chloride with sodium. Aluminum was very expensive in those days principally because sodium was expensive. Castner attacked the problem by going after an economically viable route to sodium. He succeeded and was granted a patent in 1886 for his process of using coal-tar pitch, anhydrous caustic and iron filings to make sodium, which distilled out of the reactor. Shunned by industrialists in the U.S. he sailed to England in late '86, obtaining the support of the Webster Crown Metal Co. to establish a new venture: The Aluminium Co., Ltd. By 1888 Castner had built a plant at Oldbury in Britain to produce 100,000 pounds per year of aluminum. He enjoyed a short-lived monopoly. Hall-Heroult shut him down. In a classic case of necessity spawning invention, he doggedly pursued other uses for sodium.

Finding that peroxygen bleaches were in demand for treating woolen goods and for the increasingly popular straw hat, he produced sodium peroxide by burning sodium in air. But there came an even greater, more far-reaching opportunity. At about this same time the Forrest-MacArthur process for extracting gold and silver from low grade ores (by taking advantage of the formation of cyanide complexes) began to be adopted by the world's mining companies. Patented in 1887 it gained wide usage very quickly in South Africa, Australia and the Americas, creating a first rate market for the cyanides of sodium and potassium. Castner moved fast. By reaction between ammonia and sodium he prepared sodamide, and then reduced it to cyanamide and cyanide with hot charcoal.

The Deutsche Gold- and Silber-Scheide Anstalt (DEGUSSA) were primary suppliers of potassium cyanide, which they produced by reducing potassium ferrocyanide with sodium metal. By 1891 all of DEGUSSA's needs were being met by Castner. Similarly in 1893, another prominent mining concern, the Cassel Gold Extraction Co. of Glasgow, adopted Castner's process for making NaCN and purchased sodium from his Aluminium Co., which by now was a misnomer. To keep up with demand and lower his costs, Castner turned to the electrolysis of fused caustic to produce sodium.

Electrolytic sodium became a commercial reality when Castner devised a practical cell for electrowinning sodium on a steel cathode which was separated from a carbon anode by a diaphragm made of wire mesh. Castner's design was so effective that these cells were used to produce sodium in Britain from 1891 to 1952. They were superseded by the Downs cell in Niagara Falls in 1925. Patented in 1924 (49), Downs' cell replaced Castner's electrolyte of fused NaOH with one consisting of about 59% calcium chloride and 41% NaCl, producing sodium metal at the steel cathode and dry chlorine gas at the carbon anode (50). J. C. Downs, an American, was trained as an electrical engineer. His cell reduced the energy consumption of Castner's overall process by almost 50%, thus becoming the superior route to sodium metal. Downs began his career in Niagara Falls with the Acker Process Co., but moved to the Niagara Electro Chemical Co. when the Acker plant burned down in 1907. Niagara Electro Chemical was eventually absorbed by the firm of Roessler and Hasslacher Chemical, which, in turn, became the Electrochemicals Department of the duPont Company in 1930 (51). By then there were fewer needs for NaCN and more for HCN, which was produced from methane and ammonia. But a new, large use for sodium had arrived in the form of a process to produce lead tetraethyl using a sodium-lead alloy. This use accounted for 70% of the sodium output of the electrolytic sodium cells.

Before the impact of Downs on Castner's sodium cells, Castner was looking for ways to remove impurities, especially silica, from the caustic feed to his cells. In characteristic fashion he decided to invent a means of preparing pure NaOH. It was already known in 1889 that a Canadian, E. A. LeSueur, had demonstrated a cell for the aqueous electrolysis of NaCl brine using a mercury cathode. Castner followed LeSueur's lead, but made a most significant breakthrough by eliminating the diaphragm. He devised a three compartment cell with a mercury seal across the bottom of the cell to prevent mixing of the cell liquors. The mercury was made cathodic, with carbon anodes inserted into each of the two end compartments. Brine was fed to these end compartments, and water was fed to the middle compartment to convert the sodium amalgam which forms at the mercury cathode into caustic and hydrogen. By impinging a cam drive under one end of the cell and a pivot under the opposite end, the cell was gently rocked back and forth as electrolysis proceeded. The overvoltage of hydrogen on mercury is high enough to prevent the decomposition of water at the cathode, allowing the deposition of sodium into the mercury. Wet chlorine is discharged at the carbon anodes (52). As the dilute sodium amalgam moves into the bottom of the middle compartment it decomposes the water, producing a strong solution of very pure NaOH which is continually removed from the cell. The hydrogen was vented to the atmosphere. Castner had given birth to the mercury chlor-alkali cell.

In seeking patents for his new cell in 1894, Castner found that he had been anticipated in Germany by the Austrian inventor, Karl Kellner, who was then attempting an installation near Salzburg. To avoid litigation over the interfering German patent of Kellner's, even though the latter's specifications were flawed, Castner reached an agreement with Kellner, and the Castner-Kellner Alkali Company was formed. Castner's original site at Oldbury was inadequate in that power was too expensive and there were no salt deposits in that locale. Accordingly, the new company built its first installation of mercury cells at Runcorn, England, in 1897, where there were ample salt deposits. Castner sold his other European rights to the Belgian firm of Solvay and Co., and the U.S. rights were granted to the Mathieson Alkali Works. The latter built a successful pilot plant at their site in Saltville, Virginia in 1896, and immediately established a full-scale operation at Niagara Falls near inexpensive sources of both power and brine. Incredibly, the last of Castner's rocking cells was not retired there until 1960 (53), to his credit.

But Castner was not finished. He was dissatisfied with the performance of the carbon anodes and turned toward the development of graphite electrodes. This brought him into contact with Edward G. Acheson in Niagara Falls. Castner prevailed upon Acheson's knowledge of electrothermics in graphitizing carbon to make graphite electrodes, which precipitated another great advance for industrial electrochemistry (54).

Electrothermic processes were pioneered in the laboratory by Henri Moissan in France, who received the Nobel Prize in Chemistry in 1906 for his two outstanding accomplishments: the isolation of fluorine by the electrolysis of KF in anhydrous HF, and for his development of the electric arc furnace in which previously unattainable temperatures provided the means for synthesizing exciting new materials. Fluorine did not advance in commercial significance until George Cady developed a practical cell just prior to WW(II). The need for uranium hexafluoride during the war hastened developments by Union Carbide and others, but the need remains insignificant in quantity compared to electrochemical products in general. Moissan's greater contribution in terms of its industrial impact was the arc furnace, in which he first made calcium carbide in 1892, followed by tungsten metal and steel alloys. He made an early claim to have discovered SiC as well, but finally gave the proper credit for that find to Acheson (55-56).

In his twenty-fourth year in 1880, Acheson began working with Edison at Menlo Park. He was sent to Paris in '81 to assist with the Edison exhibit at the First International Electrical Exposition, followed by a stint in Italy. In '83 he quit Edison, returned to Paris as a free agent, and then returned almost destitute to the U.S. in '84 on a boat ticket paid for by Edison. Working on his own he patented an anti-induction telephone cable for which he sold the rights to George Westinghouse. Full of ideas on generating and using electricity in new ways, he talked the small town of Monongahela, near Pittsburgh, into allowing him to establish a generating station to provide lighting for the city. He was really after a source of electricity for his experiments. Finally, in March, 1891, he did the key experiment which led to his discovery of silicon carbide, by heating a mixture of clay and powdered carbon in an iron bowl, as one electrode, with an arc-carbon inserted into the mass as the other electrode. He soon devised larger furnaces by loading mixtures of coke, silica sand, salt and sawdust around a core of carbon. In August, 1891, The Carborundum Co. was formed. The first two years were lean, but by 1894 orders were sixty days ahead of production, and The Carborundum Co. was headed for Niagara Falls (57).

By then, Acheson had observed the conversion of carbon to graphite in the hottest parts of the Carborundum furnace. More importantly, he noticed the extent of that conversion was quite dependent upon the presence of impurities in the starting coke, such as silica and iron oxide, which enhanced the conversion. He patented this electrothermic production of graphite in 1896 (58), and established the Acheson Graphite Co. in 1899. Castner made immediate gains in the efficiency of his mercury cell operations by using Acheson's graphite anodes. Soon these electrodes were standard throughout the industry to the credit of both Castner and Acheson.

Another pioneer in the application of electricity to industrial chemistry and metallurgy was the Canadian, Thomas L. Willson, whose early association with the Cowles brothers brought him to the accidental discovery of calcium carbide. Willson worked for Brush Electric in Cleveland, Ohio, in 1885 when the Cowles used that facility to operate their electric furnace (59). At the same time, J. T. Morehead, owner of a cotton mill in Spray, North Carolina, was looking for profitable new uses for the excess capacity he had in a hydroelectric installation which was used to power the mill. Morehead and Willson formed the Willson Aluminium Co. based on Willson's concept of a carbothermal route to aluminum metal using an electric furnace. Poor chemistry and lots of hard work at Spray frustrated Willson into trying a different route. Calcium was chosen as a reducing agent, analogous to the earlier sodium-based process. In the course of this work in 1892 he loaded a mixture of lime and coal into his arc furnace and got an unexpected result. When added to water, the furnace product produced a gas which burned with a very sooty flame, not the clean, barely visible flame expected had the furnace product been calcium metal and the gas hydrogen (60).

Willson sent samples of his furnace product to an outside laboratory for analysis. The solid was identified as calcium carbide and the gas as acetylene. Ever the entrepreneur, Morehead headed north with samples and demonstrations to commercialize Willson's find. Seven plants were built, of which only the one at Sault Ste. Marie, Michigan, and one at Niagara Falls survived. The Michigan plant was owned by a Chicago Co., the Peoples Gas, Light and Coke Co., whose W. S. Horry advanced the electric furnace technology very significantly, a key factor in the survival of this company compared to many of the other early adventurers in the calcium carbide/acetylene business. Horry's invention of the semi-continuous rotary arc furnace brought greater productivity to the process, supplanting Willson's batch carbide furnaces (60-61).

Morehead licensed his company's technology to others only to produce "heat, light and power". The "chemical uses" were retained by the Willson Aluminum Co. Whether clairvoyant, lucky, or both, this move by Morehead portended another smashing success, since chemical uses included metallurgical applications. Developments in alloy steels in France and Germany in the 1890s induced Morehead to use Willson's furnace to smelt ores. Early success in producing high-carbon ferro-alloys with chromium and silicon created another opportunity for the Willson Aluminum Co., which by now, like Castner's Aluminium Co., was a misnomer. Unlike Castner, the Willson operation never did produce aluminum. In terms of commercial success it didn't need to.

In the meantime Willson returned to Canada. He there established carbide operations in Merriton, Ontario, and at Shawinigan Falls, Quebec. He formed the International Marine Signal Co. to manufacture carbide-energized buoys, and applied himself to the use of the electric furnace for smelting phosphate ores in his remaining years (62). Willson died in 1915, by which time he had seen his invention produce 90,000 tons of calcium carbide annually by 1904 and 250,000 by 1910, from zero in 1892 (63). He perhaps would have been amazed to have witnessed the growth in the chemical uses of acetylene equivalent to one million tons per year of calcium carbide by 1960, produced in continuous furnaces which were 30 feet in diameter by 15 feet tall, each rated at 30,000 kw (64). Nor could he have foreseen his furnace eventually supplanted as a source of acetylene by yet another electrothermic process, the direct formation of acetylene in an electric arc used to crack hydrocarbons such as natural gas.

The success of the Chicago group at Sault Ste. Marie with Horry's modifications of Willson's process led them to seek new capital, which led to the formation of a new company in 1898: Union Carbide. The new company acquired the plants at Sault Ste. Marie and Niagara Falls, where their power contract was expanded from the original 1000 hp to 15,000 hp in 1899 (65). Solicitous of future growth as well as present optimum utilization of their power contracts, the interchangeability of the existing furnaces to produce either calcium carbide or ferro-alloys was very attractive.

In his pursuit of metallurgical advances using the electric furnace, Morehead had sought the services of Frederick M. Becket, who had organized the Niagara Research Laboratories in 1903. Becket had established himself at age 30 as a pioneer in the metallurgy of refractory metals, such as chromium, vanadium, molybdenum and tungsten. Morehead's ferro-alloy business beckoned Union Carbide, who particularly could not resist locking in the services of the talented young Becket. In 1906, Carbide absorbed both Morehead's Willson Aluminum Co. and Becket's Niagara Research Laboratories, forming them into the Electro Metallurgical Co., a unit of Union Carbide. Becket was a charter member of the Electrochemical Society, serving as its president in 1925. He also became president of the Union Carbide Research Laboratories and a vice president of Union Carbide. A graduate engineer and metallurgist with very solid training in physical chemistry, Becket brought real science to the field of industrial research, particularly in electrothermics (66-68).

The extraordinarily high temperature of the oxyacetylene flame and its usefulness in cutting and welding was known by 1901. Carbide producers rushed to capitalize on this new use. Oxygen was commercially available at that time only from the Brin process, based on BaO and air, or from the electrolysis of water. The German physicist, Carl von Linde, had pioneered the refrigeration industry and was its technical leader for twenty years when, in 1895, he demonstrated the liquefaction of air. On a visit to Cleveland in 1906 he caught the attention of a group from the National Carbon Co., and before he left town the Linde Air Products Co. was born. The National Carbon Co. itself had been established in 1886 by W. H. Lawrence, who had worked for Brush Electric. Both companies manufactured carbon rods for use in arc-lighting. Brush had found that much improved performance of the arc-carbon was achieved if they were made from petroleum coke. National Carbon expanded into using their superior grade of carbon to build another electrochemically-based business as well. In 1896 the Columbia Dry Cell was marketed. Patented by Coleman in 1893 it was the premier mobile electric power source in its day, enjoying the growing telephone industry. By the 1900s its use expanded into and grew with the automotive market. This was followed by the early days of radio. Telegraphy and railroad signaling employed primary batteries as well as the emerging secondary cells. Union Carbide was impressed with these advances made by National Carbon, particularly those which fit their own interests, as in the improving quality and performance of ever-larger carbon electrodes. In 1917 Union Carbide, including the Electrometallurgical Company, joined with Linde Air Products, National Carbon and others to form the Union Carbide and Carbon Corporation. In 1928 Acheson Graphite was added to the fold. The world's most innovative user of electricity in the field of electrothermics was now the top supplier of electrode materials for the rest of the electrochemical industry as well (69).

Beginning with arc-carbon rods for anodes and ending with graphite replacing carbon, the fortunes of the coke-based electrothermic operators and those of the practitioners of electrolysis of brine to produce chlorine, caustic soda and hydrogen were parallel. The Chlor-Alkali Industry epitomizes the achievements which applied science and sound engineering practice have made since the late 1890s. Early plants contained cells running at a few hundred amperes furnished by motor-generator sets which were difficult to control. The surroundings were primitive with wood the most often used material of construction, including for the cells as well as the armature cores in the generators. These evolved into gigantic single-train plants employing hundreds of neatly-arrayed 150,000 ampere cells being energized by 50 megawatt gas-turbine-driven co-generation power plants.

Though the advent of the titanium-based anode in the late 1960s spelled the decline of the relationship between carbon and chlorine, in the heyday of its use of graphite anodes the world's chlor-alkali plants reached a point where they were consuming upwards of a half million pounds per day of graphite. Little did Castner realize seventy years earlier that there would be that much incentive for Acheson. Today's global production capability is over 85,000 tons of chlorine, 90,000 tons of caustic soda (dry basis) and 2400 tons of hydrogen each per day (70). Some of the hydrogen is vented, most is used as fuel and in other manufactures, such as HCl, but with little finding its way any longer into the synthesis of ammonia. The latter is now also produced in large single-train plants which require much more hydrogen than those relatively small amounts co-produced in the electrolysis of brine. To achieve these production rates requires over 8500 megawatts of power, which is about eleven million horsepower. The evaporator plants for converting cell liquor to the dominant caustic product, which is a 50% solution, require about 8 million pounds per hour of steam delivered at 150 psig. About 30% of this production is in mercury cells which do not require steam for evaporation. The rest is made principally in diaphragm cells. In no small way does this branch of industrial electrochemistry owe a debt to Canada. The diaphragms separating the anode and cathode compartments in diaphragm cells are made from the asbestos found plentifully in Quebec. The inventor of the first practical asbestos diaphragm was a Canadian, E. A. LeSueur (71-73).

LeSueur was a student at M.I.T. in chemical engineering in 1890 and had gained some notoriety two years earlier for his demonstration of mercury as a cathode in electrolyzing brine, which led to his being approached by a New England group to devise for them a cell to make chlorine and caustic. His experiments with asbestos as the separator in that cell were eminently successful, including his correct perception of the need for the proper differential pressure across the diaphragm. The New Englanders capitalized on LeSueur's invention and a small plant was built in Rumford Falls, Maine in 1892. Isaiah Roberts is credited with receiving a patent for a similar cell in 1889, but LeSueur's cell was the first in production in North America.

The Siemens-Billiter cell was already in operation by the Griesheim interests in Germany. Roberts founded a company, Roberts Chemical Co., in Niagara Falls much later, in 1901. This company produced KOH from KCl, and established a first by developing a burner to combust the chlorine and hydrogen from the cells to make HCl. The German's bought Roberts out in 1910, changed the name to Niagara Alkali Company, and replaced the Roberts cells with their own Siemens-Billiter cells. The former operated with enough difficulty that the Griesheim people apparently felt more comfortable with their own cell. But WW(I) brought them too many concerns over their future in the U.S., and the plant was sold to a group called the International Agricultural Corp. (74).

In the meantime, the British bleach interests, alarmed at the rapid movements in America toward establishing competitive plants for the production of alkali and bleaching powder, decided to nip the problem in the bud by dropping the price of bleaching powder from $0.035/lb to $0.0187 in 1898. The Bleach War was on. LeSueur's backers got cold feet, closed the plant in Rumford Falls, and sold the cells to a paper company in Berlin, New Hampshire, which later became the Brown Paper Co. Any doubt as to the performance of LeSueur's cells is removed when it is realized that the Brown Company operated those cells until June of 1965 (75). The asbestos diaphragm was here to stay. Creative minds-to-come improved the diaphragm cells to perfection over the next many years. But LeSueur had left his mark, and unlike most other pioneers became totally detached from the technology for which he had been a founding father. His legacy lived on, however, in that other notables, such as E. A. Allen, H. K. Moore and J. Mercer who were LeSueur's associates in the Rumford Falls venture, went on to make names for themselves in the diaphragm cell industry. The Allen-Moore cell became quite well known in the trade. It was the first diaphragm cell to use graphite anodes instead of carbon. This cell was installed in many plants, particularly by paper companies, and was the cell of choice for the founding installation of cells in the Great Western Electrochemical Co. in 1917 in Pittsburg, California (76).

Castner had set the stage for the chlor-alkali industry in Europe with his invention of the mercury cell. The only competition then was the Billiter diaphragm cell. Unlike LeSueur's diaphragm, Billiter's was a poorer diaphragm in performance, being much less uniform than the asbestos paper used by LeSueur. So that in Europe the mercury cell reigned from the outset.

Much good work was done to improve the design and performance of the mercury cell by all of the early practitioners: Solvay, Brunner Mond (which became a part of Imperial Chemical Industries in 1926), Mathieson and others. But politics entered a strong influence in the 1930s. The Germans and Italians were preparing for war. Hitler aided Franco in the Spanish Civil War. Spain paid the debt back to Germany in the form of mercury, which was plentiful there. Anticipating being cut off from the asbestos mines of Quebec, Germany offered its industry attractive incentives for the development of mercury cells (77). The Japanese had similar incentives, and further, in Japan there is no salt. Any producer who has no local source of brine must purchase salt. Since the mercury cell operates by a constant recycle and resaturation of its feed brine, a source of solid salt is required. The fit is natural. In Italy, DeNora developed mercury cell technology which was sold the world over. DeNora has never been a producer per se, but a designer and manufacturer of cells, both diaphragm and mercury, and peripheral equipment necessary to outfit plants. The mercury cell evolved into a very effective instrument for chlor-alkali. Environmental concerns in the last twenty years have impacted negatively on that technology, particularly in Japan. But the mercury cell has its advantages and is alive and well in many plants in the world. It is especially attractive to locales rich in hydroelectric resources and poor in fossil fuels, because of its high consumption of electrical energy coupled with the lack of need for steam. No evaporation of mercury cell caustic is necessary because the cell produces 50% caustic directly. The diaphragm cell produces a 10-12% solution which requires evaporation. The early commitment of Canada to mercury cells is an example of this.

In the U.S. the early movement was to the diaphragm cell. A young entrepreneur from Cleveland, H. H. Dow, had devised an electrolytic means of releasing bromine from naturally occurring brines (i.e. those not requiring water for mining, as do salt brines). With his degree in chemistry from the Case School of Applied Science in Cleveland he traveled in search of bromide-containing brines in order to establish himself as a producer of bromine and bromides. The 24-year old Dow came to Midland, Michigan in 1890 with $375.00 to his name and the promise of backing from J. H. Osborn, a family friend and an executive of the National Carbon Co. in Cleveland. Midland had been a boom town during the lumber era. Slab wood from the mills had been used as cheap fuel to evaporate local brines to provide salt for sale. These brines were known for some time to contain bromides. Actually, Midland was a center for bromine production in the period 1878-1888 when there were no less than fourteen producers. Bitterns from the salt evaporators were chemically oxidized to release the bromine. But by 1890 the lumber was nearly gone and there was a surplus of salt, so Midland was in the doldrums. Dow was able to rent an old barn near an existing, abandoned brine well. Osborn agreed, and the partners founded the Midland Chemical Co. With his second-hand 15-volt generator riding piggyback on the steam engine of the grist mill next to his barn, Dow brought the first electricity to the small town in Central Michigan (78).

Early in 1891 Dow started up his electrolytic bromine blowing-out process, and in a continuous run of 24 hours made the first bromine known to have been produced electrolytically. Though selling his small output of product, he was constantly short of cash and plagued with equipment problems. He correctly assessed his need for greater productivity, which meant larger cells and more generating capacity. With Osborn's help in

Cleveland, more backing was secured, the partnership was dissolved, and the Midland Chemical Co. was incorporated in 1892 with Dow as general manager in Midland, and with B. E. Hellman, the principal backer, as president in Cleveland. Two months later Dow was selling KBr. But Hellman's eagerness for early financial returns led him to harass Dow. In 1893 Hellman replaced Dow as general manager. Despite this Dow remained on the payroll and turned his attention to modifying his bromine cell to electrolyze debrominated brine to recover the major constituent of that brine: chlorine. A small bleach plant was built in 1894. Energized in early '95, it ran one hour before being destroyed by an explosion, which fortunately occurred during the lunch hour, and no one was hurt. But there was great consternation at the company headquarters in Cleveland. In emergency session, the Board decided Dow needed to stick with bromine. He was ordered to retrench and to try to sell his chlorine cell technology to the Solvay Process Co. in Syracuse. But Dow convinced his Board that if he couldn't make the sale within thirty days, they would grant him the rights to his new process in return for 10% of gross.

Dow left Midland, did not sell his technology, garnered more support, and established the Dow Process Co. in Navarre, Ohio, where he worked the kinks out of his cell and his process for making bleaching powder. Returning to Midland in 1896 he built yet another electrolytic chlorine plant. The following year The Dow Chemical Co. was established, in August, 1897. The incentive to manufacture bleaching powder was the 100 million pounds per year being shipped to the U.S. by the British. As for the incentive for Dow to return to Midland it is likely he had a keen eye on the very low cost raw material for his process there. He could get debrominated brine from the Midland Chemical Co. practically for disposal cost, which in those days would have been nothing (79).

Dow's "trap cell", which pioneered his bipolar concept, was constructed of tarred wood. The electrodes were arc-light carbon rods. He was not interested in producing caustic then, so there was no need for a diaphragm except to prevent the hydrogen and chlorine from mixing and exploding. Accordingly, he inserted inverted wooden troughs around each bank of carbons to trap the chlorine, and he allowed the alkalinity which formed around the cathodic portions of the bipolar carbons to form a gelatinous precipitate of the hydroxides of iron, magnesium and calcium on the surface of the carbon, which acted as a diaphragm. There were many problems of plugging and voltage rise associated with this arrangement which required all available manpower and an ample supply of broomsticks to keep the cells rodded out and operating. The chlorine was conducted from the cells in wooden pipes made of bored-out pine logs, cooled with water, and then passed over scrap zinc to dry it sufficiently to make good bleaching powder by reaction with lime. The by-product zinc chloride was also sold. The bleaching powder was high grade product containing 40% available chlorine. Eventually there were sixteen cell buildings with two million carbon rods in service in 26,000 traps. Energy was supplied at 300 volts from a 3000 kw power plant. (Dow

Historical Files.) Production in 1897 was 9 tons of bleaching powder per day, increasing to 72 tons per day in 1902 (80). In 1900 The Dow Chemical Co. absorbed the Midland Chemical Co. Dow got control of his bromine again, as well as his destiny.

Dow had the lowest cost electrolytic bleach process, and he elected to take on the United Alkali Co. in their attempt to freeze out American competition. This British combine had grown out of the need of the Leblanc alkali producers to compete against Solvay alkali. They couldn't do it head on, but by taking advantage of their position in bleach during a seller's market they covered their losses on alkali by sustaining a high price on bleaching powder. They decided to give the upstart Dow a much needed lesson: use your own economy-of-scale and diversification to weather the storm while you continue to lower prices to freeze out the upstarts. Dow met them head on. By 1903 the price had dropped to $0.0125/lb, almost a third of the price level of five years earlier. The other U.S. producers capitulated and closed down their bleach plants. Dow Chemical was deeply in debt and overdrawn at the bank. But when the British announced that the going rate would be reduced even further in 1904 to $0.0088/lb, Dow quoted $0.0086 to his customers. The British folded. United Alkali immediately raised their price back up to $0.0125. Dow honored his contracts and lost a lot of money. But the Bleach War was over. The American chemical industry had arrived, thanks to pioneers such as Herbert Dow (81-82), whose creativity in industrial electrochemistry spawned what was to become an enterprise with global impact in many areas of technology.

Though Dow deserves credit for taking a stand at a crucial point in the history of industrial electrochemistry, his cells would not have won any prizes for engineering excellence. Nor did his cells produce recoverable alkali. Billiter, Bell, Castner and Kellner in Europe, and Roberts, LeSueur, Allen, Moore, Gibbs, Vorce and Townsend in North America all had developed cells for producing caustic as well as chlorine ahead of Dow. LeSueur's impact needs to be remembered for more than just his seminal work on the asbestos diaphragm, but for sending forth from Rumford Falls the apostles who would preach the gospel of the diaphragm cell to the world.

As mentioned earlier, chief among LeSueur's early associates were E. A. Allen and H. K. Moore. The agreement of sale of LeSueur's technology after the closing of the plant at Rumford Falls precluded him from further promotion of his cell (83), and he turned to other pursuits. Meanwhile Allen and Moore developed a cell based on LeSueur's diaphragm, but with a new twist: it was vertical and contained an unsubmerged cathode, allowing greater productivity per unit of plant floor area. In 1903 Moore established the American Electrolytic Co. to take advantage of this new cell, which was already performing in Malden, Mass., in a paper mill owned by the S. D. Warren Co. The first use of graphite anodes in a diaphragm cell was in the cells installed for American Electrolytic's plant in Glen Rock, New York. Difficulties in running this plant due to insufficient attention to plant operations peripheral to the cells, along with under-capitalization caused this venture to fail. But vertical diaphragm cells employing graphite anodes were here to stay (84).

Pennsalt, known for their practice of the Danish process to make alkali from cryolite, which in 1878 had provided the U.S. with its first domestically produced caustic, found it increasingly difficult to compete in the alkali trade after the establishment of the Solvay Process Co. in Syracuse, N.Y. in 1881. In 1902 Pennsalt decided to go into the chlor-alkali business and built a plant at Wyandotte, Michigan to take advantage of salt deposits there. A. E. Gibbs of Niagara Falls had designed a cylindrical cell which was chosen by Pennsalt for its new plant. After much difficulty in achieving satisfactory performance, L. D. Vorce was hired in 1905 to make modifications. Success in increasing the current efficiency from 70% to 93%, as well as reducing the cell voltage from 4.50 volts to 3.54 at 650 amperes (a current density of about 0.2 amps per sq. inch), led Pennsalt to a program of plant expansions which saw 2600 of these Vorce-modified Gibbs cells installed in Wyandotte by 1913. The Canadian Salt Co., later to become Canadian Industries, Ltd., purchased this cell technology in 1909 for its ten ton per day plant to be installed in Sandwich, Ontario just across the Detroit River from Wyandotte. The United Alkali Co. also purchased this technology from Pennsalt, and by 1918 this British plant had 6000 of these Gibbs-Vorce cells installed at Widnes (85).

The chief chemist at Pennsalt's Wyandotte plant during that period was F. G. Wheeler, who resigned in early 1913, designed his own version of the Gibbs cylindrical cell, and went to work for the Kimberly-Clark Co. Wheeler cells were subsequently installed in that company's paper mills in Wisconsin, as well as in the Champion Fibre Co. plant in South Carolina in 1916. In that same year the Niagara Smelting Co. installed Wheeler cells to obtain chlorine to make aluminum chloride which they intended to convert to aluminum metal by the electrolysis of the fused chloride. This process did not work and the plant was purchased by Stauffer Chemical to supply their needs for chlorine, caustic and aluminum chloride.

The last of the LeSueur alumni was F. McDonald, who patented a rectangular cell in 1902 similar to the Allen-Moore cell. Expanding its interests into chlorinated products, the Warner Chemical Company purchased McDonald's technology and installed cells in their plant in Carteret, New Jersey in 1905. Dissatisfied with its performance, H. R. Nelson and other Warner people improved the design. The Nelson cell was included in Warner's expansions and was made available to others as well. It was adopted by the U.S. Government's Edgewood Arsenal to produce fifty tons of chlorine per day for the war effort in 1918, which was a large single installation for its day. Warner developed a process for sodium hypochlorite into a commercial bleach business, and also began manufacturing carbon disulfide and sulfur monochloride as intermediates for the manufacture of carbon tetrachloride starting with coke and chlorine. To expand these operations, a plant was built in South Charleston, W. Va. in 1915 under the name Westvaco Chlorine Products Corp., near inexpensive sources of coal and brine. Nelson cells were originally installed there, but in a large expansion in 1927, Westvaco added Vorce cells to its stable (86-87).

By 1915 the shakedown cruise of the chlor-alkali industry was over. But not before Elon H. Hooker and his brothers came aboard. With his A.B. degree from the University of Rochester in 1891, and his B.S. and Ph.D. degrees from Cornell a few years later, Elon Hooker started his career as a civil engineer. In 1899 he was appointed Deputy Superintendent of Public Works of the State of New York by then governor Teddy Roosevelt. An impressive young man in a position to meet many influential people, Elon Hooker was offered a job as vice president of a venture capital company (88). In early 1903 he left their employ to start his own venture capital company, The Development and Funding Co., with no capital and no venture in mind. But with his charm and connections he soon had $60,000 in cash and considerably more promised. He and his associates examined 250 projects before deciding in favor of manufacturing chlor-alkali using a cell patented by C.P. Townsend, chemist and patent attorney, and E. A. Sperry, the pre-eminent electrical engineer who later gained fame as the inventor of the gyroscope. Townsend eventually worked for Union Carbide and Carbon as head of their Patent Dept., as well as Director of Research. The Sperry-Townsend design was not unlike other rectangular cells of its day. It included submerged cathodes arrayed on either side of a central chamber containing graphite anodes, with an asbestos paper diaphragm between the two compartments. They paid attention to LeSueur. Unique to their design was the addition of kerosene oil to the cathode compartments, supposedly to discourage back-migration of hydroxide into the anolyte compartment. Not so, but Elon liked it enough to have his brother, A. H. Hooker, a chemist, visit Townsend and Sperry in Washington to examine a working model. At the time, Townsend worked as a patent examiner for the U.S. Patent Office. A. H. Hooker's report encouraged Elon to acquire an option on the cell. Leo H. Baekeland, later renowned as the inventor of resins and plastics made from phenol and formaldehyde, was hired by Hooker as a consultant in 1904. Sperry had three 1000 ampere cells built in Cleveland and shipped to a former boiler house in Brooklyn rented from the Edison Co. A younger Hooker brother, Willard, was in charge of this pilot operation under Leo Baekeland's supervision. The maiden voyage was stormy with respect to excessive wear of the graphite. Townsend suggested the cause was unsaturated brine. Baekeland devised a means of saturating and recirculating the anolyte brine. The tests were successfully completed and Elon Hooker, with only $16,000 remaining, raised additional capital to build a plant at Niagara Falls with its plentiful supply of electricity, salt deposits, and fresh water to mine the salt. Early in 1906 the fledgling plant's 2000 ampere cells were energized, and by mid-year they were producing about five tons of caustic and enough chlorine to make eleven tons of bleaching powder per day. But as in so many of these early excursions in commercial chemistry, productivity was insufficient to generate profit. In 1907 Hooker mounted a campaign for larger cells, and by late 1909 the plant had almost quadrupled in capacity. The red ink stopped flowing.

Ready to return his attention to other ventures for his Development and Funding Co., Hooker established the Hooker Electrochemical Co. as a subsidiary of the Development Co., supposedly just one of many to come. But it was not to be. A few months later the cell house in Niagara Falls was destroyed by fire. Elon again obtained financial backing, rebuilt, and was operating again within five months with increased capacity. A few years later Hooker dissolved the Development and Funding Co. and began to devote all of his attention to the Hooker Electrochemical Co. Resolved to engineer the best technology possible, and to profit both from its use within his company as well as by selling it to others, Hooker embarked on a long campaign of improving productivity in cell technology which was at the leading edge for many years. Notable examples are C. W. Marsh's invention of the corrugated cathode assembly in 1913, and K. E. Stuart's invention of the vacuum-drawn diaphragm from a slurry of asbestos. Cells were redesigned to accommodate these improvements and to increase amperage at lower voltages, thus decreasing the total energy consumed per unit of cell products. Cell amperages increased steadily upwards to 55,000 amps in the 1960s. None of these cells bore any resemblance to Townsend's. The practice of industrial electrochemistry was enhanced considerably by Hooker Electrochemical. But enhancement was coming from other influences as well.

Three strong influences on chlorine manufacture came into the picture in 1909 and the following few years. One was the stemming of a typhoid epidemic by injecting chlorine directly into domestic water supplies, another was the blockade of German dyestuffs and organic intermediates, and the third was the liquefaction of chlorine and its transport in cylinders and tank cars. Sanitation and liquefaction had already been achieved in Europe. But these influences in North American made a stronger impact when coupled with the economic incentives deriving from the lack of trade with Germany during WW(I). Additional impetus arose as well from the introduction of liquid chlorine into the manufacture of pulp and paper after the war.

The young North American chemical industry had not been able to enter the coal-tar chemical business founded by the British because of the domination by the German cartel. But the door to this opportunity swung wide open during the war. Hooker built the first monochlorobenzene plant in the U.S. in 1915. Warner, Dow and others were already selling chlorine as carbon tetrachloride and chloroform. Dow was soon to follow with chlorobenzene, phenol, indigo and others, and they all impacted chlor-alkali consumption. The future of this important segment of the electrochemical industry was clearly in the hands of these new uses for chlorine.

No sooner was the Bleach War over and Herbert Dow found himself in a similar struggle with the German Bromine Convention, which threatened to cut prices if Dow insisted on marketing bromine products in Europe. He refused, and there followed another two years of battling. Dow won again, but the result in 1909 was an outcry from his investors who had received no dividends that year. Worse, his directors refused to allow him seed money to pursue the new cell developments he was working on. In 1907 Dow hired L. E. Ward, a young electrochemical engineer who had studied with Prof. Burgess, later of battery fame, at the University of Wisconsin, which had pioneered a program of studies in chemical engineering and electrochemistry. Ward was assigned to work on a cell to make chlorine and KOH. The latter was in very short supply, and Dow formed the Midland Manufacturing Co. in 1908 in partnership with the Fostoria Glass Co. and the Libbey Glass Co., who were interested in alternate supplies of KOH for their operations. The venture was dissolved in 1909, but it covered Dow's work on a new cell which was designed by T. Griswold, Jr. Ward had left Dow for a six year stint with the Oxford Paper Co. because he could no longer afford to be paid in stock certificates instead of regular paychecks. But Dow talked him into returning in 1917, and Ward distinguished himself in further chlor-alkali and magnesium cell developments until his retirement in 1950. Based on Ward's earlier work and his clear knowledge of the correct path to take in chlorine utilization, Dow shocked everyone by abandoning his trap cells in 1914, and by quitting the manufacture of bleaching powder (89).

The electrolytic cell is the heart of a chlor-alkali operation, and 1913 saw much activity at The Dow Chemical Co. in the construction of hundreds of new vertical, filter press cells. Wooden frames and arc-light carbons were gone, replaced by concrete and graphite (90). Dow's diaphragm was still not up to LeSueur's. But overall this was a good start in the game of catch up, and he was gambling on being the only practitioner of the bipolar, filter press type cell.

All other producers had adopted the unit cell, which refers to individual cells being connected externally to each other with cable or busbars. In the bipolar arrangement, electrical continuity is achieved internally, with external connections to the rectifier circuit being made only at the anode and cathode terminals of a series which contains a multiplicity of cells. Dow had elected to use 75 cells in a filter press series. Often referred to as bipolar, these cells are not bipolar in the strict sense that a given electrode material serves as cathode on one side and anode on the other, because steel is the cathode and graphite the anode in these cells. But the term bipolar is often used. There is a risk in squeezing together 75 cells in a filter press type array with all cells under a common cover. If one cell goes bad, and if corrective actions on the run are ineffective, the whole series must be dismantled. The advantage of the unit cell concept is that a jumper bus can be inserted around a troubled cell, which can then be taken off line for maintenance. But Dow's wisdom in cutting costs and saving floor space paid off in the long run by forcing the discipline of the filter press operation. The arrangement is used to this day, typically with 150 tons of chlorine per day and the equivalent amounts of caustic and hydrogen in a single unit, as though it were one 4.5 million ampere cell. Bold, but effective if you learn how to do it right.

From the first 1000 ampere filter press cell on, the name of the Dow game was no different than the rest of the industry: increase productivity by increasing amperage and current efficiency; and decrease voltage to decrease energy consumption. This requires a good diaphragm, especially one having a long life when you use a filter press arrangement. Dow purchased Wheeler know-how in 1918, and sent Ward off to bring back LeSueur's diaphragm (91). L. E. Ward's revisions took Dow in that direction in 1921 (92).

Dow followed Hooker's lead in implementing corrugated cathodes and drawn diaphragms in the '30s (93). But a 1912 dictum of A. H. Hooker's on the harmful effects of cell height on efficiency (94), subsequently adopted by others when they attempted to ignore it, froze the industry into a maximum electrode height of 3-4 feet until Dow broke the ice in the mid-'60s by learning how to operate efficiently with electrode heights of 6-9 feet (95). This breakthrough, coupled with Dow's operating discipline in uniquely operating bipolar series gave Dow a very competitive position. This type of cell was adopted by Pittsburgh Plate Glass and Oronzio DeNora, who teamed up to design the Glanor cell, used extensively by PPG and marketed by Oronzio DeNora for use by others.

Oronzio and his brother Vittorio had for some time made substantial impacts on the chlor-alkali industry, particularly with the DeNora mercury cell, which is used quite extensively around the world. They, as well as other mercury cell designers and users, had for some time been searching for a dimensionally stable anode. Graphite anodes slowly wear away to carbon dioxide. In mercury cell operations this dictates constant adjustment of the anode assemblies in order to minimize cell voltage. This could be avoided if the anode were constructed of a more durable material, one whose dimensions remained constant over time. Much work had been done by many researchers on using titanium metal for this purpose because it is noble in wet chlorine service. But that very fact precludes its conductivity as well. Coating with thin layers of precious metals works, but is costly because of wear rates of the precious metals. Vittorio DeNora found a lone inventor of the Edisonian type who solved the problem and changed the face of the chlor-alkali industry forever. The Belgian, Henri Beer, is a story all by himself (96-98), but suffice here to say that while the giants in the industry made incremental strides of steady, but modest improvements over the years, Beer the inventor and DeNora the entrepreneur exploded on the scene in the late-60s with the finest development in the history of the industry. Beer discovered that a solid solution of ruthenium oxide in titanium oxide is a conductor of electricity which, in thin layers on a titanium surface, has a very long life. It's elegant, it's affordable, and it made possible not only the original dream of a dimensionally stable anode for mercury cell use, as well as making diaphragm cells more efficient and less consumptive of electrical energy, but it makes possible the complete replacement of the asbestos diaphragm with another great invention of the last twenty years: an ion-exchange membrane capable of withstanding the onslaught of both chlorine and caustic at 100°C (99-103). The industry is currently being revolutionized by these two great inventions.

These two inventions are classical in that they demonstrate the value of both Edisonian and fundamental research. Both seek results, but with different approaches, each valuable in its own way. However fortuitous, it is interesting that these revolutionary developments were both begun about the same time in the late '50s; one dedicated to solving a clear and immediate problem, the other dedicated to furthering the science of polymeric membranes with commercial applications anticipated. In any event, each of them has created new excitement in that segment of the electrochemical industry which is more widely practiced than any other.

But the the story of Beer's anode is incomplete without introducing the Diamond Shamrock Co., without whose immediate commercial response to Vittorio DeNora's entrepreneurship the development may have taken longer to impact the industry. Diamond set up the Electrode Corp. to manufacture and market the dimensionally stable anode in the late '60s. Within ten years every major player had either converted from graphite or was in the process of doing so. The industry owes Diamond much respect for its perspicacity. Not only has this anode allowed substantial reductions in cell energy consumption by virtue of its essentially zero chlorine overvoltage compared to graphite, but in replacing the latter it removes sludge and particulate matter from the anolyte compartment which makes it possible to take advantage of the ion-exchange membrane.

The early work of duPont's W. G. Grot (104) in devising the chemistry needed to produce the polymers necessary to make these perfluorovinyl ether sulfonic acid and carboxylic acid membranes, tradenamed Nafion by duPont, was seminal. Japanese workers in the Asahi Chemical Co. (105), Tokuyama Soda Co. (106) and the Asahi Glass Co. (107) also deserve much credit. All three companies market membranes. Their work was accelerated by the Japanese government's plan to rid the islands of mercury cells, to which many plants there were heavily committed. But notwithstanding a few large scale commitments, such as by Akzo in the Netherlands, Vulcan in Kansas, and Occidental in Louisiana, a number in Japan, plus several small scale installations, the conversion to membrane cells has been slower than originally anticipated. The membrane cell is an ideal replacement for the mercury cell, since it produces salt-free 32% caustic requiring little evaporation. But many large scale diaphragm cell practitioners have their own power plants, and consideration must be given to the electricity to steam balance at the site.

Most recently, nickel cathodes with activated, or so-called electro-catalytic, coatings have been developed for use in membrane cells by Dow, DeNora, Oxytech and others. These developments superbly complete the metamorphosis of the chlor-alkali cell from its crude beginnings with its mediocre performance to a finely tuned, well engineered device which one day will become run of the mill for all chlor-alkali operations.

Many chlor-alkali producers have gone unmentioned here largely because they were late bloomers, but also in the interest of brevity. Similarly in electroplating, electrowinning and electrorefining there is much to be included, as for example Weston's classical work in nickel plating before the turn of the century. Nor has mention been made of the industrial electrolysis of water or of the production of chlorates and perchlorates, both of which were heavily impacted by Oldbury Electrochemical at Niagara Falls. Storage batteries are a subject unto themselves. In electrothermics there is the whole area of arc discharge in gases. Norway's Birkeland-Eyde process for fixing nitrogen is classical, though supplanted by the Haber process. As well, the electric furnace manufacture of phosphorus and silicon have not been covered. Small, but not insignificant in the history of industrial electrochemistry are the achievements of Monsanto and Asahi Chemical in commercializing Baizer's electrolytic hydrodimerization of acrylonitrile to adiponitrile, though the promise of other significant commercial applications of organic electrochemistry has proven more exciting than the fulfillment because of economics. Electroforming, electromachining, and the applications of electrophoresis all have industrial impact. Indeed, the use of electricity to conduct chemistry is an enormous field with a rich history. But Dow's struggle to bring magnesium metal to the U.S. must be covered, as it is another example of freeing North America from domination by its cousins across the Atlantic in an important area of industrial electrochemical practice.

The I.G. Farben Co. in Germany had pioneered the electrolysis of anydrous magnesium chloride in 1886 in an interesting combination of an electrothermic and an electrolytic process. Crude magnesium chloride containing oxide is obtained from the Strassfurt or Bitterfeld deposits, briquetted with coke and fed to an electric furnace to be chlorinated with cell gas and converted to the molten chloride, free of oxide. The molten chloride is electrolyzed in a eutectic bath in cells consisting of steel pots having graphite anodes and steel cathodes to produce the metal. In 1914 German magnesium was showing up on the battle field in incendiary flares. The U.S. had no supply, and the price shot up to $6.00/lb.

H. H. Dow knew he had a plentiful source of magnesium in the brine from which he derived bromine. L. E. Ward and E. O. Barstow had already worked out the details of separating all of the constituents of this brine, which contained largely calcium chloride, magnesium chloride, lesser amounts of the chlorides of sodium and potassium, 600 ppm of lithium chloride and 2500 ppm of bromide. By controlled evaporation and differential crystallization they were removing all these constituents from the debrominated brine except the lithium (108). The partially hydrated magnesium chloride which was recovered was being sold to other companies who were trying to get into the magnesium business as well. Government incentives attracted eight players in the

U.S., but by 1920 there were only two left: Dow and the American Magnesium Corp. Key in the development of Dow's electrolytic magnesium process was W.R. Veazey, a Dow consultant who was on the staff at the Case Institute in Cleveland, Dow's Alma Mater. Veazey spent summers working on Dow's plant problems in Midland, Michigan. R.M. Hunter, a young summer student who was half way through his engineering studies at Case, got on a steep learning curve during the summer of 1916 working with the group using Veazey's cell to produce Dow's first 100 lb ingot of magnesium metal. Hunter was hired by Dow in 1918 and spent his entire career of 43 years working in Dow's electrochemical areas, principally magnesium and chlorine.

The magnesium market essentially dried up after WW(I). By 1927, the American Magnesium Co. ceased manufacturing but continued marketing, with Dow their source of supply, the only one remaining on this side of the Atlantic until 1941 when war once again opened the floodgates of interest in magnesium. In the meantime, Dow subsidized his research and production of the metal, establishing a world class metallurgical laboratory in the early '20s. By 1928 four commercial alloys had been developed out of literally hundreds of compositions studied. Dow also invested heavily in fabrication methods: extrusion, rolling, sand and die casting. As a commercial enterprise magnesium had trouble standing alone, even though by 1939 the capacity was twelve million pounds per year. But the following year brought a change as the Arsenal of Democracy tooled up to supply the British. Prior to this, in 1930, H. H. Dow died and his son, Willard H. Dow, took over the company. Willard Dow continued to champion magnesium, but added a new twist: a march to the sea.

Since 1932 Dow had been interested in producing bromine and magnesium from seawater. At the same time Dow's growth in organic chemicals, especially chlorinated hydrocarbons, and its need for additional sources of chlorine, hydrocarbons and energy, required serious consideration of expanding manufacturing operations to the Gulf Coast. Several potential sites had been thoroughly studied when it was decided to locate in Freeport, Texas, one of many choice sites on the Gulf where plentiful supplies of salt, fresh water, natural gas, gas liquids, oil and seawater existed. The decision was made in 1940, and by '41 the first magnesium from seawater was produced in Freeport in a 36 million lb/yr plant. A second plant with a capacity of 72 million lb/yr was on stream in the same area by the following year. Oyster shells furnished the lime to precipitate magnesium hydroxide from the 1270 ppm of magnesium in the seawater. For its day this was a monumental operation and an achievement of the first magnitude. Dow went on to assist the U.S. Government in establishing several other magnesium plants in the country for the war effort, mostly for aircraft use. Some of these plants were operated by Dow, and some by other concerns. The U.S. goal was six to seven hundred million pounds of magnesium per annum. Space does not permit going into the many problems and technical developments which occurred during this period. But today The Dow Chemical Co. stands tall in magnesium circles for its pioneering efforts in earlier days, and continues to produce this lightweight metal in Freeport in amounts upwards of 150 million lb/yr in cells consuming over 150,000 amperes which would make H. H. Dow and L. E. Ward proud (109-110).

H. H. Dow was invited by Prof. J. W. Richards, renowned talent in electrometallurgy at Lehigh University, in 1901 to be a charter member of the American Electrochemical Society, forerunner of The Electrochemical Society. Dow took an active part and encouraged his people to do likewise. Many of them have played active roles in the history of industrial electrochemistry and continue to do so.

Literature Cited

1. Haber, L.F. The Chemical Industry During the Nineteenth Century; Oxford: London, 1958; Chapter 7, 9.
2. Williams, T.I. In A History of Technology; Singer, C., Ed.; Oxford: London, 1958; Vol. V; Chapter 11.
3. Haynes, W. American Chemical Industry; Van Nostrand: New York, 1954; Vol. I, Chapters 9, 10, 12 and 17.
4. Ihde, A.J. The Development of Modern Chemistry; Harper and Row: New York, 1964; Chapter 17.
5. Trescott, M.M. The Rise of the American Electrochemicals Industry, 1880 - 1910; Greenwood Press: Wesport, CT., 1981; Chapter 1.
6. Haber, L.F. op. cit., pp. 55-96.
7. Hou, T.P. Manufacture of Soda; ACS Monograph No. 65; Hafner Publishing: New York, 1969; 2nd Edition, p 7.
8. Haber, L.F. op. cit., p 208.
9. Hou, T.P. op. cit. p 4-5.
10. Haber, L.F. op. cit., p 93.
11. Hunter, R.M. E.A. LeSueur Memorial Lecture; The Society of Chemical Industry: Toronto, 1966; p 1.
12. Ratcliffe, J.A. In A Short History of Science; Doubleday; New York, 1959; pp 110-117.
13. Boorstin, D.J. The Discoverers; Random House: New York, 1983; p 680-1.
14. Patterson, E.C. American Scientist 1988, 76, 221.
15. Ihde, A.J. op. cit., pp 135-7.
16. Hughes, T.P. Am. Heritage of Invention and Technology 1988, 3, 59-64.
17. Ihde, A.J. op. cit., p 124-7.
18. Vinal, G.W. Storage Batteries; Wiley: New York, 1955; 4th. Ed.; pp 1-8, 352-5.
19. Clark, V.S. History of Manufacturing in the United States; McGraw Hill: New York, 1929; Vol. III, p 165.
20. Jarvis, C.M. In A History of Technology; Singer, C., Ed.; Oxford: London, 1958; Chapter 9.
21. ibid. p 191.
22. Conot, R.E. A Streak of Luck (Biography of Edison); Seaview Books: New York, 1979; p 144.
23. ibid.
24. Jarvis, C.M. op. cit., pp 192-4.
25. Clark, V.S. op. cit., pp 150-3.

26. The Encyclopedia Americana, International Edition; Grolier: Danbury, CT., 1987.
27. Allen, J. In A History of Technology; Singer, C., Ed.; Oxford: London, 1958; Chapter 22.
28. Clark, V.S. op. cit., p 2-5.
29. Elkington, J. British Patents 2 838, 1865; 3 120, 1869.
30. Chadwick, R. In A History of Technology; Singer, C. Ed; Oxford: London, 1958; pp 83-85.
31. Steele, J.M. In Industrial Electrochemical Processes; Kuhn, A.T., Ed.; Elsevier: London, 1971; Chapter 7.
32. Carr, C.C. An American Enterprise; Rinehart: New York, 1952; Chapter 1.
33. Hall, C.M. U.S. Patent 400 766, 1886.
34. Carr, C.C. op. cit., pp 53-55.
35. Haber, L.F. op. cit. p 54.
36. Haynes, W. op. cit., Vol. VI, pp 329-30.
37. Haber, L.F. op. cit., p 92.
38. Chadwick, R. op. cit., p 93-4.
39. Carr, C.C. op. cit., p 163.
40. ibid., pp 85-107.
41. ibid., pp 106-7.
42. Tilak, B.V. and Van Zee, J.W. J. Electrochem. Soc. 1987, 134, p. 283c.
43. Gilroy, D. In Industrial Electrochemical Processes, Kuhn, A.T., Ed.; Elsevier: London 1981, p 193.
44. Mantell, C.L. Electrochemical Engineering; 4th Ed.; McGraw HIll: New York, 1960, pp 371-78.
45. Carr, C.C. op. cit., p 120.
46. Mantell, C.L. op. cit., pp 379-83.
47. Williams, T.I. op cit., pp 248-51.
48. Demmerle, R.L. In The Encyclopedia of Electrochemistry, Hampel, C.A. Ed.; Reinhold: New York, 1964, pp 154-5.
49. Downs, J.C. British Patent 238 956, 1924; U.S. Patent 542 030, 1924.
50. Mantell, C.L. op. cit., p 411.
51. Walsh, C.W. In The Encyclopedia of Electrochemistry, op. cit., pp 348-9.
52. Mantell, C.L. op. cit., p 260.
53. Williams, T.I. op. cit., p 251.
54. Szymanowitz, R. Edward Goodrich Acheson; Vantage Press: New York, 1971; p 235.
55. Clark, G.L. In The Encyclopedia of Electrochemistry, op. cit., p 833.
56. Trescott, M.M. op. cit., p 25.
57. Szymanowitz, R. op. cit., pp 58-188.
58. Stansfield, A. The Electric Furnace; McGraw HIll: New York; 1914, p 283.
59. Trescott, M.M. op. cit., p 25-6.
60. Benford, E.S. In The Encyclopedia of Electrochemistry, op. cit., p 1173.
61. Haynes, W. op. cit., Vol. VI, p 430-1.
62. Benford, E.S. op. cit., p 1174.
63. Stansfield, A. op. cit., p 9-19.

64. Tinnon, J.M. In The Encyclopedia of Electrochemistry, op. cit., p 140.
65. Trescott, M.M. op. cit., p 73.
66. ibid., p 74.
67. Walsh, C.W. In The Encyclopedia of Electrochemistry, op. cit., p 96.
68. Haynes, W. op. cit., pp 431-2.
69. ibid., p 433-5.
70. Tilak, B.V. and Van Zee, J.W. op. cit., pp 280-1(C).
71. Trescott, M.M. op. cit., p 20.
72. Hunter, R.M. op. cit., p 2.
73. Vorce, L.D. Trans. Electrochem. Soc. 1944, 86, p. 70.
74. Trescott, M.M. op cit., p 79.
75. Hunter, R.M. op. cit., pp 2, 15-16.
76. Vorce, L.D. op. cit., pp 76-81.
77. Hunter, R.M. op. cit., p 10.
78. Karpiuk, R.S. Dow Research Pioneers; Pendell: Midland, MI; 1984, Era I.
79. Whitehead, D. The Dow Story; McGraw Hill: New York; 1968, pp 29-48.
80. Trescott, M.M. op. cit., p 94.
81. Whitehead, D. op. cit., pp 54-5.
82. Hunter, R.M. op. cit., p 2-3.
83. Vorce, L.D. op. cit., p 70.
84. ibid., p 71.
85. ibid., p 73-4.
86. ibid., p 75.
87. Haynes, W. op. cit., Vol. VI; p 478-9.
88. Thomas, R.E. Salt and Water, Power and People; Hooker Electrochemical Co.: Niagara Falls, NY, 1955; pp 3-51.
89. Karpiuk, R.S. op. cit., pp 28-33.
90. Griswold, T. U.S. Patents 987 717, 1911; 1 070 453, 1913.
91. Hunter, R.M. op. cit., p 4.
92. Ward, L.E. U.S. Patent 1 365 875, 1921.
93. Hunter, R.M., et. al. U.S. Patent 2 282 058, 1939.
94. Vorce, L.D. op. cit., pp 78-80.
95. Blue, R.D., et. al. U.S. Patent 3 876 520, 1975.
96. Beer, H.B. Chemistry and Industry July 15, 1978, pp 491-6.
97. Beer, H.B. In Electrochemistry in Industry, Landau, U., et. al., Ed.; Plenum Press: New York; 1982, pp 5-18.
98. Beer, H.B. Chem. Tech. 1979, 9, p 150.
99. Grot, W.G. Chem. Ing. Tech. 1972, 44, p 167.
100. Grot, W.G. U.S. Patents 3 692 569, 1972; 3 718 627, 1973.
101. Seko, M. In Ion Exchange Membranes, Flett, D., Ed.; Ellis Horwood: Chichester, 1983, Chapter 12.
102. Ezzell. B. and Carl, W. U.S. Patent 4 417 969, 1983.
103. Jackson, C. Modern Chlor-Alkali Technology Ellis Horwood: Chichester; Vol. 2, 1983.
104. Grot, W.G. In Electrochemistry and Industry, Landau, U., et. al., Ed.; op. cit., pp 73-87.
105. Seko, M. In Modern Chlor-Alkali Technology, Coulter, M.O., Ed.; Ellis Horwood: Chichester; Vol. 1, 1980, Chapter 16.
106. Motani, K. And Sata, T. ibid., Chapter 18.

107. Nagamura, M. et. al., ibid., Chapter 17.
108. Karpiuk, R.S. op. cit., pp 41-8.
109. Schambra, W.P. Trans. Am. Inst. Chem. Eng. 1944, 41, 1-16.
110. Shigley, C.M. Am. Inst. Mining and Met. Eng. 1945, No. 1845, 1-9.

RECEIVED August 9, 1988

Chapter 34

Industrial Diaphragms and Membranes

P. R. Roberge

Department of Chemistry and Chemical Engineering, Royal Military College, Kingston, Ontario K7K 5L0, Canada

In any electrolytic process, where it is necessary to control the diffusion of the products of decomposition, the most vital point to life and efficiency of the cell is the diaphragm or more recently the membrane. This tough reality, which was met with crude short lived diaphragms at the turn of the century, has forced inventors to eventually produce more stabilized materia and ion specific membranes with parallel progress in cell efficiency and quality of the electrolysis products. This paper will review the various electrochemical processes using separators in 1900 and attempt to describe the improvements achieved during the following decades.

In his 1915 description of an improved diaphragm material, C.J. Thatcher (1) summarized very well the problems associated with such development: "The lack of a satisfactory diaphragm material has probably hindered or prevented the commercial success of many promising electrolytic processes. At first thought it might seem that there should be no difficulty in obtaining cheap, porous or semi-permeable materials, suitable for prolonged commercial use as a diaphragm in any given electrolyte. Experience, or at least that of the writer, has not warranted any such optimistic view of the matter. A diaphragm should be moistened by, but not be permeable to liquids except by diffusion, so that the anolyte and catholyte, or suspended solid constituents therein, will not commingle during considerable periods. For this reason a diaphragm material should either not be porous, or should have such exceedingly fine pores as to offer a high resistance to the passage of liquids therethrough."

"But, on the other hand, a diaphragm material should offer little resistance to diffusion or migration of ions, so that the electrical resistance will not be so great as to be prohibitory. This requires that the diaphragm material should be somewhat porous."

"These two fundamental considerations limit the range of materials suitable for diaphragms to those which are somewhat, but not very porous. But this is not all that a diaphragm should be. For commercial use it should be strong, of course, so as to stand

0097–6156/89/0390–0510$06.00/0

the wear and tear of factory operations; and it should be permanent, that is, not disintegrated by the electrolytes or products of the electrolysis. If the material is not strong or permanent it should be very cheap, easily replaced and readily obtainable in all shapes and sizes."

Early days - 1900

The first industrial electrolytic process to be operated with a diaphragm started to produce chlorine at Griesheim, Germany, in 1888 (2). The electrical transmission of power was worldwide extremely limited in 1890 (3) but this mode of carrying energy was expanding very rapidly (4).

When the first issue of Electrochemical Industry (now Chemical Engineering) came out in September 1902, the American Electrochemical Society (now the Electrochemical Society) was only six months old but the electrochemical industry was booming. The feeling of optimism, then expressed by some people for the future of electrochemical processes is best described by the following quotation from J.W. Richards' article (5) on the electrochemical industries of Niagara Falls, published in this first issue of Electrochemical Industry: "No field in the whole range of the applied sciences is succeeding more signally, promising more attractively, or so pregnant with suggestions of future applications, than electrochemistry. In the hands of the masterful chemist, acquainted with the facts of chemistry and the needs of civilization, the uniting and decomposing powers of the electric current, its almost infinite control of chemical analysis and synthesis, the generation of inconceivable temperatures on a commercial scale and the methods it furnishes of torturing poor Mother Nature into new shapes and wringing from her new secrets, has brought to realization fancies which were undreamed of by the alchemists."

In fact, the developing electrochemical industry gave incentive for, and was practically the mainspring of the development of power at Niagara Falls where, by the end of year 1902, the Niagara Falls Power Co. would be developing 60,000 horse-power; of which 45000 (75%) was to be used electrochemically. Only a fraction of that electricity was to be run through a diaphragm since the Roberts Chemical Co. was the only industry to produce the chemicals that necessitated a separator. This company was all contained in a one-story frame building 60-by 200 feet and its electrical usage estimated at 500 horse power for the production of caustic potash and hydrochloric acid from potassium chloride.

Alkali-Chlorine. For the first part of the 20th century the electrolytic production of chlorine and alkali seems to have been the main process that successfully made use of diaphragms. The first commercial production of chlorine (6) in 1898 by the Griesheim Company was achieved with cells equiped with porous cement diaphragms, invented earlier (1886) by Brauer. These diaphragms were prepared by mixing portland cement with an acidified brine. The porosity was created by soaking the set cement in water to remove soluble salts. Twelve anode compartments constructed of this diaphragm material supported by angle-irons were hung into the tank illustrated in figure 1. Porous pots were provided in each anode compartment for addition of solid KCl. The cell was operated batchwise at a temperature of 90°C. The KCl solution was introduced initially and electrolysis was continued for about three days until a concentration of approximately 7% KOH was obtained. The current efficiency was low (70% to 80%) but the cell was simple, inexpensive and relatively large in capacity (2.5 kA).

The first use of a percolating diaphragm is credited to Le Sueur who put it to practice for the first time in 1890. By permitting brine to flow into the anolyte and through the diaphragm much higher current efficiency could be obtained and the process was continuous. The first cell employed at the Rumford Falls plant, U.S.A., was of an improved design (figure 2) of the original Le Sueur cell (figure 3).

The possibility of eliminating convection effects and mechanical disturbance enabled anode and cathode to be placed near to one another, and for higher current densities to be used than would otherwise have been possible without great losses through interaction between the alkali and the chlorine. This gain in current density was obtained in parallel with a gain in the alkali concentration of the catholyte.

These diaphragms could also be vertical. The respective advantages and disadvantages of the two arrangements have been reviewed by Billiter (7) in 1911. Vertical diaphragms permitted a more accessible cell construction, could be easily changed and for the same power consumption would give a far more compact cell. Further, impurities would settle on the bottom of the cell instead of on the diaphragm. On the other hand the horizontal diaphragm was completely soaked with the alkaline liquor. If the plant was to make money from its alkali production this was a double advantage. First the catholyte was free of the unavoidable losses due to the dissolved chlorine and acid coming from the anode. The second point was that is was easier to prepare diaphragms chemically resistant against alkali than against acid.

Allmand, in his first edition of Applied Electrochemistry published in 1912 (8), described the general features and merits of the various cells that were then in operation. The Hargreaves-Bird cell (figure 4) is an example of a cell with a vertical diaphragm. The Hooker Electrochemical Co. used a modified version of this design, the Townsend cell illustrated in figure 5, at the Niagara plant around the turn of the century.

Another very popular design, during that period, was the one called the Billiter Diaphragm cell (figure 6). Its diaphragm was closing each anode compartment and consisted of woven asbestos cloth, resting on an iron-wire network (the cathode) and carefully cemented all around its edges to the belljar. The upper surface of the diaphragm was uniformly covered with a buffering mixture consisting of asbestos wool and $BaSO_4$. This fine insoluble powder was used to reduce convection and diffusion through the diaphragm and to give it very uniform properties at all points. Table I illustrates the electrochemical data obtained during the normal operation of some of these cells. As a comparison the data for a mercury cathode cell is included, the Castner rocking cell (9).

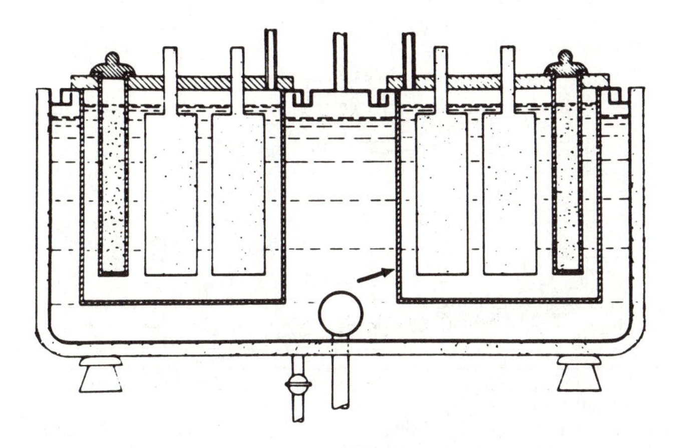

Figure 1. Griesheim non-percolating diaphragm cell with diaphragm indicated by an arrow (Reproduced with permission from Ref. 6. Copyright 1972 Robert E. Krieger).

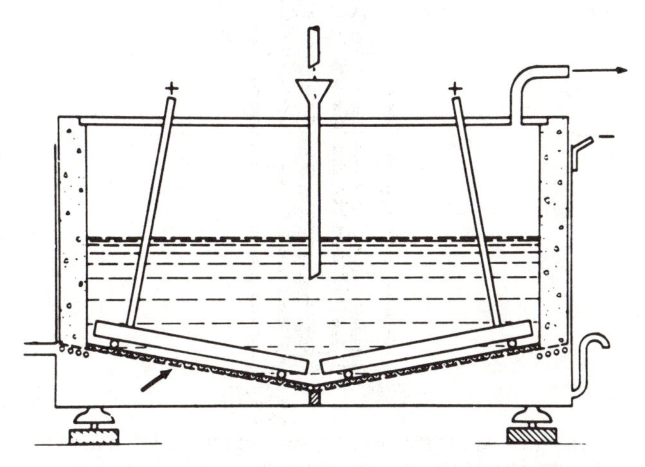

Figure 2. Le Sueur cell (Reproduced with permission from Ref. 6. Copyright 1972 Robert E. Krieger).

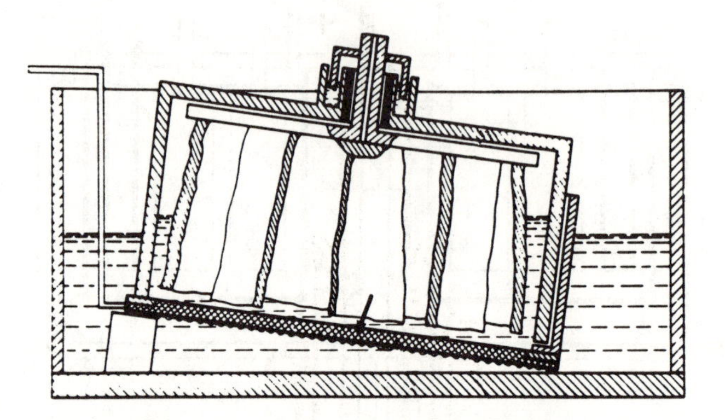

Figure 3. Original Le Sueur cell (Reproduced with permission from Ref. 6. Copyright 1972 Robert E. Krieger).

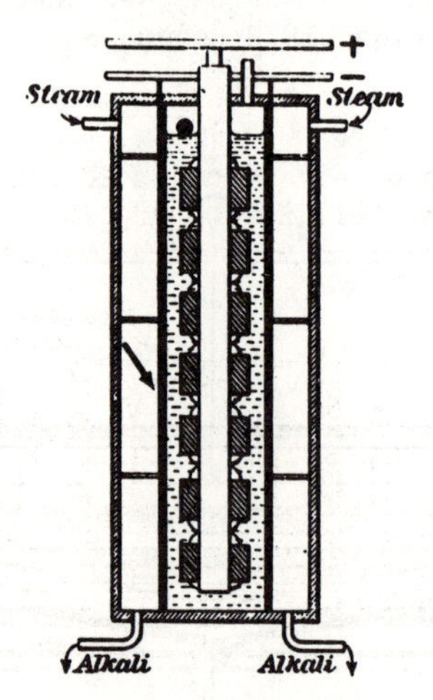

Figure 4. Hargreaves-Bird cell (Reproduced with permission from Ref. 8. Copyright 1912 Edward Arnold).

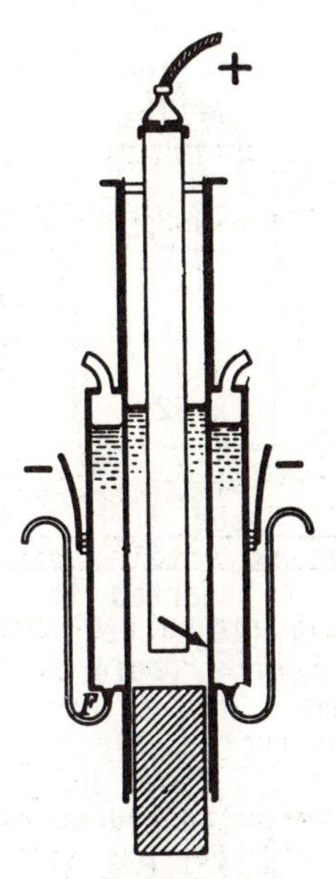

Figure 5. Townsend cell (Reproduced with permission from Ref. 8. Copyright 1912 Edward Arnold).

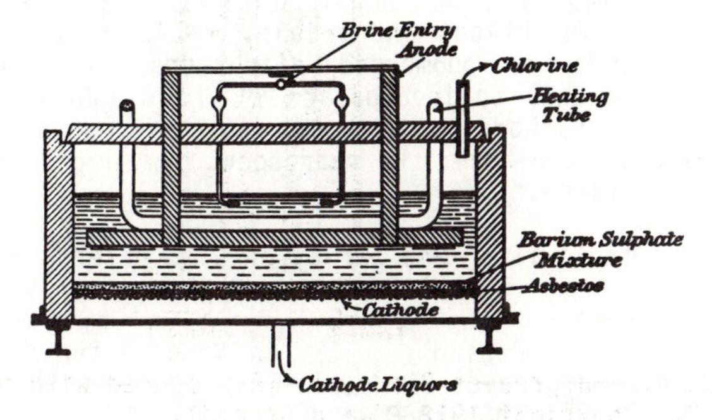

Figure 6. Billiter cell, flat cathode (Reproduced with permission from Ref. 8. Copyright 1912 Edward Arnold).

Table I. Parameters of normal operation of alkali-chlorine cells used at the turn of the century

Cell	Alkali (N)	Cathodic Current efficiency (%)	Voltage (V)	Energy efficiency (%)	KWh/ (kgNaOH)
Griesheim					
- carbon anode	1-2	70-80	3.6	41-51	3.0-3.4
- magnetite anodes	1-2	70-80	4.0	40-46	3.3-3.8
Billiter	3-4	95	3.7	59	2.6
Hargreaves-Bird	3	85	3.7	-	as Na_2CO_3
Townsend	4	94	4.8	45	3.4
Castner	5	92	4.2	50	3.1

The basic material used then to fabricate diaphragms was asbestos. This remained so until the development of sturdy ion specific membrane materials started to make a difference, in the early 1980's. J.R. Crocker (10) made a very good review of the many attempts which have been made, before 1908, to meet the various requirements of diaphragms. This author suggested the use of the following method to construct a diaphragm with an asbestos fiber material: "When the asbestos fiber has been formed as desired, it is subjected to an acid bath, sulphuric or nitric acid, for a short time, and, afterwards baked under an intense heat. The heat changes the fiber into the crystallized state, and the acid bath serves to eliminate any metallic oxide which may be in the asbestos and strengthens the same".

The first era of industrial electrolytic activities was very prolific in inventions and trials. "Claims were made for special features such as submerged diaphragms, unsubmerged diaphragms, special compounds to incorporate in the diaphragm, use of petroleum oil in the cathode compartment etc. Although considered quite important at the time, in retrospect these differences are not of great importance." (6)

Table II. 1949 Survey of U.S. patents issued on chlor-alkali diaphragm cells

Period	No of Patents
1883-1889	5
1890-1899	105
1900-1909	129
1910-1919	90
1920-1929	69
1930-1939	54
1940-1949	43

A great deal of cell development was done in the United States between 1890 and 1910 as the survey (11) of U.S. patents presented in Table II indicates.

Water Electrolysis. The production of hydrogen and oxygen by the electrolysis of water has been practiced on the industrial scale since the beginning of the century. Some important dates in the development of the technology are shown in Table III (12).

Table III. Important dates in the development of water electrolysis

Year	Development	Inventors or Company
1800	Electrolytic Decomposition of Water	Nicholson, Carlisle
1888	Experiments with Pressure Cells	Latchinoff
1890	Monopolar Tubular Cell	Renaud
1892	Load Levelling Electrolyser	Garutti
1899	Design of Filter-Press Cell	Schmidt
1902	Production of Filter-Press Electrolyser	Oerlikon
1910	Design of Knowles Cell	International Electrolytic
1912	Operation of Knowles Cell	International Electrolytic
1925	Design of Pressure Cell	Noegerath, Lavaczek
1926	Design of Filter-Press Electrolyser	Zdansky

This apparently slow industrial start was due in part to the success of the chlorine-alkali processes which also produced hydrogen. "Where one only of these gases (hydrogen and oxygen) is needed, its preparation by electrolysis is usually uneconomical. Hydrogen is produced in large quantities in electrolytic alkali works, and can be manufactured very cheaply from water-gas, whilst oxygen is best obtained by fractional distillation of liquid air. But where both gases are required, and particularly for the oxyhydrogen flame, used extensively in platinum working and in some kinds of lead burning, electrolysis is the most convenient method of preparation" (8).

The first major installation of water electrolysers was constructed in 1912 in Port Sunlight, England. It contained 200 Knowles cells that could consume 1.5 MW of power at 3.5 kA. Some of these Knowles cells have been in operation for sixty years. They were simple in design and maintenance. These tank electrolysers used paper or woven asbestos as semi-diaphragms. The separators were immersed in a concentrated alkaline solution and open at their lower edges.

From 1910 to 1940

In diaphragm cells both the cell voltage and the current efficiency depend upon:

- cell construction
- electrode materials

- electrolysis conditions
- diaphragm properties

In his second edition of Principles of Applied Electrochemistry (13), Allmand describes the virtues of a good diaphragm material before citing materials used effectively in some industrial processes:

Alkali-resisting Diaphragms
- asbestos
- Portland cement

Acid-resisting Diaphragms
- mixture of alumina/silica, LeBlanc
- mixture of kaolin/corundum, Buchner
- fused silica-electrofilters, Thatcher

The main advantage offered then by the alkali-resisting asbestos was its flexibility. The rigidity of the other materials made them more difficult to seal properly and, because of the forces involved, more easy to crack.

Chlorine-Alkali. With the exception of the Griesheim cell, the diaphragm cells have had a variety of shapes, sizes and materials of construction, but the diaphragms have always been based on asbestos used as paper, cloth or fiber. The LeSueur cell, for example, used a horizontal paper diaphragm over the horizontal cathode screen. Glass rods placed on the diaphragm served to support the anode graphite blades and maintain the proper spacing as the graphite wore away.

Where intermediate supports were required, bricks were stacked on the diaphragm (14). Another early diaphragm cell, the Townsend cell, which was developed for commercial use in 1905 with financing by the Elon H. Hooker's Development and Funding Company (later Hooker Electrochemical Co.) was of a vertical type with a long, high central row of graphite anodes. Opposing these anodes on either side of the cell was a paper diaphragm backed up with a wire screen cathode.

The original paper diaphragm of Townsend's was soon replaced with an impregnated (a colloidal mixture of iron oxide and loose asbestos fibers) asbestos cloth diaphragm developed by Leo Baekeland. In 1913 Marsh designed a cell with finger cathodes and side-entering anodes and cathodes. Asbestos paper was wrapped over this surface and sealed top and bottom with cement and putty. These putty joints provided a poor seal and thus a low cell efficiency. In order to improve the construction of a Marsh cell, K. Stuart, of the Hooker Electrochemical Company developed in 1928 a method of depositing asbestos fibre directly on the cathode by immersing it and applying a vacuum. The flexibility of design permitted by this new type of diaphragm was incorporated in the Hooker Type S series of cells. These modifications have improved the amperage capacity from 5 kA to 30 kA over the next decades.

A sign of importance of the Stuart invention is the fact that more than 90 percent of the chlorine produced in diaphragm cells in the 1970's came from cells using deposited asbestos diaphragms.

The filter-press concept, which became very popular for the electrolysis of water and of hydrochloric acid, had many attractive features for chlor-alkali producers. It required a minimum of conductor material between cells, a minimum of floor space and essentially low capital costs. The only commercial use of filter-press cells for chlor-alkali has been by the Dow Chemical Company. Early workers of this design were Kellner, Guthrie and Finally but the original patents for the bipolar cells themselves were granted to Thomas Griswold in 1911 (15) and 1913 (16).

Water Electrolysis. The demand for electrolytic hydrogen, during the same period, was still variable and mostly linked to the production of ammonia for fertilizers. In the early 1920's a few installations, made with Casale and Fauser's cells, were running in Italy for a total production of approximately 90 MW (17).

After twenty years of commercial success the old Schmidt electrolyser, built by Oerlikon in Switzerland, was superseded by a much more powerful competitor, the Pechkrang electrolyser which was built by the Hydrogène Co. in Geneva. While the biggest Schmidt electrolyser ever built could hardly produce 32 kW of hydrogen, the new unit contained up to 140 cells that could sink a current of 2.5 kA for a total power of 0.88 MW. During the first two years of its existence (1926-1927) the new electrolyser was to be installed in six European countries for a total production potential of 176 MW.

The biggest electrolyser for a long while was built during the same period (1924-1928) by E.A. Zdansky, an engineer working for the German company, the Bamag-Meguin. The largest model of this filter-press electrolyser was the Model C. It had electrodes with an individual surface area of 3 m^2 and could produce up to 2.6 MW of electrolytic hydrogen.

Various attempts were made to develop pressure water electrolysers between 1920 and the second world war, but all came short of industrial recognition: Noeggerath in 1927 (150 bar), Neiderreither in 1929 (bipolar electrodes, 5 kW), Lawaczek in 1928 (300 bar), Siemens and Halske in the late 30's (5 to 10 bar).

1940-1975, The Material Revolution

An ideal diaphragm should be (18):

1. permeable to ions but not to molecules
2. of high void fraction to minimize electrical resistance
3. of small mean pore size to prevent the passage of gas bubbles and minimise diffusion
4. homogeneous to ensure good current efficiency and even current distribution
5. non-conducting to prevent action as an electrode
6. chemically resistant to the reactants and products
7. resistant to cell operating conditions of temperature, pH, etc.
8. of some mechanical strength and rigidity
9. cheap or long lasting.

Several of these properties were in conflict and generally a compromise had to be made. The most widely used materials up to then were asbestos, either in the form of fibre cloth or paper, and ceramics. Neither material was ideal, asbestos being slowly attacked in acid or in very alkaline conditions and ceramics having a high electrical resistance whilst at the same time being brittle and thus unstable to temperature change and mechanical shock. Since the production of large ceramic diaphragms was still not possible without increasing the thickness of the material, ceramics did not invade the electrolytic scene of that period.

In order to improve the properties of asbestos diaphragms, many attempts had been made to modify the structure of the diaphragm by impregnating or combining the asbestos with various compounds. Many new materials were also starting to be fabricated into porous sheets with controlled properties for use as diaphragms for electrolytic processes. Most of these materials were to fail in chlorine electrolysis due to the high surface area exposed to the corrosive nature of the anolyte which enhances chemical degradation.

Polytetrafluoroethylene would withstand the cell conditions but the porous P.T.F.E. made then failed as a diaphragm because of its non-wettability (19). The breakthrough was to be inspired by fuel cell technologists who were then using P.T.F.E. powder (20) to limit the wettability of fuel cell electrodes.

Chlor-Alkali. Major changes in materials of construction were to happen to this half century-old electrochemical technology.

Chlorine production in the U.S. had increased about six-fold between 1950 and 1975: 7000 tons per day in the early 1950's compared to 41000 tons per day in 1976 (21). The replacement of graphite anodes by coated titanium anodes had been a great advantage to diaphragm cells. With graphite anodes, the anode-cathode gap increased as the graphite wore away. This increase in electrode gap would increase cell voltage and heat evolution, and shorten the life of the diaphragm due to clogging of the pores by graphite particles and chlorinated products. The new anodes were not only very stable but could be moved very close to the diaphragm by an adjustable feature.

The swelling and increasing porosity of asbestos diaphragms with operation were finally overcome by curing the asbestos diaphragm with special additives (chlorinated fluoro-polymers, wetting agent ec.). These additives lent dimensional stability to the diaphragm so that the anode could be moved very close to the cathode without invading the diaphragm layer. The result of incorporating the modified diaphragm and expandable anode produced a total cell voltage reduction of 400 mV at a typical current density of 2.2 kA/m^2 (22).

Diamond Shamrock Corporation had been experimenting with Dimensionally Stable Anodes (DSA) since 1966. Their cells had, by 1975, expandable anode, FRP cover and a modified asbestos diaphragm (23). These cells could produce 360% more chlorine at 2.5% less power per ton on half of the original floor space. The modified asbestos diaphragm retained the good qualities of asbestos and the troublesome properties were reduced. It also provided easier hydrogen evolution, prevented swelling, had a longer life, improved uniformity and flow properties, was easy to remove and could be stored up to 5 months before use.

Hooker Chemicals and Plastics Corporation also modified drastically the power profile of its cells by incorporating similar improvement in their construction. Table IV shows an analysis of voltage improvement in Hooker cells for two generations of electrolysers (24).

The frequency of cutout activity was reduced by a factor of from 6 to 11 times for the new bipolar PPG-De Nora Glanor cells (25)

Table IV. Analysis of voltage improvements at 1 kA/m^2 in Hooker cells

Model Date	S-3 1960	H-4 1976
Decomposition, anode	1.32	1.32
Decomposition, cathode	0.93	0.93
Overvoltage, anode	0.33	0.03
Overvoltage, cathode	0.27	0.27
Brine gap	0.49	0.27
Hardware	0.36	0.17
Diaphragm	0.30	0.16
Total	4.00 V	3.15 V

even if their diaphragm material was still of the deposited asbestos type. A life expectancy of one year was expected from the diaphragm in such cells. With additives it was claimed to become 2 to 3 times longer.

Electrolysis of Hydrochloric Acid. Development of this process was conducted initially at Bitterfield in the I.G. Farben Industrie Plant (26). The German patent application covering this innovation was dated October 15, 1942. It took 14 years before this work became known and taken up by De Nora in Milan. The original HCl electrolysis cell constructed on this concept had vertical bipolar electrodes. An extensive study made of possible diaphragm materials had resulted in a choice that gave satisfaction since the last electrolyser was operated by I.G. Farben continuously for fifteen months, but the details of manufacture of the diaphragm were never published.

On the other hand the De Nora electrolysis unit is well described in the literature (26, 27). The standard assembly was composed of 40 cells of the filter-press type (De Nora was already in the water electrolyser business). The diaphragm itself consisted of PVC cloth pressed between the ribs of graphite anodes and cathodes so as to separate the electrodes by a width of 2 mm. The life expectancy of such a diaphragm, under good operating conditions, was approximately three years.

Water electrolysis. In 1975 electrolytic hydrogen accounted only for 2% of the worldwide production of hydrogen (28). Large plants (>100 MW installed power) were scarce and they relied basically upon very cheap water power. Hydrogen production by electrolytic decomposition of water was restricted to small plants where the purity of the product and the flexibility of the process are prime qualities. It was felt that electrolytic hydrogen could become competitive with chemical hydrogen if the energy efficiency of water electrolysers reached 75 to 80%. Only three ways of doing that were then possible:

- decreasing the electrical resistance of the diaphragm
- increasing the working temperature
- improving the activation of the electrodes

Most installations were comparatively small in size, with a few notable exceptions. At the Aswan High Dam, for instance, 40,000

m^3/h^{-1} of hydrogen were produced by electrolysis, using Demag electrolysers (12).

The commercial water electrolysis units were of the types and models shown in Table V. Of the materials available to serve as diaphragms, woven asbestos-fibre cloth was still almost invariably chosen, although fine metal-mesh was also marginally used. For diaphragms of large area, asbestos was interwoven with fine nickel wire to give increased strength.

1975-2000 Advanced Membrane Materials

The availability of perfluorinated ion exchange membranes of exceptional chemical stability and electrochemical properties has opened new opportunities for the application of electrochemical technology in the chemical process industry (29).

Chlor-Alkali. The energy crisis of the 1970's combined with tougher environmental regulations accelerated drastically the integration of the newly developed membrane material by this electrochemical industry. In Japan, for example, the mercury cells that had been producing most of the country's chlor-alkali production since the second world war were legislated out of existence in the early 1980's. At the same time working with asbestos became recognized as a serious hazard for the workers handling the fiber material.

During the first four years (1978-1982) after commercialization of its Flemion membrane, Asahi Glass Co of Japan (30) has installed three plants (two in Japan, one in Thailand) for a total potential production of 41400 tons NaOH/year. During the following 4 years (1982-1986) Asahi Glass Co equipped electrolysers with its membranes for a total estimated production of 691,400 NaOH ton/year. The new process presents a real savings in energy and capital as well as improvement in the environmental area. All the major companies involved in the production of electrolysers for chlor-alkali have their own version of the original ion-exchange Nafion membrane (31). This search for innovation will itself generate new niches in the industrial transformation of chemicals (29, 32).

Water Electrolysis. The predictions, made in 1978 (33), that water electrolysers were soon to become more energy efficient to be competitive with other sources of hydrogen seem to have recently materialised (34). While the advanced hydrogen plant at Becancour, Quebec, still has asbestos for separator material, an extensive study of other possible materials for use in alkaline water electrolysers (35) has demonstrated how some replacement materials could become advantageous for higher operating temperature or efficiency. One such alternative is the use of a solid polymer electrolyte on the basis of outstanding performance and operating characteristics which have been demonstrated in systems developed for aerospace and military applications (36).

The modern alkaline water electrolysis technologies available commercially have the following common features: they operate at atmospheric pressure between 70-90°C with a current density of up to 5 kA/m^2 and energy efficiency reaching 85% (37). Some pre-commercial technologies have demonstrated higher values for these four parameters.

Table V. Operating conditions of commercial water electrolysers available in 1970

Type Model	Operating Temperature °C	Unit Capacity $Nm^3 H_2 \cdot h^{-1}$	Energy Efficiency %
Tank electrolyser			
• Knowles cell	80	2.1	70
• Stuart cell	85	2.4	60
Filter-press electrolyser			
• CJB electrolyser	80	240	60
• Demag electrolyser	80	150	65
• Oerlikon electrolyser	75	210	68
• Pintsch Banrag	80	100	65
• Moritz oxyhydrolyser	60	40	67
• De Nora electrolyser	75	18	64
Pressure electrolyser			
• Zdansky-Lonza	90	145	65
• CJB electrolyser	65	2	67
• Treadwell generator	90	0.3	40

Other Processes, Other Membrane Materials. Many other industrial processes use membranes or diaphragms. Some also operate in conditions very similar to those of the electrolysers described so far in this paper. Reverse-osmosis, batteries and fuel cells are amongst mature technologies that are forcing material developers to produce stabler, more performant and cheaper separators. The new porous materials will themselves probably inspire innovators with novel designs and applications.

Literature Cited

1. Thatcher, C.J. Met. Chem. Eng. 1915, 13, 336-8.
2. Gardiner W.C. Proc. Symposium on Selected Topics in the History of Electrochemistry, ECS 1978, pp 413-428.
3. Richards J.W. Electrochem. Ind., 1902, 1, 49-55.
4. Greene S.D. Cassier's Mag., 1895, 8, 175.
5. Richards J.W. Electrochem. Ind., 1902, 1, 11-23.
6. Kircher M.S. In Chlorine, Science J.S., Ed.; Robert & Krieger: New York, 1972; Chapter 5.
7. Billiter J. In Die Elektrochemischen Verfalnen, 1911, Vol. II, p 203.
8. Allmand A.J. In The Principles of Applied Electrochemistry; Edward Arnold: London, 1912; Chapter XXI and XXII.
9. Ornstein G. Trans. Am. Electrochem. Soc., 1916, 29, 530.
10. Crocker J.R. Electrochem. Metal. Ind., 1908, 6, 153-6.
11. Murray R.L. Ind. Eng. Chem., 1949, 41, 2155.
12. Smith D.H. In Industrial Electrochemical Processes; Kuhn A.T., Ed.; Elsevier, Amsterdam, 1971; Chapter 4.
13. Allmand A.J. In The Principles of Applied Electrochemistry; Edward Arnold: London, 1924.
14. Hubbard D.O. J. Electrochem. Soc., 1952, 99, 307C-309C.
15. Griswold T. Jr. U.S. Patent 987, 717, 1911.

16. Griswold T. Jr. U.S. Patent 1, 070, 454, 1913.
17. Aureille R. In Hydrogen Energy Progress V: Veziroglu T.N., Taylor J.B., Ed.; Pergamon, New York, 1984; pp 21-43.
18. Jackson C., Cooke B.A., Woodhall B.J. In Industrial Electrochemical Processes; Kuhn A.T., Ed.; Elsevier: Amsterdam, 1971; Chapter 15.
19. Chapman E.A. Chem. Process Eng., 1965, Aug., 387.
20. Liebhafsky H.A., Cairns E.J. In Fuel Cells and Fuel Batteries; John Wiley, New York, 1968; Chapter 7.
21. Gardiner W.C. J. Electrochem. Soc., 1978, 125, 22C-29C.
22. Sikstrom L.E., Liederback T.A. Chem. Ing. Technik, 1975, 47, 157.
23. Thomas V.H., O'Leary K.J. In Chlorine Bicentennial Symposium; Jeffrey T.C., Danna P.A., Holden, H.S., Ed.; ECS: Princeton, 1974: p 218.
24. Grotheer M.P., Harke, C.J. In Chlorine Bicentennial Symposium; Jeffrey T.C., Danna P.A., Holden, H.S., Ed: ECS: Princeton, 1974: p 209.
25. Kienholz P.J. In Chlorine Bicentennial Symposium; Jeffrey T.C., Danna P.A., Holden, H.S., Ed.; ECS: Princeton, 1974; p 198.
26. Berkey F.M. In Chlorine; Sconce J.S., Ed.; Robert & Krieger: New York, 1972; Chapter 7.
27. Prayer S., Strewe W. In Chlorine Bicentennial Symposium; Jeffrey T.C., Danna P.A., Holden, H.S., Ed.; ECS: Princeton, 1974, p. 257.
28. Mas L., Solm J.C., Damien A. Proc. 1st World Hydrogen Energy Conference, University of Miami, Florida, 1976; 6B41-6B62.
29. Grot W.G.F. Proc. Diaphragms, Separators, and Ion-Exchange Membranes; 1986 ECS Symposium, pp 1-14.
30. Sajima Y., Sato K., Ukihashi H. Industrial Membrane Process Symposium, 1985 Spring National AIChE Meeting, Houston, Texas; pp 108-113.
31. Robinson J.S. In Chlorine Production Processes: Recent and Energy Saving Developments; Noyes Data, New Jersey, 1981.
32. Ezzell B.R., Carl W.P., Mod W.A. Industrial Membrane Process Symposium, 1985 Spring National AIChE Meeting, Houston, Texas; pp 45-50.
33. LeRoy R.L., Stuart A.K., Industrial Water Electrolysis Symposium, 153rd Meeting of the ECS, Seattle, 1978.
34. Crawford G.A., Hufnagl A.F., Int. J. Hydrogen Energy, 1987, 12, 297-303.
35. Renaud R., LeRoy R.L., Int. J. Hydrogen Energy, 1982, 7, 155-166.
36. Nuttall L.J., Russell J.H., Int. J. Hydrogen Energy, 1980, 5, 75-81.
37. Crawford G.A., Benzimra S., Int. J. Hydrogen Energy, 1986, 11, 691-701.

RECEIVED August 3, 1988

Chapter 35

Gibbs, LeSueur, and Willson

Pioneers of Industrial Electrochemistry

Robert V. V. Nicholls

Department of Chemistry, McGill University, Montréal, Quebéc H3A 2K6, Canada

William T. Gibbs (1868-1910) migrated from England to Buckingham, Que. There in 1893 he perfected electrolytic methods for the production of chlorates and dichromates, and an electrothermal method for phosphorus. The latter process may have been an example of simultaneous invention. Ernest L. LeSueur (1869-1953) was born in Ottawa. While enrolled in M.I.T.'s Electrical Engineering course, he invented a diaphragm cell for the electrolysis of salt, which was widely used. Thomas L. Willson (1860-1915) grew up in Princeton, Ont. Moving to Spray, N.C., he discovered by chance in 1892 an electrothermal method for making calcium carbide, starting material for acetylene, cyanamide, etc.

Thomas Leopold Willson (1,2,3) was born at Princeton, a village between Brantford and Woodstock, Ont., on March 14, 1860. He enrolled at Hamilton Collegiate Institute. After graduation he was apprenticed to a blacksmith, John Rodgers, who allowed him to work in the loft of the smithy. There he built a dynamo, then a novel source of direct current. He used the machine to produce an arc light. So spectacular was the result that a local merchant gave him a contract to illuminate his factory. There was even some talk of lighting the public park with electricity. However, at this time John J. Wright, having solved the problem of automatic feed for the carbons, was erecting lights along Yonge and other streets in Toronto. Wright had patented his device and Willson wisely abandoned further effort.

Failing to find satisfying work in Ontario the young man emigrated to the United States in 1881. He continued to be interested keenly in the potentialities of abundant and cheap electricity. In 1891 he met Maj. James Turner Morehead, who owned a cotton mill and water rights at Spray, N.C. (4). Together they formed the Willson Aluminum Company "for the manufacture of aluminum .... in an electric furnace". Apparently their intention was to obtain the metal from aluminum oxide or a salt, such as the chloride, by reduction with calcium. Presumably, they were aware that about five years before the British Aluminium Company had built a plant to accomplish

0097–6156/89/0390–0525$06.00/0

the same result using sodium, recently made available in quantity by the Hamilton-Young-Castner Process.

They may not have given sufficient weight to the significance of Charles Martin Hall's invention just at this time of a method of producing aluminum by the electrolysis of aluminum oxide in molten cryolite, which by 1889 was being exploited by the Pittsburgh Reduction Company.

Be that as it may Willson certainly tried to devise a route for the manufacture of calcium metal in quantity by heating calcium oxide (lime) with carbon (coke). He had moved to Spray in the autumn of 1891 and built a small hydroelectric plant and an electric-arc furnace. On May 2, 1892, the initial experiment was performed. At the high temperature of the arc a brown liquid flowed from the furnace and solidified on cooling.

When brought into contact with water the solid produced a gas. Clearly it was not the hoped-for hydrogen as it would have been had the solid been calcium. The flame was sooty, suggesting that the gas was a hydrocarbon. Could it be acetylene, then still a laboratory curiosity, which had been first prepared by Friedrich Wöhler in 1862? A sample of the solid was sent to Lord Kelvin at Glasgow University. In due course his reply, dated October 3, was received. The gas was acetylene.

With a minimum of delay Willson applied for American, Canadian, British and other patents. Just in time he obtained protection for his invention, beating Henri Moissan, the eminent French electrochemist, by the narrowest of margins.

Typical of those patents was U.S. Patent 541 137, June 18, 1895. It claims "An Improved Calcium-Carbide Process". The electric furnace is described. An intimate 35:65 mixture of finely powdered coke and lime is to be heated. The calcium carbide is obtained by raising one electrode to which it adheres or by tapping the carbon-lined furnace at the bottom. The applicant emphasizes the advantage of using an alternating current, because thereby the charge readily and automatically feeds between the electrodes.

Willson sold his American patents to the Union Carbide Company, which exploited the invention by operating a large plant at Niagara Falls. It eventually prospered. Not so its competitors (5).

Willson Sells American Patents and Returns to Canada

Willson returned to Canada as quickly as possible in 1896. The Canadian Patent Act required that an invention be used within two years or the patent would be revoked. He obtained permission to harness the power of the excess water from Locks 8, 9 and 10 of the Welland Canal, at Merriton, near St. Catherines. The fall at each of these locks was 12½ feet. The potential electrical horsepower was 1,650. Using $ 90,000 of his own money he erected a power plant (housing a 400-h.p. General Electric single-phase, alternating-current generator) and a carbide plant (housing two furnaces with room for eight more). The arc of each furnace operated at 75 volts and 1,000 to 2,000 amperes. It produced four 500-pound pigs of carbide in twenty-four hours. The surfaces of the pigs were trimmed giving impure flake. The pure cores were broken into lumps.

At that time the sole use of calcium carbide was to provide acetylene as an illuminant in homes and factories, and for bicycles, carriages and trains. Willson set out with vigor to develop these and other uses. Since sad experience had shown that the gas could not be compressed safely in cylinders, it had to be made when and where required. Soon a number of cleverly-designed generators appeared on the market.

The Merriton works commenced operation on April 13, 1896. The first, 200-lb., trimmed pig was ready on August 15. The first export shipment left in December.

"Carbide" Willson incorporated a second enterprise, the Ottawa Carbide Company, in 1900 and took up residence in the Capital a year later. A plant was built on Victoria Island in the Ottawa River. Power was purchased from the nearby hydroelectric plant of Ahern and Soper's Ottawa Electric Company.

In 1903 Ernest LeSueur, of whom more below, with characteristic originality, devised a novel method of shipping acetylene. Useing the flakes chipped from the pigs at the Ottawa works he generated acetylene, compressed it to 8 p.s.i. (10 p.s.i. was accepted as the safe upper limit), and by rapid expansion converted the gas into a solid, in a manner now familiar for the conversion of compressed carbon dioxide into Dry Ice. The solid acetylene was then shipped by express train in containers insulated by feathers the 35 miles to the Town of Maxville. There the streets were illuminated in this fashion for a number of years until electricity became more economical and convenient.

At this time (1901) the Shawinigan Carbide Company was formed with J. W. Pike as president and T. L. Willson as vice-president. Three years before the Shawinigan Water & Power Company had been incorporated to exploit the hydroelectric power of the St.Maurice River at Shawinigan Falls, Que., and upstream. A year later 10,000 h.p. became available but, because there was an oversupply of carbide, the Carbide Company did not start production until two years later.

Meanwhile, a very important, new use for acetylene had been developed, oxy-acetylene welding and cutting. The oxy-acetylene became popular because it gave a temperature of 6,000 to 7,000^{o} F. contrasted with 4,000, the maximum attainable with the oxy-hydrogen flame. This application stimulated the invention of a safe method of compression acetylene to facilitate its storage and shipment. The challenge was met by the Prest-O-Lite Company, who devised a method involving the compression of acetylene to 300 p.s.i. into a cylinder packed with a solid absorbent saturated with acetone.

To give an account of the employment of calcium carbide for the fixation of atmospheric nitrogen through its conversion to calcium cyanamide and of acetylene as a versatile, raw material for organic synthses is to digress from the accomplishments of "Carbide" Willson.

Willson was not to be free from accusations of patent infringement. Particularly notable was his protracted litigation with L. M. Bullier.

In 1908 Shawinigan Water & Power purchased control of Shawinigan Carbide and three years later the Canada Carbide Company was formed to acquire all of Willson's manufacturing rights under his carbide patents. The works at Merriton, Ottawa and Shawinigan Falls were purchased outright. The Merriton plant continued in operation until 1924 when it was sold to Union Carbide. The Ottawa plant was closed in 1908 when it was found that power at that city could be more advantageously used for other purposes. The facilities of Canada Carbide were further improved and expanded.

Willson's Interest in Ferro-Alloys and Fertilizers

The large-scale manufacture of ferro-alloys in Canada originated in this century when electric-furnace capacity had expanded beyond the lagging demand for calcium carbide, This situation focused attention on alternative uses for the furnaces, including the production of these alloys. Typically, Willson became involved. His participation led to the organization of the Electro-Metallurgical Company of Canada in 1906. A year later the Company erected at Welland Canada's first plant for the production of ferro-silicon. During the Great War it consumed as much as 50,000 h.p. to manufacture 85 per cent alloy, to the order of the British Admiralty for the making of hydrogen for the inflation of balloons. At that time it was the largest in the British Empire.

Within a couple of years of taking up residence at 188 Metcalfe St., Ottawa, Willson completed a laboratory in its basement. In the words of his daughter, Mrs. Marion S. Roberts (6), "He wrote glowingly that he had equipped the finest laboratory on the continent for its purpose. It was his dream to do scientific research on nitrogen." To elucidate the cryptic phrase, "on nitrogen", will require further study. Could she have been referring to cyanamide?

The Willsons' association with the Gatineau River Valley, which is close to the Capital, began in 1904, when they rented a cottage on Meech's Lake. Within five years they had bought considerable land and built a house. What is important for our purpose is that the property included the falls on Meech Creek. There, to the consternation of some of their neighbors, Willson constructed a dam, a powerhouse and a double-superphosphate plant, which required the processing of phosphoric acid. The devising of novel methods for the manufacture of phosphatic fertilizers became his consuming passion. This passion was to lead to his financial ruin.

The enormous potential of the Saguenay River for hydroelectric-power production had captured his imagination. As early as 1900 he had secured from the Quebec Government very favorable rights. Now he needed customers for the power. Naturally, since obstacles to the long-distance transmission had yet to be overcome, he had difficulty in persuading industry to establish itself in such a remote place. Pulp-and-paper mills and carbide works were considered. I will quote from his daughter again. "The fertilizer appeared to be reaching the stage of practical, commercial production. .... In order to get the first capital he sold all his companies, and mortgaged all his patents, property and waterpower rights to James Buchanan Duke, the 'tobacco king'. He was unable to meet the time limit on the mortgages and in one sweep Duke took all Willson's assets. .... It was supposed that Willson's patents covering the entire world were gone. But it turned out that Newfoundland was not covered." He applied for the necessary protection, formed the Newfoundland Products Corporation, secured the required legislation from the Colonial Government, and by July, 1914, had raised millions on the London market. War was declared in August! Export of British capital was forbidden. While in New York City, searching for alternative support, "Carbide" Willson died of a heart attack on December 20, 1915.

Le Sueur Perfects a Diaphragm Cell

The practical application of electrolytic decomposition had its origin in the work of Humphry Davy in England. In the second Bakerian Lecture (1807) he summed up its primitive state and announced that he had accomplished in his laboratory the isolation of the two metals, sodium and potassium, by passing a current through fused soda or potash. In 1851 Charles Watt received a British patent, which described the preparation of caustic soda and chlorine from salt by a process which is essentially the one used today. For many years the exploitation of electrolytical decomposition on an industrial scale was inhibited by the lack of a source of cheap and abundant direct current. This lack was satisfied by the development of the dynamo by Siemens-Gramme in the 1870's. Malthes and Webe were the first to electrolyse brine in a diaphragm cell, which they patented in 1886. The firm of Chemische Fabrik Griesheim bought the patents, made the process continuous, improved the carbon electrodes and opened a plant at Bitterfeld in 1890. However, the electrical efficiency of the cells left much to be desired.

Ernest Arthur LeSueur was born in Ottawa on February 3, 1869 (7). He attended public school there, finishing at the Ottawa Collegiate Institute. Then he enrolled in the Massachusetts Institute of Technology and graduated with a bachelor's degree in electric engineering. (A Canadian university (McGill) was not to offer a similar degree until a year later.) In 1888, while still a student, he became interested in the electro-decomposition of salt solutions. First he built a cell with a supported-mercury cathode. There is some evidence to suggest that significant experiments were performed in Ottawa during the following summer vacation. In 1889 he turned his attention to the porous-diaphragm type and submitted the results as part of his graduation thesis. Many years later (1940), when in a reminiscent mood, during an address before the Toronto Branch of the Canadian Institute of Chemistry (8), he said "When Professor Charles R. Cross gave me permission to embark upon an investigation of the percentages of useful decomposition, under various conditions, of solutions undergoing electrolysis, he told me I should be quite free from leading strings, seeing that neither he nor anyone on the .... staff knew the first thing about such matters" ! Be that as it may LeSueur applied for his first patent in the year of his graduation, when 22 years old.

Again let LeSueur speak for himself. "After experimenting for some time I came to realize that although cathodic solution reaching the anode meant 'death' to the decomposition efficiency, it was a matter of almost indifference (at least as regards such efficiency) for the anolyte to reach the cathode. .... It seemed, therefore, that the solution of the problem would be to apply a pressure on the anode side of the diaphragm. .... Subject to the employment of this principle there is no limit to the variety of cell constructions that can be used successfully. ..... I adopted the arrangement of a horizontal or nearly horizontal diaphragm instead of the obvious and in respect to compactness, much more advantageous vertical one, because of one's ability to get the same .... pressure and transfer at all points .... a more rugged construction .... and the advantage of the anolyte pressure offsetting the tendency of the hydrogen beneath the diaphragm to bulge it upwardly."

After demonstrations during 1890-91 at Newton Upper Falls, Mass. and Bellows Falls, Vt., and with capital obtained in Boston, the youthful LeSueur formed the Rumford Falls Electrochemical Company. The site of the first plant in North America for the electro-decomposition of salt was selected in 1892, at a 130-foot fall on the Androscoggin River in Maine. Operation started a year later. Four circuits of cells consumed about 750 kw. of power (9).

In order to keep the chronology of brine electrolysis to the forefront one should mention that the Castner Electrolytic Alkali Co. started at Niagara Falls, N.Y. in 1897; the Dow Chemical Co. at Midland, Mich. in the same year; and Pennsalt Co. at Wyandotte, Mich. in 1901.

The LeSueur design was licensed to the Electrochemical Company and to several American firms, notably paper-making ones, who needed a cheap supply of chlorine for bleaching. In 1898 the Company, including LeSueur's patents, was purchased by the Burgess Sulphite Fibre Company. The plant was moved to Berlin, N.H. LeSueur was "out" He was always to regret his self-imposed exclusion from further developments in the electrochemical field. He returned to Canada and became a very successful consultant in chemical engineering. But that is another story.

Let it suffice to state that he was retained by the Consolidated Lake Superior Corporation to obtain oxygen-enriched liquid-air for steel smelting at Sault Ste.Marie, Ont.; in 1903 he produced solid acetylene (as has been recounted already); after 1905 he became interested in explosives in association with General Explosives Ltd. of Hull, Que., an interest which continued into the Great War; he designed and installed a phosphorus sesquisulphide plant for the Eddy Match Company. He continued as a very active and highly regarded practioner until his death in Ottawa in 1953.

Ernest LeSueur was a founding member of the American Society of Chemical Engineers (1908) and of the Canadian Institute of Chemistry (1920). He received the Canada Medal of the Canadian Section, Society of Chemical Industry, and that society established a medal in his honor.

And what of the original LeSueur cells? The Brown Company, successor to Burgess Fibre, operated them, and others like them, for more than fify years -- in competition with later designs of diaphragm cells, such as, the Townsend-Hooker (first used in 1906), the Billiter (1907) and the Gibbs. Of the last named more anon.

William Gibbs; Electrolytic Chlorates and Dichromates

William Taylor Gibbs was born on a farm at North Stoke, near Bath, England (10). He enrolled in the Merchant Venturers School in Bristol and in due course won a scholarship in chemistry and geology at the Royal College of Science, London. On graduation he engaged in electrolysis research in East London. In 1890 he came to Buckingham, Que., to be chemist at one of the British-owned apatite (tricalcium phosphate) mining companies. Shortly afterwards the phosphate boom in that area collapsed due to the inroads of Tennessee and Florida rock.

William Gibbs saw the possibilities of cheap, local power and began experimenting with electrochemical processes. On the Lièvre River, which flows into the Ottawa River, 25 miles below Canada's capital, there were two readily accessible falls. The nearer one had the potential of producing 10,000 h.p. Gibbs gave his attention to the manufacture of potassium chlorate via the electrolysis of the chloride in alkaline solution. He also developed an electrolytic method for the conversion of sodium chromate into dichromate, with recovery of half of the sodium as carbonate. The chlorate process was refined and patented (Can. Patent 42 429, 1893). With the financial assistance of Stanislaus P. Franchot and Alexander MacLaren, a plant was built at Masson, at the junction of the rivers. After a year of successful operation it was totally destroyed by fire. Gibbs, Franchot and others formed the National Electrolyte Company and, using a later design of cell, established a works at Niagara Falls, N.Y.,for the manufacture of chlorates and dichromates. Franchot, an American citizen and co-director with Gibbs, moved to superintend its operation.

Gibbs and Electrothermal Phosphorus and Ferro-Chromium

Gibbs, rather surprisingly, decided not to accompany Franchot. Instead he saw an opportunity to use electricity to exploit the local phosphate rock. With a partially-built 10,000-h.p. powerhouse nearby and awaiting a purchaser, he had no great difficulty in persuading Walter A. Williams to become a partner with him. He invented a method for the production of yellow phosphorus, which involved the heating of a mixture of mineral phosphate, carbon and silica. He successfully adapted a furnace of the Siemens type by using jointed,carbon electrodes, to insure their complete utilization and continuous operation. Phosphorus distilled off and a calcium-silicate slag drained away. Using 500 h.p. production began in 1896.

Requiring further capital, Gibbs and Williams secured it a year later from the Anglo-Continental Gold Syndicate, with headquarters in London. A corporation, Electric Reduction Company, was formed in November with an authorized capital of £ 40,000. The Syndicate held a 50-percent share; Gibbs and Williams the remainder. (One should note that Gibbs retained his interest in National Electrolyte.) (Hambly, Fred T. _A History of the Electric Reduction Company of Canada, Ltd., Buckingham, Quebec, 1897-1951_, unpublished manuscript)

At the beginning of the enterprise the furnace was charged with calcined alumina phosphate (imported from the West Indies), lime phosphate (obtained locally), coke and sand. After about 1900 local phosphate was replaced by ore from Tennessee and then Florida. Alumina phosphate was dispensed with after 1910.

A decision was reached by the British shareholders to extend the power facilities and, pending the development of an enlarged market for phosphorus, to use the power for another purpose. International interest in the manufacture of stainless steel had recently arisen and this incentive, coupled with the existence of chrome-iron ore at Black Lake, Que., suggested the appropriateness of embarking upon the production of ferro-chromium. Experiments were successful and in 1899 an alloy averaging 63 percent chromium was made in a carbide-type furnace. Upwards of 1,000 tons of the alloy were produced and exported to Sheffield, England, up to 1906, when mining operations were suspended at Black Lake due to the exhaustion of the ore. From experience gained from the design and frunning of the ferro-chromium furnace, it was found possible to tap the molten alloy directly. This improvement had an interesting sequel. When Gibbs became technical director of Shawinigan Carbide, acting jointly with R. A. Witherspoon, he applied this continuous method of discharging to carbide production.

Apparently, Gibbs was unaware that a method for the manufacture of yellow phosphorus, very similar to his own, had been patented by J. B. Readman in England in 1888. The Readman patent (together with similar Parker-Robinson one) had been acquired and exploited (in 1890) by The Phosphorus Company, formed for that purpose. After negotiations Albright's bought The Company and within two years had moved the operation from Widnesfield to Oldbury. Albright's had enjoyed a virtual monopoly of white-phosphorus manufacture in Britain and a lucrative oversea trade for many years, using a batchwise procedure (heating of the mixture in iron retorts by coal as first then by coal gas). So advantageous was the new, continuous electrothermal process that Albright's terminated the use of retorts in 1895 (11).

Gibbs did not learn of the Readman patent until Fred J. Hambly came from Aberdeen in 1898 and directed his attention to Readman's papers in the 1890 issues of the Journal of the Society of Chemical Industry. In 1900 Albright's filed suit in a Canadian court alleging infringement of their Canadian patent on the part of Electric Reduction. After giving consideration to many factors Albright's withdrew their suit and purchased the controling share held by the Syndicate. Gibbs declined to enter into this out-of-court settlement. He held his share until his death in November, 1910 (from a heart attack following upon a hunting accident). Williams held his until his retirement in 1914.

Arthur Gibbs and Another Diaphragm Cell

This paper ends with a "footnote". William Gibbs' brother, Arthur E., had joined the Pennsylvania Salt Manufacturing Company (Pennsalt). William recommended to Arthur that he undertake research directed to improving the diaphragm cell, designed for the electrolysis of brine to produce caustic soda and chlorine. I do not know whether William's help went beyond this initial advice. I do not know when the work was performed. I do know that it was very fruitful. From it emerged the Gibbs cell. It was patented in many countries (Can. Patent 110 604, 1908). Pennsalt had established an electrolysis works at Wyandotte, Mich., in 1898, where it eventually used Gibbs cells. They were first used in Canada by the Canadian Salt Company in 1911, adjacent to its mine at Sandwich, Ont. The cell was of the vertical, cylindrical diaphragm type.

Literature Cited

1. Langford, M. W. M.A. Thesis, Carleton University, Ottawa, 1977
2. Warrington, C. J.; Nicholls, R. V. V. A History of Chemistry in Canada; Pitman: Toronto, 1949, p 167
3. Willson, T. L. Letters, pamphlets, articles, patents, clip-Public Archives of Canada, Ottawa; MG 30, A85
4. Morehead, J. Motley James Turner Morehead; An address delivered before the International Acetylene Association, Chicago, October 27, 1922
5. Willson, T. L.; Suckert, J. J. J. Franklin Inst. 1895, 139,21

6. Roberts, M Up The Gatineau, Historical Society of the Gatineau, 1976, 2, 16
7. Warrington; Nicholls. Ibid, p 344
8. LeSueur, E. A. Can. Chemistry Process Industries 1940, 24, 113
9. LeSueur, E. A. Trans. Electrochemical Soc. 1933, 63, 188
10. Warrington, Nicholls. Ibid, p 80
11. Threlfall, R. E., The Story of 100 Years of Phosphorus Making, 1851-1951; Albright & Wilson: Oldbury, 1951

RECEIVED August 12, 1988

Chapter 36

History of Electrochemistry in Mexico

Silvia Tejada

Departamento de Fisicoquímica, División de Ciencias Básicas, Facultad de Química, Universidad Nacional Autónoma de México, Ciudad Universitaria, 04510 México, D. F., México

The development of electrochemistry in Mexico, including some interesting prehispanic discoveries, the teaching of electrochemistry, and some examples of the industrial application in Mexico since the beginnings of Mexican electrical energy, will be presented.

Practically everything dealing with the history of metallurgy in the New World (1) differs from the way in which the Old World utilized metals, and the differences are astonishing. The discovery of metals in America occurred much later than in the Old World. Developmental techniques and scales of values were quite different. Copper, for example, was important at the outset of metallurgy in the Old World, since it was used for the creation of useful objects. In the New World, however, copper remained relatively forgotten. The metal smiths on the American Continent preferred to work with gold; this choice greatly influenced the roles that metals would play in pre-Columbian American societies.

When the Spanish conquerors melted gold and silver objects, they were surprised to see that the bars obtained were actually rather impure; the objects, apparently of gold or silver, were in reality made of alloys of those metals and copper. The smiths produced these alloys and employed a process to give finished objects the appearance of pure gold or silver.

The smiths who invented and used this method were members of societies of the central region of the Andes, such as the Chavin culture (about 800-400 B.C.), the Moche culture (100 B.C.-800 A.D.), and the Chimu culture (1150-1416 A.D.) in Peru. Spanish archives indicate that this technique was known in Mexico, (2) at first in Oaxaca (300-900 A.D.) by the Mixtecs, who later spread it to other places such as Tenochtitlan, which was founded by the Aztecs approximately in the year 1325 A.D. The method used by the natives of Mexico commences with the making of an alloy of various proportions of copper and silver or of copper and gold or of all three metals (3).

The most important alloy, known as Tumbaga, was that of copper and gold. When gold and copper are melted together and cooled,

0097-6156/89/0390-0534$06.00/0

they form a wide range of solid solutions. Some of these alloys contain 12 per cent of gold by weight. The color of the alloy depends on its composition. Tumbaga with high concentration of copper is red or pink. Tumbaga with high concentration of gold is yellow. Gold was sometimes added to an alloy of silver and copper; nevertheless, the product was still called Tumbaga.

Several cycles of annealing and hammering of a plate of Tumbaga causes copper at the surface to be converted into copper oxide, so that the surface is enriched with precious metal. To remove the copper oxide that was formed after each annealing the natives may have used old urine (the urea of recent urine is degraded to ammonia), corrosive minerals, or sour juices of certain plants. In the region of Oaxaca, it is thought that the juice of a particular species of the Liana vine was used while the Aztecs may have used Mansteras and Anthuriums, which have a high acidic content, mainly in the form of oxalic acid (4). To lower the silver content of Cu-Au-Ag alloys, the workers may have utilized natural mixtures of corrosive minerals to treat the surface of the plate.

Fray Bernardino de Sahagun, in his General History of the Objects of New Spain (5), written in Nahuatl approximately in the year 1548, mentions: "When the artifact has been melted, it is polished with a rough, uncut stone, and when polished, a bath in alum is given; in this way the melted gold is mixed. Fire is applied once more, and the artifact is heated, and when it has been taken from the fire, once again it is given a bath and mixed with what is called "The gold remedy" which is merely like yellow earth. Some salt is mixed, and in this way it is perfected as the gold achieves a brilliant yellow. Then it is stripped and rubbed, by which actions it becomes very beautiful with great brilliance, and splendor and brightness".

Tumbaga was used in Mexico to mold three-dimensional objects, using the method of "melted wax" (6). Objects of this kind have been found in the valley of Oaxaca; These objects date from two hundred years before the arrival of Cortes in Mexico. Mixtecan craftsmen handed on to the Aztecs all the secrets of their art, and indeed a great part of Moctezuma's Treasure taken by Cortes was really Mixtecan. When the Spanish missionary Toribio de Motolinia wrote that the Aztecs were able "to melt a bird together with head, tongue, feet, and mobile wings and to place any insignificant object on the wings in such a way to make it look as if it were dancing", he is undoubtedly describing an ornament invented by the Mixtecs. It should be pointed out that the Mixtecs used soldering as a normal technique in repairing metal pieces, this enabled them to achieve filagree (7). Sahagun (8) describes this: "If a part of a work of art is broken or damaged, it is repaired by cementing with solder and it is scraped with an adz, with which it is polished. It is placed once more where the alum is applied, it is cleaned and burnished and left completely clean".

The Beginnings of Teaching Electrochemistry in Mexico.

The antecedent of teaching electrochemistry in Mexico is the Royal Seminary of Mining (9). This was created by Joaquin Velazquez de Leon, by appointment of the viceroy Antonio Maria

Bucareli to support the proposition expressed by Francisco Javier Gamboa (1760) in reference to the need of qualification of personnel in mining.

Due to this, the Spanish authorities sent their best chemists to Mexico, to participate as teachers in this Seminary.

Fausto and Juan Jose de Elhuyar greatly promoted this project and got it into practice. They are considered to be the creators of the first institution dedicated to the teaching of science in Mexico. These two famous scientists discovered tungsten in the Patriotic Seminary, Vergara (Spain) in 1783 (10). Their discovery was published with the title: "The chemical analysis of wolfram and an examination of a new metal, which makes part of its composition".

The Royal Seminary of Mining was opened in January, 1786, with the initiation of its courses in 1789. The formal inauguration took place in January 1792, with two professors and eight students.

Andres Manuel del Rio (11) (1795) later took charge of the Mineralogy courses. In the year 1801 this scientist discovered a new metal while examining some minerals of "dark lead", from the mine "La Purisima", located in Zimapan, now known as the state of Hidalgo, Mexico. He first named the metal PANDROMO, due to the range of colors of its oxides, solutions, salts and precipitates. He later named it ERITRONIO, because the salts turned red with heat or with acids.

Four years later, in 1805, the Collet-Descotils examined what was supposed to be the metal, stating it was just an impure chromium oxide. Del Rio apparently accepted this conclusion. In 1830 the Swedish chemist N.G. Seftroem discovers a new metal which he named VANADIUM. This metal was described by Berzelius in 1831, as follows:
" This metal, recently discovered by Professor Seftroem in an iron original from Taberg minerals in the Smaland, had already been found in Mexico in the region of Zimapan by Mr. del Rio in 1801". The unstable political situation which prevailed in Mexico from 1810 to 1867 greatly hindered all teaching institutions, in particular the Colegio de Mineria. However, this experience led to the institutionalizing of teaching. An example was the proposal of Valentin Gomez Farias to create six Higher Education Foundations such as the Medical Science Foundation, and the Physics and Mathematics Foundation which was the continuation of the Mining School.

Later, Ignacio Ramirez founded the "Special School of Mining". At the same time, he proposed that all professors should write a yearly memoires' of every subject taught, arising the events and advances of the year. These documents had surveys of the most important Mexican and European publications dealing with the subjects. This is a clear evidence of the need to keep records of all teachings and of the knowledge acquired by each and every professor.

Jose Manuel de Herrera, the chemistry professor at the School of Mining and at the Mining Seminary of the Mexican capital, presented a brilliant dissertation on "The Electrochemical Forces" on October 29, 1840.

Following the Mexican Government Reform, several scientific institutions and associations were founded during the late 19th and the early 20th centuries. Examples are: The Mexican Society of Natural History (1868), The Geographic Exploitation Commission (1877), and The Antonio Alzate Scientific Society. One of the activities of such institutions was to publish compilations of data and descriptive matter; very little research was included.

By 1900, most of the textbooks were European. Accordingly, academic activity was limited to the teaching of what was already well known.

In 1916, the first school of Chemistry is established in Mexico. It is worth noting that this school was founded during World War I (1914-1918); an interesting fact is that the Chemistry Institute was founded during World War II (1939-1945). The creation of these institutions was due to the scarcity of raw materials originated by these wars (12).

The first Chemistry School was founded by Don Juan Salvador Agraz, who was at the head of his graduating class at the University of Paris, where he obtained his bachelor's degree in Science. Later, he registered at the Applied Chemistry Institute in France and then he obtained his PhD in Chemistry at the University of Berlin (13).

Dr. Agraz suggested the founding of the School of Chemistry to President Francisco I. Madero in January, 1913. Madero favored his idea but he was assassinated a month later.

It was not until December 24, 1915 that Dr. Agraz's project was approved by President Venustiano Carranza. Felix F. Palavicini then Secretary of Education and Fine Arts, then named Dr. Agraz as director and founder of the first School of Chemistry of Mexico.

The School of Chemistry began in an old and abandoned building with smoky walls and classrooms without floors. The building had been a fort during the Zapatista Revolution.

On April 3, 1916 classes for 70 students began with no ceremony at all. Three career programs were offered: Industrial Chemist (4 year), Expert in Chemical Industry (2 year), Chemist Technician (1 year).

On February 5, 1917, the National School of Industrial Chemistry was incorporated into the University of Mexico, the American Continent's oldest and first University (14).

Both Spanish-American Universities and Spanish Universities had papal and monarchical origins. The Church had been involved in the evolution of previous universities and was naturally interested in subjects of an ecclesiastical nature.

The foundation of the University of Mexico has double documentation, the royal letters patent of April 30, 1547, signed by Carlos V and the papal bulla authorized by Pope Paul IV dated 1555. Once the National School of Industrial Chemistry was incorporated into the University of Mexico, the formal studies of Chemical Engineering were established in 1918. The Electricity and Electrochemistry course was taught during the fourth year. Laboratory courses in Electrochemistry were not started until 1927.

In 1921 Electrochemistry was taught for the first time to the Chemical Technicians and Metallurgists. The professional studies in Metallurgical Chemistry were opened in 1967. Electrochemistry, Corrosion, and Metal Protection are included, the latter subject as an elective. During this year, Electrochemistry was included in the Chemistry program.

In 1972, Electrochemistry was given a new impulse. Analytical Chemistry was taught in place of the traditional "chemical analysis". Electrochemistry is then taught to Pharmaceutical Chemists, Chemists, Chemical Engineers and Chemical Metallurgical Engineers. During the period 1916 to 1986, the estimated number of graduates from the School of Chemistry in the four areas mentioned is 13,000.

Through the initiative of Francisco Diaz Lombardo, Director of the National Chemistry School, this school became a Faculty in 1965, with Manuel Madrazo Garamendi as Director.

The Chemistry School of the National Autonomous University of Mexico is not the only institution in Mexico where electrochemistry is taught. There are other centers, such as The National Polytechnic Institute, founded in 1937 by the initiative of Lazaro Cardenas, President of Mexico at that time. The purpose was to create a basic substructure for the industrialization of the country. Electrochemistry is also part of the curriculum of some state universities, regional schools of higher education, and private universities as well as technical schools, where chemistry and related courses are offered. There are about 50 institutions where electrochemistry is taught in Mexico.

Some 367 dissertations that involve electrochemistry were registered in Mexico during the period 1926 to 1980 (15). The first, in the UNAM, had the title "Project of a Caustic Soda Factory With a 5 Ton-a-Day Capacity Using Sodium Chloride as Raw Material, Working by Electrolyte Process" and was by Edmundo de Jarmy. Since then, more and more dissertations on electrochemistry have been registered. In the Polytechnic Institute, the first dissertation, "An Electrolyte Plant for the Production of Caustic Soda" was presented by Agustin Cid Suarez.

The first regional university to register a dissertation on electrochemistry was the University of Guadalajara in 1953. Then Guillermo Fernandez Ruiz presented "The Installation of a Plant for the Production of Electrolyte Sodium Chlorate".

Additionally, Mexican students have been sent to other countries to pursue Masters and Doctorates dealing with electrochemistry. If we consider that the total number of graduates who worked in the electrochemistry field is approximately 200, this means that there is one electrochemist for every 400 thousand Mexicans. This is evidence of our small number of researchers and of research groups in electrochemistry. The groups are mainly of young scientists.

The following table registers the number of graduates in each master and doctorate from their beginnings until 1988.

Master (16) (17)	Institution (1)	Graduates
Physical Chemistry	UNAM	35
	CIEA	22
	ITESM	14
Analytical Chemistry	UNAM	21
	U. Reg.	2
Chemistry (Electrochemistry optional)	UAM-I	2
Metallurgical Engineering	ESIQIE	12
Metallurgy of Ferrous Processes	ITS	8
Metallurgy	UNAM	6
Metallurgy and Science of Materials	UMSNH	4
Iron and Steel Industry	ITM	7
TOTAL		133

Doctorate	Institution	graduates
Physical Chemistry	UNAM	12
	CIEA	3
Chemistry (Electrochemistry optional)	UAM-I	1
TOTAL		16

(1) Abbreviations:
UNAM, National Autonomous University of Mexico.
CIEA-IPN, Center of Investigation and Advanced Studies, National Polytechnic Institute.
ITEM, Technological Institute of Superior Studies of Monterrey.
U. Reg., Regiomontana University.
UAM-I, Autonomous Metropolitan University at Iztapalapa.
ESIQIE, Superior School of Chemical Engineering and Extractive Industries.
ITS, Technological Institute of Saltillo.
UMSNH, University of Michoacan at San Nicolas of Hidalgo.
UANL, Autonomous University of Nuevo Leon.

The main themes referring to electrochemistry in Mexico from 1926 to 1980, deduced from articles published in Mexican journals, abstracts of papers presented in Mexican meetings, theses registered in Mexican institutions and Mexican patents are as follows:

Theme	Number of publications	% of total
Corrosion	144	23.7
Electrodeposits	129	21.2
Electrolysis	119	19.6
Batteries	70	11.5
Electrophoresis	43	7.0
Other themes	101	17.0
TOTAL	606	100.00

From 1929 to 1974, 49 patents involving Electrochemistry have been registered at the General Office of Industrial Property. In the entire country, 367 theses involving Electrochemistry were registered from 1926 to 1980. Of these 252 were from UNAM, 48 from the IPN. The rest are from State Universities and Institutes or from other private schools.

The publication of periodicals did not begin until the middle of the 19th century. In the Seminario de la Industria Mexicana appears "Electrochemistry, its progress and applications to Arts and Metallurgy" (1841 pp 55-59). Examples in the Sociedad Cientifica Antonio Alzate (1887-1920) concern the decomposition of salts etc., by the electric current (18), the electrochemistry of gold and silver minerals (19), electro-sinu-caustic surgery (20), industrial electrochemistry (21), and the theories of ions (22).

After the demise of the Antonio Alzate Scientific Society, there was practically no permanent chemical periodical, until the appearance of the "Revista de la Sociedad Quimica de Mexico". Systematic articles dealing with electrochemistry appear in this. For example 13 papers and 59 abstracts are registered as works participating in different congresses from 1959 to 1980.

Naturally papers written by Mexican scientists have appeared in foreign publications. One of these was an important contribution towards the improvement of the Pourbaix diagram (23). However, publications other than in Mexico are beyond the scope of the present survey.

Installation of the First Electricity Plants in Mexico.

The first electricity plants were installed in Mexico towards the end of the 19th century (24). The government of President Porfirio Diaz welcomed foreign investments. The first purpose of electric plants was to illuminate mines and to operate looms to increase fabric production.

In 1879 the first thermoelectric plant was installed in Leon, Guanajuato for a textile factory. In 1889 the first hydroelectric plant, with a capacity of 22.38 KW, was installed in Batopilas, Chihuahua. Thereafter hydroelectric, thermoelectric, and geothermic plants began to appear in different parts of Mexico for mining and the extraction, casting, and refining of metals. Fabric factories, flour mills, glass factories, etc., were then electrified.

The Mexican Light and Power Company was founded in Ottawa, Canada in 1902, obtaining grants to provide with electric energy to Mexico City and, later on, to the states of Hidalgo, Puebla, Michoacan, and Mexico State.

Later, the Electrical Company of Chapala and the American and Foreign Power Company took part in the electrification of Mexico. However, these companies were not very efficient, and there were many consumers' complaints. For these reasons the Federal Commission of Electricity was created in 1937, during the term of Lazaro Cardenas. Its goals were to organize and direct a national system of generation, transmission, and distribution of electric energy as a non-profit public service organization.

The Mexican government bought this foreign company in 1960, during the term of Adolfo Lopez Mateos. By 1977 its capacity reached 12 million KW.

Concurrently with the development of the electrical energy production, electrochemistry flourished. Today, among the 500 most important companies in Mexico, several use one or more electrochemical processes (25).

From the point of view of economic importance, the Mexican Electrochemical Industry can be subdivided as follows: examples of the firms that are involved are given:

1. Metal extraction and refining: Industrias Peñoles S.A. de C.V., Grupo Industrial Minero de Mexico S.A. de C.V., Industriales Nacobre S.A. de C.V., Mexicana de Cobre S.A. de C.V., Cobre de Mexico S.A., Compañia Minera Autlan, S.A. de C.V., Aluminio S.A. de C.V., etc. It is worth mentioning that Mexico has been the first silver productor in the capitalist world, for several years.
2. The Chlor-Alkali Industry: Cloro de Tehuantepec S.A., Penwalt S.A., Cydsa S.A., etc.
3. Other inorganic electrolytic processes: Cydsa, S.A., Quimica Fluor, S.A. de C.V., Tetraftalatos Mexicanos S.A., Electroquimica Mexicana S.A. de C.V., etc.
4. Metal Finishing: Cromo de Mexico, S.A., Galvanizadora Latinoamericana, S.A., Nacional de Galvanoplastia, S.A., Plasticos Especializados Mexicanos S.A., Anodizados Especiales de Mexico S.A., Galvanostegia Industrial de Mexico S.A. de C.V., etc.
5. Batteries: Acumuladores Globe S.A. de C.V., Acumuladores Insuperables S.A. de C.V., Ray-O-Vac de Mexico S.A. de C.V., Duracell S.A. de C.V., etc.

This survey is by no means a complete and critical study. It demonstrates that Mexico has contributed both ideas and happenings to the history of electrochemistry.

Acknowledgments.

I want to express my gratitude to the following persons from the School of Chemistry: Lucia Alvarez, head of the Language Department, for her support in the translation of this paper. Jaime Perez, Susana Torres and Veronica Napoles, for helping in the search of bibliography and Dr. Joan Genesca for his suggestions.

Literature Cited.

1. Knanth, P. El Descubrimiento de los Metales. Ed. Offset Multicolor, S.A.: Mexico, D.F., 1979; Chapter 6.
2. Sanchez, F.R. Historia de la Tecnologia y la Invencion en Mexico; Salvat Mexicana de Editores, S.A. de C.V.; Mexico, 1980; Chapter 1.
3. Lechtman, H. Sci. Am. 1984, 250, 56-63.
4. Scott, D.A. Hist. Metall 1983, 17, 99-115.
5. Sahagun, B. de Historia General de las Cosas de Nueva España; Ed. Porrua, S.A.; Mexico, 1985; p. 522-23.
6. Leon-Portilla, M. Historia de Mexico; Salvat Mexico, 1978; Vol. 4, p 736-38.
7. Grinberg, D.M.K. de Metalurgia Moderna 1987, 3, 45-48.
8. Sahagun, B. de Historia General de las Cosas de Nueva España; Ed. Porrua, S.A.: Mexico, 1985; p 524.
9. Flores, L.R. ICYT 1985, 7, 13-17.
10. Bargallo, M. Tratado de Quimica Inorganica; Ed. Porrua, S.A.: Mexico, 1962; p 843.

11. Valle, H.R. Historia Mexicana, 1954, 4, 115-23.
12. Estrada, O.H. Cuadernos de Posgrado 20, UNAM; Mexico, 1986; Chapter 1.
13. Garcia, H. Historia de una Facultad; UNAM; Mexico, 1985; Chapter 1,2.
14. Garcia, S.C. Sintesis Historica de la Universidad de Mexico; UNAM; Mexico, 1978; p. 39.
15. Garcia, V.E. Thesis, UNAM, Mexico, 1982.
16. Dominguez, O.; Garritz, A. Ciencia y Desarrollo 1987, 161-75.
17. Fonseca, J. Ciencia y Desarrollo 1987, 143-53.
18. Industrias e Invenciones de Barcelona. Memorias de la Sociedad Científica Antonio Alzate, 1890, 11, 110-112.
19. Laguerenne, T. Memorias de la Sociedad Científica Antonio Alzate, 1901, 16, 179-190.
20. Jofre, R. Memorias de la Sociedad Cientifica Antonio Alzate, 1901, 16, 161.
21. Guerrero, H. Memorias de la Sociedad Cientifica Antonio Alzate, 1904-1905, 22, 193.
22. Leon, G.L. Memorias de la Sociedad Cientifica Antonio Alzate, 1906-1907, 5.
23. Markovic, T.; Leon, L.E.G. Metaux. 1973, 48, 50-53.
24. Alonso, M. Evolucion del Sector Electrico en Mexico. Comision Federal de Electricidad: Mexico, 1977; Chapter 2,3,4.
25. Equipo de Expansion. Expansion. 1983, 15, 112-87.

RECEIVED August 9, 1988

Chapter 37

The Search for Portable Electricity

History of High Energy-Density Batteries

Sankar Das Gupta

Department of Metallurgy and Material Science, University of Toronto, Toronto, Ontario M5S 1A4, Canada and The Electrofuel Manufacturing Company, 9 Hanna Avenue, Toronto, Ontario M5S 1A4, Canada

This paper reviews the history of rechargeable batteries starting from Volta's pile to the recent development of high energy density batteries such as sodium/sulphur and LiAl/FeS. The development of batteries such as Pb/acid and Ni/Cd has been reviewed and it has been pointed out that most of the present battery technology has stemmed from work done during the last century, and modern research has done rather little for rechargeable battery development, unlike the progress achieved by the primary batteries.

The current world market for batteries is estimated to be in excess of $23 billion, of which about $12 billion is the market for larger batteries (mainly automotive and UPS systems) and about $11 billion for the smaller sized batteries. The interesting feature of this industry is that the presently dominant technologies were all invented in the last century. Billions of dollars have been invested in the research of new battery systems with little positive effects. Newer forms of batteries have promised structural changes in major industries such as transportation and power, with concommitant improvement in air quality and environmental pollution. However we still await any such spectacular developments.

Reports of the usage of batteries,in prehistoric times, have been proposed by archeologists. Electrochemical power was possibly used for medicinal purposes and for the plating of gold. Mysterious earthenware vessels have been found in various Mesopotamian sites which might have been electrochemical power sources. These vessels have a copper tube along with a central iron rod, the latter being cemented in position with asphalt. An electrolyte such as organic acids would certainly convert that vessel into a battery.

Two thousand years passed before Volta's experiments ushered in the modern age of batteries and electrochemistry. Volta had

0097–6156/89/0390–0543$06.00/0

shown interest in Galvani's experiments and thereafter within a period of 30 years, progress was swift and the foundations of the science of electrochemistry were established. Throughout the history of electrochemistry, it is interesting to note that major discoveries came in groups all within a short period and then years would pass without any significant development.

Luigi Galvani was a lecturer in anatomy at the University of Bologna in 1790 when he (or was it his wife) noticed the twitching in the legs of a frog when touched with a copper probe; the frog had been hung from a iron hook. Count Alessandro Volta was a professor of physics at Pavia University, and in his laboratory the modern world's first battery was born in 1800 (1).

The first battery was a pile of discs alternating between silver and zinc interleafed with a separator soaked in an electrolyte. Soon Volta had improved the system with a multicell battery called 'couronne de tasses', whereby it was possible to draw electric current from that multicell battery at a controlled rate (2). After Volta's discovery in 1800, Nicholson and Carlisle in the same year (1800) decomposed water into hydrogen and oxygen by an electric current, and Cruickshank also in 1800 deposited metals from solutions. In 1803, Hisinger and Berzelius showed that when solutions are decomposed by electric current, acids, oxygen and chlorine are deposited at the positive pole, and alkalis, metals and hydrogen at the negative pole. All this within three years of Volta's discovery.

The first theory of electrolysis was developed by Grotthuss in 1805 and Davy in 1807 isolated alkali metals by electrolysis. Volta, thereafter developed the electrochemical series (he called them the contact series) and in 1811, Berzelius extended the electrochemical series to include non metals, and correctly established some of the scientific basis of inorganic chemistry. Volta not only produced the world's first battery and the electrochemical series, but he qualitatively differentiated between groups such as metals, carbon and some sulphides which conducted electricity without undergoing chemical change, this he called 'conductors of the first class' and others such as salt solution which is decomposed by electric current, which he called 'conductors of the second class'. Volta's electrochemical series, correctly identified the flow of current from the more electropositive to the less electropositive metal in his series, when connected with each other and a 'conductor of the second class'.

It then remained for Michael Faraday between 1832 and 1833 to use a similar apparatus as Volta's 'couronne de tasses' and derive the quantitative laws of electrochemistry and open up the floodgates of technology and the subsequent developments of chemistry, modern physics, thermodynamics and ofcourse electrochemistry. Today, battery research is a billion dollar phenomena and the battery industry is the largest industry in the field of electrochemistry, employing the largest number of researchers in electrochemistry. Despite its long history, this ancient area of electrochemical science promises to provide the solution to newer problems of pollution, scarce energy resources, better transportation and essentially a better method of storing and transporting electrical energy. If any of latter promises are fulfilled,then the growth of the newer battery systems is predicted to far outstrip the present multi-billion dollar business. Count Volta would surely have been impressed by the growth of his brainchild and possibly more so by the massive research efforts being now made to produce better energy storage systems with the high energy density batteries.

High Energy Density Battery

Electrochemical techniques of energy conversion can be classified as batteries (primary and secondary), chemical storage through electrolysis and electrochemical convertors such as fuel cells. Of these three technologies, only secondary batteries will be discussed here, as this technology is not only the most advanced for electrochemical energy conversion and storage but also shows the most promise for further development. Protogonists for fuel cells and electrolysis/hydrogen economy may disagree, but it is unlikely that the pre-eminent position of batteries in the electrochemistry world will be abrogated in the near future.

The use of batteries as portable electrical power source has increased and to some extent technology has not been able to keep pace with the demands for such power sources. Electric vehicles are demanding longer lifetime and higher volumetric energy densities, load levelling applications are more sensitive to cost than to the gravimetric or volumetric energy densities, computers and other electrical and electronic gadgets are demanding higher safety and shelf life while the space station requires enormous amounts of power storage capacity in a small volume and weight. The defense industries require extreme battery performance while a major target, the internal combustion engine suprises all by the rapidity of its improvement in efficiency.

The ideal high energy density battery has to meet all these new demands on power and energy density. Only time will tell, whether improved batteries will evolve to meet these stringent requirements.

Science Versus Technology.

Recent studies done by Rustom Roy and others have shown how far Science has lagged behind Technology in the Materials development area. In the field of Batteries, a similar viewpoint has been alluded to by Kordesch (3), who has pointed out that the efforts of 'pure scientists' have been largely unimportant in any major development in this field. An important reason why we examine the 'history' of electrochemistry or any other field of technological importance, is to understand the underlying factors which nurtured technological growth and innovation, such that future effort for further growth can be channelled to those important underlying factors. As we note,in battery development, most of the significant growths in technology came from the empirical scientists, and indeed battery systems which were developed a century ago is today still the dominant technology.

The billions of dollars spent on modern research has provided rather scanty outputs in newer rechargeable battery technology, and better understanding of quantum mechanics and solid state theory has not provided any breakthroughs. The above is mentioned, not as some polemical statement, but as a genuine concern, recognising the intense requirements for improved battery characteristics from major economic sectors such as environmental pollution, scarce resources and energy conservation and recognising that the electrochemistry community has not met the challenge with improved systems inspite of major increases in research efforts over the last fifty years.

The electrochemistry of batteries is a complicated phenomenon and thermodynamic data is only applicable for equilibrium conditions and does not define overall kinetics, including charge transfer, overpotential and rates. To date quantum mechanics and solid state theory has provided no understanding in defining the behavior of real batteries or cells. Battery research needs an interdisciplinary approach of electrochemistry, material science, and various disciplines of engineering and above all a 'can-do' innovative spirit which is somewhat missing in some of today's big science approach where publications are the highest form of achievement.

Golden Age of Battery Development.

The golden age of batteries, and indeed electrochemistry, started with Volta's experiments, and by the end of the 19th century tremendous progress had been made. Table I, shows the main milestones in the progress of battery evolution, and essentially by the end of the 19th century most of the rechargeable battery landmarks had been reached. These includes the dry cells, the lead acid battery(4,5), the nickel cadmium battery(6), elements of the fuel cell (7) and metal air batteries(8), nickel iron (6,9) and even the nickel zinc (10) battery concept. Progress has been slow ever since.

Table I. Golden Age of Battery Development

L.Galvani	1790	Electrochemical phenomena.
A.Volta.	1800	World's first battery.
M.Faraday	1834	Quantitative laws of electrochemistry.
N.R.Grove	1839	Hydrogen/oxygen fuel cells.
A.Smee	1840	Metal air cells.
N.J. Sindsted	1854	Lead acid battery.
G.Plante	1859	Lead acid battery.
G.Leclanche	1868	Zinc manganese dioxide primary.
T. deMichalowski	1899	Nickel zinc battery.
W. Jungner	1901	Nickel cadmium Battery.
W. Jungner	1901	Nickel iron battery.
T.A.Edison	1901	Nickel iron battery.

Even the conceptual design on zinc-bromine battery is about 100 years old, and it has been said that during the Franco/Prussian war, French balloonists used static zinc bromine batteries to light 'Nernst glowers' in order to illuminate their maps as they ballooned over the Prussian lines (R.J. Bellows and P.G.Grimes). Then, after a long hiatus,came the energy crisis of the 70's and the zinc/bromine and other zinc/halogen batteries were rediscovered. Similar stories abound with the various metal air systems, and of course with the hydrogen-oxygen fuel cells.

The Lead Acid Battery

About 40% of the world's lead production goes into the manufacture of the lead acid battery, and it is by far the dominant technology in rechargeable battery systems with market size of about $12 billion. The earliest reference on this battery is from 1854 (5), while the real commercial breakthrough came with Plante's experiments in 1859(4). Gaston Plante had been investigating the effect of polarization on different metals and he noticed the unique behavior of lead plates in dilute sulphuric acid as the electrolyte, and the rest is history.

The secondary lead acid battery was at that point ahead of the electrical industry, and the only way Plante could charge his battery was to use the Bunsen types of primary cells in series. Bunsen, in 1850, had been producing cells with a carbon cathode and zinc anode with open circuit voltage of about 2.0V. This cell was actually a modified Grove cell (Platinum cathode and zinc anode). Sir William Grove was an extremely practical scientist, who had earlier invented in 1939 the hydrogen/oxygen fuel cell and had devised working gas electrodes and operating fuel cells. As Barak mentions (2), Grove showed remarkable scientific insight in correctly identifying the platinum electrodes as being both the current collector as well as the catalyst for the reaction. Moreover, he identified that the reaction took place at the three phase interface of gas-liquid-solid and also noticed the effect of surface area on the fuel cell process. It was a remarkable piece of scientific work.

Plante's seminal work on the electrochemical power storage was taking place at about the same time Werner von Siemens (1857) was producing a practical electromechanical generator or dynamo and thus the electrical industry was born.

The work by Plante, Faure and others during the next thirty years laid the foundations for startup companies which have since grown to become the major battery companies in the world.

Today, most of the commercial electric vehicles operate with the lead acid battery, despite the efforts to develop newer types of rechargeable batteries.

The lead-acid battery being used for electric vehicles is somewhat different in design from the standard SLI battery, and has been modified to provide longer cycle life under deep discharge conditions.

Table II highlights the various milestones in the evolution of the lead acid battery.

Table II: Lead Acid Battery Milestones

1854	Sindsted's concept of the lead acid battery.
1859	Plante producing the first practical battery.
1881	Faure produced the pasted plate.
1882	Sellon: antimony lead alloys for grids.
1882	Gladstone and Tribe: double sulphate theory.
1890-1900	Wood separators.
1914-1920	rubber separators.
1935	lead calcium grid alloys.
1948-1950	synthetic fibre separators.
1965	maintainance free batteries.
1980's	plastic components, gel electrolyte.

Note: Adapted from Reference 11.

Nickel Cadmium Battery

Nickel cadmium and nickel iron are the prime examples of alkaline rechargeable batteries, which were invented again within years of each other by Waldemar Jungner in Sweden and the ever inventive Thomas Edison. Here again both the inventors laid the foundation of major battery businesses.

Jungner, in 1896, had developed the pocket electrodes to hold the finely divided electrode powders, while Edison used tubular electrodes for the positive and flat plates for the negative. The cell reaction is given below:

$$2NiOOH + Cd + H_2O = 2Ni(OH)_2 + Cd(OH)_2$$

Nickel cadmium batteries are extensively used in rechargeable portable electronic and electrical devices. Some efforts were made by DOE and others to produce higher energy density batteries based on sintered high surface area electrodes, but till now there has been no commercial success for this development program.

Nickel Iron Battery

Nickel iron batteries were developed by Edison, and were commonly known as the Edison cells. Of all batteries, perhaps the Nickel Iron can claim to be the most rugged and reliable as well as being virtually indestructible. In the 1920's and 30s these batteries were found everywhere, and when the need arose for heavy water during the Manhattan project, the electrolyte from thousands of these batteries were used as an enriched source. The cell reaction is given below:

$$2NiOOH + Fe + H_2O = 2Ni(OH))_2 + Fe(OH)_2$$

The Edison cells have ceased production in North America, and USSR today has the largest production capability in these batteries.

Recently, the NiFe cells were evaluated as potential batteries for the electric vehicle program, and a development project launched. The target was to produce a sintered iron electrode along with a sintered nickel electrode, improve energy density and current efficiencies. Till now, this has remained an experimental program. We believe, that the NiFe batteries show extraordinary performance with respect to cost and cyclelife, and this battery should be evaluated as low cost power storage for solar and wind energy, amongst other applications.

Other Alkaline Batteries

Other alkaline batteries have been commercially developed, the principal system being the silver-zinc which was developed about 50 years ago. With an energy density of 150 wh/kg, this battery has possibly still the highest energy density of any rechargeable batteries. The lifecycle of the zinc electrode is poor and the cost of silver electrodes is high for any large scale use of this battery. Table III outlines the performance characteristics of the presently available rechargeable batteries.

Table III

Characteristic performance of commercial rechargeable batteries

Battery		Energy Density wh/kg	Power Density w/kg	Cycle life cycles
Pb/acid	1859	20 - 40	20 - 175	300 - 500
Ni/Cd	1896	25 - 45	250 - 600	300 - 2000
Ni/Fe	1899	20 - 54	60 - 104	5000
Ag/Zn	1930	50 - 175	200 - 450	100

History Of The New Rechargeable Batteries

It is intriguing that in a sophisticated technology area such as electrochemistry, where billions of recent dollars have been spent in the search for newer batteries, the commercial rechargeable cells are essentially those invented in the last century. The demand for better power sources has been increasing over the last few decades. The demand has come from numerous places, especially the space program, the defence requirements including SDI, the electric vehicle movement, the clean air/environmental issue, the energy and oil crisis, computers and power tools, load levelling and the list goes on.

The military requirements have been mainly in the primary batteries area and spectacular newer systems have been developed with enormous power densities, wide temperature ranges of operation and very high shelf life. The same progress is not the case for secondary batteries. Some of the interesting development programs in secondary batteries are mentioned in this following sections.

The Zinc Electrode

The zinc electrode has been a constant lure for electrochemists over the last 150 years and in every case it has been the death knell of the project. With its high energy density, low costs, and high reactivity, efforts have been continuously made to produce a 'good' rechargeable system based on this electrode.

Shortly after Grove's work on fuel cells in 1839, Smee in 1840 introduced the original concept of the metal air batteries and the last 150 years have seen discontinuous developments in the metal air batteries. The zinc/air battery had been prominent but generally unsuccessful due to both problems in the rechargeability of the zinc electrode as well as problems with the gas electrodes.

The zinc halogen systems (both bromine and chlorine) were rediscovered during the last two decades and the efforts to commercialize these systems have been enormous. Madison Avenue orchestrated public relations hype certainly raised the awareness of the general public towards Zinc/Chlorine batteries but other than improved share prices there were no long term technological gains.

The zinc/bromine battery showed good promise, especially the Exxon work using recirculating electrolytes, liquid bromine complexing agents, low cost conductive plastic electrode structure and good system design. However after about 15 years of development, commercialization is not yet in sight.

The nickel/zinc battery, with a theoretical energy density of 375 Wh/Kg lured another large group of electrochemists. Major automobile companies pronounced the battery to be 'adequate', and the largest auto company of all pronounced that by the mid 80's full 25% of its production will be electric cars operating on nickel/zinc batteries. The battery development work has been instead a glorious failure. Here again the rechargeability of the zinc electrode became the stumbling block of this enterprise.

An interesting twist to this saga is being given by a startup Toronto company. This is developing a rechargeable alkaline manganese battery (initially for small scale batteries) and promises commercial production in 1989.

It is estimated that the total development costs for the zinc based rechargeable batteries has been around $300 million.

Sodium Sulphur System

The Sodium/sulphur battery concept dates from 1966, when Ford's initial patents were filed. Since then, it is estimated that over $500 million has been spent on the development of this battery, of which $250 million has been spent by one group alone (DOE, personal communication).

Unlike the other systems mentioned above,this is a true'high technology' battery born out of the modern 'materials age'. The reactants are molten sodium and sulphur, separated by a sodium ion conducting ceramic (beta-alumina), and the battery operates at temperatures of about 350C. The battery shows excellent power densities and good energy densities. A plethora of companies had been involved in this battery technology, including Ford, General Electric and Dow, and they have all left the field, leaving Brown Boveri and Chloride to be the only major players. Problems includes propensity of the battery to blowup, as can be seen in the etchings of the ceilings in the laboratories concerned and more recently with blownup vehicles. Thermal cycling of the batteries remain unsolved while systems engineering grows more complex as attempts are made to improve safety.

Variations of the sodium/sulphur battery include systems with sodium/beta-alumina/sodium tetrachloroaluminate/nickel chloride or iron chloride. The early results are encouraging and the system suffers less from the problems of the sulphur electrode, such as rechargeability, high vapour pressure, various polysulphide formations etc.

Lithium Rechargeable Ambient Temperature

The lithium primary batteries have been one of the bright spots of battery development in this century. With the advent of better materials and better understanding of the electrochemistry of lithium in organic electrolytes, a number of successful lithium based primary batteries are now in commercial production.

The first commercial rechargeable lithium battery was a button cell produced by Exxon in 1979 based on the $LiAl/TiS_2$ couple. This product was later discontinued.

Other lithium rechargeable concepts are being actively developed, and prominent amongst them includes a battery being developed by Moli since 1976, where advantage is taken of lithium intercalation in materials such as molybdenum disulphide. The Moli cells (especially AA) are now (1988) available in a limited way and full production is promised for 1989.

The safety considerations for pure lithium batteries are such that it is unlikely that they will ever be used for large battery systems, as identified by the electric vehicle market. The lithium ambient temperature rechargeables contain metallic lithium along with combustible organic electrolytes.

Another variation of these batteries is the solid state lithium system, which is still in the laboratory bench scale stage of development.

Lithium Rechargeable: Molten Salt Electrolyte

The most prominent lithium rechargeable battery concept is that of the LiAl/FeS system using molten salt electrolyte. The latter electrolyte theoretically provides very low internal resistance, such that batteries of enormous power densities can be visualised. The LiAl/FeS battery is a derivation of the lithium/sulphur battery, where both the anode and the cathode have been made less energetic thus lowering safety and materials problems. This family of batteries (higher sulphur activity can be used to generate higher energy densities), was conceptualised in the mid 1960's and development was carried out essentially at the Argonne National labs. The performance of this battery is similar to that of the sodium/sulphur, but with lower safety and thermal cycling problems.

A number of companies have been developing this battery including Gould and Electrofuel, and large sized batteries are expected to be demonstrated by 1990. In addition to Argonne, the British Admiralty laboratories has been developing this battery for submarine applications while other potential applications include electric vehicles and load levelling.

Table IV outlines the potential near term capabilities of the rechargeable batteries which are undergoing serious developments.

TABLE IV

Projected performance of the Future Batteries

Battery	Specific Energy		Specific Power		Cycle life
	wh/l	wh/kg	w/l	w/kg	cycles
Zn/Br	74	65	156	138	400
Ni/Zn	-	42-68	-	70-130	200
Na/S	135	150	130	145	1000+
LiAl/FeS	150	100	225	150	1000+

Conclusions

The need for high energy density batteries has intensified, as the demand for portable electricity is coming from all sectors of the economy. The development of rechargeable batteries has not as yet kept pace with the demand; however there are certain interesting systems which may be a success in the near future.

The developments in batteries have evolved mainly from the work of 'experimental scientists' and owes very little to 'pure scientific work'. Progress in materials has possibly been the major factor for progress in battery development.

General progress in rechargeable battery development has been poor during the last 80 years, and developments in other areas of technology could make much of battery technology obsolete. With the advent of High Temperature Superconductors, power storage in superconducting magnets could soon become a commercial reality (12), and is not only a threat to the batteries now undergoing development, but could replace large sections of the present battery markets.

Literature Cited:

1. Volta, A. Phil.Trans. 1800, 90, 403.
2. Barak, M. Electrochemical Power Sources; Institution of Electrical Engineering,. London, 1980; p 23.
3. Kordesch, K. Electrochemical Energy Storage, In Comprehensive Treatise in Electrochemistry, Plenum: N.Y., 1981; V3, p 149.
4. Plante, G. Compt. Rend. 1859, 49, 402.
5. Sindsted, N.J. Ann. Physik Chem. 1854, 92, 21.
6. Junger, G. Swed. Patent 15 576, 1901.
7. Grove, N.R. Phil.Mag. 1839, 14, 127.
8. Smee, A. Phil.Mag. 1840, 16, 315.
9. Edison, T.A. Ger.Patent 157 209, 1901.
10. deMichalowski, T. Brit. Pat.15 370, 1899.
11. Weissman, E.Y. Batteries,In Lead Acid Batteries and Electric Vehicles; Kordesch, K.V., Ed.; Marcel Dekker: N.Y., 1977, Vol. 2, p 200.
12. Das Gupta, S. Opportunities in Superconductor Research; Ministry of Energy, Toronto, Ontario, 1988.

RECEIVED August 18, 1988

Chapter 38

Evolution of Electrochemical Reactor Systems for Metal Recovery and Pollution Control

Bernard Fleet

TRSI Incorporated, #14, 14 Tasken Drive, Rexdale, Ontario M9W 5M8, Canada and Chemistry Department and Institute for Environmental Studies, University of Toronto, Toronto, Ontario M5S 1A1, Canada

A review of the development of electrochemical reactor systems for metal recovery in pollution control and chemical waste management applications is presented. After reviewing the historical development of electrolytic cells for metal recovery (electrometallurgy), which led to the introduction of planar cathode and rotating electrode cell designs in the mid-1960's, the paper focusses on newer concepts of high surface area cathodes and high performance electrochemical reactors. Finally, some concepts and strategies for applying these systems for treating toxic metal bearing waste streams from the electronics products, metal finishing and gold milling industries are presented.

The electrochemical recovery of metals (electrometallurgy) is an old art (1,2). The earliest literature reference to an electrochemical phenomenon is found in Pliny's treatise on chemical subjects (3) and refers to the protection of iron with lead plating. However, it is known that various forms of metal plating were practised much earlier and remains of primitive storage batteries have been found associated with ancient Assyrian civilizations dated from around 2000 B.C. (1,4).

The first recorded example of the application of electrometallurgy dates from the mid-17th century and occurred at a mine known as Herrengrund, in the then Austro-Hungarian empire, now located at the small town of Spania Dolina in the Slovakia region of Czechoslovakia (5). This involved the recovery of copper from cupriferous mine waters by electrochemical replacement with iron and was used to produce a popular form of decorative copper plated ironware known as Herrengrund Ware. Dr.Edward Brown, in his account (5) to the Royal Society in 1670 writes "There are also two springs of a Vitriolat water, which are affirm'd to turn iron into copper. They are called the Old and the New Ziment. These springs lye deep

0097–6156/89/0390–0554$07.00/0

in the mine. The Iron is ordinarily left in the Water 14 dayes. They make handsome Cups and Vessels out of this Salt of Copper".

Although there was considerable interest in electrochemistry in the years following Nicholson, Davy and Faraday's pioneering works, most of this was confined to the fashionable uses of electrotyping in decorative art forms. However, by the 1850's more industrial applications of electrotechnology were being introduced (6) and electroplating was quite widely practised on a commercial scale. This was given further stimulus by the introduction of the first practical generators in the mid 1860's (7) and in 1865 the Birmingham electroplating firm, Elkington's built the first large scale electrolytic refinery in Swansea (8). Until 1880 this plant produced an average of six tons of copper per week for use as telegraph wire using cathode designs not too dissimilar from those used in some modern electrorefineries (2).

Since the focus of this paper is on pollution control applications of metal recovery, the complex and as yet incompletely told story of the early development of electroplating, surface finishing and early electrowinning techniques will not be discussed further. The development of electrolytic cells for pollution control applications of metal recovery dates from the mid-1960's when several major advances in electrochemical engineering took place. Advances in potential and current distribution theory, mass transfer processes, coupled with the introduction of new materials, created a stimulus for the introduction of novel cathode designs with improved mass transfer characteristics.

There are some major differences between electrochemical engineering and classical electrochemistry. In conventional electrochemistry the mechanism of the electrode process and its kinetics are often the factors of major concern whereas in electrochemical engineering the actual mechanistic details of the process are usually less important than its specificity or process efficiency. The rate of the process defined either as current efficiency or as a general measure of reactor efficiency, the space-time yield are the main performance criteria. This latter factor determines whether a process is economically or commercially viable since it can be used to compare performance of different electrode designs as well as comparing an electrochemical process with the space-time yields for alternate non-electrochemical technologies.

Another difference between classical electrochemistry and electrochemical engineering lies in the size of the electrode. Conventional electrochemistry most commonly employs micro electrodes of well defined area operating under carefully controlled current and mass transfer conditions. Conversely, electrochemical engineering typically employs large surface area electrodes, where, moreover, the surface area and electrode activity varies constantly as metal is deposited. In addition, there are usually difficulties in maintaining uniform potential control and current distribution over the electrode surface. It is also necessary to consider the reverse stripping process of recovering the metal after collection.

Electrochemical engineering theory is now well established with the major parameters in reactor design and operation, potential and current control, potential and current distribution, mass-transfer characteristics, role of electrocatalysis and the role of electrode material all being well defined (9-17). The keystone undoubtedly has been the development of convective diffusion theory by Levich (10) which has led to the proper evaluation of concentration profiles in a wide range of electrolytic systems. The concept of the boundary layer defines the concentration profiles of electroactive species in the bulk solution, in the boundary diffusion layer and at the electrode surface. The role of convective diffusion in both laminar and turbulent forced convection modes has also been defined and has led to the development of mass-transfer relationships for a wide range of electrode-cell geometries.

Electrochemical Reactor and Process Design

In the design of an electrochemical metal recovery process two main groups of factors need to be considered. The first group relates to the design of the reactor system and the second to the process. The design or selection of the electrochemical reactor system is dependent on the the type and concentration of metal in the process stream, the desired level of removal efficiency and any other process constraints such as corrosivity of the medium or presence of impurities. Although simple laboratory scale experiments will usually confirm or disprove the feasibility of a given process, much supporting information on metal deposition processes is available from the literature including sources such as Pourbaix diagrams (18) which define pH-potential equilibria.

There are four major design criteria in the selection or design of an electrochemical reactor. These are the overall cell design and geometry including electrode materials of both cathode and anode and the requirement, if any, for a diaphragm or separator, the mode of control of potential for the working electrode, the control of current distribution within the working electrode surface and finally the definition and control of the mass-transfer characteristics of the system.

Potential Control and Current Distribution

The electrode potential of the working electrode is a complex parameter involving current distributions within the working electrode and the solution, conductivities of the bulk electrode and the process solution and the concentrations of all reactants and products for all of the possible electrode processes. Potentiostatic control in classical electrochemistry is achieved by electronically comparing the voltage difference between the working and reference electrodes with the required operating potential (set point) and feeding the difference back to a rectifier such that the rectifier provides an output through the counter electrode to maintain the desired condition. In electrochemical engineering applications, however, potentiostatic control suffers from a number of practical problems. The first is the problem of ground loops from cell currents which can drive destructive currents through the reference electrode, effectively destroying its performance. There are also problems of locating the reference electrode, where the normal arrangement of a Luggin capillary probe situated close to the working electrode surface is only able to monitor potential at that specific site. This approach clearly has little relevance to controlling potential in a large area or three-dimensional electrode, although in some instances it may find use for the mapping or plotting of electrode potential distribution.

Current distribution through the electrode which controls the current efficiency is also dependent on the mass-transfer characteristics of the system and on the control of potential over the working electrode surface. Thus in bulk, porous or three-dimensional electrodes it is usually mass-transfer characteristics which control most situations. The Nernst diffusion model (Fig. 1) gives a simplified picture of the electrode-solution interface conditions. This simplified picture, however, does not account for real operating conditions since the stationary diffusion layer thickness is strongly dependent on the solution flow characterics. Most modern hydrodynamic treatments take these factors into account.

Process Control in Electrochemical Reactors

Although voltage control would seem to provide an obvious alternative to potentiostatic control in electrochemical reactors, in actual practice this approach is also of limited value. This is because the voltage-current relationships of most cells are dominated by inter-electrode conditions, electrolyte resistance, diaphragm resistance and mode of operation (monopolar or bipolar) so that typical cell voltages lie in the range of 4 to 10 volts. The corresponding working electrode potentials, however, rarely exceed 1000 mV.

For most electrolytic metal recovery processes from dilute solution, constant current control is the most useful operating approach from both a practical and theoretical standpoint. Since the electrode potential cannot be controlled in a "potentiostatic" sense it can be controlled by a combination of controlled current and controlled mass-transfer (B.Fleet and C.E.Small., Process Control Systems in Electrochemical Engineering, in Electrochemical Reactors: their Science and Technology, Ismail M.,(Ed), Pergammon, Oxford, in press, 1988). Since mass-transfer characteristics are so important in controlling electrode potential, for any given electrode-cell geometry there is a well defined range of mass-transfer and imposed current conditions that will achieve the desired electrode process efficiency. In some cases the control of mass transfer may also pose problems. This is particularly true with high surface area, high performance electrodes where metal deposition alters the porosity and hence the hydrodynamic characteristics of the electrode.

Electrochemical Engineering Parameters

Several definitions are often used to define the performance of an electrode or electrochemical reactor. These are the current efficiency, process efficiency and electrode/cell space-time yield. In some cases, particularly with high surface electrodes an

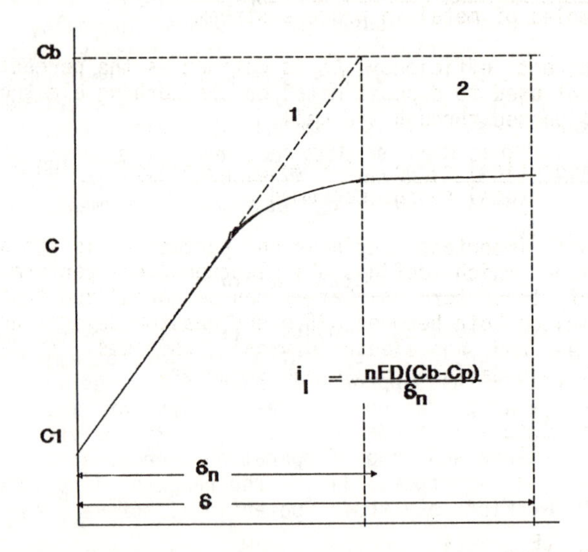

Fig.1 Diagram of the Nernst Boundary Layer (conc. of reacting species vs. distance from the electrode): 1. Nernst model concentration profile, 2. true concentration profile, δ_n = Nernst diffusion layer, δ = true diffusion layer

additional parameter, the percentage conversion or removal per pass may also be specified. These parameters are important since for a given process they define the size and hence the capital and operating costs of the required reactor system. The first of these terms, ϕ the process efficiency is defined as the number of moles of metal removed as a function of the initial metal concentration in the input stream.

$$\phi = \frac{\text{moles of metal removed}}{\text{moles of metal in process stream}} \cdot 100 \qquad (1)$$

The current efficiency, β is defined as the percentage ratio of the current used to deposit metal on the working electrode to the total current passed through the cell,

$$\beta = \frac{\text{theoretical current for metal deposited}}{\text{total charge consumed}} \cdot 100 \qquad (2)$$

The most important engineering parameter is the space-time yield parameter which defines the electrode/cell performance. The importance of this term is that it can be normalized thus allowing direct comparison both between different reactor designs and between electrochemical and non-electrochemical processes. The space-time yield term Y_{st} is defined for an electrode as;

$$Y^{E}{}_{ST} = \frac{A^{s}\beta}{czF} \qquad (3)$$

while for a cell,

$$Y^{C}{}_{ST} = Y^{E}{}_{ST} \frac{1}{1 + V_A/V_B} \qquad (4)$$

where A^s is the specific electrode area, β the current efficiency, C the concentration change during one solution pass through the reactor and V_A and V_B the volumes of anode and cathode compartments. Whereas most pollution control metal recovery processes are not subject to the same economic constraints as are, for example, electrosynthetic processes, the space-time-yield for a given cell design/removal process is often the deciding factor. In fact it has only been since the development of high surface area cathodes in the mid-1970's that reactor systems have become available with sufficiently high space-time-yields to provide technical and economically feasible processes for pollution control.

ELECTROCHEMICAL REACTOR SYSTEMS FOR METAL RECOVERY

The major application areas of electrochemical reactors in metal recovery are electrowinning of metals from ores and primary sources, electrorefining of metals from aqueous solutions or molten salts and electrochemical pollution control or detoxification/removal of metals from industrial effluents and other liquid wastes. Although cell designs for electrorefining of metals are usually quite specific, many cell designs used for both electrowinning and electrochemical effluent treatment are often very similar and in many cases the applications overlap. Apart from the major electrolytic processes for the production of aluminium, magnesium and sodium, the most widely used electrowinning processes are the recovery of copper, nickel and the precious metals. Electrolytic processes have been also been described for a wide range of metals (19) but due to the world depression in metal values only a small fraction of these processes are presently either being developed or operated on a commercial scale.

Electrochemical Reactor Design Classification

Several attempts have been made to classify the plethora of electrochemical reactor designs(12,15,20,21). One approach has been based on defining the motion of process solution in relation to the direction of current flow through the working electrode (Fig. 2). This definition describes three main modes of process flow; a "flow-by" mode where solution flows past the surface of the electrode and two flow-through modes, "flow-through parallel to current" and "flow-through perpendicular to current". A simpler classification (21) is based on working electrode geometry and describes two main classes of reactor; two-dimensional, usually planar electrode designs and the three-dimensional extended or high surface area designs. These two major categories of electrochemical reactor should be viewed as the extremes of reactor design. In between is a broad spectrum of intermediate reactor and working electrode designs, from which the optimum reactor and process design can be selected for a given application. A brief review of the development of electrochemical reactor systems which have either found application for metal removal from industrial wastewaters or which show future promise is presented below.

TWO-DIMENSIONAL REACTOR SYSTEMS

Classical electrochemical reactor designs invariably evolved from direct scale-up of simple laboratory electrolysis experiments. The most common example of this concept is the tank cell where an array of electrodes is immersed in a plastic or metal tank. More sophisticated versions involve a variety of approaches to enhancing convection, by rapid stirring, rotating or moving electrodes or improving geometry with plate and frame or filter-press-type cells.

Tank cells are one of the simplest and most popular designs of cell for both electro-organic and inorganic processes. Commercial designs which are available in a wide range of sizes and electrode areas can operate both in the monopolar or bipolar mode. They usually function as an undivided cell with a single electrolyte/process stream; the incorporation of membrane/separators is convenient and in cases where cathode and anode chambers need to be separated a plate and frame construction is usually preferred. They have found application in pollution control and metal recovery, mostly for electrowinning applications. Primarily based on planar cathode designs (22,23) with cathode areas ranging from $0.5m^2$ upwards, the main applications for this type of cell are in the electrowinning of high concentration process streams such as spent

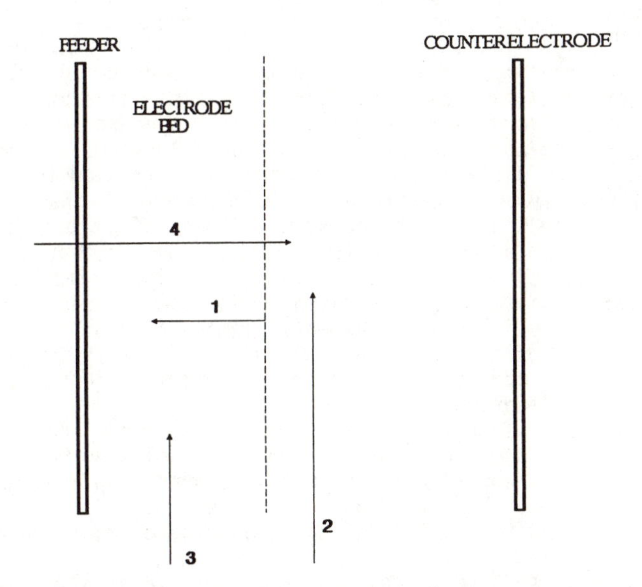

Fig.2 Mass-Transfer Modes in Electrochemical Reactors, 1- direction of current flow, 2- flow-by, 3- flow-through perpendicular to current flow, and, 4- flow-through parallel to current flow

plating baths, etchants, and ion-exchange eluates. Typically these cells can economically treat process streams to provide effluents in the 200-500 ppm range after which they may require further processing by a high surface area reactor. Another example of a tank cell used as a resource recovery system is the Capenhurst Cell (24), which has been designed for regeneration of etchants in the printed circuit board industry. These spent etchants usually contain the dissolved Cu(II) as well as the reduced form of the oxidant etchant, typically Fe(II) from $FeCl_3$ or Cu(I) from $CuCl_2$. In the Capenhurst process the cell recovers copper at the cathode while in the anode compartment the active etching species is regenerated.

A bipolar version of the tank cell design, the Bipolar Stack cell (25), consists of an assembly of parallel, planar electrodes separated by insulating spacers. Flow of electrolyte between electrodes may be either by natural or forced circulation. This design of cell is easy to construct since it simply comprises a stack of alternating electrodes and spacers with electrical connections being made to the two end electrodes. The number of electrodes in a stack typically ranges from from 10 to 100. Process flow is usually by gravity feed so that the cell has none of the complex hydraulics and plumbing features of plate and frame cell designs. One example of this design, the bipolar trickle tower reactor consists of a regular array of bipolar perforated carbon discs separated by an insulating mesh. The original work on this design described simultaneous metal removal and cyanide destruction in cyanide containing rinse streams; despite the high demand for this type pf process, no industrial demonstrations of the concept have yet been reported.

Although plate and frame cells represent one of the most widely used and developed commercial electrochemical reactor concepts (26), their use in metal recovery applications has been very limited. This has primarily been due to the problems of stripping deposited metal which requires dismantling of the cell assembly (26,27).

The limited metal removal rates for planar electrodes has led to a variety of approaches for enhancing mass-transfer in this type of cell. These have included mechanical stirring, solution forced flow, and gas sparging. Rotating the working electrode has also been a popular route for enhancing mass transfer. Other approaches have tried to increase active electrode area by surface roughening or by using expanded mesh in place of planar electrodes; these variants approach the domain of three-dimensional electrodes.

Although not strictly a two-dimensional cell design, the Chemelec Cell (28,29) developed by Bewt Engineering uses an array of expanded mesh cathodes with alternate noble metal coated, DSA type planar anodes in an undivided cell arrangement (Fig.3). Based on the original concept (29) developed at the Electricity Research Council (UK), the unique feature of this design is that it uses a fluidized bed of inert glass ballotini to promote mass-transfer. This cell has been successfully applied to a range of metal recovery and pollution control applications in the metal finishing, printed circuit board and photographic industries.

Rotating the working electrode has been one of the most obvious ways of increasing mass-transfer rates, particularly in metal recovery applications. One attractive feature of this approach is the ability to adjust the cathode rotation speed to the shear stress of the electrode to dislodge the deposited metal in powder form. Mechanical scraping has also been used in addition to direct shearing of dendritic deposits. By analogy to the analytical rotating disc and ring-disc electrodes, this configuration of cell should also have application where short-lived intermediate species are involved in the process, for example in the electrolytic generation of oxidants or reductants for effluent treatment. Although the ability to produce high purity metal powders is an attractive feature of this design, in routine industrial practice rotating electrodes are subject to mechanical problems.

A rotating cylinder cathode has featured in several commercially developed sytems for metal recovery. The Eco Cell (30) was one of the earliest designs based on this concept, where the rotating drum cathode was scraped by a wiper to dislodge deposited metal which is then passed through a hydrocyclone to produce a metal powder. A cascade version of this cell with six chambers separated by baffles was also demonstrated and was claimed to reduce a 50 ppm copper input stream to 1.6 ppm in the output (31). Commercial development of this design is now being carried out by Steetley Engineering. A laboratory prototype of an ingenious design of rotating disc electrode cell has been described (J. Tenygl, personal communication) for the production of hydrogen peroxide for potable or industrial wastewater sterilization. By using a partly submerged rotating disc array, a thin film of electrolyte is maintained in contact with air, thus enabling a much higher concentration of hydrogen peroxide to be formed than in the alternative homogeneous solution route. The peroxide in the thin solution film is continuously fed into the bulk solution by the rotation process.

The pump-cell concept (32) devised by Jansson and coworkers in Southampton is another variant of the rotating cell. In the simplest version (Fig.4) the process stream enters the thin layer between the rotating disc cathode and the stationary cell body. The electrolyte is accelerated to high mass transfer rates and the cell becomes self priming. In metal recovery applications the deposited metal film is discharged from the cell in the form of fine powder. It is also relatively straightforward to scale-up this cell design to produce a multiplate bipolar stack; a 500-amp version of this cell has been tested.

Another version of the rotating electrode concept has been developed by Gotzellman KG, in Stuttgart, West Germany. In this design, known as the SE Reactor Geocomet Cell (33) a concentric arrangement of rod-shaped cathodes rotates inside inner and outer anodes. The deposited metal is dislodged by the rotating rods and falls to the bottom of the cell. The anodic destruction of cyanides in metal plating process rinses can also be carried out with this cell.

THREE-DIMENSIONAL REACTOR SYSTEMS

The development of high surface area electrochemical reactor systems has been one of the most active research and development areas in electrochemical engineering. The demand for systems with high space-time yields has been driven by the need for economic metal

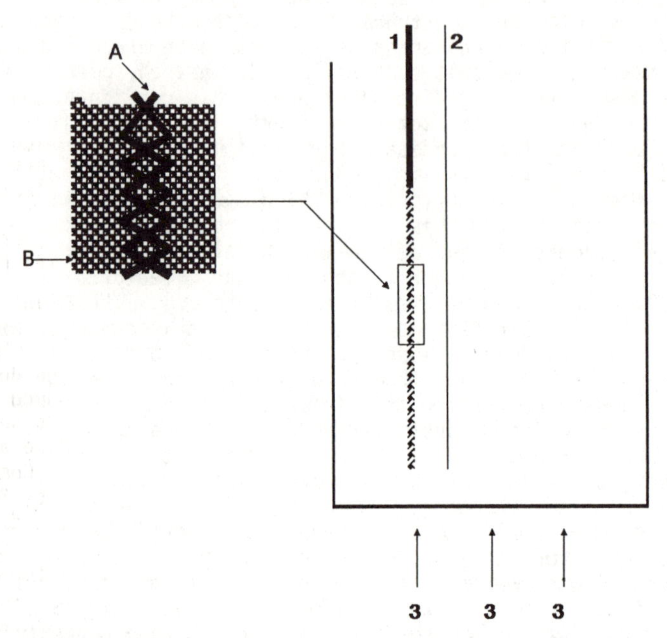

Fig.3 Chemelec Cell, 1- expanded steel or titanium mesh cathode, 2- DSA Ti/RuO_2 coated anode, 3- process steam inlet. A and B expanded view of cathode showing, A - expanded mesh and B -inert ballotini used for fluidised mass-transfer enhancement

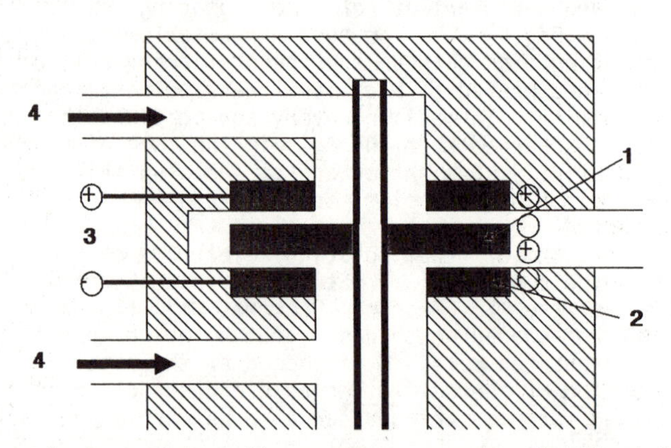

Fig.4 Bipolar Pump-Cell; employs a magnetically driven rotor on ceramic shaft; 1- rotor, 2- stator, 3- current connections, and 4- flow direction

recovery systems for both pollution control and resource recovery applications. The common feature in all of the reactors of this type is the use of an extended surface cathode, where the cell chamber is filled or partly filled with the working electrode material such as carbon or metal fibres or mesh, porous materials and packed particles.

The Fluidized-Bed Electrode

The invention of the fluidized bed cell by the University of Newcastle group (34) marked both an important landmark and also created a major stimulus to the field of electrochemical engineering. Despite the elegance of this cell design, it was the inherent limitations of the fluidized-bed concept that provided a focus for the development of alternative high surface area reactors. The principle of the fluidized-bed cell is shown in Fig.5; the conductive electrode particles are contacted by a porous feeder electrode, while the process stream causes fluidisation of the electrode bed. The main limitation to the design is that fluidization of the bed causes loss of electrical contact between the particles, resulting in an extensive ohmic drop occurring within the cell so that uniform potential and current distribution is virtually impossible to maintain (Figs. 5a and 5b). The original design by Goodridge et al (33) shows why the cell was initially so attractive. The electrode had a specific area of 200 m^2/m^3 and was thus able to support large currents at an effective current density of around 0.01 A/cm^2. Despite these advantages, it was more than 15 years from the initial concept before any significant commercial applications of the concept were demonstrated.

Early commercial development of the fluidized-bed cell was carried out by Constructors John Brown Ltd., in the U.K. and by Akzo Zout Chemie in the Netherlands (35). The Akzo cell design overcame some of the scale-up limitations of the original design by the use of a large number of rod feeders to the cylindrical, 0.35m diameter, cathode bed. The cell also contained six symmetrical rod anodes encased in cylindrical diaphragms. Applications to copper removal from chlorinated hydrocarbon waste and mercury from brine have been demonstrated by Akzo. The technology is currently licensed and is being developed by the Billiton Group of Shell Research, also in the Netherlands. The Chemelec Cell described earlier (28,29) is also an example of the use of the fluidized-bed concept to promote mass-transfer. In this case, however, the working electrode is an expanded metal mesh and the inert fluidized bed simply acts to enhance mass-transfer.

Contiguous and Packed-Bed Electrode Designs

Many of the problems of fluidized-bed cells appear to have been overcome by the development of three-dimensional contiguous-bed reactors (36,37) which mainly evolved from the porous electrode designs used in battery or fuel cell applications. These electrode designs are characterised by very high specific surface areas and space time yields. At the same time the ability to control

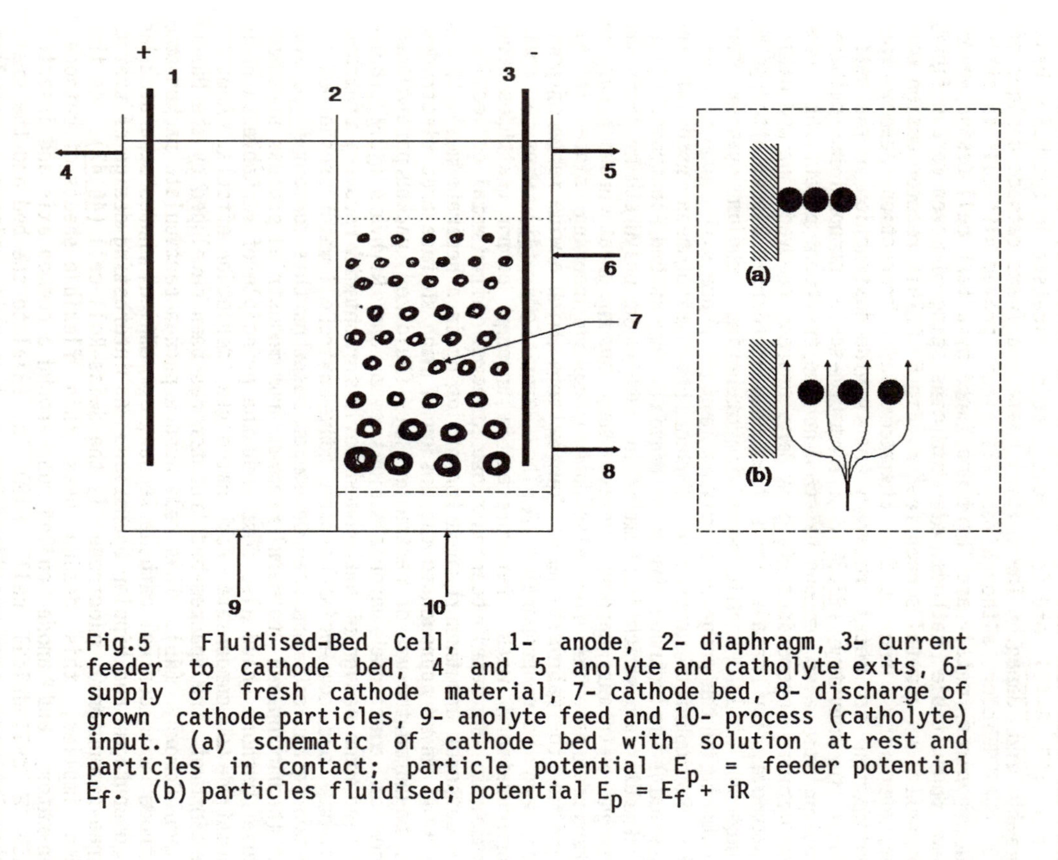

Fig.5 Fluidised-Bed Cell, 1- anode, 2- diaphragm, 3- current feeder to cathode bed, 4 and 5 anolyte and catholyte exits, 6- supply of fresh cathode material, 7- cathode bed, 8- discharge of grown cathode particles, 9- anolyte feed and 10- process (catholyte) input. (a) schematic of cathode bed with solution at rest and particles in contact; particle potential E_p = feeder potential E_f. (b) particles fluidised; potential $E_p = E_f + iR$

potential and current distribution through the electrode bed is far superior to the earlier fluidized bed designs.

The first demonstration of the use of this electrode principle in metal recovery and pollution control applications was the use of carbon fibers as a high performance cathode (37). Subsequently this electrode concept was developed commercially by HSA Reactors in Canada (27,38-40) and is currently being licensed in the U.S., West Germany and Japan. The initial plate and frame cathode designs of the HSA Reactor, although very efficient, proved difficult to strip of deposited metal and were replaced by a tank cell design which used demountable cathode elements consisting of woven carbon fiber supported on a metal screen feeder (Fig.6). This reactor design was evaluated by the U.S. Environmental Protection Agency for application to metal recovery and cyanide destruction in the metal finishing industry (27,39). A pilot-scale carbon fiber cathode reactor was also demonstrated for the treatment of effluent from Gold Mining operations (40). This process removed zinc and other trace impurity metals from the effluent (Barren Bleed) after gold recovery while simultaneously liberating sodium cyanide for recycling to the pregnant stream extraction process.

Reticulated vitreous carbon has also been used as an alternative electrode material for contiguous bed electrodes (41). This material, which is prepared by pyrolysing polyurethane foams to give a vitreous carbon surface has shown great promise as an electrode material for electrochemical engineering since it is possible to fabricate electrode material with a wide range of pore size and void proportion. The Retec Cell (41) uses a tank configuration with reticulated vitreous carbon cathodes for treatment of metal bearing rinse streams. After metal collection, the cathodes are then removed for stripping in a separate cell.

A third approach to the design of high surface area electrodes is based on the use of restrained, non-fluidized, packed particulate beds of electrode materials. Kreysa and Reynvaan (42,43) have studied the design of packed-bed cells for optimal recovery of trace metals. A commercial system, the EnviroCell which uses a bed of carbon granules, has been developed based on this concept (44), and has been applied to metal removal from industrial process streams. These authors have also described the principle of variable cathode geometry to compensate for catholyte depletion effects. Another version of the packed-bed reactor has been developed by the Nanao Kogyo Company (45). This cell uses a packed particulate packed-bed cathode with central cathode feeder, symmetrical anodes separated by a diaphragm and bipolar operation. An interesting design of static three-dimensional electrode is the Swiss-Roll cell (46,47). As the name implies, this design uses thin flexible sheets of cathode, separator, and anode rolled up around a common axis and inserted into a cylindrical cell. Flow is axial to the bed and the small interelectrode gap provides for a low cell voltage. A similar version of this concept was developed by DuPont (47) and was known as the extended surface electrode or ESE Cell. This system was demonstrated at the pilot scale for metal removal from wastewaters. A modified plate and frame cell design which uses a similar type of construction to the HSA reactor (Figure 5) has also been described by Simonnson (48), using the Swedish Electrocell concept. However, unlike conventional practice with packed-bed electrodes which use a flow-through approach this system employs the flow-by mode.

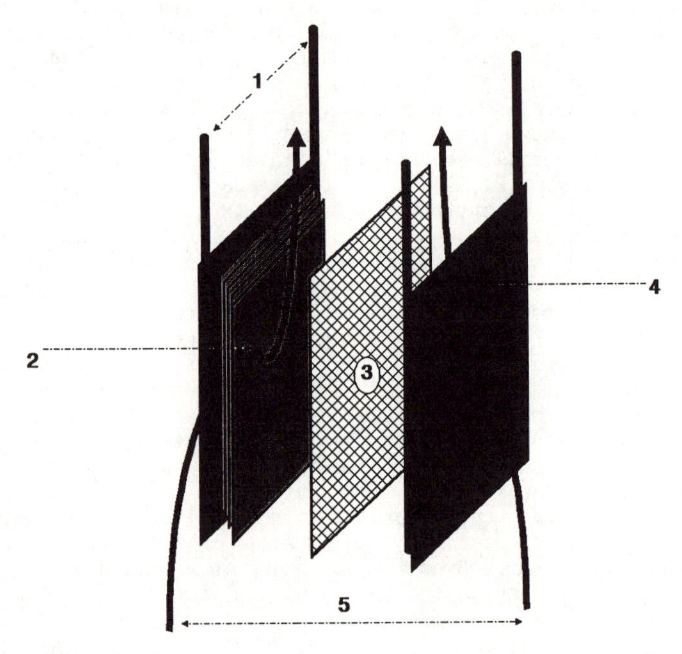

Fig.6 High Surface Area Reactor, 1- cathode current feeders, 2- layered carbon fibre cloth, 3- anode, DSA type or steel, 4- stainless steel or titanium mesh cathode support/feeder, 5- process solution flow.

The enhancement in mass-transfer characteristics obtained with these designs in comparison to conventional two-dimensional planar reactors has had dramatic consequences for the commercial development of electrochemical engineering. First, the orders-of-magnitude increase in space-time yields obtained with many of these designs has resulted in substantial decreases in capital and operating equipment costs for a given process. This in turn has made the electrochemical processes more competitive with alternative, non-electrochemical technical routes. Second, it has opened up important new application areas, such as low level metal recovery, which were either technically or economically impractical with the far less efficient two-dimensional cells. Some trade-offs were involved, particularly in the area of potential control and uniformity of current distribution but the leverage obtained in terms of improved space time yields more than compensates for this limitation. With further research in the areas of better computer modelling techniques and advanced process control systems even this limitation may eventually be overcome.

APPLICATIONS TO METAL RECOVERY/POLLUTION CONTROL

As there is a very wide range of applications of electrochemical reactors in metal removal and pollution control, it is important first of all to attempt to identify the main approaches or strategies for their use (49,50). The first of these is known as "point source" or the treatment of a pollutant at its point of origin. The second is known as "end of pipe" and describes the use of a treatment technology to process a combination of waste sources, generally before being discharged from the plant into a sewer or local watercourse. Three examples or case studies will be presented which illustrate these applications.

Metal Finishing

Electrochemical reactors have shown great promise for treating waste streams from metal finishing operations. This industry generates a highly toxic and complex waste stream containing a range of metals including copper, cadmium, lead and chromium, cyanides and some organics. Because of the lack of uniformity of of most metal finishing operations it is difficult to define an optimum treatment strategy. One of the most generally useful treatment approaches for this industry is based on the use of electrochemical reactors used for point-source treatment units. In this process they are linked in a "closed-loop" configuration on the rinse tank immediately following plating operations (Fig.7). In this situation, the rinse tank R_1 is a drag-out or non-flowing rinse so that the metal plating solution adhering to the plated parts is captured. The electrochemical reactor which is of the high surface cathode type, continuously removes the accumulated metal from this "drag-out" tank and thus maintains the metal concentration at a suitably low level, typically in the range 10-100 ppm. Thus, as the plated parts move to the next rinse station the carry-over of metal solution is minimized as is the discharge from this tank to the sewer. In most

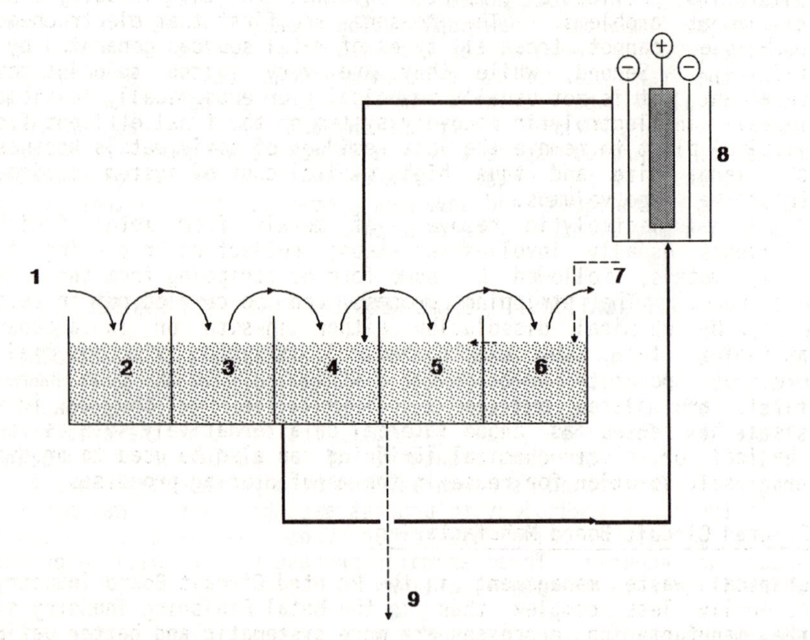

Fig.7 Point-Source Treatment of Electroplating Rinses; 1- direction of plating work flow, 2- pre-plating, 3- plating bath, 4- static or drag-out rinse tank, 5- first running rinse, 6- second running rinse, 7- process water input, 8- electrolytic recovery cell, 9- discharge to sewer or waste treatment

cases this approach can ensure that the effluent from this particular source is well below the discharge limit. Demonstrated applications have included the treatment of drag-out rinses after acid copper, cyanide copper, cadmium, nickel, tin, lead, gold, silver and platinum plating, cyanide destruction and some few examples of plating bath control.

Although electrolytic metal removal is a valuable pollution control tool for the metal finishing industry, in general it is unable to provide a complete solution to the industry's waste management problems. The reasons are first that electrochemical techniques cannot treat all types of metal sources generated by the industry. Second, while they are very suited to point source treatment, it is not usually technically or economically feasible to install an electrolytic recovery system on the final effluent from a plating plant to remove the last residues of toxic metals because of the large size and thus high capital cost of system required to treat the large volumes.

The electrolytic recovery of metals from metal finishing effluents usually involves two steps; collection or plating of the heavy metals, followed by some form of stripping from the working electrode. The stripping process can be carried out in several ways; by chemical dissolution either in-situ or in a separate stripping tank, by electrochemical stripping via reversal of electrode polarity or by galvanic stripping. In the case where the metals are plated out as a solid metallic sheet they may in some cases be reused as anode material or alternatively sold as scrap. Chemical or electrochemical stripping can also be used to produce a metal salt solution for reuse in the manufacturing processes.

Printed Circuit Board Manufacturing

Chemical waste management in the Printed Circuit Board industry is generally less complex than in the Metal Finishing industry since the manufacturing processes are more systematic and better defined. This industry generates toxic wastes comprising acid and chelated copper from circuit board plating/etching, tin/lead from solder coating and masking and nickel and gold from edge connection preparation. Although a number of waste management strategies have been described for treating this industry's waste management problems (49,50), one of the most promising would seem to be based on a combination of ion-exchange and electrolytic recovery (Fig.8). In this approach the dilute metal-bearing process rinse streams are first segregated and treated by an ion-exchange process. The higher concentration waste sources (spent process baths, circuit board etchants etc) are treated directly with an electrowinning cell using planar stainless steel cathodes in a tank cell with a membrane isolated anode and turbulence promotion. This electrowinning process can reduce metal concentrations from initial values in the 5 - 20 g/L range down to 200-500 ppm. The residual solution after treatment is then recycled through the ion-exchange process to remove the remaining metals. Finally, by passing the treated effluent through a demineralisation step, over 90% of the plant process water can be recycled as high purity, deionized water. There are clearly significant benefits to this approach in elimination of

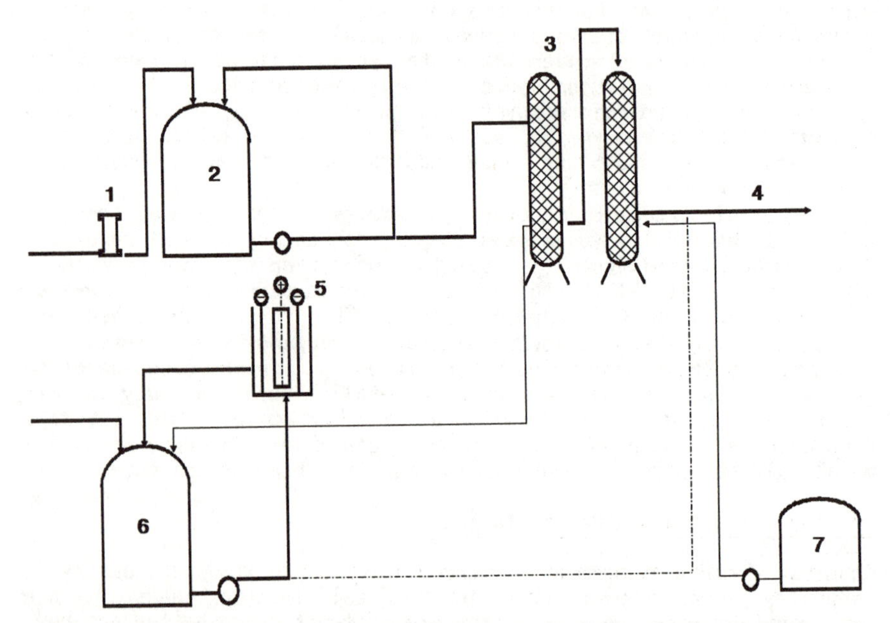

Fig.8 Integration of Ion-Exchange and Electrolytic Recovery; 1- filter, 2- collection tank for segregated dilute metal stream, 3- ion-exchange columns, collector and guard, 4- discharge to sewer or demineralization for recycling to process, 5- electrowinning cell for treating high concentration metal sources and ion-exchange eluate, 6- high concentration metal sources storage, 7- eluent reservoir

offsite waste transport and disposal costs and process water savings. Despite the higher capital cost of resource recovery systems, the economics of this approach are very favourable for medium to large size plants with total daily process water usage over 200,000 litres per day. For smaller plants some alternate strategies may presently be more cost effective although with increasing constraints on land disposal of toxic wastes and more realistic costs being imposed for water and sewer use, on-site resource recovery and zero discharge may eventually be the only available option. Furthermore, this type of waste management strategy involving integrated ion-exchange, electrolytic recovery and water recycling is readily transferable to a range of metal bearing waste streams generated by industries such as semiconductor, integrated circuit or cathode ray tube manufacturing.

Mining

While electrolytic methods have been developed for a range of mining and primary metal recovery processes, in most cases the currently depressed economic state of the metals industry has prevented their commercial exploitation (19). One area where electrochemical processes still offer considerable promise, however, is in gold mining. At the present time this industry relies very heavily on the Merrill-Crowe process which consists of adding powered metallic zinc to a cyanide extract of the gold ore. This causes precipitation of the gold while at the same time generates a highly toxic waste stream containing free cyanide along with zinc and other metal cyanide complexes. An alternative process which is normally used for lower grade gold ores (typically 0.015 to 0.15 oz/ton) is known as the Heap-Leach process. In this approach, shown schematically in Fig.9, a cyanide process stream is percolated through a mound of coarsely crushed ore and the eluate collected in a plastic lined tank. This stream is then passed through a carbon adsorption column which collects the gold by adsorbing the gold cyanide complex. The column is then washed with a concentrated cyanide stream which elutes the gold as a concentrate prior to the final gold processing step.

Electrolytic cells have long been used for final electrorecovery of gold after conventional processing and cyanide extraction using either the Merrill-Crowe or carbon-in-pulp processes. The cell designs used to date have not generally adopted recent advances in electrochemical engineering and it seems clear that the full potential of electrochemical techniques has yet to be realized in this industry. There are several possible conceptual process schemes incorporating electrolytic recovery which could both improve the economics of conventional gold processing and which, in most cases, would simultaneously eliminate many of the hazardous waste problems faced by this industry. The first of these is the direct electrowinning of gold from gold cyanide process streams. Electrolytic treatment of the pregnant leach stream to directly recover gold has not been applied commmercially due partly to the lack of availability of a suitable design of high performance electrochemical recovery cell as well as the significant process problems caused by the corrosive, high particulate streams.

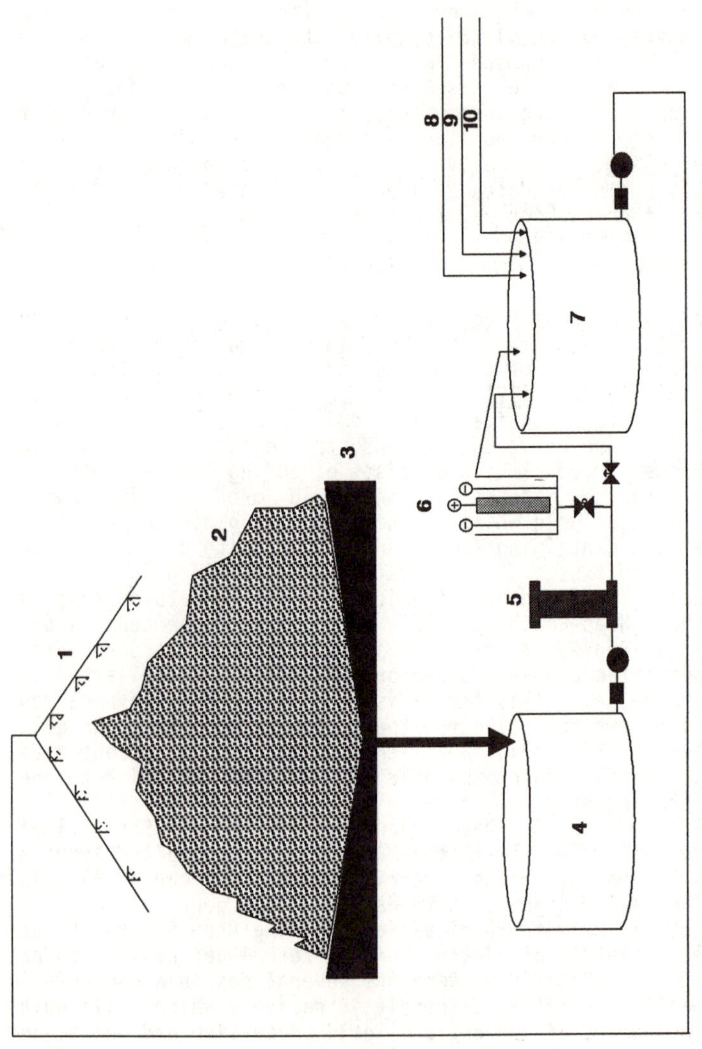

Fig.9 Heap-Leach Gold Recovery; 1- cyanide sprays, 2- crushed ore, 3- asphalt or plastic lined pad, 4- collection tank, 5- carbon filter collector for gold and metal cyanides, 6- electrolytic cell, 7- process return tank, 8- process water, 9- caustic and 10- cyanide feed

The treatment of Barren Bleed streams after conventional Merrill Crowe precipitation of gold with zinc, has been successfully demonstrated at pilot scale (40). The HSA reactor cell was used to remove impurity metals from the Barren stream, thus liberating free cyanide, which could then be recycled back to the extraction process. This approach elegantly demonstrates the concepts of resource recovery and pollution control; sodium cyanide, a valuable process chemical is recovered which would otherwise have to be subjected to waste treatment.

Electrolytic treatment of Heap-Leach gold mill process streams is another potential application for electrochemical reactors. The approach shown schematically in Figure 9 summarises one application which has been demonstrated at pilot scale. This is the removal of copper which is often present as an impurity in the process. Unless it is removed, the copper is recovered along with the gold and results in dilution of the gold bullion. Another attractive application is the use of high performance cathode cells for the direct electrowinning of gold from heap-leach extract thus eliminating the need for an intermediate carbon adsorption recovery and pre-concentration step.

Conclusions

The safe management of hazardous wastes is one of the most crucial issues facing modern society. Pollution by toxic metals including cadmium, copper, chromium, lead, mercury, nickel and zinc is generated by a wide range of manufacturing industries and most of these, if they use any form of waste management at all, rely on chemical treatment, "sludge" generation and land disposal of their wastes. There is now indisputable evidence (51) that land disposal of toxic wastes is only a temporary solution to the problem. All of these storage or disposal sites leak to varying degrees with the result that many of the toxic species find their way into natural watercourses and subsequently into the biological food chain. In the US, the results of 50 years of industrial waste mismanagement are now apparent. The EPA has identified over 17,000 hazardous waste dump sites; Congress under "Superfund" has allocated over $9 billion for cleanup. Other sources claim that the real number of toxic chemical dump sites could exceed 250,000 and that the realistic cost for cleanup to include nuclear and military waste dumps will be in excess of $100 billion.

Environmental legislation in most developed countries as well as some newly industrializing regions is now focussing on two main themes; waste reduction or minimization and onsite recovery and recycling. Electrochemical technologies will almost certainly play a major role in providing practical recovery and recycling technologies for the future. Already they have proven their utility in metal removal, detoxification of organic effluent streams and the electrogeneration of chemical sterilizing agents. In conclusion I would like to acknowledge the contributions of my coworkers, Dr. Colin Small and Bruce Cardoza at TRSI Inc., and especially my colleague and friend, the late Dr. Jiri Tenygl whose work contributed so much to the environmental applications of electrochemistry.

Literature Cited

1. Dubpernel, G. In Selected Topics in the History of Electrochemistry; The Electrochemical Society: Princeton, 1978, p 1.
2. Smith, C.S. ibid., p 360.
3. Bailey, K.C. The Elder Pliny's Chapters on Chemical Subjects, Part II, Edward Arnold: London, 1932; p 60.
4. Gray, W.F.M. J.Electrochem.Soc., 1963, 110, 210c.
5. Brown, E. Philosophical Transactions, London, 1670, 5, 1042. R. Pribil, Jr., personal communication.
6. Smee, A. Elements of Electro-Metallurgy or the Art of Working in Metals by the Galvanic Fluid; London, 1841.
7. Wilde, H. Philosophical Transactions, 1867, 157, 89.
8. Percy, J. Metallurgy,Silver and Gold; London, 1880, p 48.
9. Selman, J.R. Amer.Inst.Chem.Eng. Symposium Series, 1981, 204,77.
10. Levich, V.G. Acta Physicochem.(URSS)., 1942, 17, 257., Physicochemical Hydrodynamics; Prentice-Hall: New York, 1962.
11. Newman, J.S. Electrochemical Systems; Prentice-Hall: New York, 1973; In Electroanalytical Chemistry, Bard, A.J.,Ed.; Marcel Dekker: New York, 1973, p 187.
12. Pickett, D.J. Electrochemical Reactor Design; Elsevier: Amsterdam, 1979.
13. Kuhn, A.T.,Ed.; Industrial Electrochemical Processes; Elsevier: Amsterdam, 1971.
14. Tutorial Lectures in Electrochemical Engineering and Technology; Amer.Inst.Chem.Eng. Symposium Series; 1981, 204, 77.
15. Pletcher, D. Industrial Electrochemistry; Chapman and Hall: London, 1983.
16. King, C.J.H. in Ref 14, p 46.
17. Fahidy, T.Z.; Mohanta, S. In Advances in Transport Processes, Mujumdar,A.S.,Ed.; Halstead, 1980, p 83.
18. Pourbaix, M. Atlas of Electrochemical Equilibria in Aqueous Solutions; Pergamon: Oxford, 1966.
19. Barbier, M. In Electrotechnology Volume 1; Wastewater Treatment and Separation Methods; Ouellette, R.P.; King, J.A.; Cheremisinoff, P.N.; Eds.; Ann Arbor Science: Ann Arbor; 1978, p 239.
20. Marshall, R.J.; Walsh, F.C. Surface Tech., 1985, 24, 45.
21. Kalia, R.K.; Weinberg, N.L.; Fleet B. Impact of New Electrochemical Technologies on the Demand for Electrical Energy; Contract CEA 410-U-479; Canadian Electrical Association: Montreal, 1986.
22. Lancy International Inc., Zelionople, PA.
23. Fleet, B. In Electrochemical Technologies for Profit and Progress, Canadian Electrotechnologies Association, Toronto, 1988.
24. Hillis,M.R. Trans. Inst. Met. Finish., 1979, 57, 73.
25. Ehdaie, S.; Fleischmann, M.; Jansonn, R.E.W.; Alghaoui A.E. J.Appl. Electrochem., 1982, 12, 59.
26. Carlsson, L.; Sandegren, B.; Simonsson, D.; Rihovsky, M. J.Electrochem. Soc., 1983, 130, 342.

27. Evaluation of the HSA Reactor for Metal Recovery and Cyanide Oxidation in Metal Plating Operations; EPA/600/2-86/094, U.S. Environmental Protection Agency, Cincinatti, OH, 1986.
28. Tyson, A.G. Plating and Surface Finishing, December, 1984.; also BEWT Water Engineers, Alcester, UK.
29. Lopez-Cacicedo, C.L. Br. Patent; 1,423,369; 1973.
30. Gabe, D.R.; Walsh F.C. J. Appl. Electrochem., 1983, 13, 3.
31. Walsh, F.C.; Gardner, N.A.; Gabe, D.R. J. Appl. Electrochem., 1982, 12, 229.
32. Jansson, R.E.W.; Tomov, N.R. Chem. Eng., 1977, 316, 867.
33. Kammel, R. Metalloberfleche, 1982, 5, 194.
34. Backhurst, J.R.; Coulson, J.M.; Goodridge, F.; Plimley, R.E.; Fleischmann, M. J. Electrochem. Soc., 1969, 116, 1600.
35. Van der Heiden, G.; Raats, C.M.S.; Boon, H.F. Chem. and Ind.(London), 1978, 13, 465.
36. Sioda, R.E.; Keating, K.B. In Electroanalytical Chemistry, Bard A.J.; Ed.; Marcel Dekker: New York, 1982, p 12.
37. Fleet, B.; Das Gupta, S. Nature 1976, 263, No. 5573, 122.
38. Kennedy, I.F.T.; Das Gupta, S. First Annual Conf. on Advanced Pollution Control for the Metal Finishing Industry, EPA-600/8-78-010, U.S. Environmental Protection Agency, Cincinatti, OH.,1978, p 49.
39. Roof, E. Third Annual Conf. on Advanced Pollution Control for the Metal Finishing Industry, EPA-600/2-81-028, U.S. Environmental Protection Agency, Cincinatti, OH., 1982.
40. Mohanta, S.; Fleet, B.; Das Gupta, S.; Jacobs, J. Evaluation of a High Surface Area Electrochemical Reactor for Pollution Control in the Gold Industry, Department of Supply and Services, Contract # DS 0477K204-7-EP58; Ottawa, 1981.
41. Eltech Systems Corporation, Chardon, OH.
42. Kreysa, G.; Reynvaan, C. J. Appl. Electrochem., 1982, 12, 241.
43. Kreysa, G. Chem.Ing.Tech., 1970, 50, 332.
44. Deutsche Carbone Akttiengesellschaft, Frankfurt, West Germany.
45. Nanao Kogyo Co. Ltd., Yokohama, Japan.
46. Robertson, P.M.; Ibl N. J.Appl.Electrochem., 1977, 7, 323.
47. Williams, J.M. U.S.Patent; 3,859,195 ;1975.
48. Simonsson, D. J. Appl. Electrochem., 1984, 14, 595.
49. Fleet, B.; Small, C.E.; Cardoza, B.; Schore G. Performance and Costs of Alternatives to Land Disposal, Oppelt, E.T.; Blaney, B. L.; Kemner, W.F. Eds.; APCA Publications; Pittsburgh, 1987, p 170.
50. SCADA Systems Inc.; Cal-Tech Management Associates: Waste Reduction Strategies for the Printed Circuit Board Industry, California Department of Health Services; Contract 85-00173, Sacramento, CA., 1987.
51. Piasecki, B.L., Ed.; Beyond Dumping: New Strategies for Controlling Toxic Contamination, Quorum Books: Westport, 1984.

RECEIVED August 18, 1988

Chapter 39

Electrochemical Machining

Development and Application

J. A. McGeough and M. B. Barker

Department of Mechanical Engineering, University of Edinburgh, Edinburgh EH9 3J1 Scotland

Electrochemical machining (ECM) is a metal removal process based on the laws of electrolysis. The need for ECM has stemmed from the recent development of high-strength heat-resistant alloys which are difficult to machine by conventional techniques.

With the alternative ECM technique, hard metals can be shaped electrolytically; and the rate of machining does not depend on their hardness. Moreover the tool-electrode used in the process does not wear, and therefore soft metals can be used as tools to form shapes on harder workpieces, unlike conventional practice. The bases of ECM will first be described in the paper. Then some unsolved research problems will be discussed. Industrial applications, which illustrate how the process is used in practice, will be investigated.

Conventional machining practice relies on the mechanical forces exerted by a tool, in order that a workpiece made of a softer material can be cut to a required shape. To that end, various tool materials and formations of tools are used in different types of machine tools to produce a large range of items in a wide variety of materials.

However the production of tough, heat-resistant alloys has made machining by established methods increasingly difficult. Alternative methods therefore had to be sought. Electrochemical machining (ECM) has been developed primarily to tackle these hard alloys, although any metal can be so treated. Although ECM is well-known to engineers, the technique may not be familiar to electrochemists. A bibliography is contained therefore in the Appendix.

ECM is founded on the laws of electrolysis namely that the rate at which metal is dissolved is proportional to the product of the chemical equivalent of the anode-metal and the current.

0097–6156/89/0390–0578$06.00/0

The rate of evolution of the gas generated at the cathode can be similarly determined. In principle, therefore, we have an alternative way of removing material from an anodic workpiece. Thus if a nickel anode were used, it would dissolve, yielding nickel hydroxide. If brass or steel, for example, is substituted for copper as the cathode, the reaction at that electrode is still hydrogen gas generation.

Observations relevant to ECM can now be made. Firstly, the rate of metal removal from the anode is not affected by the hardness or other mechanical properties of that electrode. Secondly, the shape of the cathode remains unchanged during the electrolysis, since gas evolution is the only reaction that occurs there.

In the development of ECM as a metal removal process, safety and power limitation restrict the voltages that can be used to about 10 to 20V (normally d.c.). Now aqueous electrolytes such as 20% sodium chloride solution have a specific conductivity of the order of only 0.1 ohm^{-1} cm^{-1}, at ambient temperatures of about 19°C. Therefore, metal dissolution rates that are comparable with those of established machining processes can be achieved only by narrowing the inter-electrode gap to about 0.5 mm, in order to reach the currents of the order of 100 to 10,000A that are needed for the electrolytic action. (This gap is usually kept constant at an equilibrium width by mechanically feeding one electrode, the cathode say, towards the other at a fixed rate, typically at 1 mm/min, such that the rate at which the anode surface recedes through dissolution is matched by the forward movement of the cathode-tool).

The accumulation of the products of the reactions at both electrodes, the metal hydroxide and gas bubbles, is undesirable. Being insulators, they reduce the specific conductivity of the electrolyte. The electrolyte is also likely to boil, since it is rapidly heated by the electrical power transmitted across the gap. Boiling of the electrolyte is also unwanted in ECM, since it can lead to premature termination of machining. To wash away the electrolytic debris and keep the temperature of the electrolyte cool (usually between 20 and 40°C) in the machining zone, the solution is pumped through the gap at velocities which are typically as high as 30 m/s; the corresponding electrolyte pressures are about 700 kN/m^2. Breakdown in machining is thereby avoided.

The characteristics of the ECM process can now be summarized. A cathode-tool is cut from a soft metal, such as brass or copper, to a shape which is the image of that required on the anode-workpiece, which typically would be a tough metal, such as a nickel alloy or titanium. A solution of electrolyte, for example 20% sodium chloride or sodium nitrate, is pumped between the two electrodes. When d.c. of about 10V is applied between them, the inter-electrode gap tends to an equilibrium width, if the tool is moved mechanically towards the workpiece in order to maintain the ECM action and a shape, complementary to that of the tool, is reproduced on the workpiece.

The main advantages of ECM are therefore: (i) no tool wear; (ii) the rate of machining is not affected by the hardness of the workpiece; (iii) complicated shapes can be machined with a single tool, which can be softer than the workpiece.

Dynamics of ECM

The production of complicated shapes by ECM can be more fully understood if we derive some expressions for the variation in gap width with machining time. Consider a set of plane-parallel electrodes, with a constant voltage V applied across them, and with the cathode-tool driven mechanically towards the anode-workpiece at a constant rate of f.

In this analysis, the electrolyte flow is not expected to have any significant effect on the specific conductivity, K_e, of the electrolyte, which is assumed to stay constant throughout the ECM operation. Also, all the current that is passed is used to remove metal from the anode, i.e., no other reactions occur there. Under these conditions, from Faraday's law, the rate of change of gap width h relative to the tool surface is

$$\frac{dh}{dt} = \frac{AJ}{ZF\rho_a} - f \qquad (1)$$

where A, Z are the atomic weight and valency respectively of the dissolving ions, F is Faraday's constant, ρ_a is the density of the anode-workpiece metal, and J is the current density.

From Ohm's law, the current density J is given by

$$J = K_e V/h \qquad (2)$$

where h is the gap-width between the electrodes. On substitution of equation (2) into (1) we have

$$\frac{dh}{dt} = \frac{AK_eV}{ZF\rho_a h} - f \qquad (3)$$

Two practical cases are of interest in considering solutions to equation (3):

(i) Feed-rate $f = 0$, i.e., no tool movement. Equation (3) then has the solution for gap h(t) at time (t).

$$h^2(t) = h^2(0) + \frac{2AK_eV_t}{ZF\rho_a} \qquad (4)$$

where $h(0)$ is the initial matching gap. That is, the gap width increases indefinitely with the square root of machining time t. This condition is often used in deburring by ECM, when surface irregularities can be removed from components in a few seconds, without the need for mechanical movement of the electrode.

(ii) Constant feed-rate, f_e, i.e., the tool is moved mechanically at a fixed rate towards the workpiece. Equation (3) then has the solution

$$t = \frac{1}{f}\ h(0) - h(t) + h_e \ell n_e \frac{h(0) - h_e}{h(t) - h_e} \qquad (5)$$

Note that the gap width tends to a steady state value, h_e, given by

$$h_e = \frac{AK_eV}{ZF\rho_a f} \qquad (6)$$

This inherent feature of ECM, whereby an equilibrium gap width is obtained, is used widely in ECM for reproducing the shape of the cathode-tool on the workpiece.

Rates of Metal Removal

It is now useful to indicate how Faraday's laws can be employed to calculate the rates at which metals can be electrochemically machined. His laws are embodied in the simple expression.

$$m = \frac{AIt}{ZF} \qquad (7)$$

where m is the mass of metal electrochemically machined by a current I (amp) passed for a time t(s). The quantity A/ZF is called the electrochemical equivalent of the anode-metal.

Table 1 (McGeough, 1988) shows the metal machining rates that can be obtained when a current of 1000 A is used in ECM; (currents from 250 to 10,000 A are common). Note that the metal removal rates are given in terms of volumetric machining rates as well as mass removal rates. The former is often more useful in practice.

Table 1 assumes that all the current is used in ECM to remove metal. Unfortunately that is not always the case. Some metals are more likely to machine at the Faraday rates of dissolution than others. The factors, other than current, that influence the rate of machining are multifarious. They involve electrolyte type, the rate of electrolyte flow, and other process conditions. For example, nickel will machine at 100 per cent current efficiency (defined as the percentage ratio of the experimental to theoretical rates of metal removal) at low current densities (e.g., 25 A/cm^2). However, if the current density is increased (say to 250 A/cm^2) the current efficiency is found to be reduced to typically 85 to 90 per cent. This is the result of other reactions at the anode, such as oxygen gas evolution which becomes increasingly preferred as the current density is raised.

If the ECM of titanium is attempted in sodium chloride electrolyte, very low current efficiencies, of about 10 to 20 per cent, usually result. When this solution is replaced by a mixture of fluoride-based electrolytes, higher efficiencies can be obtained. Admittedly, the additional measure of higher voltages (roughly 60 V) is still needed to break down the tenacious oxide film that forms on the surface of this metal. (The corrosion-resistant character of titanium, which, together with its toughness and lightness, makes this metal so useful in the aircraft engine industry, owes much to the formation of this film).

If the rates of electrolyte flow are kept too low, the current efficiency of even the most easily electrochemically machined metal is reduced; with insufficient flow the products of machining cannot be so readily flushed from the machining gap. The accumulation of debris within the gap impedes the further dissolution of metal, and the build-up of cathodically generated gas can lead to short-circuiting between tool and workpiece, causing termination of machining, with both electrodes damaged. Indeed, the inclusion of proper flow channels to provide sufficient flow of electrolyte, so that machining can be efficiently maintained, remains a major exercise in ECM practice. When complex shapes have to be produced, the design of tooling incorporating the right kind of flow ports becomes a considerable problem, requiring skill and great expertise on the part of the design engineers.

Although Table 1 provides data on machining rates for pure metals, various expressions have been derived from which the corresponding rates for alloys can be calculated. All these procedures are based on calculating an effective value for the chemical equivalent of the alloy. Thus for Nimonic, a typical nickel alloy used in the aircraft industry, a chemical equivalent of 25.1 may be derived from the expression:

$$\text{Chemical equivalent of alloy} = 100 \left[\frac{X_A}{A_A/Z_A} + \frac{X_B}{A_B/Z_B} + \ldots \right] \quad (9)$$

where A, B, ... denote the elements in the alloy. (A typical Nimonic alloy has the following constituents by weight percent (X): 72.5 Ni, 19.5 Cr, 5.0 Fe, 0.4 Ti, 1.0 Si, 1.0 Mn, 0.5 Cd).

Electrolytes

From the preceding paragraphs we note that the current efficiency achieved with a particular metal depends greatly on the choice of the electrolyte solution.

In ECM, the main electrolytes used are aqueous solutions of (i) sodium chloride, and (ii) nitrate, and occasionally (iii) acid electrolytes. These solutions would have a typical concentration and density of 400 g/l, and 1100 kg/m^3 respectively; the electrolyte will have a kinematic viscosity of about 1 mm^2/s. The solution would normally be operated at temperatures between about 18°C and 40°C. Temperatures above ambient are often preferred because the electrolyte solution warms during ECM due to electrical heating caused by the passage of current. The machining action is often found to be easier to control if the electrolyte is maintained at a higher temperature from the outset. This is usually achieved by heating the electrolyte in its reservoir to the required temperature, before it is pumped into the machining zone between the tool and workpiece.

The two main electrolytes mentioned above, sodium chloride and sodium nitrate solutions, exhibit different machining characteristics for the same metals. For example, in the ECM of most steels and nickel alloys, sodium chloride solutions show only a very slight decrease in current efficiency from the value of 100 per cent, when the current density is increased. (Occasionally, efficiencies higher than 100 per cent are obtained, when actual grains of metal are dislodged by the traction forces of the electrolyte flow.) With sodium nitrate electrolyte, the current efficiency rises from comparatively small values at low current densities, to maximum values usually below 100 per cent. The efficiency only very slowly increases thereafter, with further rise in current density.

Although sodium chloride electrolyte has generally a higher current efficiency than sodium nitrate over a wide range of current densities, the latter electrolyte is often preferred in practice, because closer dimensional accuracy of ECM is obtained with it. The superior machining performance obtained with sodium nitrate becomes particularly relevant in hole-drilling by ECM. (This technique will be explained more fully later).

Other electrolytes that are used include mild (about 5 per cent) hydrochloric acid solution; it is useful in fine-hole drilling, since this acid electrolyte dissolves the metal hydroxides as they are produced. Like NaCl electrolyte, the current efficiency is about 100%. Sodium chlorate solution has also been investigated. However industry has been reluctant to employ it as an ECM electrolyte, owing to its ready combustibility, even though this electrolyte is claimed to give even better throwing power and closer dimensional control than sodium nitrate solution.

Surface Finish

As well as influencing the rate of metal removal, the electrolytes also affect the quality of surface finish obtained in ECM, although other process conditions also have an effect. Depending on the metal being machined, some electrolytes leave an etched finish, caused by the non-specular reflection of light from crystal faces electrochemically dissolved at different rates. Sodium chloride electrolyte tends to produce an etched, matte finish with steels and nickel alloys; a typical surface roughness would be about 1 μm Ra.

In many applications, a polished finish is desirable on machined components. The production of an electrochemically polished surface is usually associated with the random removal of atoms from the anode-workpiece, the surface of which has become covered with an oxide film. This is governed by the particular metal-electrolyte combination being used. (Nonetheless, the mechanisms controlling high-current density electropolishing in ECM are still not completely understood.)

For example, with nickel-based alloys, the formation of a nickel oxide film seems to be pre-requisite for obtaining a polished surface; a finish of this quality, of 0.2 μm Ra, has been claimed for a Nimonic (nickel alloy) machined in saturated sodium chloride solution. Surface finishes as fine as 0.1 μm Ra have been reported when nickel-chromium steels have been machined in sodium chlorate solution. Again, the formation of an oxide film on the metal surface has been considered to be the key to these conditions of polishing.

Sometimes the formation of oxide film on the metal surface hinders efficient ECM, and leads to poor surface finish. For example, the ECM of titanium is rendered difficult in chloride and nitrate electrolytes, because the oxide film formed is so passive. Even when higher voltages (e.g., about 50 V) are applied to break the oxide film, its disruption is so non-uniform that deep grain boundary attack of the metal surface can occur.

Occasionally, metals that have undergone ECM are found to have a pitted surface, the remaining area being polished or matte. Pitting normally stems from gas evolution at the anode-electrode; the gas bubbles rupture the oxide film causing localized pitting.

Process variables also play a significant part in the determination of surface finish. For example, as the current density is raised, generally the smoother becomes the finish on the workpiece surface. A similar effect is achieved when the electrolyte velocity is increased. For instance, tests with nickel machined in HCl solution have shown that the surface finish improves from an etched to a polished appearance when the current density is increased from about 8 to 19 A/cm^2, the flow velocity being held constant.

Applications of ECM

ECM can be used to shape metals in three main ways:

(1) Deburring (smoothing of surfaces). Its simplest and most common application is smoothing, of which deburring is a simple illustration. A plane-faced cathode-tool is placed opposite a workpiece, which carries irregularities on its surface. The current densities at the peaks of the surface irregularities are higher than those in the valleys. The former are therefore removed preferentially, and the workpiece becomes smoothed, admittedly at the expense of stock metal (since metal is still machined from the valleys of the irregularities, albeit at a lower rate). Electrochemical smoothing is the only type of ECM in which the final anode shape can match exactly that of the cathode-tool.

Electrochemical deburring is a fast process; typical process times are 5 to 30s for smoothing the surfaces of manufactured components. The technique has many applications. Owing to its speed and simplicity of operation, electrochemical deburring can be performed with a fixed, stationary cathode-tool. The process is used in many industries and is particularly attractive for the deburring of the intersectional region of cross-drilled holes.

(2) Hole-drilling. Hole-drilling is another well-known way of using ECM. A tubular electrode is used as the cathode-tool. Electrolyte is pumped down the central bore of the tool, across the main machining gap, and out between the side-gap formed between the wall of the tool and the hole electrolytically dissolved in the workpiece. Considerable improvement in machining can often be obtained as the electrolyte flow is reversed, as will be explained below.

The main machining action is carried out in the inter-electrode gap formed between the leading edge of the drill-tool and the base of the hole in the workpiece. ECM also proceeds laterally between the side-walls of the tool and component. The current density in that region is lower than that at the leading edge of the advancing tool. Since the lateral gap width becomes progressively larger than that at the leading edge, the side ECM-rate is lower. The overall effect of the side-ECM is to increase the diameter of the hole that is produced. (The local difference between (i) the radial length between the side-wall of the workpiece and the central axis of the cathode-tool, and (ii) the external radius of the cathode, is known as the "overcut"). The amount of overcut can be reduced by several methods. A common procedure involves the insulation of the external walls of the tool, which inhibits side-current flow.

Another practice is the choice of an electrolyte like sodium nitrate solution, as discussed earlier. With this kind of electrolyte, the current efficiency is greatest at the highest current densities. In hole-drilling these high current densities occur between the leading edge of the drill and the base of the workpiece. If another electrolyte such as sodium chloride solution were used instead, then the overcut could be much greater. Its current efficiency remains steady at almost 100 per cent for a wide range of current densities. Thus, even in the side gap, metal removal proceeds at a rate which is mainly determined by the current density, in accordance with Faraday's law.

A wide range in hole-sizes can be drilled or trepanned by ECM; diameters as small as 0.05 mm to as large as 75 mm have been achieved. For fine holes of 0.5 to 1.0 mm diameter, depths of up to 110 mm have been produced. Drilling by ECM is not restricted to round holes; the shape of the workpiece is determined by that of the tool-electrode.

Finally, some significant differences between drilling and smoothing are noteworthy. With the former technique, forward mechanical movement of one of the electrodes, say the tool, towards the other is usually necessary in order to maintain a constant equilibrium gap width in the main machining zone between the leading face of the drill and the workpiece. A typical feed-rate in ECM-drilling would be 1 to 5 mm/min., to drill holes ranging from 0.5 to 20 mm in diameter. In deburring, mechanical drive of the tool is often unnecessary. The times of deburring are typically 10 to 30s for burrs of about 0.10 to 0.15 mm high.

A well-used application of drilling by ECM is the production of cooling holes in gas turbine blades for aircraft engines.

(3) ECM shaping Here a three-dimensional shape is formed on a workpiece. The successful use of this technique requires that a constant equilibrium gap be maintained between the two electrodes, by use of a constant rate of mechanical feed of one electrode towards the other. In order to achieve the required dimensional accuracies, electrolytes such as sodium nitrate are commonly employed. Nitrate solutions enable superior tolerances to be achieved than with their chloride counterparts as the current density is more sensitive to electrode gap distance with nitrate solutions and so they are more effective in the forefferential removal of high spots.

Electrolyte flow plays a very significant role in contour shaping. Careful design of tooling is necessary to provide the right entry and exit ports for the electrolyte, which usually has to be maintained at high pressures in order to rapidly flush away the products of machining before they can interfere with the machining action. Indeed, adjustments often have to be made to the profile of the cathode-tool to take account of the effects of the cathodic gas bubbles upon the specific conductivity of the solution. These bubbles can cause a tapering in the profile. The effect arises in contour shaping; the problem is rendered even more complicated by the electrical heating of the electrolyte. This causes the specific conductivity to rise, with a consequential widening taper in the downstream direction. The two effects, gas bubbles and heating, act in conflicting ways but, unfortunately, do not eliminate each other.

Careful design of the tooling is therefore necessary, and can account for about 20% of the cost of the entire machining. The complexity of the shapes needed means that empirical methods of design are still largely used.

Contour shaping is well known for the production of turbine blades in the aircraft engineering industry. Current densities of 100 to 200 A/cm^2 being commonly used, with machines of current capacity of 50,000 A.

The Present and Future of ECM

High-rate anodic electrochemical dissolution has been found to be a practical method of smoothing and shaping hard metals, without wear of the cathodic tool, and by employment of simple aqueous electrolyte solutions. ECM can therefore prove attractive to the production engineer, since it can offer substantial advantages in a wide range of cavity sinking and shaped hole production operations.

The control of the ECM process is improving all the time, with more sophisticated servo-systems, better insulating coatings and so on. But even now there is still a clear need for basic information on electrode phenomena at high current densities and electrolyte flows.

Tool design continues to be of paramount importance in any ECM operation. The ingenuity of the tool designer will be tested continually as he endeavours to optimize electrolyte flows and metal removal rates, to produce the required size and shape of

The advent of new technology for controlling the ECM process and with the development of new and improved metal alloys which are difficult to machine by conventional means, ECM has an assured future.

APPENDIX

Bibliography of Electrochemical Machining

Bellows, G. Non-traditional Machining Guide 26 Newcomers for Production; Metcut Research Associates, Inc.: Cincinnati, Ohio, 1976, pp. 28,29.

Bellows, G.; Kohls, J. D. 1982, American Machinist, 178-83.

Clifton, D.; Midgley, J. W.; McGeough, J. A. 1987, Proc. Inst. Mech. Eng. 201,B4, 229-231.

Crichton, I. M.; McGeough, J. A.; Munro, W.; White, C. 1981, Precision Engineering 3(3), 155-60.

De Barr, A. E.; Oliver, D. A. Eds; Electrochemical Machining, MacDonald Press: London, 1968.

De Silva, A.; McGeough, J. A. Proc. Inst. Mech. Engrs, 1986, 200 (B4), 237-46.

Drake, T.; McGeough, J. A. Proc. Machine Tool Design and Research Conf. Macmillan: New York, 1981; pp. 362-9.

Elltofy, H.; McGeough, J. A. (1988) Evaluation of an Apparatus for Electrochemical Arc Wire Machining. "Trans. Am. Soc. Mech. Eng. J. Eng. Industry" (in press).

Ghabrail, S. R. et al. (1984) Electrochemical Wirecutting. "Proc. 24th Int. Machine Tool Design and Research Conf.", Department of Mechanical Engineering, University of Manchester in association with Macmillan Press, pp. 323-8.

Graham, D. The Production Engineer, 1982, 61(6), 27-30.

Jain, V. K.; Nanda, V. N. Precision Engineering, 1986, 8(1), 27-33.

Kaczmarek, J. Principles of Machining by Cutting, Abrasion and Erosion", Peter Peregrinus: Stevenage, 1976; 498-513.

Khayry, A. B. M.; McGeough, J. A. Proc. Roy. Soc. A. 1987; 412, 403-29.

Kubota, M. Mechanique; 1975, 303, 15-18.

Kubota, M.; Tamura, Y.; Omori, J.; Hirano, Y. J. Assoc. of Electro-Machinery, 1978; 12(23), 24-33.

Kubota, M.; Tamura, Y.; Takahashi, H.; Sugaya, T. J. Assoc. of Electro-Machinery, 1980; 13 (26), 42-57.

Landolt, D. (1987) "Experimental Study and Theoretical Modeling of Electrochemical Metal Dissolution Processes Involving a Shape Change of the Anode", 172nd Meeting of the Electrochemical Society, Honolulu, Abstract No. 544.

Mao, K. W. J. Electrochem. Soc., 1971; 118, 1870-9.

Mao, K. W.; Mitchell, A. L.; Hoare, J. P. J. Electrochem. Soc., 1972; 119(4), 419-27.

McGeough, J. A. Principles of Electrochemical Machining; Chapman and Hall: London, 1974.

McGeough, J. A. Advanced Methods of Machining, Chapman and Hall: London, 1988.
Newton, M. A. (1985) "Stem Drilling Update", Tech. paper, Society of Manufacturing Engineering, Paper MR85-381.
Wilson, J. F. Theory and Practice of Electrochemical Machining; John Wiley & Son: London, 1971.

RECEIVED October 7, 1988

Author Index

Affiliation Index

Subject Index

Production by Joyce A. Jones
Indexing by Deborah H. Steiner

Elements typeset by Hot Type Ltd., Washington, DC
Printed and bound by Maple Press, York, PA